Management in the Built Environment

Series editor
Low Sui Pheng
Department of Building, National University of Singapore, Singapore, Singapur
Singapore

The aim of this book series is to provide a platform to build and consolidate a rigorous and significant repository of academic, practice and research publications that contribute to further knowledge relating to management in the built environment. Its objectives are to:

(1) Disseminate new and contemporary knowledge relating to research and practice in the built environment
(2) Promote synergy across different research and practice domains in the built environment and
(3) Advance cutting-edge research and best practice in the built environment.

The scope of this book series is not limited to "management" issues per se because this then begs the question of what exactly are we managing in the built environment. While the primary focus is on management issues in the building and construction industry, its scope has been extended upstream to the design management phase and downstream to the post-occupancy facilities management phase. Management in the built environment also involves other closely allied disciplines in the areas of economics, environment, legal and technology. Hence, the starting point of this book series lies with project management, extends into construction and ends with facilities management. In between this spectrum, there are also other management-related issues that are allied with or relevant to the built environment. These can include, for example cost management, disaster management, contract management and management of technology.

This book series serves to engage and encourage the generation of new knowledge in these areas and to offer a publishing platform within which different strands of management in the built environment can be positioned to promote synergistic collaboration at their interfaces. This book series also provides a platform for other authors to benchmark their thoughts to identify innovative ideas that they can further build on to further advance cutting-edge research and best practice in the built environment.

If you are interested in submitting a proposal for this series, please kindly contact the Series Editor or the Publishing Editor at Springer:

Low Sui Pheng (bdglowsp@nus.edu.sg) or
Ramesh Premnath (Ramesh.premnath@springer.com)

More information about this series at http://www.springer.com/series/15765

Leni Sagita Riantini Supriadi • Low Sui Pheng

Business Continuity Management in Construction

Springer

Leni Sagita Riantini Supriadi
Department of Building
National University of Singapore
Singapore, Singapore

Low Sui Pheng
Department of Building
National University of Singapore
Singapore, Singapore

ISSN 2522-0047 ISSN 2522-0055 (electronic)
Management in the Built Environment
ISBN 978-981-10-5486-0 ISBN 978-981-10-5487-7 (eBook)
DOI 10.1007/978-981-10-5487-7

Library of Congress Control Number: 2017945380

Printed on acid-free paper

This Springer imprint is published by Springer Nature
The registered company is Springer Nature Singapore Pte Ltd.
The registered company address is: 152 Beach Road, #21-01/04 Gateway East, Singapore 189721, Singapore

Contents

1 Introduction 1
- 1.1 Background 1
  - 1.1.1 BCM Overview 2
  - 1.1.2 Tools for Decision Making Process During Crises 3
- 1.2 Motivation for Research 4
- 1.3 Research Problems 5
  - 1.3.1 Knowledge Gap 7
  - 1.3.2 Research Questions 9
- 1.4 Objectives 9
- 1.5 Scope 10
- 1.6 Research Process 10
- 1.7 Research Significance 11
- 1.8 Structure of the Book 11

2 The Management of Crisis 15
- 2.1 Introduction 15
- 2.2 Management 15
- 2.3 Organizational Management 18
  - 2.3.1 Organization Overview 18
  - 2.3.2 Traditional Organizational Theories 19
  - 2.3.3 Systems Theory 22
  - 2.3.4 Contingency Theory 24
  - 2.3.5 Complexity Theory in Organizations 25
  - 2.3.6 Change in Organization 27
- 2.4 Crisis Management 29
  - 2.4.1 Definition of Crisis 30
  - 2.4.2 History of Development 31
  - 2.4.3 Main Concept of Crisis Management 32
- 2.5 Summary 38

3 Business Continuity Management (BCM) . . . 41
3.1 Introduction . . . 41
3.2 Background . . . 41
3.2.1 BCM Definition and Development . . . 41
3.2.2 BCM and Other Related Concepts . . . 44
3.3 BCM as a Management System . . . 47
3.4 Main Principles of BCM . . . 50
3.5 Business Continuity Planning (BCP) . . . 52
3.6 BCM Implementation . . . 55
3.6.1 Legislation and Standards Relating to BCM . . . 55
3.6.2 BCM Level of Preparedness . . . 57
3.7 Reviews of BC Plan . . . 62
3.7.1 BC Plan from Financial Services Sector . . . 62
3.7.2 BC Plan from Education Institutions: A Case Study . . . 64
3.7.3 BC Plan for Influenza Pandemic: A Review . . . 66
3.8 The Need for BCM . . . 68
3.8.1 Benefits of BCM . . . 69
3.8.2 Challenges in BCM . . . 70
3.9 Summary . . . 72

4 Organizational Culture and Institutional Forces . . . 75
4.1 Introduction . . . 75
4.2 Culture in Organization . . . 75
4.2.1 Culture Overview . . . 75
4.2.2 Definitions of Culture . . . 76
4.2.3 Organizational Culture Dimensions . . . 77
4.2.4 Managing Organizational Culture . . . 93
4.2.5 Benefits of Identifying Organizational Culture . . . 94
4.3 Institutional Theory . . . 94
4.3.1 Defining Institutions . . . 95
4.3.2 Three Pillars of Institutions . . . 96
4.4 Summary . . . 100

5 Mainstream Theories: Implementation by Contractors . . . 101
5.1 Introduction . . . 101
5.2 Mainstream Theories for Implementation by Contractors . . . 101
5.3 Organizational Management in Contractors . . . 102
5.4 Traditional Organizational Theories . . . 104
5.4.1 Systems Theory . . . 107
5.4.2 Contingency Theory . . . 108
5.4.3 Complexity Theory . . . 108
5.4.4 Change in Organization . . . 109
5.5 Crisis Management Implementation by Contractors . . . 110
5.6 BCM in Construction Firms . . . 111
5.7 Culture in Organization . . . 112

5.8 Institutional Theory 115
5.9 Summary 116

6 The Indonesian Construction Industry 117
6.1 Introduction 117
6.2 Profile of Indonesia 117
6.2.1 Geography 117
6.2.2 Demography and Resources 118
6.2.3 Form of State 119
6.2.4 Economy 119
6.3 The Indonesian Construction Industry 127
6.3.1 Construction Activities and National Contribution 127
6.3.2 Role of Construction Administration by Government . . . 130
6.3.3 Role of Construction Industry Associations 131
6.3.4 Regulation and Management Systems of Construction Projects 134
6.3.5 Technology 137
6.3.6 Research and Development 137
6.3.7 Lessons Learned and Future Developments 138
6.4 Indonesian Construction Firms 139
6.4.1 Types of Firms 139
6.4.2 Firms' Characteristics 140
6.4.3 Indonesian Contractors 142
6.4.4 Business Activities of Contractors 143
6.5 Review of Crises Faced by Indonesian Contractors 145
6.5.1 Crisis Overview 145
6.5.2 Possible Crises Faced by Contractors 146
6.5.3 Crises Experienced by Indonesian Contractors 148
6.5.4 Current Findings 152
6.6 Summary 154

7 Knowledge Based Decision Support System (KBDSS) 155
7.1 Introduction 155
7.2 Decision Making Process 155
7.3 Decision Making Tools 156
7.3.1 Decision Making Tools Using Fuzzy Approach 157
7.4 Decision Support System (DSS) Overview 158
7.4.1 Definition and Development of DSS 159
7.4.2 DSS Implementation 160
7.5 KBDSS Overview 163
7.5.1 KBDSS Concept 164
7.5.2 Review of KBDSS Implementation 166
7.6 KBDSS Development 167
7.6.1 Formulation 167
7.6.2 System Development 168

7.6.3 Validation 171
7.7 Summary 173

8 Conceptual Framework 175
8.1 Introduction 175
8.2 Indonesian Contractor's Knowledge About BCM 175
8.3 Relationships Between BCM, Organizational Culture and Institutional Forces 176
8.3.1 Defining Organizational Culture Dimensions 176
8.3.2 Relationships Between OC Dimensions and BCM Principles 179
8.3.3 Defining Institutional Forces Attributes 186
8.3.4 Relationships Between IF Attributes and BCM Principles 187
8.3.5 The Importance of Organizational Culture and Institutional Forces in Adopting BCM 188
8.4 Developing BCM for Indonesian Contractors 193
8.5 Development of KBDSS for BCM 199
8.5.1 Benefits of KBDSS for Management Process and Construction 199
8.5.2 The Application of KBDSS for BCM 201
8.6 Conceptual Framework 207
8.7 Summary 207

9 Research Design and Methodology 209
9.1 Introduction 209
9.2 Research Framework 209
9.2.1 Research Questions and Hypotheses 210
9.3 Research Design 212
9.3.1 Research Design for the Study 212
9.3.2 Sampling 213
9.4 Methods of Data Collection 215
9.4.1 Methods of Collecting Data 216
9.5 Methods of Data Analysis 217
9.6 Summary 219

10 Data Analysis: Surveys 221
10.1 Introduction 221
10.2 Pilot Study 221
10.3 Questionnaire Survey 222
10.3.1 Crises Response 225
10.3.2 BCM General Knowledge 227
10.3.3 BCM Principles 229
10.3.4 Institutional Forces (IF) that Support BCM Principles Implementation 229

10.3.5 Organizational Culture Attributes (OC) Implemented for BCM Principles 234
10.3.6 Perceived Importance and Implemented for OC Attributes 240
10.3.7 Survey Validation 240
10.4 Summary 252

11 Data Analysis: Case Studies 253
11.1 Introduction 253
11.2 Case Studies 253
11.2.1 Case Study 1: Firm A 253
11.2.2 Case Study 2: Firm B 265
11.2.3 BCM Preparedness 275
11.3 Interviews with Experts on BCM Implementation 283
11.3.1 The Benefits of Implementing BCM for Contractors 283
11.3.2 The Drawbacks of Implementing BCM for Contractors 284
11.3.3 The Need for BCM in Contractors' Firms 285
11.3.4 BCM Certification 286
11.3.5 Important Elements (BUs, CBFs, MBCOs, and Significant Crises with High Impacts) in Contracting Firms that Relates Mostly with BCM 287
11.3.6 Additional Recommendations for BCM Implementation by Indonesian Contractors 288
11.4 BCM Implementation Guidelines Development 291
11.4.1 BCM Implementation Guidelines Framework 291
11.4.2 Levels of Preparedness Development 291
11.4.3 Assessment Phase 292
11.4.4 Action Plans per BCM Level of Preparedness 298
11.5 Summary 299

12 Data Analysis: BCM-KBDSS Development 301
12.1 Introduction 301
12.2 BCM-KBDSS Development 301
12.2.1 Knowledge Base (KB) Development 301
12.2.2 Rules and Logics for BCM-KBDSS 302
12.2.3 Design User Interface and BCM-KBDSS Compilation 309
12.2.4 BCM-KBDSS Prototype 311
12.3 KBDSS Validation 314
12.3.1 BCM-KBDSS Validation Results 322
12.4 Summary 328

13 Findings and Discussion . . . 329
13.1 Introduction . . . 329
13.2 Findings and Discussion for RQ 1 . . . 329
13.2.1 Findings from Surveys . . . 329
13.2.2 Findings from Case Studies . . . 349
13.2.3 BCM Implementation Guidelines Development . . . 361
13.3 Findings and Discussion for RQ 2 . . . 361
13.3.1 Findings from KBDSS Development . . . 361
13.3.2 Findings from KBDSS Validation . . . 362
13.4 Summary . . . 362

14 Conclusions and Recommendations . . . 365
14.1 Introduction . . . 365
14.2 Summary . . . 365
14.2.1 Research Problems . . . 366
14.2.2 Knowledge Gap . . . 367
14.2.3 Research Objectives and Research Designs Revisited . . . 367
14.3 Conclusions of the Research Questions . . . 369
14.3.1 Research Question 1 . . . 369
14.3.2 Research Question 2 . . . 375
14.4 Conclusions of the Research Hypotheses and Propositions . . . 376
14.4.1 Testing Hypotheses H1, H2 and H3 . . . 376
14.4.2 Propositions for RO4 and RQ2 . . . 378
14.5 Academic Contributions . . . 379
14.6 Practical Contributions . . . 380
14.7 Limitations of the Research . . . 380
14.8 Recommendations for Future Studies . . . 381

Appendix A . . . 383

Appendix B . . . 387

Appendix C . . . 417

Appendix D . . . 435

Appendix E . . . 445

Appendix F . . . 465

Appendix G . . . 505

References . . . 511

List of Abbreviations

AHP	Analytical hierarchical process
BC	Business continuity
BCI	Business Continuity Institute
BCM	Business continuity management
BC	Plan business continuity plan
BCP1	Developing the detailed BC plan
BI	Business intelligence
BIA	Business impact analysis
BIA1	Conducting business impact analysis
BIA2	The involvement of experts, employees from related BUs and key staff members in BIA
BU	Business unit
CBF	Critical business function
CM Plan	Crisis management plan
CMT	Crisis management team
COA	Center of area
CSR	Corporate social responsibility
DML	Decision making logic
DRP	Disaster resources partnership
DSS	Decision support system
DW	Data warehousing
EFA	Exploratory factor analysis
EIA	Environmental impact assessment
EIS	Executive information system
EIU	Economist intelligence unit
EMS	Environmental management system
ENS	Emergency notification system
EOC	Emergency operations centre
EPC	Engineering, procurement and construction
ERRG	Emergency response resource group
FLS	Fuzzy logic systems

GDP	Gross domestic product
GLM	General linear model
GLOBE	Global leadership and organizational behaviour effectiveness
GSS	Group support systems
ICA	Indonesian Contractors Association
ICT	Information communication technology
IF	Institutional forces
I/O	Input/output
KB	Knowledge base
KBDSS	Knowledge based decision support system
LISP	List processing
MBCO	Minimum business continuity objective
MBMS	Model base management system
MIS	Management information systems
NBCSD	National Board of Construction Service Development
NSS	Negotiation support systems
OC	Organizational culture
OCAI	Organizational culture assessment instrument
PDCA	Plan-do-check-act
PDSS	Personal DSS
PM	Programme management
PM1	Analysing the ongoing efforts and activities to maintain the effectiveness of its BCM, including providing systematic training and awareness programmes to staff members
PM2	Conducting BCM training and awareness programmes for all staff and related external parties
PROLOG	Programming in logic
QSHE	Quality, Safety, Health and Environment
R&D	Research and development
RA	Risk analysis
RA1	Conducting risk analysis and cost benefit analysis
RA2	The involvement of experts and BCM committee in risk review
RA3	Conducting a detailed risk review that examines and assesses the availability of critical equipment, technology, and facilities for BU/CBF
RFI	Requests for information
RFP	Requests for proposals
RPO	Recovery point objective
RTO	Recovery time objective
RO1	Research objective 1
RO2	Research objective 2
RO3	Research objective 3
RO4	Research objective 4
RO5	Research objective 5
RQ1	Research question 1

RQ2	Research question 2
SA	Strategy analysis
S1	Conducting strategy analysis for maintaining the operations of CBFs that cover pre-incident preparedness, response and recovery
S2	Determining staff members to support the recovery strategy and providing training and awareness programmes
SAW	Simple additive weight
SIT	Site implementation team
SME	Small and medium-sized enterprise
SOC	Satellite Operations Center
SOP	Standard operating procedure
TE	Tests and exercises
TE1	Providing periodic tests and exercises to ensure that the BC plan is viable and workable

List of Figures

Fig. 1.1 Research process. Source: Adapted from Tan (2008) 10
Fig. 1.2 Relationships between the chapters 13

Fig. 2.1 The organization as an input-process-output system. Source: Adapted from Naoum (2001) 23
Fig. 2.2 Contingency model of management. Source: Adapted from Carlisle (1976) 25
Fig. 2.3 Three pillars to successful response to crises. Source: Adapted from Geraldi et al. (2009) 36

Fig. 3.1 Old and new BCM approach. Source: Adapted from Herbane et al. (1997) 44

Fig. 6.1 GDP growth of Indonesia. Source: SCA (2012) 120
Fig. 6.2 Annual end-period exchange rate Rupiah:$1 (USD). Source: EIU (2012) 121
Fig. 6.3 Indonesia's population in selected years. Source: SCA (2012) 121
Fig. 6.4 Indonesia's annual current-account balance. Source: World Bank (2012) 122
Fig. 6.5 Indonesian construction industry: value. Source: SCA (2012) 128
Fig. 6.6 Indonesian construction industry: real growth. Source: SCA (2012) 129
Fig. 6.7 Indonesian construction industry: percentage of GDP. Source: SCA (2012) 129
Fig. 6.8 Total permanent workforce in the Indonesian construction industry. Source: SCA (2012) 130
Fig. 6.9 Indonesian construction industry support infrastructure. Source: Adapted from NBCSD (2004) 132
Fig. 6.10 Value of construction projects. Source: SCA (2012) 142

Fig. 7.1 DSS characteristics. Source: Adapted from Raymond (1990) 161

Fig. 7.2 Main components of KBDSS. Source: Adapted from Arain and Low (2006) 165

Fig. 8.1 Framework of IF–BCM–OC relationships 190
Fig. 8.2 BCM–IF variables 191
Fig. 8.3 BCM–OC variables 192
Fig. 8.4 BCM implementation guidelines framework 199
Fig. 8.5 KBDSS process: BCM level of preparedness assessment 203
Fig. 8.6 KBDSS components 204
Fig. 8.7 BCM–KBDSS framework 205
Fig. 8.8 Conceptual framework 206

Fig. 12.1 A fuzzy logic system. Source: Adapted from Mendel (1995) and Ahmed and Capretz (2006) 303
Fig. 12.2 BCM-KBDSS decision situation scenario flowchart 305
Fig. 12.3 Structured situation diagram (conceptual) 305
Fig. 12.4 Structured situation diagram (coding) 306
Fig. 12.5 BCM-KBDSS dependency diagram 306
Fig. 12.6 Linguistic variables 308
Fig. 12.7 Two-input and one output rule base 309
Fig. 12.8 BCM-KBDSS introduction page 311
Fig. 12.9 User information page 312
Fig. 12.10 BCM introduction page 312
Fig. 12.11 Assessment page—BCM principle 1 313
Fig. 12.12 Assessment page—BCM principle 2 313
Fig. 12.13 Assessment page—BCM principle 3 314
Fig. 12.14 Assessment page—BCM principle 4 314
Fig. 12.15 Assessment page—BCM principle 5 315
Fig. 12.16 Assessment page—BCM principle 6 315
Fig. 12.17 Assessment result page 316
Fig. 12.18 Level description page 316
Fig. 12.19 Recommended action plans page—BCM principle 1 316
Fig. 12.20 Recommended action plans page—BCM principle 2 317
Fig. 12.21 Recommended action plans page—BCM principle 3 317
Fig. 12.22 Recommended action plans page—BCM principle 4 318
Fig. 12.23 Recommended action plans page—BCM principle 5 318
Fig. 12.24 Recommended action plans page—BCM principle 6 319
Fig. 12.25 Additional recommendations page 319
Fig. 12.26 Summary page 320
Fig. 12.27 Feedback page 320
Fig. 12.28 End page 321
Fig. 12.29 BCM-KBDSS result of Firm A 322
Fig. 12.30 BCM-KBDSS result of Firm B 323

Fig. 13.1 Findings of the study 363

List of Tables

Table 3.1	Comparison between Risk Management and BCM [adapted from BCI (2005, p. 6)]	45
Table 3.2	The main aspects of BCM principles	58
Table 3.3	Marsh BCM preparedness level	61
Table 3.4	BCM distinction with other related concepts	70
Table 3.5	BCM benefits	71
Table 4.1	Characteristics of Hofstede's cultural dimensions	81
Table 4.2	Cultural dimensions from other scholars related to Hofstede's dimensions	84
Table 4.3	Organizational culture factors	89
Table 4.4	Strategic responses to environmental pressures	96
Table 4.5	The three pillars of institutions	97
Table 4.6	Institutional pillars and carriers	99
Table 5.1	Project management competence for construction in UK	105
Table 5.2	Research on cultures in construction industry	113
Table 6.1	The key functions of Indonesian construction industry support infrastructure	113
Table 6.2	Number of contractors in 33 provinces of Indonesia	143
Table 6.3	Value chain of contractor	144
Table 6.4	Unstructured list of crises	146
Table 6.5	Possible crises faced by contractors	149
Table 7.1	DSS framework subsystems	163
Table 8.1	BCM knowledge variables	177
Table 8.2	Organizational Culture (OC) dimensions	180
Table 8.3	Institutional Forces (IF) attributes	187
Table 8.4	BCM steps for developing implementation guidelines	194
Table 8.5	BCM preparedness criteria	196

Table 8.6 Firm Y—State owned: Relationships between business units and value chain ... 197
Table 8.7 Firm Z—Private owned: Relationships between business units and value chain ... 198
Table 8.8 Variables for BCM–KBDSS model ... 205
Table 10.1 Respondent details ... 222
Table 10.2 Respondent feedbacks ... 223
Table 10.3 Respondents' job levels ... 225
Table 10.4 Years working in the company and construction sector ... 225
Table 10.5 Crises that occurred in the firm within 5 years ... 226
Table 10.6 Types of crises that occurred in the firm within the last 5 years ... 226
Table 10.7 Impacts from crises ... 227
Table 10.8 Crisis SOP elements ... 227
Table 10.9 Sources of BCM knowledge ... 228
Table 10.10 Reasons for implementing BCM ... 228
Table 10.11 Effectiveness of BCM implemented in a firm [based on their agree-disagree perspectives; Disagree-Agree (1–5)] ... 228
Table 10.12 BCM principle implementation in the firm (whether this principle has been implemented or not; Scale: 1 = strongly disagree, 2 = disagree, 3 = neutral, 4 = agree, 5 = strongly agree) ... 230
Table 10.13 Correlations between private and state-owned firms ... 231
Table 10.14 Most important institutional forces factors (IF) that support risk analysis and review (variable orders by size of loadings) ... 231
Table 10.15 Most important institutional forces factors (IF) that support Business Impact Analysis (variable orders by size of loadings) ... 231
Table 10.16 Most important institutional forces factors (IF) that support strategy analysis (variable orders by size of loadings) ... 232
Table 10.17 Most important institutional forces factors (IF) that support BC plan development and tests and exercises (variable orders by size of loadings) ... 232
Table 10.18 Most important institutional forces factors (IF) that support programme management (variable orders by size of loadings) ... 233
Table 10.19 Most implemented organizational culture attributes for risk analysis and review (variable orders by size of loadings) ... 235
Table 10.20 Least implemented organizational culture attributes for risk analysis and review (variable orders by size of loadings) ... 235
Table 10.21 Most implemented organizational culture attributes for business impact analysis (variable orders by size of loadings) ... 236

Table 10.22 Least implemented organizational culture attributes for business impact analysis (variable orders by size of loadings) 236
Table 10.23 Most implemented organizational culture attributes for strategy analysis (variable orders by size of loadings) 236
Table 10.24 Least implemented organizational culture attributes for strategy analysis (variable orders by size of loadings) 237
Table 10.25 Most and least implemented organizational culture attributes for BC plan development (variable orders by size of loadings) 237
Table 10.26 Most and least implemented organizational culture attributes for tests and exercises (variable orders by size of loadings) 238
Table 10.27 Most implemented organizational culture attributes for programme management (variable orders by size of loadings) 238
Table 10.28 Least implemented organizational culture attributes for programme management (variable orders by size of loadings) 239
Table 10.29 P-value range of OC attributes (between the perceived importance and implemented) for each BCM principle 240

Table 11.1 Documented suppliers in 2012 (Firm A, 2012) 257
Table 11.2 Firm A's BUs and CBFs 258
Table 11.3 Firm A's MBCO for each CBF 260
Table 11.4 Firm A's internal and external coordination overview 261
Table 11.5 Firm A crises response 264
Table 11.6 Firm B's BUs and CBFs 269
Table 11.7 Firm B's MBCO for each CBF 271
Table 11.8 Firm B's internal and external coordination overview 272
Table 11.9 Firm B crises response 274
Table 11.10 Practices that have been implemented in the initiation phase 276
Table 11.11 Hindrances in implementing the initiation phase (in the context of its current condition) 276
Table 11.12 Practices that have been implemented for risk analysis and review 277
Table 11.13 Hindrances in implementing risk analysis and review (in the context of its current condition) 277
Table 11.14 Practices that have been implemented for BIA 278
Table 11.15 Hindrances in implementing BIA (in the context of its current condition) 278
Table 11.16 Practices that have been implemented for BC strategy 279

Table 11.17 Hindrances in implementing BC strategy (in the context of its current condition) ... 279
Table 11.18 Practices that have been implemented for BC plan development ... 280
Table 11.19 Hindrances in implementing BC plan development (in the context of its current condition) ... 281
Table 11.20 Practices that have been implemented for tests and exercises ... 281
Table 11.21 Hindrances in implementing tests and exercises (in the context of its current condition) ... 281
Table 11.22 Practices that have been implemented for programme management ... 282
Table 11.23 Hindrances in implementing programme management (in the context of its current condition) ... 282
Table 11.24 BCM-KBDSS level of preparedness ... 293
Table 11.25 BCM practices for assessment (per BCM principle) ... 297

Table 12.1 Triangular membership function ... 308
Table 12.2 Truth value rules ... 309
Table 12.3 Results from questionnaire's criteria (11 criteria) ... 324
Table 12.4 Additional feedback (interviews) ... 327
Table 12.5 Questionnaire results from field testing ... 328

Table 13.1 The reasons and motivations for conducting risk analysis and cost benefit analysis ... 336
Table 13.2 The reasons and motivations for involving experts and BCM committee in risk review ... 337
Table 13.3 The reasons and motivations for conducting a detailed risk review ... 337
Table 13.4 The reasons and motivations for conducting BIA ... 337
Table 13.5 The reasons and motivations for involving experts, employees from related BUs and key staffs in BIA ... 338
Table 13.6 The reasons and motivations for conducting strategy analysis for maintaining the operations of CBFs ... 338
Table 13.7 The reasons and motivations for determining staff to support the recovery strategy and providing training and awareness programme ... 338
Table 13.8 The reasons and motivations for developing a detailed BC plan ... 339
Table 13.9 The reasons and motivations for providing periodic tests and exercises to ensure that the BC plan is viable and workable ... 339
Table 13.10 The reasons and motivations for conducting BCM programme management ... 339
Table 13.11 The reasons and motivations for conducting BCM training and awareness programmes for all staff and related external parties ... 340

Table 13.12 OC dimensions that have been implemented and support the BCM principles 344
Table 13.13 Organization characteristics of Firm A and Firm B 350
Table 13.14 Additional recommendations for Indonesian contractors to implement BCM 360

Chapter 1
Introduction

1.1 Background

An organization engaged in business activities requires actions and operational procedures that must be completed in order to attain its mission and goals. A business is termed as an integrated set of activities and assets that is capable of being conducted and managed for the purpose of providing either a return to investors or dividends, lower costs, or other economic benefits directly and proportionally to owners, members or participants (Council on Corporate Disclosure and Governance 2005, p. 3). Nowadays, the business world is experiencing an increasing level of uncertainty in its environment, where it can lead to adverse financial implications, damage to corporate reputation, viability and integrity (Marsh 2007). Furthermore, throughout the business process, there are times when an unexpected situation or event may occur. Events like crises or emergencies can occur unexpectedly during the course of conducting business activities, and these cannot be overlooked.

Organizations may fail or suffer from many events or crises such as corporate collapses, acts of terrorism, war, and natural disasters. These events can affect the organization directly and indirectly through supply chain issues and loss of customers. This eventually will have negative impacts on employees, their families and the wider community (AS/NZS 2004). According to some studies, whilst bombs, fires and floods capture the headlines, almost 90% of business-threatening incidents were "quiet catastrophes" which were unreported in the media but could have a significant impact on an organization's ability to function. Many of the causes are outside of an organization's control (BCI 2007a). Therefore, as mentioned by Barton (1993), these events are considered as abnormal situations that threaten operations, staff, customers, or the reputation of the organization.

A crisis may give various consequences to an organization, whether financial, legal, or operational consequences. It may disrupt the business process from a few

L.S.R. Supriadi, L. Sui Pheng, *Business Continuity Management in Construction*,
Management in the Built Environment, DOI 10.1007/978-981-10-5487-7_1

minutes up to several months or years in extreme cases. These consequences can impact the business process, and hence may threaten the firm's sustainability.

In managing any event, including overcoming crises, a successful outcome is judged by both the technical response and the perceived competence of the management. However, thoughts like "It won't happen to us", "We will cope—we always do", "We are too big to fail" and "We are not a terrorist target" are frequent responses by companies when questioned about their preparedness toward these threats. Others would believe that their insurance company will pay for everything, and most would also think they do not have the time to prepare for something that will never happen. Various companies that have failed due to an incident found that these responses are based on false assumptions (BCI 2007a). In addition, Knight and Pretty (1996) found that an organization that had successfully dealt with a crisis experienced an increase in their share value in the long-term, while those who were perceived not to have managed the crisis well had experienced the opposite, and after a year had still not recovered.

In order to overcome a crisis and to continue business as usual, an organization or firm must first have systematic ways and approaches in place. Although some organizations survive such events due to perseverance, but continuity of a business is primarily due to planning and preparation. There are various management concepts that can be adopted to overcome crises, such as crisis management, emergency management, risk management, and disaster recovery concepts. In addition to these, there is one concept that is viewed as a unifying process where the abovementioned concepts are included. The concept is Business Continuity Management (BCM), where it is not only focusing on overcoming any crises that occurred, but also considering thoroughly on how to sustain the business in order to obtain its goals and mission (Smith 2003). BCM provides a method for managing any disruption to ensure continuity of service when there is a disruption of business. Moreover, business continuity means to conduct activities that are needed for keeping the business operations running during a period of displacement or interruption (VMIA 2007).

1.1.1 BCM Overview

Historically, BCM was developed many years ago, where this concept is an evolution of a disaster recovery approach in a firm. Elliott et al. (2002) developed on these theories in more details explaining that the evolution of BCM has progressed from a focused technical aspect such as protecting the corporate computer systems to a broader strategic organizational requirement which focused on the needs of the business. The latter viewed business continuity as the integration of social and technical systems which together enable effective organizational protection (Swartz et al. 1995). Therefore, BCM not only protects but is also seen to contribute to the value adding process through more efficient systems or providing value-adding benefits to customers through higher responsiveness, security and reliability.

The Australian National Audit Office (2000) stated that the main deliverable of a BCM process is the Business Continuity Plan (BC Plan). The BC Plan documents the approach for dealing with a disruption to business and includes the steps required to be undertaken to recover from the loss in business function. Fundamentally, various plans such as contingency plans, disaster recovery plans, and business resumption plans are integrated in the BC Plan. Other plans which are usually already in place, such as emergency response procedures, evacuation plans, communication strategies and media liaison strategies, are also important elements of a comprehensive business continuity plan. The BC Plan is activated when a risk event or crisis occurs that causes business interruption. Often these business interruptions is categorized as an extraordinary event, which has a high impact on the organization (VMIA 2007).

Regarding its benefits, BCM is able to help firms to have a response for major disruptions that may threaten their business activities, assist in addressing some key risks in the firm, and help them achieve compliance. Also, BCM can be used as a competitive advantage to gain new customers and to improve margins by using it as a demonstration of "customer care" (BCI 2007b).

1.1.2 Tools for Decision Making Process During Crises

In managing a crisis, decision making is considered as an important part of the process. Critical decisions such as task assignment, resource allocation, guideline to long-term decisions, training and the control capabilities of the organization are necessary for this situation (Yoon et al. 2008). As part of a decision making process in responding to crises or unexpected events, BCM can be designed into an effective model, which is using computer application for providing faster and reliable decision. Based on the development of advanced computer programming technology nowadays, BCM can be automated by using a Decision Support System (DSS) (Eom and Min 1999).

The important roles computer-based information systems play in supporting managers in their semi-structured or unstructured decision making activities have been recognized since the early 1970s by the scholars in the management information systems (MIS) or decision support systems (DSS) areas. Since then, there has been a growing amount of research performed in the area of DSS (Eom and Min 1999). DSS can be defined as interactive computer-based systems that help decision makers utilize data and models to solve unstructured problems (Janakiraman and Sarukeshi 2001). A DSS can also be developed into a system which utilizes knowledge as its based information, where a knowledge base consists of groups of knowledge from experts which provide any information related to the focused problems (Mockler,1989). The latter description is called Knowledge Based Decision Support System (KBDSS).

KBDSS have been proposed as an important technology in managing decision making within businesses over the past few decades. These assist businesses to deal with basic and/or complex problem solving and provide the most suitable alternative for implementation in a real time situation. The broad benefits from this system are that it provides a quick access for the user to all relevant information, the process is direct and personalized and the problem models that are evaluated in the system can be integrated into a logical framework (Singh et al. 2008). Moreover, Yoon et al. (2008) found that KBDSS have been developed for emergency response and management. A KBDSS for emergency response was developed to assess the preparedness of response to emergencies, to provide guidelines for emergency response, and to empower employees in the decision making process. However, before using KBDSS for management process, it is essential to understand that KBDSS is not designed to make decisions for users, but rather it provides relevant information in an efficient and easy-to-access format that allows users to make more informed decisions (Arain and Low 2006).

1.2 Motivation for Research

One of the industries that should implement BCM is the construction industry, where it has an important role in a country's economic growth and development. Considering its characteristics and complexities, overcoming crises and threats in order to continue business in this industry is necessary. Construction is typically a complex, crisis-prone activity carried out in an environment which is relatively uncontrollable compared to the manufacturing industry (Galbraith 1973). Hillebrandt (1988) acknowledged that the structure of this industry is complex because of the large range of contractors' types and professional firms involved, including main contractors and sub-contractors, one-man firms and international firms, low-technology firms and sophisticated specialist firms, builders, and civil engineers and a whole range of additional professionals connected to the industry. The projects delivered in the construction industry have a life cycle consisting of several phases. Various parties are involved in the project, and they have different responsibilities at each stage in the life cycle, thus more interfaces between parties are needed. These imply that the project has a high level of fragmentation in nature. Also, this eventually contributes to increasing complexity and high uncertainties.

The Indonesian construction industry is one of the important sectors in Indonesia. Its role can be seen from the major usage of domestic goods and services that contribute some significant amount in the country's total GDP. This industry is also supported by and connected with a broad spectrum of the nation's legislation and agents. The industry has so far been developed by the government and coordinated by a special construction board for establishing good governance in the sector. In delivering construction projects, a vast network of relationship between many parties is involved in the process, supported by regulation and management systems (NBCSD 2004).

The country's regulation on construction services has provided provisions about the types of services that can be carried out by the construction firms, types of firms, and the business area of the services. Recently, non-residential projects, roads and bridges, and residential projects are still the highest value projects that have been delivered in Indonesia. Regarding its contractors, they are owned either by the government (state-owned) or private parties (private firms). Until now, the state-owned and large contractors are still dominating the construction market in Indonesia (Sutjipto 1991; NBCSD 2002).

There had been an analysis of the existing conditions that are considered as problems faced by the Indonesian contractors. Larasati and Tsunemi (2009) described that the production process in the construction industry in Indonesia does not run smoothly. This is indicated by problems in project delivery systems such as the lack of appropriate materials and the necessity to waste valuable resource on rework. Moreover, low skill index and experience of construction workers make the business face difficulties in undertaking new concepts and technologies. Although there are still weaknesses and threats that are faced by Indonesian contractors, the strength and opportunities of these firms are quite significant. Moreover, the growth of the country's economic performance and investment opportunities in the infrastructure sector create a promising environment for Indonesian contractor's business. BCM in the context of Indonesian contractors is therefore an important issue to address for the construction industry to continue to play an important role in the economic growth plans of Indonesia.

1.3 Research Problems

As firms located over a vast geographical area, which is known as the world's largest archipelago (Raftery, Chiang and Anson 2004), Indonesian contractors have also experienced various threats or crises that have significant impacts on their business activities. Various crises have recently occurred in Indonesia, such as the financial crises in 1997 and 2008, natural disasters that occurred frequently (earthquakes, floods, tsunami), the political and financial instability, terrorism issues, and other internal events (IFRC 2004; Saparini 2009; Sutardi 2006; Tirtosudarmo 2005; UNR/HC 2005).

The crises stated above had resulted in various levels of impacts, where it can start from disruptions to business activities, loss of potential markets, loss of productivity and profitability, to the extreme case such as bankruptcy of a firm. As for the crisis responses, the contractors appeared to have reacted differently for different types of crises. Most of the contractors have provided relevant emergency responses for evacuating people during the crises (in external man-made events and natural disasters). However, a detailed recovery procedure for their business to resume after the crisis does not appear to have been planned in advance. The firms would have created the recovery team and developed steps to resume operations based on the management's further decisions. Moreover, crises related to natural disasters (such as earthquakes, tsunami and floods) caused the contractors to apply emergency responses (site or office evacuation procedures), and further coordination with the local government. The processes after the disaster appeared to be mostly dependent on the government's aid. These included temporary shelters, health aid facilities, and surveys of damage around the area provided by the government. From these cases, it seems that these firms have not developed their crisis responses into a holistic management approach in the organization, and there is a lack of detailed responses for their business stakeholders (Agustinus and Luhur 2008; Firdausy 2002; Herlijanto 2004; ICG 2002; Kartasasmita 2000; Lee 2009; MiyamotoINTL 2007; PTX 2008; Tambunan 2006; UNR/HC 2005).

Furthermore, considering the types of crises and the severe impacts that have occurred and have been experienced by Indonesian contractors, the existing responses made for these crises were not fully effective for safeguarding the business continuity of these firms. To become resilient and capable of providing an effective response to such threats, Indonesian contractors should start to adopt a systematic management concept in their organizations. BCM provides this framework, where based on its definition, it builds resilience and the capability for an effective response that safeguards the interests of its key stakeholders, reputation, brand and value creating activities (Elliott et al. 2010).

Ferre (2000) stated that BCM is not only focusing on death avoidance or survival, which has a low level of preparedness in the system, but more of preparing to be proactive in facing the disruption. This high-level of preparedness will provide better resilience and continuity for the business, hence supporting the improvement of its business performance. BCM is different from other emergency response process. It is not used only in response to a failure or crisis, but it emerges with a functional approach that integrates the whole components in the organization to prepare, respond, recover and restore from crises.

From these observations, it can be concluded that BCM is needed for implementation by Indonesian contractors because there are many threats by virtue of the nature of the business of a contractor, particularly in Indonesia. Secondly, the impact of crises can interrupt the firm's business activities from the low level impacts to the most severe impacts, such as bankruptcy. Last but not least, the existing crisis responses of the firms are still ineffective, and hence BCM is needed in the organization for safeguarding the firm's business continuity during any disruptions.

There are several benefits that can be gained by Indonesian contractors in adopting the BCM concept. BCM will help to improve resilience to disruptions. With BCM, in the event of a major incident, the contractors will manage to continue their business with little or minimum disruption, which is important to protect the livelihoods of all employees and those in the supply chain. BCM helps businesses to better understand the threats and activities at risk, and structured the process for implementing measures to protect against these. Also, having BCM in the firm will increase the client's confidence levels and help in gaining new markets. Finally, developing BCM in an automated form as a KBDSS can assist the firm in reaching an efficient decision (Tinston 2010; BSI 2010; Singh et al. 2008).

1.3.1 Knowledge Gap

Currently, studies on Indonesian contractors adopting BCM have yet to be found. Nonetheless, there are some studies that can be used for consideration in filling the knowledge gap. These studies described that BCM have been implemented by other organizations in other sectors, how some construction firms from other countries have adopted BCM in their organizations, the importance of organizational culture and institutional perspectives in adopting a concept in an organization, and the benefits of KBDSS.

BCM is found to be widely used in various types of firms, particularly in banking, telecommunication, oil and gas, and retail industries. These firms had developed BCM in their management systems. The BCM development is based on their business strategies and activities. Moreover, these firms developed different procedures for overcoming different types of crises, where some of them had also focused not only on their business continuity, but the service continuity to their customers as well (Elliott et al. 2002).

Herbane et al. (2004) noted that all of the organizations that adopt BCM had recognized that in the face of internal and external threats to the continuity of operations, a socio-technical approach, which is more than IT-focused disaster recovery, is necessary in order to improve business recovery from crises. These firms have strategically linked BCM to their essential business functions of their operations. This study was observed from firms in the financial service industry, vehicle maintenance services, gas supplies, water utilities, supermarkets, and local authorities.

As for the construction industry, the BCM concept seems to be relatively new for contractors, especially in the Asian region (e.g. Singapore, China and Hong Kong). Most of the large construction firms in China, Hong Kong and Singapore have not implemented BCM in their organizations due to a lack of awareness (Low et al. 2008b). Particularly in Singapore, BCM is far from being fully embraced by construction firms. Although the importance and usefulness of BCM in the construction industry is clear, the receptiveness of BCM among the construction firms is far from ideal (Low et al. 2010).

Furthermore, this situation also occurred in the United Kingdom. A study reported that although construction firms had identified threats, they had done little to prepare for resilience. Less than 50% of the construction firms had a business continuity plan in place, and the drivers of developing the plan were mostly due to regulations or requirements from central government, insurers or auditors (Broughton 2005).

Adopting a new concept like BCM is not a straightforward process. There are issues to consider before implementing the concept into the firm. Previous studies had shown that the immediate motivation for a firm in adopting a concept or system comes from institutional forces and organizational culture. Institutional forces emanating from the environment and transmitted through operational channels can strongly affect firms taking on a new concept. Moreover, organizational culture is the key to many change initiative. In adopting any concept or management system into a firm, it must redefine its culture to some extent and that success in implementing a concept can depend on the organizational culture. A firm is more likely to adopt a system if the values embedded in the system fit its organizational culture (Liu et al. 2010).

According to Low et al. (2010b), Scott's (2001, 2004) Institutional Compliance Framework could be adopted to examine how construction companies manage impending crises and the drivers that would spur them to implement BCM. It can investigate the motivation and current situations of implementing BCM from an institutional perspective. This framework offers a sound platform to explain why construction companies do or do not wish to implement BCM from the factors that may influence compliance and the reasons for compliance. This framework is based on the Institutional theory that examines the processes and mechanisms by which structures, schemas, rules and routines become established as authoritative guides for acceptable social behavior (Scott 2001, 2004). Different components of the Institutional theory explain how these elements are created, diffused, adopted, and adapted over space and time; and how they fall into decline and disuse. Collectively, this theory appears to provide a framework to garner a reasonable interpretation of the corresponding implementation issues.

Furthermore, establishing a BCM culture in the organization is crucial to ensure that continuity is embedded in the company (Business Continuity Institute 2002, 2007a). A culture that supports BC planning is the end result of successful efforts to engage business process owners in the planning process. Creating an organizational culture that supports BCM does not imply that the entire organizational culture needs to change for BC planning to be effective. Rather, those implementing BCM need to be aware of the organization's culture, seeking to change only that which is within that leadership's sphere of influence. By identifying what the existing organizational culture is serves to facilitate successful BC planning. Culture determines the degree to which resistance to change affects the rollout of any program. Culture also determines degrees of accountability and ownership of business processes. Similarly, the organization's tolerance for risk, a cultural outcome, helps to determine BC strategy. Lastly, the ability to get things accomplished within the organization is a characteristic of culture (Goldberg 2008).

Thus, institutional forces and organizational culture may work together and interact with each other to affect concept or system adoption (Liu et al. 2010). Based on these considerations, before implementing BCM, the Indonesian contractor's organizational culture and institutional forces should be identified in order to determine whether these elements support or do not support BCM implementation.

Before adopting and implementing a concept, an organization should also analyze its level of preparedness towards the concept. This assessment helps to identify strengths and areas for improvement. Based on these analyses, the organization can further invest its resources to implement the concept accordingly (McKinsey 2013). This process can be developed in a form of a KBDSS. A KBDSS can be used as a supporting tool to assess the organization's level of preparedness and to provide the knowledge needed by the management team in developing BCM, where the knowledge base can be updated regularly (Arain and Low 2006; Sudarto 2007).

According to McKinsey (2013), the insights from this type of supporting tool provide an effective foundation for conversations and decision making in the organization. This supporting tool provides benefits such as (McKinsey 2013):

- Personal understanding of the organization and what improvements can be made;
- Sharing opinions and building alignment amongst leadership and staff;
- Agreeing on areas of focus for organizational improvement; and
- Access to the knowledge needed for the organization.

1.3.2 Research Questions

Based on the research problems and the knowledge gap, BCM is needed for implementation by Indonesian contractors in order to prepare and overcome crises or threats. Although there are no studies yet on Indonesian contractors adopting BCM in their organizations, some studies on BCM implementation in other sectors and other construction firms in other countries may provide relevant findings for this study. Furthermore, the roles of organizational culture and institutional forces were found to be important in adopting a concept such as BCM, and developing BCM in a form of a KBDSS can assist the firm in understanding its level of preparedness towards the concept. Therefore, the research questions for the study presented in this book are:

1. What are the BCM implementation guidelines for different levels of preparedness for Indonesian contractors?
2. How can the BCM implementation guidelines for Indonesian contractors be automated as a KBDSS?

1.4 Objectives

The research presented in this book aims to study BCM implementation for Indonesian contractors. The research objectives are:

1. To identify Indonesian contractors' knowledge about BCM.
2. To identify the significant drivers and hindrances from institutional forces for implementing BCM.
3. To identify the significant drivers and hindrances from organizational culture dimensions for implementing BCM.
4. To develop BCM implementation guidelines for different levels of preparedness for Indonesian contractors.
5. To automate BCM implementation guidelines for Indonesian contractors into a KBDSS.

1.5 Scope

In the study presented in this book, Indonesian contractors which are state-owned and private-owned enterprises were chosen for analysis. The large type of contractors was chosen because these firms are mostly involved in major construction projects and are dominating the construction market in Indonesia (NBCSD 2002; ICA 2013). In addition, all of them are members of the Indonesian Contractors Association (ICA).

1.6 Research Process

In general, a research process consists of steps which are to (1) identify the research problem or question, (2) review the literature to develop a hypothesis, (3) determine an appropriate research design to test the hypothesis, (4) devise appropriate methods to collect data, (5) collect data, (6) analyze the data and finally (7) conclude the study (Tan 2008).

Figure 1.1 illustrates the research process for the study presented in this book. The conceptual framework based on a comprehensive literature review was the starting point to develop the research designs. Following that, data were collected and analyzed for the development of the KBDSS. Validation was conducted for the KBDSS and findings were generated to reach the conclusion of the study presented in this book.

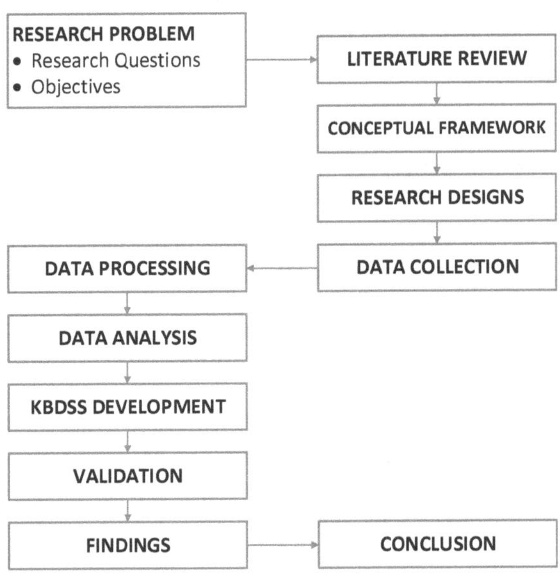

Fig. 1.1 Research process. Source: Adapted from Tan (2008)

1.7 Research Significance

The study presented in this book provides the following contributions to knowledge and practice, particularly to the improved understanding of BCM implementation for Indonesian contractors. The academic contributions consist of: (a) providing knowledge about adopting and implementing BCM in construction firms; (b) addressing perspectives on how the institutional theory can be a constructive platform for explaining the reasons to implement (or not implement) the BCM principles; (c) providing knowledge about organizational cultural studies in the context of BCM implementation in construction firms; (d) developing a conceptual framework for an organization to adopt BCM. The framework collates various schools of thoughts that relate to the study's topic; and (e) developing an automated system (KBDSS) for the management's decision making process when implementing BCM.

Furthermore, with respect to its practical significance, the study presented in this book provides the following: (a) knowledge for Indonesian construction firms on implementing BCM based on the technical and non-technical aspects; and (b) an effective BCM level of preparedness assessment process in the form of a KBDSS. Further details on these contributions can be found in the final chapter of this book.

1.8 Structure of the Book

There are fourteen chapters that describe the study in this book, which are:

- Chapter 1: This chapter introduces the background and motivation of the study. It will be followed by addressing the research problems (including its knowledge gap and research questions), research aims and objectives, scope and organization of the study presented in this book.
- Chapter 2: This chapter reviews the literature on the management of crisis in organizations. It describes the theories about management, organizational management, and crisis management.
- Chapter 3: This chapter reviews the literature on BCM, starting from its definition and development, its relationships with other concepts, BCM as a management system, and its main principles. Further, the reviews will describe Business Continuity Planning (BCP) and BCM's implementation (relating to standards and levels of preparedness). Finally, reviews of BC plans from various sectors or types and the need for BCM are elaborated.
- Chapter 4: This chapter reviews organizational culture and institutional forces as important aspects in adopting a concept. There are reviews on culture in organization, its dimensions, and benefits of identifying organizational culture. Institutional theory will be described that focus mainly on its three pillars of institutions.

- Chapter 5: This chapter reviews the mainstream theories implementation by contractors. It starts with describing organizational management, traditional organizational theories, and crisis management in contractors. The next sections are reviews on BCM in construction firms, organizational culture and institutional theory's implementation for contractors.
- Chapter 6: This chapter portrays the Indonesian construction industry, starting from general profile of Indonesia, its construction industry overview, and Indonesian construction firms. The last section reviews crises faced by Indonesian contractors.
- Chapter 7: This chapter describes the KBDSS that will be used in the study. The background of the KBDSS is its decision making process and tools. The review will be followed by discussing DSS and KBDSS. The last section will review KBDSS development, based on its formulation, system development, and validation.
- Chapter 8: This chapter explains the conceptual framework developed in the study. Indonesian contractor's knowledge about BCM is the first section that will be described, followed by relationships between BCM, organizational culture, and institutional forces. The next sections are developing BCM for Indonesian contractors, automating BCM through KBDSS, and the conceptual framework.
- Chapter 9: This chapter elaborates the research design and methodology for the study. It consists of research framework, research design, methods of data collection, and methods of data analysis.
- Chapter 10: This chapter describes the data analysis process from the pilot study and surveys. It explains the analyses conducted to provide the results for this phase of the study.
- Chapter 11: This chapter is the continuation of the previous chapter, which elaborates on the data analysis process from the case studies and BCM implementation guideline development.
- Chapter 12: This chapter focuses on the BCM-KBDSS development. It elaborates the Knowledge Base development (based on the BCM implementation guideline from the previous chapter), rules and logics development, synthesizing the Knowledge Base and rules, KBDSS finalization and validation.
- Chapter 13: This chapter discusses the findings derived from the data analyses results. It explains the findings for each research question of this study.
- Chapter 14: This is the final chapter that elaborates the conclusions and contributions of this study. In addition, it highlights the limitations of the research and provides recommendations for future studies.

Figure 1.2 illustrates the relationships between the chapters. The literature reviews sections are described in Chaps. 2, 3, 4, 5, 6 and 7. As shown in the figure, Chaps. 3, 4, and 5 are further reviews on the literatures related with Chap. 2, which are organization and management of crisis. Chapter 6 is a general portrayal of the Indonesian construction industry. Review about KBDSS (that provides an overview about KBDSS in construction) is described in Chap. 7. Chapter 8 presents the

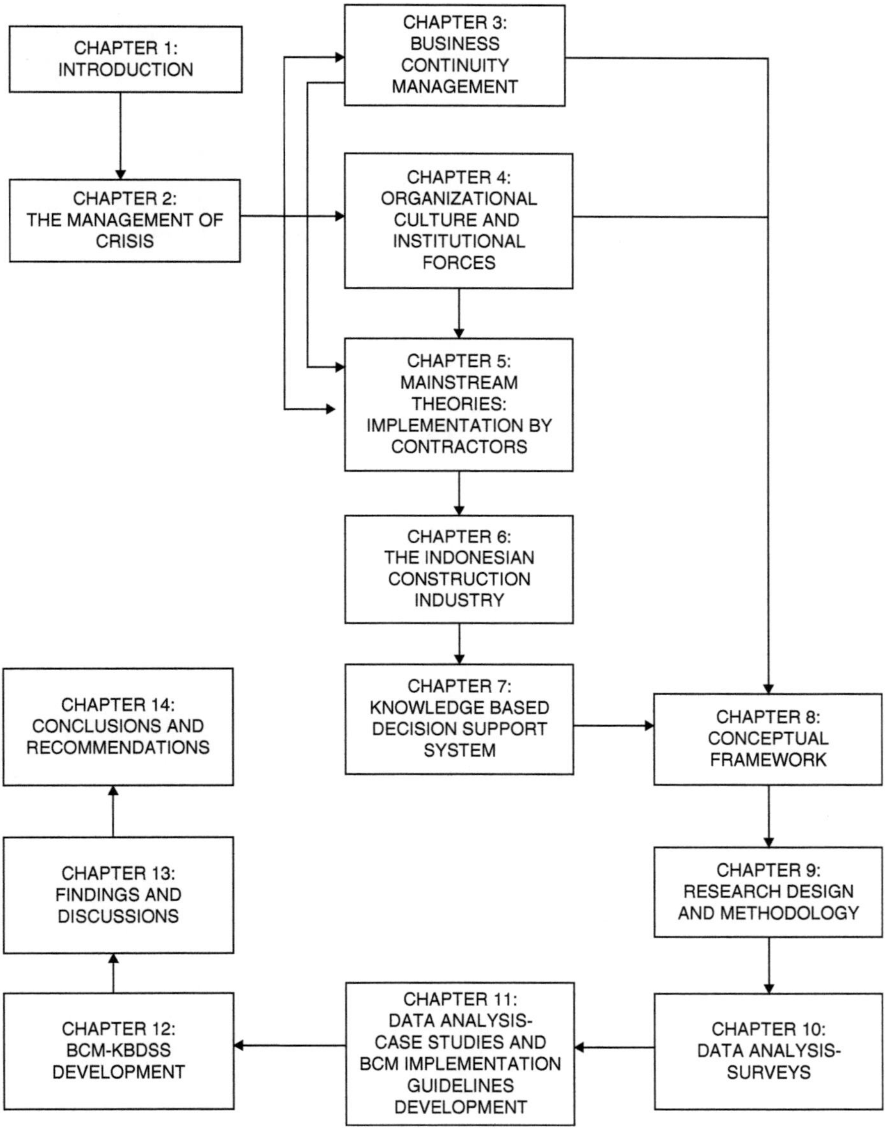

Fig. 1.2 Relationships between the chapters

conceptual framework of the study that compiles the reviews from previous chapters. Following that, Chap. 9 explains the study's research design and methodology. The study's results, findings, conclusions, and recommendations are described in Chaps. 10, 11, 12, 13 and 14.

Chapter 2
The Management of Crisis

2.1 Introduction

This chapter reviews the theories relating to the concept of crisis management in organizations. The first section describes an overview of management concepts in general. It is then followed by a review of mainstream theories in organizations which had evolved from the traditional organizational theories into systems and contingency theories. Other essential concepts relating to the organization such as complexity, and change are also discussed in this section. Furthermore, the chapter continues to describe crisis management, its definition, history of development and its main concepts.

2.2 Management

Management, which is regarded as an art by some scholars, has its aspects documented since the beginning of the industrial period. Management started to become a specific study in the nineteenth century, where the first major contribution to its definition was made by Henri Fayol. His definition of management was in terms of five functions (Lavender 1996; Naoum 2001):

1. To forecast and plan.
   This involves selecting missions, objectives and actions to achieve them. It requires decision-making, such as choosing future courses of action from among alternatives.
2. To organize.
   This function establishes an intentional structure of roles for people to fill in an organization, where all tasks necessary to accomplish goals are assigned to people. In other words, this function involves building the structural, material and human aspects of the undertaking.

L.S.R. Supriadi, L. Sui Pheng, *Business Continuity Management in Construction*, Management in the Built Environment, DOI 10.1007/978-981-10-5487-7_2

3. To command.
   This function maintains the activities among the personnel in the organization. It also influences people so that they will contribute to organization and group goals, where motivation, leadership styles and communication approach are part of the elements needed in this function.
4. To coordinate.
   This involves binding together, unifying and harmonizing all activities and efforts. Moreover, differences in approach, timing, effort or interest should be reconciled to contribute to organizational goals.
5. To control.
   This function measures and corrects activities of subordinates to ensure that events conform to plans. In details, it measures performance against goals and plans, shows where negative deviations exist, and, by providing actions to correct deviations, helps ensure accomplishment of plans.

Koontz and Weihrich (1990) further defined management as the process of designing and maintaining an environment in which individuals, working together in groups, efficiently accomplish selected aims. Moreover, several scholars on organization and management studies added social and cultural factors to Fayol's definition of management, where they relate the organization to the external environment and responding to society needs. They also opined that in management, there is a need to develop an organizational climate where people can accomplish their individual and collective goals (Naoum 2001).

Based on these definitions, it can be viewed that management refers to the development and operation of organization that derives its importance from the need for strategic planning, coordinating, directing and controlling various and complex decision-making processes. The key areas which are essential in management are (Olum 2004):

- Problem solving
  Management is mostly about solving problems that occur in an organization. This process is supported by problem identification, analysis and the implementation of remedies to the problems.
- Administration
  This is the process of developing and following procedures in the organization to achieve its goals and objectives.
- Human resources management
  Human resources will be managed based on strategic integration, assessment of workers and relationships between shareholders and workers.
- Organizational leadership
  This aspect should be developed along with interpersonal relationship, teamwork, self-motivation to perform, emotional strength and maturity to handle situations, and personal integrity.

Since the nineteenth century, the studies of management have developed into modern management thought, which was established by F. W. Taylor, and named as scientific management. This school of thought focused on a production-orientated view with units of labour feeding into the production process in a passive manner. Later on, the studies focused more significantly on the human element in production and in the post-1945 period, the authors have considered a wider variety of functions and contexts in an organization (Lavender 1996).

According to Tengblad (2012), the most recent management studies (since 1990 until 2012) increasingly emphasized on the emotional, political and symbolic aspects of managerial work. Along with focusing on classic activities such as planning, decision-making, organizing, and controlling, these aspects should not be ignored. Hill (1992) argued that the manager requires a mental transformation in order to learn to think, feel, and value as a manager. Such transformation can be acquired from learning through the on-the-job training where jobholders learn from their experience (such as successes, failures, and insights). They learn to cope with complexity, ambiguity, fragmentation, emotional stress, conflict and the importance of handling the symbolic aspects of management. Watson (1992) also described the chaos, uncertainties, ambiguities and contradictions that surround managers. Concepts like strategy and culture are presented as important empirical artefacts and rhetorical devices.

Mintzberg (2009) also acknowledged the highly complex, fragmented, hectic, and often chaotic nature of management. Mintzberg's (2009) study presented the required competencies of good managers, which reveal the complexity of the role of management:

- Personal competencies: managing oneself (reflecting, managing time, prioritizing, and agenda setting);
- Interpersonal competencies: leading individuals and groups (selecting, mentoring/coaching, inspiring, team-building, and resolving conflicts); administering and linking the organization/unit (allocating resources and delegating);
- Informational competencies: communicating verbally and non-verbally, and analyzing information;
- Actionable competencies: designing and mobilizing (planning, visioning, negotiating, politicking, and managing change).

Mintzberg (2009) also stated that management practice develops from three major human spheres: art (the imaginative, creative and insightful), science (analysis and systematic evidence), and craft (experience and practical learning). Management at its best is insightful, engaging, and mindful; at its worst, it is disconnected from reality, demotivating, and disorganized.

2.3 Organizational Management

2.3.1 Organization Overview

Daft (1998) and Kirst-Ashman (2000) have defined organizations as social entities that are goal directed, designed as deliberately structured and coordinated activity systems and are linked to the external environment. Social entities in this context mean that organizations are made up of people who receive responsibilities in doing their jobs, supported by their own values and personalities. Thus, patterns of behavior develop in organizational environment. Regarding the term goal-directed, this means that an organization exists for some specified purposes. It must then clearly define its goals in order to evaluate the extent to which it achieves these goals. Furthermore, organizations are considered as deliberately structured and coordinated activity systems, where such systems are guided by a technology to achieve desired ends. These systems are structured by policies for how the organization should be run, hierarchies of how the personnel are supervised and different units working in various ways to help the organization function. The linkage of an organization to the external environment refers to the constant interaction with other systems in the social environment including individuals, groups, other organizations and communities.

In line with the definition stated above, Naoum (2001) also described that organizations come in various sizes and shapes in performing specific functions to achieve certain objectives through a collection of people and other resources. These resources are coordinated by a set of procedures and integrated by a form of organizational structure. The ways in which the objectives are planned and how the resources are managed differ considerably among organizations. However, there are six common elements in an organization, which are the operation (task and technology), objective (tangible and intangible products), resources (human and non-human), structure (formal and informal), management (strategic and operational), and environment (internal and external).

Naoum (2001) further identified three main sub-systems that integrate the operation and resources within the organization, which are:

1. The managerial sub-system—including setting goals, planning a strategy, structuring, coordinating and administrating.
2. The technical sub-system—including planning and control techniques, production methods, facilities and equipment.
3. The social sub-system—including leadership, personnel management work groups, motivation and culture.

The success or failure of the organization depends on the clarity of the operation and the objective, the quality of the people employed, the availability of the resources and the suitability of the structure and the management system adopted.

In order to work within organizations, understanding the major theories regarding how organizations operate and functions is beneficial. The concepts of organization and management theory are not completely distinct and unrelated. They have been developed and evolved from views derived by practicing executives, administrators, scientists, sociologists and economists, who observed and set general guidelines for others. The current management practices are somehow influenced, either consciously or subconsciously by these theories. The evolution of these theories can be viewed in four main stages (Naoum 2001; Lavender 1996; Wertheim 2001; Coffey 2010):

1. The classical view (1900–1910s)
   This view emphasized the division of labour and the importance of machinery to facilitate labour. Scientific management and the development of basic principles for specialization of work, unity of command, scalar chain of command and coordination of activities were established.
2. The human relations view (1920s–1960s)
   This view focused on the importance of the attitudes and feelings of workers, and informal roles and norms which influence performance of individual worker and groups. Theories about group dynamics, bureaucracy, leadership and decision theory were further developed in this era.
3. The development of modern systems theory (1960s–1970s)
   After developing theories about work groups and technology, scholars then focused on understanding a work system, which also described the existence of mechanistic and organic structures and stated their effectiveness with specific types of environmental conditions and technology. They developed systems theory which represented organizations as open systems with inputs, process, outputs and feedback.
4. The contingency concept (1980s)
   This concept emphasized the fit between organization processes and characteristics of the situations in its environment.

2.3.2 *Traditional Organizational Theories*

Traditional organizational theories consist mainly of two schools of thoughts, which are the classical view and the human relations view. Brief descriptions about the two views are given below:

- Classical school of thoughts

Lawrence and Lorsch (1986) found that the classical school of thoughts was pioneered by Fayol, Mooney, Urwick, Gracunias and Gulick, which had developed a highly formalized structure with a directive or authoritarian leadership style that was considered to lead a high performance in a certain, homogeneous and stable environment. In details, classical theory looked into organizational management in terms of its purpose and formal structure. There are three distinctive pillars to classical theory; these are: traditional universal management (administrative management), quantitative management (scientific management) and formal structuring (the bureaucratic model) (Naoum 2001).

As mentioned by Naoum (2001), the traditional universal management (administrative management) was identified by Fayol, who not only described the five functions of management, but also established the fourteen principles of management, which are:

1. Division of work. The principle of specialization of labour in order to concentrate activities for more efficiency.
2. Authority and responsibility. Authority is the right to give orders and the power to exact opinion.
3. Discipline. Discipline is absolutely essential for the smooth running of business, and without discipline no enterprise can prosper.
4. Unity of command. An employee should receive orders from one superior only.
5. Unity of direction. One head and one plan for a group of activities have the same objectives.
6. Subordination of individual interests to general interests. The interests of one employee or a group should not prevail over that of the organization.
7. Remuneration of personnel. Compensation should be fair and, as far as possible, afford satisfaction to both personnel and the firm.
8. Centralization and decentralization. Centralization is essential to the organization and is a natural consequence of organization.
9. Scalar chain. The scalar chain is the chain of superiors ranging from the ultimate authority to the lowest tank.
10. Order. The organization should provide an orderly place for every individual. A place for everyone and everyone in their place.
11. Equity. Equity and a sense of justice should pervade the organization.
12. Stability of tenure of personnel. Time is needed for the employees to adapt to their work and to perform effectively.
13. Initiative. At all levels of the organizational ladder, zeal and energy are augmented by initiative.
14. Esprit de corps. This principle emphasizes the need for teamwork and the maintenance of interpersonal relationships.

At the same time when Fayol identified the traditional universal approach for management, Frederick Taylor, Frank Gilbreth and Henry Gantt were tackling the problem of efficiency in a scientific way. The aim of scientific management (quantitative management) was to identify universal principles on which production could best be organized. It was assumed that there was a best way of doing things, and it was the task of management to determine what it was. These principles were derived into:

1. The development of a true science for each person's work and not the old rule-of-thumb.
2. The scientific selection, training and development of workers, unlike in the past when they chose their own work and trained themselves as best they could.

3. Cooperation with the workers to ensure work is carried out in the prescribed way.
4. The division of work and responsibility between management and the workers.

Later on, these principles were carried forward throughout the twentieth century and, in about 1960, the scientific approach was characterized by the techniques of operational research, where various mathematical models were developed to solve decisional problems such as CPM, PERT and linear programming. However, this approach then turned away from an emphasis on narrow operational research techniques towards a broader perspective of management science. The management science approach incorporates quantitative decision techniques and model building as in operational research, but it also incorporates computerized information systems and operations management (Lawrence and Lorsch 1986; Naoum 2001; Lavender 1996).

The third pillar in the classical theory was created by Max Weber, who developed the bureaucratic model structure (or formal structuring). It was the beginning of the formal organization where rules and procedures were designed to coordinate and direct people towards organizational goals. The characteristics of Weber's model are as follows (Blau and Scott 1966; Naoum 2001):

1. Organization tasks are distributed in the clear cut division of labour, creating specialisms and expertise of staff emphasizing technical qualifications.
2. Job roles are organized hierarchically, in most cases in a pyramid structure where authority and subordination are clearly seen.
3. A formal set of rules exists to govern decisions and actions.
4. Officials are accepted to assume impersonal organization to clients and individuals as cases. Here, formal behavior is encouraged.
5. An employment and career structure using qualifications, experience, seniority, etc. as a rational basis for advancement is in operation.

- Human relations school of thoughts

This school of thoughts was focusing more on the human beings in the organizations, where this path moved towards a more broadly based organizational behavior. According to Lawrence and Lorsch (1986), scholars such as Mayo, Roethlisberger, Lewin, and McGregor stressed the importance of viewing organizations as systems of relationships. Furthermore, March, Weber, Blau, Gouldner and Crozier developed decision theory and bureaucratic theory that relate the interaction between the human being and the formal organization.

The most popular studies in the human relations school were those carried out by Elton Mayo. The studies were focusing on increasing productivity and efficiency, but concluded that motivation was essential and found to be greatly enhanced when workers felt to be a part of a group. Thus, social interaction was regarded as beneficial. Other findings from these studies were that informal organizations develop within the formal, and therefore it should be encouraged where it can result in better performance for the organization. Human relations studies go beyond scientific management, where they viewed people not as mere machines who only have to be paid a certain rate to motivate them to work. Nonetheless, these studies still regarded workers as units of labour where they are still subject to direct management control and have little role in giving feedbacks to the organization (Naoum 2001; Lavender 1996).

2.3.3 Systems Theory

The systems approach to management attempted to integrate the earlier approaches, which are the classical and human relations theories. This approach had evolved due to the increasing complexity of organizations and as a response to a rapidly changing environment (technical, economic, social and governmental) (Naoum 2001). Griffith (2011) found that systems theory was developed by Hegel, who suggested that in a system of any kind the whole adds up to be greater than the sum of the parts, or produces synergy; the whole determines the characteristics of the parts; the parts cannot be fully understood if seen in isolation from the whole; and the parts are interrelated and therefore interdependent. These attributes of systems are used in the configuration of almost all organizations (Bertalanffy 1968; Checkland 1981; Kirst-Ashman 2000; Cleland and King 1983). Hamilton (1997) further stated that a systems perspective enables an organization to develop:

- Its way of thinking (philosophy).
- The design and operation of the organization as a whole (management).
- Its method of analysis and problem solving technique.
- Logical and regular consideration (systematic thinking).

Therefore, systems theory can provide a framework for the managed direction of organizational activity through the provision of all its managed parts with a focus on and benefit to the core business it undertakes, which creates synergy and a holistic perspective (Griffith 2011).

Systems theories focus on how organizations take resources and process them into some kind of product or service. They emphasize on how all parts of the organization (as subsystems) are interrelated and function together to produce output (Holland 1995; Hodge, Anthony and Gales 2003; Holland and Petchers 1987). Moreover, a system can be looked at as having inputs (e.g., resources such as raw materials, money, technologies, people), processes (e.g., planning, organizing, motivating and controlling), outputs (products or services), and outcomes (e.g., enhanced quality of life or productivity for customers/clients, productivity). Systems share feedback among each of these four aspects of the system. Figure 2.1 illustrates an example of an organization that uses a systems approach, involving the four aspects of the system.

This theory had a significant effect on management science and understanding organizations. As a collection of parts unified, if one part of the system is removed, the nature of the system is changed as well (Koontz 1980; Olum 2004). In practice,

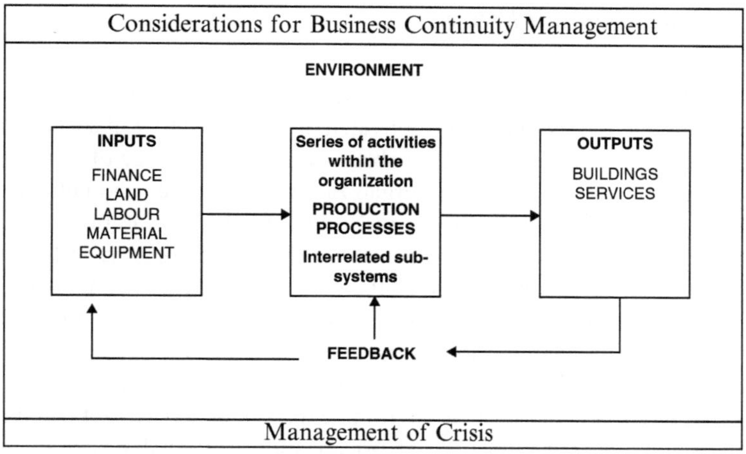

Fig. 2.1 The organization as an input-process-output system. Source: Adapted from Naoum (2001)

Coffey (2010) described that the analysis of an organization using the systems approach would involve steps as follows:

- Defining the problem and the scope of system to be studied.
- Breaking down the system into basic components.
- Gathering data about each component.
- Identifying/evaluating alternative solutions and selecting the best one.
- Evaluating interactions among components of alternatives.

Generally, there are two types of system, which are the open and closed system. In an open system, the organization is highly affected by the external environment and the four factors (inputs, process, outputs and feedback). The environmental influences on an organization are mainly from the economical, social, technical, and political aspects. In a closed system, the environment does not play an important part in the business processes. A closed system focuses on optimizing the internal efficiency of the operation rather than looking outside the organization for effective actions (Naoum 2001). According to Lavender (1996), most organizations are open systems, where they interact with their environment. Thus the organization as a system takes in resources from the outside environment, processes them and then sends them out. From this view, it can be seen that applying systems theory in management helps managers to look at the organization more broadly, and enable them to interpret patterns, various parts, interrelations of the parts and events in the workplace (Koontz 1980; Olum 2004).

2.3.4 Contingency Theory

Naoum (2001) mentioned that while interaction between the organization system and its environment is essential for survival, a firm's probability of success depends more on the ability of management to obtain an optimum degree of fit between the complex and sometimes conflicting organizational objectives and culture; culture and structure; structure and strategy; individual employee's ability and expectations; type of work and external environment. These principles are in line with the contingency theory, which states that there is no one best way to structure the organization, to lead a team, and to design a system. All are contingent upon the situation or problem at hand. Moreover, contingency theory suggest that organizational variables are in a complex inter-relationship with one another, with conditions in the environment and that environmental contingencies act as constraints and opportunities which influence the organization's internal structures and processes (Lawrence and Lorsch 1986).

Scott (1992) also opined that contingency theory is considered a dominant, theoretical, rational, open system model at the structural level of analysis in organization theory. The basic assertion of contingency theory is that the environment in which an organization operates determines the best way for it to organize, with basic assumptions that there is no one best way to organize, and any way of organizing is not equally effective (Galbraith 1973). This theory is mainly concerned with adapting to change, since it recognizes that no universal approach to management is possible. The most appropriate form of management will differ between organizations, and will differ over time within the same organization. Contingency theory has some common threads and overlaps with systems theory. The similarity with systems theory is that there is some emphasis on the environment in which the organization exists. In the case of contingency theory, the environment is one of the major influences for change. Organizations exist to prosper in a changing political, regulatory and financial marketplace where prevailing circumstances must be grasped as catalysts for change and actions responsive to the needs of change (Lavender 1996; Griffith 2011; Lawrence and Lorsch 1967).

In applying the theory, Carlisle (1976) had developed a contingency model, which analyzes a situation and determines what variables influence the decision of which is being concerned. Figure 2.2 shows the contingency model, where the center circle represents the agency. The primary internal contingency on which management depends is the agency's purpose or goals. The people hired, technology used, tasks performed and organizational structure are heavily influenced by an agency's goals. Moreover, the agency is also influenced by its environmental forces, to process the inputs in order to obtain the outputs.

From this model, it shows that when managers make a decision, they must take into account all aspects of the current situation and act on those aspects that are essential to the situation at hand. Alternatively, this situational or contingency approach can also be considered as an "it depends" approach (Olum 2004).

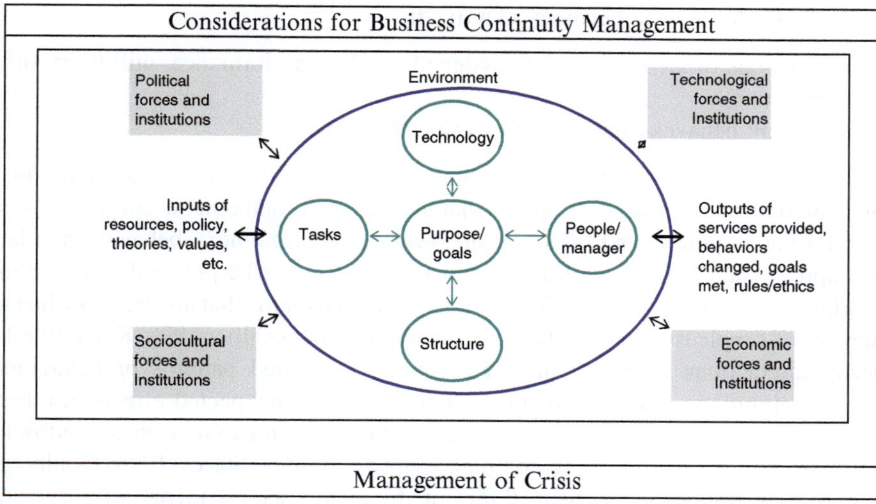

Fig. 2.2 Contingency model of management. Source: Adapted from Carlisle (1976)

2.3.5 Complexity Theory in Organizations

Complexity in organization has been discussed and considered by many practitioners and scholars. Even though there is no universal definition of complexity theory, its principles have been considered by many academics and practitioners in organizational studies and business management (Mitleton-Kelly 2004). Developed principally in the fields of physics, biology, chemistry and economics, complexity theory arises from chaos theory, in that it shares chaos theory's focus on the sensitivity of phenomena to initial conditions that may result in unexpected and apparently random subsequent properties and behaviors. The philosophy of chaos states that physical and lawful do not mean predictable, controllable or completely knowable. Furthermore, chaos theory suggests that even a very slight degree of uncertainty about initial conditions can grow inevitably and cause substantial fluctuations in the behavior of a particular phenomenon (Mason 2008; Griffin 1996; Goerner 1994; Doll 1987).

Generally, complexity theory is mainly focusing on "the study of the dynamics of complex adaptive systems which are non-linear, have self-organizing attributes and emergent properties" (McMillan 2006, p. 25). Holland and Miller (1991) defined a complex system as:

1. A network of interacting agents.
2. Exhibiting dynamic aggregate behavior that emerges from the individual activities of the agents.
3. Aggregate behavior can be described without a detailed knowledge of the behavior of an individual agent.

The complex system becomes adaptive if:

1. The actions of an agent can be assigned a value (performance, utility, payoff, fitness, etc.).
2. The agent behaves to increase this value over time.

Moreover, Stacey (1996, p. 10) defines complex adaptive systems as consisting of "a number of components (agents) that interact with each other according to sets of rules that require them to examine and respond to each other's behavior in order to improve their behavior". These interactions can either be physical or relate to sharing information, where the interactions develop patterns that are created when a number of simple rules are applied over many iterations (Cillier 1998; Kelly 1999). Many interactions in a system can produce unexpected patterns or behaviors because stimulating one part of the system can have unexpected effects in other parts of the system. Such unexpectedness is due to the nature of non-linear feedback networks and the nature of the complex adaptive system (Stacey 1996; Goldberg and Markoczy 2000). Morrison (2002, p. 6) puts it as, "a theory of survival, evolution, development and adaptation". It concerns itself with environments, organizations, or systems that are complex in the sense that very large numbers of constituent elements or agents are connected to and interacting with each other in many different ways. Also, complexity theory offers some useful insights into the nature of continuity and change (Mason 2008).

According to Pascale et al. (2000) and Santos (1998), generative complexity takes place in the boundary between rigidity and randomness, which can be applied to the production process. The boundaries are where change is managed and these are set out in procedures. Procedures are defined as organizational design statements, and capture the methods to execute a task (Rogers 1995). Organizations write procedures in order to manage aspects of operations, where many of them today are still written through hierarchical command-and-control structures with a high level of rigidity. A rigid, rule bound structure, although providing management with a sense of control, is incapable of adapting to meet new situational requirements. These rigid procedural structures also often depersonalize the social elements and practices that had been developed. If applied processes are too rigid, an organization will fail owing to a lack of creativity, too random and there are numerous examples of failed organizations which could not find a balance (Stacey 2000; Rogers 1995; Brodbeck 2002; Mercer 1999).

Scholars like Anderson et al. (1999), Harald et al. (1999) and Sherman and Schultz (1998) argued that complexity theory can influence the design and development of procedures which focus more on the impact of natural human behavior, that is considered as the natural drivers to "get the job done". Based on this thought, procedures could be developed to promote self organizing frameworks utilizing natural laws which would drive simplicity and generate greater influence without the need for "force" or detailed bureaucratic approach. The behavior of individuals is self-organizing when people (or agents using the language of complexity) are empowered, free to associate with others and cross organizational boundaries to pursue their goals (Coleman 1999).

Lissack (1997) stated that complexity theory identifies the gatekeepers of an organization as those that stand at its boundaries and translate information between the internal and external world. The nature of these gatekeepers that interact with other business units, suppliers and customers should be standardized without resort to the application of rigid procedures that would destroy the evolutionary nature of the system. Therefore, by embedding a system which is capable of undergoing continuous metamorphosis in order to respond to a dynamic business landscape, through creating the capability of continuously adapting and co-evolving within the environment, a competitive advantage may be gained in the organization (Brodbeck 2002; Lewin et al. 1999).

2.3.6 Change in Organization

According to Lawrence et al. (1976), there are two main objectives of an organization change, which are (a) changes in an organizations' level of adaptation to its environment; and (b) changes in the internal behavioral patterns of employees. Organizations are continuously struggling to adapt themselves to their external environment. Due to the inability of the management to fully control its environment, internal organizational changes have been introduced which allow them to cope more effectively with new challenges such as increased competition, advances in technology, new government legislation, and social demands. The most frequent organizational changes are due to environmental pressures. In some cases, changes are made to anticipate future pressures. For this latter course, although it is more difficult to pursue because employees may not recognize its direct importance, it is a standard that can often be applied to organizations that lead their industries. These types of organizations can be considered as proactive, where they engage in attempting to change their environments and themselves. Any organizational change, whether it is introduced through a new structural design or a training program, is principally trying to get employees to adopt new patterns of behavior and ground rules for relating to each other and their jobs. For the changes in the organization to have significant effects, these new behavior patterns must surface within superior-subordinate relations, work groups and also include larger subsystems (such as departments and divisions) of the whole organization.

A visible crisis faced by an organization can be an important force for triggering behavioral change, although such change may have costs derived from it. Essentially, such crisis has an unfreezing impact on the members of the organization, causing them to review and analyze their current attitudes and behavior patterns. During this period, the organization is in a fluid state and people are more apt to accept new ways of thinking and acting. However, such crises do not fully affect the awareness of organization members, unless the top management communicates the need for change. Whether the means for communicating the importance of change is a top-management pronouncement, the entry of selected consultants, or from the new top management, the force for the change is fundamentally at the top of the organization. Alternatively, the approach on how change should be brought is through involving a wider segment of the organization, such as small group meetings, where the process of establishing and diagnosing the need for change are discussed. Either one of the approaches or some fusion of the two can be used with focus on the entire change process, the plan and how to implement them (Lawrence et al. 1976).

The process of introducing change in an organization involves several steps (Lawrence et al. 1976):

1. Creating an awareness of the need for change.
2. Analyzing the situation which creates the need for change and determining the direction for change.
3. Communicating the change to the people involved and affected.
4. Monitoring the change and making adjustments if needed.

Moreover, Lawrence et al. (1976) also described several factors that indicate which approaches for changes can be most effective, either top-down or wider segment approach. Firstly, the person who has the relevant data to determine the need for change, the direction of change and the likely consequences should plan and direct the change. If such knowledge is sourced from the top management, then they can be the one who plan and direct the process. Secondly, the existing norms about involvement of subordinates in decision making can also have an important impact. If the members of the organization expect that top management will make important decisions and communicate them to the organization, then the top-down approach to the change process will be more suitable. Thirdly, the approach selected must be consistent with the leadership style of the top managers in the firm. This means that whatever strategies top managers use must be compatible with their own leadership styles. Last but not least, the size of the organization can also affect management's thinking about its use of power to introduce the change. For example, even in situations where other factors had suggested a power sharing approach (a wider segment approach), the large size of an organization can make such an approach seems impractical.

In managing change, if it is conducted successfully, it will be one of the key factors in organizational effectiveness. The reason for this is because diagnosing and solving organizational problems involves the interaction of a multiplicity of factors influencing an organization's ability to change and its proper mode of change (Naoum 2001). Lawrence et al. (1976) stated that there are four areas that should be considered by the manager involved in order to successfully manage the change process:

- Diagnosing organization problem

  This phase is deciding on what are the specific problems to be corrected?; what are the determinants of these problems?; what forces are likely to work for and against change? This will be followed by a variety of diagnostic methods to use such as meetings, conferences, consultants, task forces, interviews, surveys, etc.

- Planning for change
  This stage is translating a thoughtful diagnosis into an appropriate action plan. This is implemented by deciding about the overall goals for change, selecting the basic approach for reaching these goals, and planning a sequence of detailed steps for implementing the basic approach.
- Launching organization change
  This is the stage when the plan will be put into action. The manager's interpersonal skills will be at its fullest test in this phase.
- Following up on organization change
  Monitoring of any change is essential, particularly in terms of getting proper and accurate feedback on the change process. A systematic evaluation should be taken to monitor and control the process.

Leavitt (1964) discussed an applied organization change in industry, which relates to structural, technical and human approaches. According to Leavitt (1964), organizations as multivariate systems, have at least four interacting variables looming, which are variables of task, structure, technology, and actors (usually people). These variables are interacting with each other and highly interdependent, so that a change in any of the variables will most likely result in compensatory (or retaliatory) change in others. As an example, the introduction of computers (as new technological tools) may affect changes in the organization's structure (e.g. in communication system or decision map of the organization), changes in people (their numbers, skills, attitudes, and activities), and changes in task performance or even task definition, since some tasks may now become feasible of accomplishment for the first time.

Regarding change in organization, Naoum (2001) concluded that an innovative firm is usually more successful to change when their environments change. An innovative firm has common characteristics such as a flexible organization structure, open communications, informality, and participative decision-making. As a whole culture, it innovates to meet changes in the needs of its customers, the skills of competitors, the mood of the public and forces of international trade or government regulations.

2.4 Crisis Management

Considering that the current dynamic and high-velocity business environment is characterized by discontinuity and continuous change, crises are regarded as more of the norm rather than exception in organizations (Paraskevas 2006; Zhong and Low 2009). The term "crisis" has been defined differently by different management writers. A crisis is an unexpected event in an organization's life which represents a significant threat to its high priority values and demands a time-pressured response. It is a situation faced by an individual, group or organization which they are unable to cope with by the use of normal routine procedures and in which stress is created by sudden change (Loosemore 1998a). Moreover, a crisis is defined as a turning point in the course of anything; uncertain time or state of affairs; moment of great danger or difficulty (Longman 1978). Often, there is substantial agreement on three conditions that are deemed necessary for a crisis to exist, which are severe threat, high levels of uncertainty, and urgent need for action (Drennan and McConnell 2007).

Organizational crises are the manifestations of unexpected risks that develop into decisive periods of acute difficulty which threaten the viability of an organization, its business units or key products (Loosemore and Teo 2006; Fink 1986). These crises are considered as low probability, high-impact events of complex and ambiguous cause and effect, and they need critical and rapid analytical decision-making skills (Pauchant and Mitroff 1992). Therefore, crisis management is crucial for all organizations, where effective crisis management helps to ensure the continuous wellbeing of an organization. Furthermore, crisis management is much more than a simple matter of setting up contingency plans and avoiding risk (Low et al. 1999).

According to Devlin (2007), examples of a crisis can include when an organization experiences a product failure, a product safety issue, product tampering, a product market-shift, and incident that results in a poor image or negative reputation, an international incident that negatively affects the organization, and a financial problem—especially a fuzzy accounting problem. In addition, the various types of disasters would also come to mind such as fires, floods, tornadoes, earthquakes, terrorist bombings and so forth.

Drennan and McConnell (2007) stated that a crisis always has the potential for bad consequences for organizations. Potential scenarios may include the following:

- Human costs—loss of life, scarred survivors, family trauma.
- Critical infrastructure costs—failures in networks such as transport, IT, water, energy.
- Policy costs—the failure of core policies.
- Economic costs—loss of revenue, loss of markets, job losses.
- Political-symbolic costs—damage to organizational/governmental legitimacy, damage to strategic policy direction.
- Personal costs—possible investigations, damage to reputation and loss of employment.

Therefore, crises are characterized by an essential sense of urgency of the organization's coping capabilities in preparing and overcoming them (Loosemore and Teo 2006).

2.4.1 Definition of Crisis

Crisis management is defined as the ability of an organization to deal quickly, efficiently and effectively with contingency operations with the goal of reducing the threat to human health and safety, the loss of public or corporate property, and adverse impact on continued normal business or operations (Gigliotti and Ronald 1991). In the corporate sector, the term crisis management refers to the successful management of public and stockholder opinions in the midst of a disaster. This definition can be extended to cover not just sociotechnical but also natural disasters, with the emphasis on successful management and coordination of various entities affecting an event in order to be prepared (Sriraj and Khisty 1999).

Crises affect many parties. In formulating its crisis management policies, stakeholders in the organization such as individuals, interest groups and institutions who are affected by the crisis must be considered. Their diverse perceptions must be taken into account (Low et al. 1999). Moreover, a crisis is also regarded as a stimulus to which certain kinds of behavior, which can be both helpful and destructive. At one extreme, a crisis can cause closer integration and innovation within an organization, while at the other end, it can bring about destructive behavior compromising an organization's viability (Pauchant and Mitroff 1992). Loosemore (1998a) stated that based on research in the behavioral sciences, crisis behavior can be explained by reference to the structure of people's communications during a crisis, to the way they cope with change and to the increased psychological pressures which characterize such periods.

2.4.2 History of Development

The knowledge of crisis management has evolved for the last three decades, stemming from research about disaster management. The evolution of crisis management research was provided by international political instability, rapid technological advances, and then by an increasingly hostile business environment. During a crisis, power configurations, interests, values, perceptions, bargaining and decision-making processes are highlighted by being focused upon a single well-defined issue. Due to the multitude of forces that interact during a crisis, it provides a context for the integration of theory (Loosemore 1998c).

According to Hallgren and Wilson (2008), there are options in dealing with crises. A crisis management plan and response team was suggested, along with some exercises that could be done in preparation. Methods such as risk analysis, contingency plans, logic charts and table top exercises can be used as tools in preparing for crises. Considering that crises are unimaginable and thus would become impossible to prevent, therefore prudent steps can be taken to deal with their occurrence.

Sriraj and Khisty (1999) studied that there are two views about crisis situations. The first one takes a simple systems point of view, where individuals are held responsible for the disasters. The second views disasters as stemming from the complex interaction between the various parts of the organization and its environment in terms of people, infrastructure, competitors, and so forth. This approach is systemic in nature and focuses on the interaction of different parts of the system.

This latter view is encouraged to be adopted by organizations in order to better equip them in facing any emergencies and/or disasters. Moreover, it is recommended that organizations should be crisis-prepared rather than reacting to a crisis.

The systemic approach in facing a crisis, suggested by Sriraj and Khisty (1999), stems from systems science that helps us to understand and manage complexity. The study suggested to adopt the critical systems thinking. It is based around three areas, which are critical awareness—examining and reexamining taken-for-granted assumptions of planning; ensuring that planning takes into consideration concepts of emancipation and power; and believing in methodological pluralism to address complex technical, socioeconomic and environmental problems (Ulrich 1983; Sriraj and Khisty 1999).

2.4.3 Main Concept of Crisis Management

In crisis management, there are four variables that need to be considered, which are (Mitroff and Pearson 1993):

1. Types

   This refers to the scope of the crisis management plan. It should answer the following questions: what crisis should the organization prepare for? What kinds of crisis can be neglected safely? What is the criterion or rationale for deciding which crises to prepare for or neglect?
2. Phases

   This variable addresses the activities involved and the management issues at each phase of a crisis.
3. Systems

   This variable examines the causes of crisis. There are five sub-variables that have critical influence in the origin of crisis, which are technical factors, organizational infrastructure, human factors, organizational culture and emotional factors.
4. Stakeholders

   This refers to the parties who may affect or be affected by the organization's crisis.

Furthermore, Low et al. (1999) found that regardless of the type of crisis, effective crisis management involves managing the five distinct phases through which all crises pass. The failure to manage any one of these phases might be responsible for the occurrence of a crisis in the first place, and then for its escalation. These five phases are as follows (Low et al. 1999; Alpaslan et al. 2009; Drennan and McConnell 2007):

1. Signal detection
   This means the sensing of early warning signals that announce the possibility of the crisis.
2. Preparation and prevention
   This involves doing as much as possible both to avert crises and to prepare for those that do occur. In the preparation phase, organizations aim to identify and interact with stakeholders and/or potential victims to prevent crises from happening and affecting stakeholders. In this phase, organizations should conduct risk analysis, threat assessment, mitigation strategies, contingency planning, simulation, training and education for facing the crisis.
3. Damage containment
   It is intended to mitigate the effects of a crisis and keep it from spreading to unaffected parts of an organization. This phase can also be considered as the response phase, where emergency working, operational deployment of resources and communications are essential.
4. Recovery
   This is the phase when organizations develop and implement tested, short-term and long-term programs designed to help them resume normal business operations.
5. Learning
   This concerns the continual study and re-examination of critical lessons learned from the organization's own experiences and from others to improve what has been done in the past.

Particularly in the damage containment or response phase, the crisis management team should be activated. This means that the crisis management team has been given the authority to take actions necessary to manage the crisis in an effective and timely manner. The crisis management team (CMT) consists of executives with specific expertise that will be needed to support business units and management during the crisis. In general, they are representatives of departments such as public relations, human resources, facilities, security, finance, insurance, purchasing and transportation. The responsibilities of the members of the CMT are:

- Take charge quickly: The CMT needs to take charge quickly, or the crisis will end up dictating the actions that will be taken, rather than having the team dictating the actions.
- Establish the facts: The CMT should reconstruct the events that led to the crisis. They should determine which employees were directly involved in the incident and speak with those employees about what happened. Getting good information about what happened is not easy. Much of what is being reported is tainted by emotion. Information that is available is also subject to differing interpretations.
- Tell your story: Make contact with all of the important segments of your public, such as the media, the general public, customers, shareholders, vendors and employees.

- Fix the problem: This is the period where tough decisions must be made and should be made fast. The goals are to recoup losses, to evaluate the organization's performance, and to make any changes that were identified as needed.

If the CMT follows these steps, it will be able to minimize the damage, be it physical damage to the assets or a crisis causing damage to the perception of the organization. The top management in the organization will assist the members of the CMT where needed, but their main responsibility will be to continue running the organization (Devlin 2007).

2.4.3.1 Communication in Crisis Management

Scholars have recognized the important role of communication in effective crisis management (Barton 1993; Coombs 2007; Millar and Heath 2003; Zhong and Low 2009). According to Grunig (1992), when an organization is threatened by crises, the need for communication increases to some level where the use of communication management to assist in transformation and relationship with the environment is essential.

Immediate and appropriate communication decision is crucial, particularly during the crisis response stage, which is characterized by short decision time, stress, complexity and uncertainty. The communication system structure that emerges in response to a crisis affects reaction efficiency of crisis management. The crisis response communication management is challenging, because during the crisis response phase, the organization's normal communication systems are under pressure as the information overload and channel bottlenecks. These situations can lead to a communication system break down. Therefore, the organization's crisis management plan should include a series of checklists or a template of procedures which can help the organization goes into auto-pilot in communicating proper information to respond to key stakeholders (Thayer 1998; Loosemore 1998a; More 1995; Quarantelli 1988; Zhong and Low 2009).

Loosemore and Teo (2006) also stated that effective communication during a crisis is essential, but difficult to implement. It was found that firms with a track-record of effective communication as an intrinsic part of their daily organization life are most likely to survive. In addition, such an organization is usually considered as a crisis-prepared organization, where it has a well-developed and comprehensive crisis management plan. It is the state of knowledge in these areas that represents the fundamental difference between crisis-prone and crisis-prepared organizations.

2.4.3.2 Organizational Embeddedness in Response to Crises

According to Hermann (1963), crises can be regarded as devices of change, where they will involve significant social and monetary change in an organization and may end up in physical change to the nature of its products. Furthermore, it is the difficulties that people have in adapting to these changes that create the behavioural, psychological and sociological problems that characterize crises periods (Applewhite 1965). Resistance to change can be seen in many forms, ranging along a continuum from passive disagreement to positive hostility. Its level in response to change is likely to depend on its contentiousness, magnitude, power balance between potential beneficiaries and losers and the manner in which it is introduced. In this situation, open mindedness, flexibility and creativity become of greatest value for responding to change during a crisis (Loosemore 1998a).

An organization's behavior towards stakeholders during the response phase of a crisis may range from denial (and hence no preparation), forced compliance, and voluntary compliance to going beyond legal expectations and making extra efforts (Shrivastava and Siomkos 1989). This type of responses is also consistent with the typologies described by other scholars, which are: deny responsibility; admit responsibility but fight it; accept responsibility; and anticipate responsibility (Clarkson 1995).

Geraldi et al. (2009) found that the characteristics present in the successful response to unexpected events such as crises resonate some of the principles proposed by the post-bureaucratic organizations. This type of organization replaces the hierarchic, centralized and formalized bureaucratic organization into a flat, decentralized organization, emphasizing flexibility rather than rule following, which resembles features of organic organizations. Nowadays, post-bureaucratic organizations are understood as a trend that encompasses changes, including the rhetoric of rapid change, globalization and competition in which survival depends on the adaptation of organizations and flexibility, autonomy and commitment of employees. Thus, flexibility, autonomy and commitment were identified as core constructs for successful responses (Alvesson and Thompson 2006). Figure 2.3 shows that there are three pillars that support successful responses to crises, which are responsive and functioning structures, good interpersonal relationship and competent people. Careful attention by organizations in allowing empowerment supported by these three pillars could have helped avoid or reduce adverse consequences (Geraldi et al. 2009).

Based on Low et al's. (1999) study, the level of crisis preparedness of organizations can be categorized into five stages. In general, each subsequent stage incorporates the crisis management capabilities of previous stages, as well as additional strengths. The stages are as follows:

1. Stage one: crisis prone

   These organizations have virtually no early warning systems in place for detecting major crises. Planning for damage containment rarely occurs before the crisis hits, and recovery systems are not established. These organizations do not learn from their past mistakes because they do not conduct formal review sessions. Moreover, crisis-prone organizations tend to be characterized by a culture of managerial invincibility and fatalism toward crises and a short-term, skeptical attitude toward the benefits of investing in the development of crisis

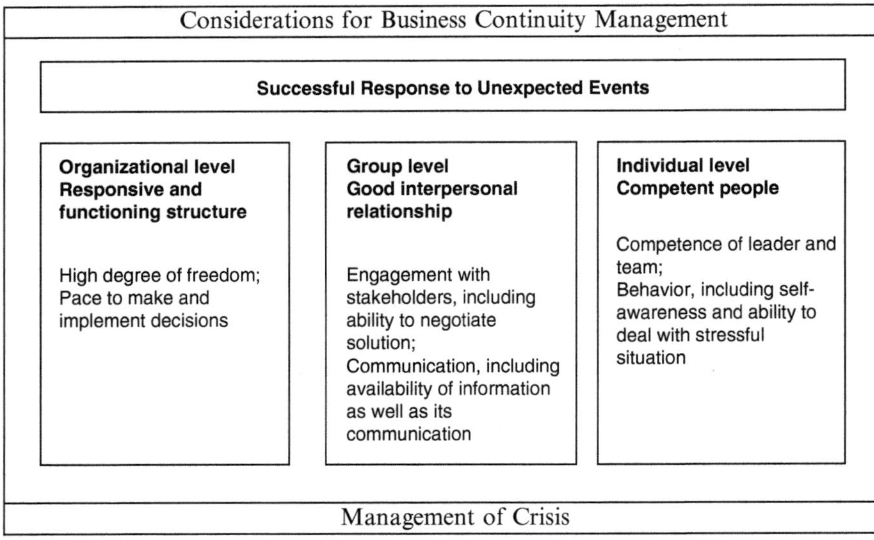

Fig. 2.3 Three pillars to successful response to crises. Source: Adapted from Geraldi et al. (2009)

management plans. Such organizations tend to have inflexible, formal structures and penal, exploitative, and task-orientated cultures that consistently stress the importance of profits over other corporate goals. People in crisis-prone organizations also believe that they are insulated from their environment and that crisis management is someone else's responsibility. Indeed, crisis management plans are considered a sign of weakness because crises are seen as a sign of managerial failure and most crises are considered to resolve themselves in time and to turn out to be unimportant. Furthermore, it is believed that each crisis is so unique that it precludes preparation and learning. In essence, crisis-prone organizations have an inappropriate structure and culture in relation to their risks and do not dedicate sufficient resources to crisis management activities. Consequently, crisis management plans represent little more than defensive routines and have minimal impact upon day-to-day organizational practices and attitudes (Loosemore and Teo 2006).

2. Stage two: crisis susceptible

   These organizations are better prepared, but are still very vulnerable to a variety of crises. They are likely to have a comprehensive programme for natural and human-directed disasters of all kind, but are not likely to plan or prepare for other types of crisis such as external economic attacks or external information attacks.

3. Stage three: crisis adjusted

   These organizations are likely to have in-depth plans and procedures for a limited number of breakdowns, such as computer malfunction, serious operator errors or major security breaches. However, they still do not appreciate the complex relationships that will contribute to a crisis.

4. Stage four: crisis braced

   These organizations have formalized their crisis management efforts toward the proactive mode. They have created a crisis management team which will be responsible for facilitating and formalizing crisis management efforts.

5. Stage five: crisis prepared

   These organizations are likely to develop plans and procedures that explicitly take into account all the critical systems that cause and prevent major crises. They do not see the causes of crises as purely technical; they are sensitive to human, organizational and staff feedback as well. As a result, they are much more likely to have explicit programs that address human factor issues. These organizations have a greater awareness of the underlying organizational culture and how it contributes, positively as well as negatively to crisis management. Moreover, they have a culture of awareness and sensitivity to organizational risks and of their social and financial responsibilities to stakeholders and the wider environment. Crisis management planning is systematically incorporated into strategic planning processes so that it is an integral part of organizational life at all levels. In this sense, senior executives provide the drive and support for crisis management by providing sufficient resources and clear statements of fundamentally held, core beliefs and attitudes relating to organizational priorities. Another characteristic of crisis-prepared organizations is their flexibility and willingness to "let-go" of formal, standardized systems and procedures that serve them well in normal times but which become restrictive and counterproductive during a crisis (Loosemore and Teo 2006).

Regarding crisis management planning in organizations, some studies had found that organizations have had a form of a crisis management plan for a number of years, but they have had a different name and several different functions. One of the examples is the corporate communications plan, where it is used to brief the news media when an organization is facing a crisis. This plan, however, is not a crisis management plan. It is a major element in the crisis management plan, but does not include other roles of other key executives. Still, some organizations use the term crisis management plan to describe their security plans, or their emergency response plans. The security plans, the emergency response plans, or the corporate communications plans are a part of the overall crisis management plan that are not integrated as a whole definitive crisis management plan (Devlin 2007).

Drennan and McConnell (2007) opined that in order to prepare and prevent future crises, and in developing crisis management within an organization, there are certain key characteristics that the organization should possess, which are:

- Preoccupation with failure.

  This is done by treating any lapse as a symptom that something is wrong with the system, encouraging reporting of errors, learning lessons from near misses and being cautious of complacency.

- Reluctance to simplify interpretations.
  Knowing that the world is complex, unstable and unpredictable, they encourage individuals to look beyond their own boundaries and to be skeptical towards received wisdom.
- Sensitivity to operations.
  This is conducted by scrutinizing normal operations in order to reveal deficiencies in supervision, safety procedures and training, hazard identification, and encouraging continuous adjustments that will prevent errors from accumulating and enlarging, and encouraging people to speak out about their concerns.
- Commitment to resilience.
  This is done by developing capabilities not only to detect problems but also to be able to continue working when things go wrong.
- Deference to expertise.
  Decisions are delegated to those on the front line and with the most expertise in that field.

Moreover, they have also found that a belief in the wider benefits to be gained from attempting to manage risks in a holistic, enterprise-wide manner and the development of plans to deal with a crisis situation, is now gaining ground and becoming an attention to organizations in various sectors.

2.5 Summary

This chapter reviews theories in organizations that are related to the crisis management concept. The first section briefly portrays the concept of management in general. This concept was first developed by Henri Fayol in which it became the basis for organizational theories in the following periods. In order to work within organizations, understanding the major theories regarding how organizations operate and functions is an advantage. The concepts of organization and management theory are not completely distinct and unrelated. They have been developed and evolved from views that focus on mechanization and production, to school of thoughts that highlighted the importance of human relations in organizations, and further consider the significance of the systems approach and contingency perspective in firms nowadays.

Other than the main concept and development of the management school of thoughts, other essential concepts such as complexity and change in an organization are also discussed. Scholars viewed organization as a complex system that needs ways to survive, evolve, and adapt due to its very large numbers of constituent elements or agents which are connected to and interacting with each other in many different ways. As a complex system that needs to adapt and evolve, this means that change in an organization is inevitable.

It has been found that crisis management had relations with the organizational mainstream theories and concepts that were discussed in the previous section. This concept views crises as more of the norm in organizations, due to its dynamic and high-velocity business environment which is characterized by discontinuity and continuous change. Moreover, the crisis management phase stems from the systemic approach that comprehends and manages complexity and uncertainties. Along with the defined five phase of crisis management, communication is crucial in this concept.

Chapter 3
Business Continuity Management (BCM)

3.1 Introduction

This chapter elaborates on a review of BCM. As the background, it describes the historical development of BCM and its relationships with other concepts. It will be followed by reviews on BCM as a management system, BCM's main principles, and Business Continuity Planning overview. The next section will describe the implementation of BCM, related with regulations or standards that support the concept and the development of BCM level of preparedness. Several reviews on BC plans from various sectors are elaborated in the final part of the chapter, followed by reviewing the need for BCM in organizations based on its benefits and challenges.

3.2 Background

3.2.1 BCM Definition and Development

The Business Continuity Institute (Business Continuity Institute 2007b) defines Business Continuity Management (BCM) as an act of anticipating incidents that will affect mission-critical functions and processes for the organization, and ensuring that it responds to any incident in a planned and rehearsed manner. Moreover, the Singapore Standard for BCM (SPRING 2008) looked at this concept as a holistic management process that identifies potential impacts which threaten an organization and provides a framework for building resilience and the capability for an effective response that safeguards the interests of its key stakeholders, reputation, brand and value-creating activities. Foster and Dye (2005) similarly viewed BCM as the process of developing advance arrangements and procedures that enable an organization to respond to an event in such a manner that critical business functions continue with planned levels of interruption or essential change. In this

L.S.R. Supriadi, L. Sui Pheng, *Business Continuity Management in Construction*, Management in the Built Environment, DOI 10.1007/978-981-10-5487-7_3

context, top management must take the lead in driving organizational BCM with a view to garnering the collective efforts of all individuals within the organization for this purpose (Low et al. 2008a).

The main objectives of developing and implementing a BCM in an organization are (O'Hehir 1999; Health 1999):

1. To enable a focused approach in developing a business continuity plan (BCP), using a well structured and comprehensive methodology.
2. To develop a pragmatic, cost effective, and operable recovery plan, to enable the firm to achieve critical business processes during a major disruption to the firm's operations.
3. To minimize the impact of the crisis on the firm's operations.

Moreover, Smith (2003) stated that an effective BCM strategy should be to ensure the safety of staff, maximize the defense of the organization's reputation and brand image, minimize the impact of business continuity events (including crises) on customers or clients, prevent impact beyond the organization, demonstrate effective and efficient governance to the media, markets and stakeholders, protect the organization's assets, and meet insurance, legal and regulatory requirements.

Historically, BCM was developed many years ago, where this concept is an evolution of a disaster recovery approach in a firm. Its roots lie in Information Systems (IS) protection although it is argued that it has grown a long way since then. Elliott et al. (2002) developed on these theories in more details explaining that the evolution of BCM has progressed from a focused technical aspect to a broader strategic organizational requirement. They also described the evolution as being linked to three mindsets within organizations which are technology, auditing and value based mindsets. The key features of these mindsets are:

a. Technology mindset in the 1970s—The focus was on the protection of computer systems, principally hard corporate main frame systems. During the 1970s, a common assumption was that business disruptions were triggered by a technology failure; thus priority was placed on protecting hard systems such as corporate main frame systems (Prithchard 1976; Broadbent 1979; Kuong and Isaacson 1986).
b. Auditing mindset in the 1980s—Technological changes in the 1980s which moved the IT element away from main frame to end user PC responsibility, brought with it regulations, corporate legislation and policies. Auditing was needed to ensure compliance. The major focus of the auditing perspective is still on the technology, the plan itself, and how continuity can be established through protecting essential business activities.
c. Value mindset in the 1990s—This described the value-based mindset as being focused on the needs of the business, where BCM is considered to have the potential to add value to the organization. The value-based perspective departs from the technology and auditing perspectives in the assumptions that were made about the scope and purpose of BCM. The scope is perceived as constituting the entire organization including employees, who are regarded as

presenting the biggest challenge in terms of implementation and management of the business continuity process. Organizational stakeholders are regarded as being the most important driver for change and BCM. The fundamental approach in this perspective is that business continuity is regarded as the integration of social and technical systems which together enable effective organizational protection (Swartz et al. 1995). Therefore, BCM not only protects but is also seen to contribute to the value adding process through more efficient systems or providing value-adding benefits to customers through superior responsiveness, reliability, and security.

According to Foster and Dye (2005), after the September 11 2001 attacks, an event that hit the World Trade Centers in New York City, many companies had realized that the world is now full of many unknown threats, requiring that business continuity plans be much broader than in the past. Significant threats are now not only confined in the categories of fire, natural disasters and some infrastructure breakdown. Threats such as terrorism, cybercrime, reliance on third-party vendors and suppliers have also become significant. Therefore, business continuity planning should require more robust prioritization efforts for business recovery, proactive development of new and innovative recovery strategies, and a greater dependence on the testing of plans. Furthermore, considerations that need strategic thinking are not only on the location decisions of a company's own facilities, but also the location decisions of a business partner (such as supplier). All of these environmental changes take BCM into a higher level, which is more focused on building resilience.

Smith (2003) also argued that BCM is not only about disaster recovery or responding to a crisis. It should be a business-owned and driven process that unifies a broad spectrum of management disciplines. In addition, crisis and risk management are part of the fundamentals used for developing a BCM concept.

Figure 3.1 shows the difference between the old and new BCM approach. Herbane et al. (1997) described the continuum of standard and better practice of BCM and identified a number of dimensions against which practice might be assessed. The first two dimensions refer to the types of staff employed in continuity projects and to the scope of their work. Standard practice is concerned with IT systems and employs only IT staff while better practice organizations employ staff from various backgrounds on a project which is business wide in scope. In standard practice, there was little need for new structures because IT could deal with continuity. In better practice cases, new structures of coordinators were identified with responsibility for the continuity process being delegated to each business unit and the dedicated continuity team providing a supporting role. The final group of dimensions relates to the strategy. Better practice saw continuity as a strategic issue both in terms of protecting its place in the supply chain and in marketing activities.

Based on these reviews, it shows that BCM has developed and evolved into a more holistic approach. It has progressed into a broader strategic organizational mindset which focuses on its business values. In the context of definition, it appears that SPRING's (2008) definition of BCM has incorporated all of these aspects and

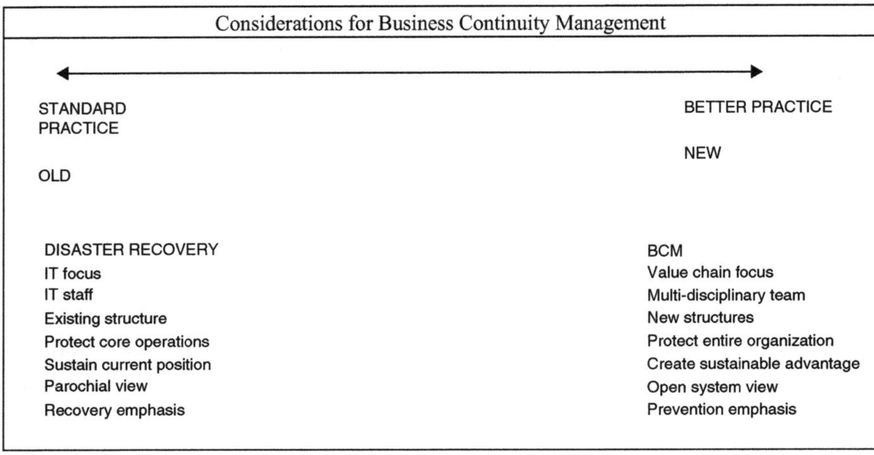

Fig. 3.1 Old and new BCM approach. Source: Adapted from Herbane et al. (1997)

represents the latest BCM mindset. Other BCM definition from BCI (2007b), Foster and Dye (2005), and Smith (2003) provide similar meanings of the BCM concept, which focuses on the keywords of: processes/procedures for the organization; response to incidents/threats/events; critical functions; and a planned and rehearsed manner. However, SPRING (2008) defined BCM's critical functions in more detailed aspects which include key stakeholders, reputation, brand and value-creating activities. Moreover, it specified the management process as holistic and the responses to threats/incidents are developed as a framework for building resilience.

3.2.2 BCM and Other Related Concepts

BCM has been considered as part of other concepts for overcoming crisis. There are relationships between BCM and these concepts, such as risk management, crisis management, and disaster recovery.

3.2.2.1 BCM and Risk Management

There are differences between risk management and BCM. Risk management focuses on a thorough organization-wide identification and assessment of risks and evaluating risks in relation to their likelihood and impact before identifying an appropriate risk response. BCM is concerned only with events that cause a significant business disruption, where it is not mainly concerned with probability but with the impact of an event and the time required for an organization to return to normal business operations (Collier 2009). Moreover, Goh (2010) mentioned that the relationship between risk management and BCM can be partially explained by referring to the Australian Standard for risk management. BCM efforts focus on addressing those risks which are deemed not acceptable to the organization. Subsequent BCM activities are aimed at establishing the appropriate measures to address these risks. It relegates BCM as part of risk treatment. Business Continuity has been defined "to safeguard the interests of an organization and its key stakeholders by protecting its critical business functions against predetermined disruptions" (BCI 2010, p. 3). The numbers and types of critical business functions in an organization would depend on the nature of the business and its mission as reflected in its Minimum Business Continuity Objective (MBCO). Risk management in BCM should be restricted to those instances where it affects the MBCO of the organization. It is also important to note that BCM is focused on identifying vulnerabilities within organizations, especially those linked to the underlying value they support and understanding the impact of their non-availability over time on the organization (BCI 2010; Hiles 2007). Table 3.1 summarizes the comparison between risk management and BCM.

3.2.2.2 BCM and Crisis Management

BCM has strong links with crisis management through the incident management component. In the BCM context, incidents come in different shapes and sizes and will typically invoke the BCM plan. Crisis management is often seen as the domain of communication and public relations (PR) practitioners with the BCM practitioner in a support role, if involved at all. Crisis management is also seen as responding to non-physical as well as physical events such as financial performance and reputation tarnishing incidents (BCI 2010).

Table 3.1 Comparison between Risk Management and BCM [adapted from BCI (2005, p. 6)]

	Risk management	BCM
Key method	Risk analysis and assessment	Business impact analysis
Key parameters	Impact and probability or likelihood	Impact and time
Type of incident	All types of events	Events causing significant business disruption
Size of events	All sizes and costs of events	For strategy planning: survival-threatening incidents only
Scope	Focus primarily on risks to core business objectives	Mostly outside the core competencies of the business
Intensity	All from gradual to sudden	Sudden or rapid events (although response may also be appropriate if a slower-moving incident becomes severe)

Source: Drennan and McConnell (2007)

Moreover, BCM considers any disruption holistically and determines how an organization will respond to the disruption, continue its activities and recover. BCM practitioners consider the media response to an incident or crisis to be an integral part of a full business continuity (BC) programme. Regarding emergency planning that is usually included in incident management, BCM views that this planning is not only seen as the domain of services from police, fire, ambulance and local authorities, but also for the organization in general. The company that adopts BCM would have a specific emergency response team that will coordinate with other external emergency response agencies (BCI 2010).

Other relationships between BCM and crisis management were also mentioned by Elliott et al. (2002), where BCM provides principles that use a crisis management approach. A crisis management approach may be defined as one that:

- Recognizes the social and technical characteristics of business interruption (organizations are socio-technical systems).
- Emphasizes the contribution that managers may make to the resolution of interruptions (the importance of the human response element).
- Assumes that managers may build resilience to business interruptions through processes and changes to operating norms and practices.
- Assumes that organizations themselves play a major role in "incubating the potential failure" (early detection is vital).
- Recognizes that, if managed properly, interruptions do not inevitably result in crises (the importance of preventative measures).
- Acknowledges the impact, potential or realized, of interruptions upon a wide range of stakeholders (think beyond the impact on the organization itself) (Elliott et al. 2002).

Some studies had made a distinction between BCM and crisis management. BCM refers to the planning and implementation of systems and procedures to enable an organization to sustain normal operations in the event of a disaster or other potential interruption. It is the process of developing advance arrangements and procedures that enable an organization to respond to an event in such a manner that critical business functions continue with planned levels of interruption or essential change. Crisis management is viewed to be a process by which an organization deals with major unexpected events that have already happened. Crisis management focuses on the immediate activities which need to be considered when the incident occurs. At most, the crisis management planning phase deals with the first couple of hours of the incident occurring, detailing who the key decision makers are, who will talk to the customers/clients/regulators and when this will be conducted (Smith 2003; Devlin 2007; Foster and Dye 2005). In addition, BCI (2007a) defined crisis management as the role that senior management have during an incident. It includes the high level command and control aspects of identifying a crisis situation, deciding how and when to respond, communicating both internally and externally, and leading and directing the recovery process.

3.2.2.3 BCM and Disaster Recovery

According to Elliott et al. (1999), the difference between disaster recovery and BCM is primarily based on its scope. Disaster recovery is a focus on technology-based problems triggered by external factors. BCM focuses more on adding value, creating an attitudinal change throughout the organization and considering its associated stakeholder groups. It is more concerned with the continuance of the whole business in the face of any unusual or unforeseen event. Moreover, disaster recovery is the implementation of a response capability to a specific type of event that impacts the continuity of the business. BCM is responsible for the overall identification of potential events, the likelihood of the occurrence of the event, and the predicted impact on the organization. BCM puts in place plans to deal with such occurrences. Disaster recovery is essentially a plan, with supporting infrastructure, which is enacted in the event of a disaster. In this way, disaster recovery is a subset of BCM, as is contingency planning, high availability planning, and the like (McCrackan 2005).

3.2.2.4 BCM and Business Resilience

BCM is a relatively newcomer to the business disciplines; however, aspects of BCM may have always been present in organizations, under different names. The vulnerabilities in the business and operating model of an organization can be considered in seven areas, which are reputation, supply chain, information and communication, sites and facilities, people, finance and customers. The nature of the BCM approach is to provide the framework to understand how value is created and maintained within an organization and establishes a direct relationship to dependencies or vulnerabilities inherent in the delivery of that value. This approach is conducted in a holistic and cross-functional manner. A successful BCM implementation would increase an organization's resilience, where it is defined as the ability to absorb, respond and recover from disruptions. This will eventually contribute to higher corporate performance (BCI 2010).

3.3 BCM as a Management System

BCM is a system that develops a framework of protocols and sets of procedures and instructions which give structure, order and stability to the particular function being managed. It is in line with the definition of a management system, stated by Griffith (1999), that sets out and describes, for a particular management function, the organization's policies, strategies, structures, resources and procedures used, within the firm to manage the processes that delivers its products or services (Griffith

2011). Based on its theory development and main principles, it can be seen that BCM adopts several management mainstream theories.

In its implementation, BCM adopts the Plan-Do-Check-Act (PDCA) methodology for achieving continual improvement. The BCM policy, objectives, processes and procedures are planned, implemented, assessed, and reviewed regularly (SPRING 2008). PDCA is a key attribute within standards-based management systems that is widely used nowadays. It was established by Deming, who propounded the view of quality management within a cycle of plan-do-check-act. The theories underpinning quality management have influenced systems development and continue to form component parts of systems applications. Historically, quality management was developed from a range of traditional organizational theories such as scientific, human and classical schools of thought. These theories are also pertinent to the evolution, development and implementation of management systems (Griffith 2011).

BCM also adopts the view of complexity theory, where an organization consists of a number of components (agents) that interact with each other according to sets of rules that require them to examine and respond to each other's behavior in order to improve their behavior (Stacey 1996). According to Griffith (2011), due to the extensive and complexity in the arrangement of business activities, processes and resourcing, a management system in an organization should establish an effective framework of responsibilities at various organizational levels. Parts of BCM principles are determining various responsibilities to the BCM members.

Based on its definition, BCM is developed and implemented in a holistic approach. The holistic perspective has much in common with systems theory. This theory viewed management system as a central part that directly supports the core business of the organization. Moreover, it is considered that a management system focuses not only on itself but also for the greater contribution that it can make to the organization (SPRING 2008; Griffith 2011; Checkland 1981).

According to Lawrence and Lorsch (1967), contingency theory suggests that organizational variables are in a complex interrelationship with one another, where environmental contingencies act as constraints and opportunities which influence the organization's internal structures and processes. Moreover, decision making are made through considerations of all aspects and situational approach (Olum 2004; Carlisle 1976). In BCM, this approach is adopted by implementing risk analysis and business impact analysis. The consideration of risk is viewed as a key element of the system (BCI 2010).

The BCM methodology has strong links with crisis management. Crisis management is often viewed as responding to non-physical as well as physical events such as financial performance and reputation tarnishing incidents. Furthermore, the domain of communication and public relations are important in crisis management. BCM considers any disruption holistically and determines how an organization will respond to the disruption, continue its activities and recover. BCM practitioners also viewed that communication and response to public are part of a full business continuity programme (BCI 2010).

Regarding change management, it is also part of crisis management. Lawrence et al. (1976) stated that a visible crisis faced by an organization can be an important force for triggering behavioral change, although such change may have costs derived from it. Essentially, such crisis has an unfreezing impact on the members of the organization, causing them to review and analyze their current attitudes and behavior patterns. Managing change in an organization should be conducted in orderly phases which are diagnosing the problem, planning the change, launching the change, and following up on the change in the organization. In this matter, it appears that these phases are similar to the PDCA approach which is adopted by BCM (SPRING 2008; Lawrence et al. 1976).

In accordance with Griffith (2011), a general approach to planning, delivering and implementing any management system consists of the following key considerations, which BCM also provides:

- The needs of the customer and other stakeholders.
- The policies and objectives of the organization.
- The organizational processes necessary to fulfill the policies and objectives.
- The assignment of responsibilities to manage processes towards the objectives.
- The provision of resources to attain the objectives.
- The establishment of procedures and instructions to manage the processes.
- The monitoring of processes to determine their efficiency and effectiveness.
- The identification and elimination of non-conformities in the processes.
- The encouragement of continual improvement in management of the processes.
- The audit and review of systems to improve the overall management approach.
- The feedback on performance to improve provision to customers through improved policies and objectives.

Furthermore, the highly influential factors to be considered in implementing a management system are as follows (Griffith 2011):

- Organizational culture. Instilling a trusting and cooperative workforce is vital to embedding the system.
- Involvement, which is bottom-up involvement from grassroots level in system development is essential, as is inviting contribution and feedback to management.
- Resources, which are trained and capable managers, supervisors and workforce are essential and, as such, investments in training and system ownership should be a priority.
- Flexibility. The system should be allowed considerable flexibility in performance upon system establishment, incrementally becoming more demanding as familiarity with its operation is developed.
- Shared commitment. Management must develop a blame-free culture where learning and improvement are preferred to difficulty and blame.

These factors should be embedded in an organization for its BCM implementation effectiveness.

3.4 Main Principles of BCM

To implement BCM, each organization must identify the threats and assess their resulting impacts. BCM needs to address issues and concerns in six broad areas in the following order (SPRING 2008):

1. Risk analysis and review: The threats to an organization can be identified through a risk analysis and review of its internal operations and external operating environment.
2. Business Impact Analysis: The potential impact of these threats on an organization and its ability to continue business operations and service can be obtained by conducting a business impact analysis. This would include, where possible, the loss impact from both a number of days of business disruption and financial consequences.
3. Strategy: The organization determines the appropriate strategies to safeguard its interests. These strategies can be preventive or pre-emptive in nature.
4. Business Continuity Plan (BC Plan): A detailed business continuity plan should be formulated to indicate the resources and capabilities required of the organization to prepare, respond, and recover from potential threats.
5. Tests and exercises: An established BC plan shall be validated by implementing tests and exercises. These are done to highlight errors or omissions and verify if the resources committed are accessible, available and adequate for efficient and effective recovery. It also verifies whether the staff is familiar with recovery procedures, and whether the BC plan meets its recovery objectives.
6. Program management: The organization will demonstrate commitment in maintaining the currency of its plan through regular and systematic review of its risks and business impacts, regularly reviewing its BCM strategies and revalidating its BC plan. Program management serves to validate the capability of the BC plan to fulfill the plan's objectives. Validation aims to uncover flaws in the plan design, for example any inaccuracies and incompleteness of the design of the plan.

There are four main components that must be considered in implementing BCM in an organization, which are (SPRING 2008):

- Policies: Senior management must stipulate policies to guide BCM efforts by the staff. The policies should set out the organization's aims, principles and approach specifying what is to be achieved or delivered, and will serve as the rationale and support for all BCM areas. In addition, policies provide the rationale for establishing the processes, people and infrastructure to support BCM on an ongoing basis.
- Processes: The set of activities with defined outcomes, deliverables and evaluation criteria to attain the objectives of the BCM policies. They include formal change control and documentation processes.
- People: Participation from various business units in the firm should be established to oversee BCM efforts and the skill sets of participants are crucial

to the success of BCM. The roles and responsibilities of staff involved in the organization's BCM efforts should be clearly defined.
- Infrastructure: The organization should allocate resources to support critical business functions against potential risk events. This consistently requires a good understanding and application of available technology and equipment, and physical facilities to respond to risk occurrences.

Generally, BCM has four main processes which are developed in an organization. The processes are the initiation process (initiating the BCM concept in the firm), planning for business continuity [which produces a business continuity plan (BC Plan)], implementation (implementing the BC Plan through testing and exercising), and lastly the operational management process (maintaining and updating the BC Plan). These four processes can be divided more comprehensively into six phases which are (Pitt and Goyal 2004; Elliott et al. 2002; BCI 2010):

1. Phase one—Project initiation

   The fundamental critical activity required prior to the establishment of a BC Plan is obtaining senior management approval, support, and commitment. Having obtained management approval, the initial phase of the BC Plan will include establishment of the BC Plan objectives and requirements of the plan. A business continuity steering committee would normally be established. This committee is likely to be made up of senior staff within the organization who have the relevant strategic view of the firm's operations. It is important that they also have nominated deputies who are suitably briefed and have an in-depth understanding of the BCP process.
2. Phase two—Risk assessment/business impact analysis

   The principal objectives of phase two relate to data gathering and review of alternative courses of action. The identification and evaluation of this information will then allow senior management to make decisions on the critical aspects of the core business. Having identified the risks, a business impact analysis should then be carried out. Karakasidis (1997) identified this as a key step in protecting an organization, and identified some of the minimum objectives as being:

   - Determine critical requirements and resources and the effects a disaster may have on the people, place, process, and premises.
   - Estimate anticipated target recovery time for each core business function and service.
   - Establish core business recovery priorities.
   - Identify key personnel, equipment, and facilities needed to support core functions.
   - Estimate costs of extended business disruption.
   - Identify resources required to develop, test, and implement BC Plan.
3. Phase three—Design and development of the BC Plan

   Essential issues to be addressed at this stage include detailed scope strategy and objectives of the plan, administration procedures, formation of business continuity committee and downstream business recovery teams, lines of communication, escalation notification and plan activation, scenario setting for plan execution, establishing BC Plan records, storage, access, and its budget.

4. Phase four—Creation of the business continuity plan

   This phase basically deals with the creation of the BC Plan. The key issues to be addressed include:

   - Emergency response procedures covering evacuation, decanting access to work areas, and access to documentation.
   - Emergency control center establishment, command and control procedures.
   - Detailed procedure for communications, delegation or designation of authority, and key stakeholders.
   - Detailed resumption, recovery, and restoration procedures.
   - External support, vendor contracts, contacts, and resources.

5. Phase five—Testing and exercising BC Plan

   In order to establish the effectiveness of BC Plan, it is essential to implement a regular testing and exercise program. The key activities to be established during the testing and exercising stage will include preparation of exercise program and objectives, the details of exercise scenarios and monitoring and recording procedures, and identification of training requirements, communication channels, and induction of new staff.

6. Phase six—Maintenance and updating

   Having established the need for testing and the degree of probability that a substantial number of plans might fail following the testing exercise, it is essential that the lessons learned and shortfalls documented are incorporated into the plans. The key issues to be addressed during this phase include:

   - BC Plan review criteria and objectives
   - Schedules and program of review
   - Plan distribution and security

In responding to the changing environment of a business from time to time, the maintenance and updating process should be done in a regular and continuous basis.

Based on this review, it is considered that BCM has evolved from a simple reactive disaster recovery planning, to crisis management principally driven by information technology, and finally to a more proactive comprehensive approach.

3.5 Business Continuity Planning (BCP)

The main process of BCM is Business Continuity Planning (BCP). BCP refers to the identification and protection of critical business processes and resources required to maintain an acceptable level of business, protection of such resources, and preparation of procedures to ensure the survival of the organization in times of business disruptions. Fundamentally, it seeks to mitigate the impact of a disaster by ensuring alternative mission-critical capability is available when disaster strikes. The process seeks to preserve the organization's assets in the event of a disaster, which are its capability to achieve its mission, its operational capability, its reputation and image, its customer base and market share, and its profitability (Low et al. 2008; Hiles 2007). This is regarded as the main process due to its vital output for the firm in handling disruptions and overcoming crises. This planning process will be followed by regular monitoring and updates.

Before formulating the BCP framework, the following issues have to be considered thoroughly (Low et al. 2008a; O'Hehir 1999; Eternity Business Continuity Consultants 2007; Civil Contingencies Secretariat 2007):

1. Policy—formulating a policy statement at the managerial level to signify the company's attitude towards a particular risk and prescribing the objectives of such a policy.
2. Methodology—analyzing the assessment processes involved in evaluating a crisis, and promoting greater commitment for the company to proceed with the plans.
3. Accountability—establishing individual accountability for managing the risk and ensuring that the nominated person has the appropriated technical expertise and authority to manage the risk.
4. Management support—determining the company's current managerial attitude or process towards assessing and managing the risk, without which the company will not have the initiative to implement BCM in the organization.
5. Dependencies—defining the scope of the BCP clearly, so that every individual is aware of the dependencies involved, whether this is external or internal (key supplier, personnel, operating system, etc.) to successfully mitigate the specified crisis.
6. Being realistic—educating the management that a crisis brings about certain risks and to mitigate the effects, certain costs are involved. The management should be ready to accept certain risks and should be prepared to spend the necessary funds to mitigate the risks involved.
7. Future actions—determining the appropriate business processes to be implemented or to be refined, to reduce the risk to an acceptable level, and assigning responsibilities and milestones.
8. Performance measures—establishing measurement indicators to enable assessment, and monitoring the effectiveness of risk management which can be proactive or reactive. Proactive action is recommended to prevent occurrence.
9. Independent expert—appointing an internal or external, qualified, independent expert to determine the adequacy of the response to the crisis, such as through regular meetings, and reporting to higher management to signify the importance of BCM.
10. Contingency plan—establishing an alternate plan for the unforeseen circumstances not being provided for.

According to Vancoppenolle (1999) and Elliott et al. (2002), the respective elements are included in the operational flow of a company's operations, which are: (1) Business processes (how the products and services are delivered to the client); (2) Participants (who the participants are, in the execution of the business process); and (3) Infrastructure and resources (what is used in the execution of the business process). These elements are necessary to be reviewed when analyzing a crisis during BCP.

Furthermore, upon the occurrence of a crisis, many parties could be affected (Elliott, Swartz and Herbane 2002). It could be the company management or interest groups like investors, suppliers, etc., who have direct or indirect investments in the company. The occurrence of a crisis, if not appropriately mitigated, could lead to adverse consequences such as withdrawal of funds, which is an external factor. Even though investors are not directly involved in the company's operations, they have an indirect influence on the growth of the company. Therefore, the requirements of the various stakeholders in the organization should also be considered, which include the following (Singapore Business Federation 2003):

- The ways and means of the employees' livelihood protection.
- The defined time lines for the resumption of support and services and transparency of operations in a crisis, which relate to customers and suppliers.
- The control of the situation, cost effective solutions to handle the impact of the crisis and the effects on business resumption, and transparency of operations by managers.
- Good corporate governance, protecting the image of the organization, and sharing of the company's profits that linked strongly to what investors will review on the company.

Hiles (2007) stated that the company's BCP should not be driven by eliminating risks according only to their probability, but rather be based on the effects and impacts on the business if an unexpected event were to occur. Such classification according to effects could be:

- Failure of an individual infrastructure element, including single points of failure.
- Longer-term interruption of a critical information flow.
- Longer-term interruption of a critical business activity chain or business process.
- Local longer-term business interruption.
- Complete business interruption.

These effects from an unexpected event may cascade into larger impact levels. Some examples of these effects are damages to infrastructure elements and resources supporting the business operations. The damage can result in impacts such as unavailability of infrastructure elements or resources or loss of information. Loss of information due to a disaster is not limited to data in computers. All of the information stored in binders, folders (with, for instance, customer information), contracts, property deeds, the archives, the legally required vital records, the paper client files, the business knowledge spread over the place, and others can be lost too.

Other than impacts on business operations, the long-term impacts of such crises or events may also arise, even after the business has been resumed and operations have returned to normal. The examples of long-term impacts are: loss of market share; lower share price; lower credit rating; loss of brand value; loss of company image, public confidence and credibility; and loss of key staff. Furthermore, the rippling effects of a business interruption should never be underestimated, particularly for companies that are an integral component of a wider supply chain. When a company participating in a supply chain is hit by a disaster, this could ripple down throughout the supply chain (Hiles 2007).

3.6 BCM Implementation

Nowadays, BCM is widely used in various types of firms. Firms in banking, telecommunication, oil and gas, and retail industries had developed a BCM concept in their management systems. BCM is developed based on their respective business strategies and activities. Due to the different business environments, the firms developed different procedures for overcoming different types of crises. Some of them had also focused not only on their business continuity, but the service continuity to their customers. This shows that they had developed the program based on the value mindset (Elliott et al. 2002).

Herbane et al. (2004) also found that BCM has evolved to encompass wider participants, threats, techniques and responses. It has been applied in the financial service industry, vehicle breakdown services, gas suppliers, water utilities, supermarkets, and local authorities. All of these organizations recognize that in the face of internal and external threats to the continuity of operations, a socio-technical approach (beyond IT disaster recovery) is essential to improve business recovery from crises. They also have linked BCM to strategically important dimensions of their operations.

When implementing BCM for the first time in an organization, project management practices should be adopted. The practices of project management that may usefully be employed include the identification of deliverables, timescales and deadlines, and budget and work effort control. Other knowledge in project management such as communications, risks, procurement and human resources management are also needed for establishing effective BCM components (Business Continuity Institute 2007a).

3.6.1 Legislation and Standards Relating to BCM

Elliott et al. (2010) elaborated that the earliest legal provisions to influence disaster recovery and business continuity (BC) ideas can be found in the 1977 Foreign Corrupt Practices Act, which is the US financial services sector's provision. It is often cited as an important development in firm's reorientation of the perceived threats and impacts. Since then, the US financial services industry has developed various regulations and legal requirements to impose greater requirements on BC provisions. Although the acts do not refer specifically to BC, they specify the importance of countering the increasing risk of external threats to digital resilience, which is one of the dependencies on BCM.

Moreover, the introduction of BCM-specific regulations in the financial services sector is not only applied in the US. The Australian Prudential Regulation Authority (APRA) Standard on BCM APS 222 (for deposit taking institutions) and GPS 222 (for general insurers) published in April 2005 (APRA 2005a, 2005b) requires Australian financial institutions to implement a whole of business approach to BCM. Elsewhere, the Reserve Bank of India (RBI) set out a requirement for Indian banks to fully implement BCP, presents a planning methodology, and further specifies a template for plan content. Banks are required to submit recovery time objectives for critical systems to RBI's Department of Banking Supervision at the end of each financial year and to report major failures and response activities or prevention measures on a quarterly basis (Parthasarathi 2005; Elliott et al. 2010).

In several countries such as United Kingdom (UK), United States of America (US), Switzerland, Australia, New Zealand and Singapore, BCM had been developed into a national standard, where every firm from various sectors is encouraged to have this system in its organization (Elliott et al. 2010). In Singapore, the SS540:2008 standard has been formally used as the standard for implementing BCM in a firm. This Singapore Standard is applicable to all organizations regardless of their size. This standard emphasizes resilience and protection of critical assets, in the human, environmental, intangible and physical domains. It focuses on continuity management and recovery of critical business functions (SPRING 2008). Up to now, Singapore is the only country in Asia that has established a BCM standard, whereas other BCM standards came from Europe, North America, and Australia (Elliott et al. 2010).

In the UK, the Business Continuity Institute (BCI) has developed a certification standard for business continuity practitioners. Besides that, a BCM standard (BS25999:1-2006) as a Code of Practice for Business Continuity Management was also published by the British Standards Institution and can be viewed as an implementation guide and a definitive text for those intending to understand BCM principles and practices in a more comprehensive manner (Business Continuity Institute 2007a). Moreover, the American Chapter of the Business Continuity Institute (BCI) and BSI America have joined forces to help businesses better prepare for disasters by encouraging the adoption of BS 25999 (Business Continuity Institute 2009). This standard is also in line with US's national standard for business continuity, which is NFPA 1600:2007 (National Fire Protection Association 2007).

Furthermore, ISO has officially launched ISO 22301, "Societal security—Business continuity management systems—Requirements", the new international standard for Business Continuity Management System (BCMS). ISO 22301 has been developed in 2012 to help organizations minimize the risk of business disruptions (St-Germain et al. 2012). This standard is similar to the previous BCM standards, but it has some improvements for BCM implementation such as (St-Germain et al. 2012; SPRING 2012):

- Greater emphasis on setting the objectives, monitoring performance and metrics;
- Clearer expectations on management; and
- More careful planning for and preparing the resources needed for ensuring business continuity.

According to Goh (2010) and St-Germain et al. (2012), the standards from various countries have similar contents. The differences are on how the standards develop the detailed components in the BCM planning process. In general, each standard has the same BCM planning methodology, which are: Risk analysis and review; Business impact analysis (BIA); Recovery strategy; BC plan development; Testing and exercising; and Programme management (some standards incorporate project management in this phase). All of the above standards have the common objectives, which are to guide the users to recover from any disasters that have occurred in their business environment and still continuously focus on the continuity of their business processes. Furthermore, the standards also help the users in identifying the potential impacts of various disruptions to the firm and be able to prioritize the efforts in aiming to achieve resilience. Table 3.2 illustrates the main aspects of the BCM concept being grouped into six categories. These aspects are summarized from various standards.

3.6.2 BCM Level of Preparedness

Regarding implementing BCM in an organization, several agencies from various countries had developed assessment levels of BCM preparedness. These levels are useful to assess whether an organization has adopted a complete BCM concept or not. From understanding the position of the company within these levels, the organization gains feedback from its current BCM preparedness level and may increase its effort for a better BCM maturity level.

Levels of preparedness assessments have been proven to be an effective evaluation method (Scott 2007). In general, this type of assessment can help the organization to verify what they have achieved relative to the topic assessed. The organization's current achievement can also be determined by describing their current activities. In addition, it can assist the organization in prioritizing the necessary improvement based on their assessment results (Peng et al. 2011; Stevanovic 2011).

The Ministry of Finance in British Columbia, Canada (MOF-BC 2007), had developed the BCM maturity assessment for every financial agency in the province. There are three levels of criteria involved, which are:

- High maturity. This level demonstrated strong executive support for BCM, the establishment of an organization-wide structure supporting the activity, and staff responsible for BCM had a strong awareness of and compliance with core policy requirements, guidelines and procedures for BCP. BC plans for mission critical processes and business priority areas were developed and updated, and testing/exercising was ongoing, with results used to make changes. Monitoring and

Table 3.2 The main aspects of BCM principles

No.	BCM principles	Description
1	Risk analysis and review	• Examine internal and external risk events and impacts (qualitative and quantitative) that can affect the critical operation's continuity • Using Risk Analysis (RA), Business Impact Analysis (BIA), and Cost benefit analysis (justification for initial treatments to prevent or reduce the effects of risks and potential losses)
2	Business impact analysis	• Examine the impact to the organization (assesses the potential impact of loss from an internal perspective), qualitatively and quantitatively, due to a disruption of business operations and processes • BIA must be conducted on a periodic and systematic basis to assess the impact of losses if the corresponding business operations and processes are disrupted in view of proposed changes
3	Strategy development	Examine the possible strategies for maintaining the operation of Critical Business Functions (CBFs). This should cover pre-incident preparedness, response and recovery
4	BC Plan development	Examine the BC plan(s) which is an action plan that guides the response and recovery actions of the organization when disaster occurs. It includes an emergency response to stabilize the situation following a disaster, the set up and operation of an Emergency Operations Centre (EOC), and specifies CBFs to be recovered within their established Recovery Time Objectives (RTOs) and Recovery Point Objectives (RPOs) when a disaster occurs. RTO is the period of time in which functions must be recovered after a disruption has occurred, and RPO is the point in time at which systems and data must be recovered after a disruption has occurred.
5	Test and exercises for BC plan	• Ensure that the BC plan drawn up and implemented by the organization is viable and workable • Tests are intended to verify the capability of the BC plan to attain specified objectives or established criteria • Exercises are intended to train and condition BC team members to improve their coordination and performance in executing the BC plan. Exercises also serve to highlight any weaknesses in the operation and effectiveness of the BC plan, with establishing generic corrective actions if the result falls below assessment criteria
6	Program management	• Examine the ongoing efforts and activities of the organization to maintain the effectiveness of its BCM. BCM involves firm commitment of organization's efforts and resources to safeguard the interests of its key stakeholders, reputation, brand and value-creating activities on a continuous basis. Assessment of an organization's BCM efforts should therefore be dynamic • The BC plan is operated by staff of the organizations. Staff in the organization should be familiar with the plan via appropriate awareness and training programs prior to any test/exercise of the plan. Periodic and systematic training and awareness programs should be conducted to familiarize employees to the operation of the BC plan.

Sources: Adapted from SS540:2008 (SPRING 2008), NFPA1600:2007 (National Fire Protection Association 2007), BS25999:2006 (BSI 2006), ANZ5050:2009 (Standards Australia 2009; Elliott et al. 2010), SS ISO 22301: 2012 (SPRING 2012)

reporting processes were effective and efficient, and pandemic planning had been undertaken.
- Moderate maturity. This level demonstrated strong executive support and a level of coordination within the organization to ensure progress is made towards BCM objectives, although roles and responsibilities may not be adequately defined to ensure all recovery staffs were clear on their expectations in a business interruption. Compliance with core policy was low, and BC plans for mission critical processes and business priority areas were either under construction or in need of updating. Monitoring and reporting processes were largely ad hoc and pandemic planning may have been in the commencement phase.
- Low maturity. This is the lowest level of preparedness, where typically the organization had a lower level of executive support and BCM may not have been considered a high priority. These organizations exhibited a low level of awareness of policies and guidelines and of roles and responsibilities. Compliance with core policy was also low, and BC plans were either not developed or in need of significant updating. Pandemic planning may have been initiated, although activities to date were limited to those driven by existing OHS committees.

The Australian National Audit Office (2009) had also developed characteristics of better BCM preparedness for public sector entities. There are two levels, which are (1) Basic level, that is generally found in small, non-complex or less time-critical entities and (2) Mature level which is found in large, complex, geographically dispersed or critical entities. The characteristics that are described and assessed in each level are:

- A BCM framework is in place.
- Training and awareness of BC has been conducted.
- A risk assessment has been conducted.
- A BIA has been conducted.
- Preparatory controls have been implemented.
- The entity has documented and the executive has endorsed, its BC plans and framework.
- BC testing and exercises have been conducted.
- The entity monitors BC.

Also in Australia, Lansley and McAtee (2009) had established a six-level BCM preparedness model for companies, which are:

- Level 1—Self-governed: BCM has not yet been recognized as strategically important by senior management.
- Level 2—Supported self-governed: At least one business unit (BU) or corporate function has recognized the strategic importance of BC and has begun efforts to increase executive and enterprise-wide awareness.
- Level 3—Centrally-governed: Participating BUs and departments have instituted a basic governance program, mandating at least limited compliance to standardized BCM policy, practices and processes to which they have commonly agreed.

- Level 4—Enterprise awakening: All critical business functions (CBFs) have been identified and continuity plans for their protection have been developed across the enterprise.
- Level 5—Planned growth: BC plans and tests incorporate multi-departmental considerations of critical enterprise business processes.
- Level 6—Synergistic: All BUs has a high degree of BCP competency. Complex business protection strategies are formulated and tested successfully.

Smit (2005) had studied and defined another BCM maturity model that can be applied to organizations. According to the study, there are six level of BCM maturity, described as follows:

1. BCM initiated. An organization has initiated BCM if there is formal management commitment to the organization of BCM. The responsibility for BCM is covered at a sufficiently high level within the organization and an explicit BCM policy is in effect. The deliverable of the initiated stage is BCM as an initiative.
2. BCM planned. An organization reaches the stage planned if it has performed all necessary analyses and has written all relevant plans. Therefore, this stage is characterized by a BC analysis and a BC plan. The deliverable of the planned stage is BCM as a blueprint.
3. BCM implemented. Implemented stage is reached as soon as not only the measures to assure BC are planned, but also realized. This means BCM facilities have to be realized, services have been contracted and BCM tasks have to be assigned to the right people. The deliverable of the implemented stage is BCM as an implemented project.
4. BCM embedded. On the first three stages, BCM is a project. As soon as an organization reaches the embedded stage, BCM has turned into a process instead of a project. This stage is reached as soon as a maintenance process is designed; hence a maintenance plan is developed, the plan is known and available within the organization and there is awareness regarding the importance of BCM within the organization. The deliverable of the embedded state is BCM as a process.
5. BCM controlled. At the stage of BCM embedded, an organization has developed a maintenance plan and probably formulated some BCM exercises and tests. In the next stage, BCM controlled, this maintenance process is also executed as it should and exercises are done as planned for. In addition to that, the existing BCM is audited and controlled. The deliverable of the controlled stage is BCM as business as usual. If an organization has reached stage 5, it controls its existing BCM. For some organization, a BCM process that is controlled is sufficient. However, other organizations will strive for stage 6.
6. BCM optimized. If an organization has optimized its BCM, it can use its BCM as a strategic instrument, for example to gain a commercial advantage or strive for operational excellence as a business strategy. For this, a strategic approach of BCM is a requisite. Furthermore, the organization should strive for continuous improvement of their BCM and the deliverable of the optimized stage is BCM as a strategic instrument.

Furthermore, other BCM preparedness level model from a risk consulting firm in Canada (Marsh Risk Consulting 2010) had been developed. The level of preparedness with its label, overview of the preparedness level description, and the organization's ability to respond can be seen in Table 3.3.

Last but not least, the Singapore Business Federation (2011) provided a BCM preparedness assessment, based on the company's level of understanding about business continuity. Red level shows that the organization has a minimal understanding of BC, whereas Yellow level shows the organization has a basic understanding of BC, and finally Green level describes the organization has an advanced understanding of BC. The assessment are conducted through rating the firm's understanding and preparedness towards risk analysis and review, BIA, strategy development, BC plan development, tests and exercises, and programme management.

According to a study from New York University (2006), most businesses, particularly small and medium sized ones, are lacking formal BCM programs. Only one-quarter of the companies surveyed have formal, written continuity plans. Moreover, only four in those companies provided BCM training to their employees. These four companies had prepared the concept within their

Table 3.3 Marsh BCM preparedness level

Preparedness level	Label	Overview	Organization's ability to respond
Level 5	Optimizing BCM	BCM driven by corporate strategy is subject to continuous improvement and is integrated into the overall risk management and operational strategy	Organization has sustained ability to respond to and survive strategic threats and crises—both anticipated and unanticipated
Level 4	Integrated BCM	BIA is done at divisional level and value/supply chain dependencies are understood and protected	Organization understands its business processes and has the ability to deal with crises and recover processes across sites and into the supply chain
Level 3	Established BCM	Emergency response, crisis management and BC plans are completed and linked. Training and exercising embedded in the organization	BCM response is integrated and BCM capabilities can be sustained
Level 2	Formalizing BCM	Corporate policy driving a consistent approach at site level. BIAs are done for sites and recovery strategy agreed	Key location(s) have built the ability to respond to a localized emergency and recover business
Level 1	Undeveloped BCM	Ad hoc and reactive approach—not a systematic BC	Minimum legal/regulatory requirements met providing protection for people and facilities

Source: Marsh Risk Consulting (2010)

organization due to regulatory forces, which are risks to employees and business operations, legal liability, and insurance requirements. From this study, it is recommended that an organization should analyze its own case for BCM preparedness and invest accordingly.

3.7 Reviews of BC Plan

Various sectors have developed their BC plans based on the functions of their business and impacts that may occur from certain crises. There are general principles that can be gained from these plans that may provide insights on developing a BC plan.

3.7.1 BC Plan from Financial Services Sector

As mentioned before, the financial services sector is the pioneer of developing and implementing BCM. In general, the main principles that are established in their BCM policy are as follows (Monetary Authority of Singapore (MAS) 2003; Bank *Van De Nederlandse Antillen* (Central Bank) 2010):

1. Board of Directors and Senior Management should be responsible for their institution's BCM.

   The responsibility for the state of BC preparedness of an institution lies with the Board of Directors and senior management. Senior management is responsible for steering BCM with policies and strategies necessary for the continuation of CBFs. In addition, they should demonstrate that they have sufficient awareness of the risks, mitigating measures and state of readiness by way of a confirmation to the Board of Directors.
2. Institutions should embed BCM into their business-as-usual operations, incorporating sound practices.

   Depending on the scale and complexity of the businesses, institutions could adopt sound BCM practices that include the following components:

   - Clear BCM policy, strategy and budget.
   - Well-defined roles and responsibilities for the BCM programme.
   - BC plan comprising of detailed tasks and activities.
   - Succession plans for critical staff and senior management.
   - BIA or similar process.
   - Programme for the development, implementation, testing and maintenance of BC plan.
   - Programmes for training and awareness.
   - Emergency responses.
   - External communications and crisis management coordination programmes.
   - Coordination with external parties (including authorities, interdependent parties, etc.).

3. Institutions should test their BC plan regularly, completely and meaningfully.

   It is essential to regularly test its functionality and effectiveness. Tests will also familiarize staff with the location of the recovery site, as well as the recovery procedures. Senior management and staff should participate in these exercises and be familiar with their roles and responsibilities in the event of activation. Exercises may include:

   - Desk-top-walk-through exercise to full system test.
   - Staff call-tree activation (with and without mobilization).
   - Back-up site to back-up site exercise (including with external service providers).
   - Alternative arrangements of shared services.
   - Back-up tape restoration.
   - Retrieval of vital records.

4. Institutions should develop recovery strategies and set recovery time objectives for CBFs.

   The establishment of recovery strategies enables institutions to execute their BC plan in an orderly and predefined manner that minimizes disruption and financial loss. Recovery strategies form the basis for defining recovery time objectives of CBFs. Without these clear markers, scarce resources may be inappropriately diverted to less important activities. This may adversely affect the institutions' reputation and survivability. Recovery time objectives may range from minutes to hours. The transparency and sharing of recovery time objectives would help improve service level expectations and understanding among institutions and further contribute towards the mitigation of interdependency risk.
5. Institutions should understand and appropriately mitigate interdependency risk of CBFs.

   When planning for the BC of CBFs, institutions should take into account the interdependencies of these business functions, and the extent to which they depend on other parties. Institutions should also understand the business processes of these parties that support their critical functions, including their BC preparedness and recovery priorities.
6. Institutions should plan for wide area disruptions.

   These financial services look to institutions to demonstrate that they have planned and catered for a wide-area disruption in their BCM. Some planning parameters that institutions may consider include the geographical concentration of institutions, transactional processing activities and dependencies on internal or external service providers. Institutions are responsible for deciding on the need to cater for multiple zones outage scenarios, taking into consideration their respective levels of critical business activities and prudent risk management policies. In addition, they should also consider broadening and deepening their BCM scope to cater for prolonged operational disruptions.
7. Institutions should practice a separation policy to mitigate concentration risk of CBFs.

Critical staff and information are important assets that are difficult to replace quickly. Many institutions assume that the same pool of staff would be available to recover their CBFs at the recovery sites. This may not always be true as disruptions may result in the unavailability of critical staff. Also, identifying alternates to critical staff may not always reduce the risk, especially if both the primary and alternate critical staffs are housed in the same location or zone. It is important, therefore, to find the right balance between mitigating concentration risk and not losing the efficiencies gained from the centralization of business processes and critical staff.

3.7.2 BC Plan from Education Institutions: A Case Study

On April 16, 2007, Virginia Polytechnic Institute and State University (Virginia Tech) experienced one of the most horrific events in American university history. A double homicide had occurred, followed by a mass shooting that left 32 students and faculty killed, with many others injured, and many more scarred psychologically. Families of the slain and injured as well as the university community have suffered terribly from this event. One of the main recommendations from the tragedy is to update and improve the university's emergency response plan. It is recommended that the plan should be more systematic, including conducting risk analysis (threat assessment) in advance and choose a level of security appropriate for the campus. Along with that, the university should update and enhance the plan where students, faculty and staff should also be trained annually about responding to various emergencies (Tridata Division 2009; Flynn and Heitzmann 2008).

In 2010, the school had developed a comprehensive emergency response and continuity plan. The brief description of the plan is as follows (Virginia Polytechnic Institute and State University 2010):

- General purpose

  The plan outlines procedures for managing major emergencies that may have threatened the health and safety of the campus community or disrupt business operations on the local campus. It identifies individuals and departments that have a direct or supporting role in emergency response, and it provides a management structure for coordinating and deploying university resources to handle the event.

  This plan consists of the basic plan, the appendices, and the emergency support function and incident annexes. The basic plan provides an overview of the university's approach to emergency response and operations. It explains the policies, organization and tasks that would be involved with the response to an emergency. The annexes and appendices give definition to the terms and acronyms used throughout the basic plan, and are the location for any supporting figures, maps and forms. The emergency support function appendices focus on detailing the specific responsibilities, tasks and operational actions to complete a specific emergency operations function, while the incident annexes focus on any additional special planning or response needs beyond the basic response plan for particular event scenarios.

- Scope

  This plan applies to all of the university's students, facilities, staff and visitors. Surrounding community in addition to the campus may be impacted by major emergencies, and if this happens, the university will further cooperate with local, state, and federal officials in their delivery of emergency services. Categories of emergencies or hazards are identified through risk assessment with significance ranking that are most likely to impact the university.
- Priorities

  The plan's response priorities are (1) to protect life safety; (2) to secure critical infrastructure and facilities (in priority order: buildings used by dependent population; buildings critical to health and safety; facilities that sustain the emergency response; classroom and research buildings; administrative buildings); (3) to resume teaching and research programs.
- Response phases

  The university response to a disaster or emergency will generally involve the following phases:

  1. Planning and mitigation. The process of evaluating exposures and developing or refining response plans that will assure an orderly and effective response to an emergency, and for identifying and mitigating areas of vulnerability.
  2. Response. The reaction(s) to an incident or emergency in order to assess the level of containment and control activities that may be necessary.
  3. Resumption. The process of planning for and/or implementing the resumption of critical business operations immediately following an interruption or disaster. During this phase, more in-depth forecasts of the impact will be available, and university-wide priorities for program resumption will be determined.
  4. Recovery/restoration. The process of planning for and/or implementing recovery of non-critical business processes and functions after critical business process functions have been resumed, and for implementing projects/operations that will allow the university to return to a normal service level.

- Emergency notification systems protocols

  The university provides an Emergency Notification System (ENS) which is intended to rapidly circulate emergency information on an incident, and give instructions to the campus population.
- Emergency operations command structure

  The university's emergency response and continuity plan had been coordinated with the town's agencies, local government and organizations. The functional groups in delivering the response and continuity process are:

1. The policy group, which is composed of lead administrators. It establishes policies and procedures as needed to support emergency operations, and determines business recovery and resumption priorities.
2. The Emergency Response Resource Group (ERRG) directs resources in support of emergency response operations, assures the continuity of critical business functions, and implements business recovery and resumption activities. The ERRG convenes at the Emergency Operations Center (EOC).
3. Satellite Operations Centers (SOCs), located in the administrative headquarters. Deans, Vice Presidents and Vice Provosts, gather emergency impact data from their constituent departments, account for their personnel, transmit reports to the EOC, disseminate emergency instructions to constituents, and develop and implement business continuity, resumption and recovery plans.

In addition to these groups, there are also essential roles who will direct these groups, supported by essential personnel.

- Business recovery

  Even when emergency response activities are nearing completion, business recovery activities may continue for weeks or months after the event. Business recovery activities include reestablishing complete services and functions following a major incident and recovering extraordinary costs caused by the event. Furthermore, recovery priorities should be established as follows:

  1. Immediate recovery (true continuity) is essential;
  2. Recovery required within 24 hours;
  3. Recovery required between 24 and 72 hours;
  4. Recovery not required within 72 hours.

- Exercises and training

  Trained and knowledgeable personnel are essential for the prompt and proper execution of the plan. All personnel will be provided with the necessary training to execute those responsibilities in an effective and responsible manner. Training on university-level emergency response roles and the incident command system will generally be coordinated by the Director of Emergency Management.

  Exercises will be conducted as needed which allow all persons involved in emergency response to practice their roles and to better understand emergency operations and their responsibilities under emergency conditions. University-wide exercises will be held at least once per year, and will consist of tabletop, practical and full-scale staged events as deemed appropriate.

3.7.3 BC Plan for Influenza Pandemic: A Review

A pandemic is an epidemic or outbreak of infectious disease that spreads through populations across a large region; for instance a continent, or even worldwide. A flu pandemic could occur when a new flu virus emerges and starts spreading as easily as normal seasonal flu. As the virus is new, the human immune system will have no pre-existing immunity. This makes it easier for people to contract the new flu and experience more serious symptoms than that caused by normal seasonal flu. Current viruses that had spread across a large region (particularly in Asia) are the influenza A (H1N1), the SARS incident in 2003, and the avian flu (H5N1) (SPRING 2009).

According to some studies, no one could predict when a flu pandemic will occur. When it does occur, the impacts may be felt in various ways. Regarding its possible general impact, public gatherings may be discouraged, people with flu-like symptoms may not be allowed in public places, public transport may be disrupted and regular updates and clarifications may be necessary. As for the business impact, supplies may be disrupted, the number of customers may drop, likely increase of electronic communications use which may lead to overloaded communication systems and some staff in any organization may be absent from work (SPRING 2009).

Based on these likely impacts, companies are encouraged to ensure their business remain viable in the event of an outbreak. BCP should be developed with further considerations on how to operate their business with minimal face to face contact between staff, staff and customers, and with suppliers; how to operate business effectively with key members of staff being absent from work; and how to operate if supply chains are disrupted. Moreover, the key risks to the company that need to be addressed in BCP are (SPRING 2009):

- Employees
- Processes and business functions (e.g. production, sales and marketing, etc.)
- Business infrastructure (e.g. offices, shops, factories, equipment, etc.)
- Stakeholders (shareholders, suppliers, customers, etc.)
- Communications, both internal and external

The Singapore government had proactively taken an approach to overcome this crisis through initiatives such as the Flu Pandemic Guide for small and medium-sized enterprises (SMEs) in 2006. The BC guideline developed by a Singapore standards agency provides these contents particularly for handling flu pandemic (Low et al. 2010a; Singapore Business Federation 2006; SPRING 2009):

3.7.3.1 Annex section

This section describes:

- Information about personal hygiene awareness, as an example: correct hand washing procedures; basic information on sanitization such as disinfectants, recommended use and their precautions.
- Contact list of key customers, key suppliers/vendor/contractors and others.
- Contact list of key personnel and key organizations for information and assistance on flu pandemic.
- Description about roles and responsibilities of the Flu Manager.

- Procedures upon detection of visitors and staff who are unwell. These include procedures of (1) Visitor detection and isolation; (2) Staff unwell at workplace; (3) Staff unwell outside workplace and (4) Contact tracing.
- Forms such as temperature screening, notification form (for suspected flu case at work), and body temperature monitoring log.

3.7.3.2 BC Plan for Flu Pandemic Contents

- Description about the alert level code. There are five levels of codes, which consist of:

  1. Green—isolated overseas or local cases of animal-to-human transmission. Threat of human-to-human infection remains low.
  2. Yellow—slight human-to-human transmission. A small risk of it being imported here, but has not resulted in sustained spread.
  3. Orange—evolves into human disease. WHO confirms several outbreaks in one country, spreading to other countries. Deaths are expected. Local confirmation of new cases and evidence of more than one transmission has occurred.
  4. Red—widespread infection. Increase in deaths has occurred. Healthcare system likely to be overwhelmed and essential services are added to ensure full operational capacity.
  5. Black—high death rates reported. Economic activities are severely disrupted, as panic sweeps through the community.

- Description of recommended actions for companies

  1. Priority tasks for various levels:

     (a) Green—to set up a team to oversee BCP.
     (b) Yellow—appoint a Flu manager.

  2. Action plans are written for every alert level.

3.8 The Need for BCM

According to a survey on trends in business continuity, it was found that BCM has become mandatory to maintain customer confidence and a competitive edge. The threat of interruption and the need to respond promptly has manifested itself, where a vast increase in regulatory requirements and a mandate from customers for BC plan development has occurred. Organizations are expected to manage the BC process more collaboratively, be driven to complete their BC plans and include it in Requests for Proposals (RFP) and Requests for Information (RFI) (BUCORIM 2008).

There are several sources of external influence that are encouraging an increased focus on business continuity. According to respondents questioned for a report conducted by the Economist Intelligence Unit (EIU 2007), customers are the stakeholder that is viewed as most important in driving decisions about business continuity, with 59% citing them as a significant influence. Moreover, in the supply chain relationships that are getting complex and more dependent, customers will most likely ask about a detailed scope of BC plan, whether the supplier has it in place and would request evidence of compliance with particular policies.

In addition to customers, pressure from regulators is also becoming more distinct. Regulators are viewed as the second most important external influence over decisions about BC, with 58% seeing them as significant in the regard. This figure rises to 72% from respondents who are in the financial services sector (EIU 2007).

3.8.1 Benefits of BCM

Previous section of this chapter had described the relationships between BCM and other concepts. Table 3.4 summarizes the distinction between these concepts based on their main focus and key methods.

Whilst BCM is able to help firms to have a response for major disruptions that may threaten their business activities, the Business Continuity Institute (2007a) found that there are other benefits that can be gained by embracing BCM as a management discipline in an organization. Firstly, BCM will help address some key risks in the firm and help them achieve compliance. Secondly, BCM can be used as a competitive advantage to gain new customers and to improve margins by using it as a demonstration of "customer care". Thirdly, a thorough review of the business through Business Impact Analysis (BIA) can highlight business inefficiencies and focus on priorities that would not otherwise have come to light. And last but not least, firms providing services or goods recognize that keeping customers through a more reliable service is cheaper than tempting back the deserters after an interruption. Other studies have also found various benefits of implementing BCM in an organization. Table 3.5 shows the BCM benefits from various studies. In addition, the table shows that BCM's main focus and key method of conducting Business Impact Analysis plays an important role and provides positive implication for an organization that implements BCM.

Table 3.4 BCM distinction with other related concepts

	BCM	Risk management	Crisis management	Disaster recovery
Main focus	BCM is concerned only with events that cause a significant business disruption, where it is mainly concerned with the impact of an event and the time required for an organization to return to normal business operations	A thorough organization-wide identification and assessment of risks and evaluating risks in relation to their likelihood and impact before identifying an appropriate risk response	Crisis management focuses on the immediate activities which need to be considered when the incident occurs. At most, the crisis management planning phase deals with the first couple of hours of the incident occurring, detailing who the key decision makers are, who will talk to the customers/clients/regulators and when this will be conducted	Disaster recovery is a focus on technology-based problems triggered by external factors
Key method	Business impact analysis; and identifying critical business function (CBF) and minimum business continuity objective (MBCO)	Risk analysis and assessment; identifying risk response	Risk analysis and contingency planning; the sensing of early warning signals that announce the possibility of the crisis	Contingency planning; emphasize on recovery of the core operations

Sources: Collier (2009), Drennan and McConnell (2007), BCI (2007a), Foster and Dye (2005), Devlin (2007), Smith (2003), Elliott (1999), McCrackan (2005)

3.8.2 *Challenges in BCM*

Although BCM is considered as necessary to be implemented in organizations, there are several issues regarding the challenges of its implementation. Robinson (2009) viewed that the recent economic recession would be a challenge in implementing BCM. Recession has delayed or reduced BCM uptake; with top management viewing it as a discretionary spend. Moreover, only a minority will recognize that recession increases the need for BCM, with cutbacks reducing operational resilience and scarce liquidity eroding financial tolerance. Nonetheless, when a senior management team still has a strong commitment in sustaining its business resilience, and perceiving the recession-BCM link being strong enough, these can be a strong contributory factor to maintain its BCM. Moreover, Molinier (2009) opined that these economic conditions should be viewed as an opportunity to demonstrate how the companies can provide resilience whilst streamlining processes and adopting a cost-benefit approach that demonstrably support business objective.

Table 3.5 BCM benefits

Description	References
Firms that invest in developing a BC tend to create value for the firm, particularly maintaining their stock price. For global 1000 firms, there is a high probability of a crisis resulting in substantial decline of stock price during any 5 year period	INTERCEP (2007), FM Global (2003)
Effective BCM by corporate management can actually lead to an increase in shareholder value	Knight and Pretty (1996), Knight and Pretty (2005)
Corporate resilience will be a competitive advantage in the twenty-first century. Globalization, technological complexity, interdependence, terrorism, climate and energy volatility, and pandemic potential are increasing the level of risk that societies and organizations now face. Risks also are increasingly interrelated; disruptions in one area can cascade in multiple directions. The ability to manage emerging risks, anticipate the interactions between different types of risk, and bounce back from disruption will be a competitive differentiator for companies and countries alike in the twenty-first century. Moreover, it is a contributor to profitability, shareholder value and competitiveness	Van Opstal (2007), Council on Competitiveness (2006)
Implementing a BC plan may also have legal significance for a corporation. Because BC recognizes risk and mitigates it, the creation and implementation of such a plan may help a corporation discharge its corporate governance responsibilities to customers and shareholders alike. BC is a strategic investment, and its dividends will be evident during an attack, and economically and legally, in the aftermath of a terrorist event	Directors and Boards Magazine (2006)
The business impact of crises can run into the billions. The 1990 Wall Street Blackout and the 1992 Chicago flood are two examples. The article argues that initiating a BIA can have positive implications for the bottom line, especially in the event of a disaster	Watkins (1997)
Rewards of corporate resiliency through BCM • Increased productivity and innovation often supported by more effective internal communications, streamlined processes, more adaptive workplaces, better workflows and increased employee morale • Protected revenue flows as a result of plans to protect key assets—Inventory, property/plant, equipment and intellectual property—as well as sustain core operations • Expanded customer base and increased customer retention, as both individual consumers and organizations place an increasing focus on safety, security and preparedness • Lower operating expenses as a result of lower insurance and legal costs, less theft, reduced employee turnover and more competition among suppliers • Reduced cost of capital as both equity and debt markets (including key rating agencies) increasingly evaluate	Raisch, Statler and Burgi (2007)

(continued)

Table 3.5 (continued)

Description	References
corporate preparedness and resiliency • Stronger reputation, as a result of both the application and communication of resilience • Better regulatory compliance and governance both internally and in terms of external review	
When made known to insurance companies, a corporate preparedness program can result in relatively lower insurance premiums and better policy terms	Raisch and Statler (2006)
BCM can help to avoid losses of important business data, which can result in significant losses in terms of both existing and future business as well as liabilities to customers, investors and legal authorities. IT downtime costs can range from $1 million to over $6 million annually for companies that focus on database in its business	Hinton (2000)

In accordance with Continuity Central's survey to BC professionals (Continuity Central 2011), the biggest challenge in implementing BCM was lack of resource for the implementation. The second biggest challenge was the difficulties in obtaining senior management support and input. Thirdly, getting the wider organization to buy-in to BC and to provide support to the process was another challenge that needs to be considered. Following these top three challenges, other reasons are: organizational cut backs and changes; technology issues; testing and exercising issues; compliance, regulations and auditing; and culture change. These findings provide important feedbacks to those who have implemented BCM and who are in the phase of initiating it.

3.9 Summary

This chapter provided a review on BCM, starting from its historical development, its relationships with other concepts, its main principles and methodology, to its implementation in various sectors that shows the necessary need of the concept in an organization.

As an act of anticipating incidents that will affect mission-critical functions and processes for the organization, and ensuring that it responds to any incident in a planned and rehearsed manner, BCM has evolved from a technology-based disaster recovery approach to a value-based drive for business resilience. It is also viewed as a unifying process that includes various concepts for overcoming crises.

BCM is considered as a management system that, similar with other management systems, needs influential factors such as organizational culture, involvement, resources, flexibility and shared commitments for its effectiveness. Moreover, these approaches are embedded in its main principles and methodology.

Currently, BCM is widely adopted in various firms from various sectors. Regulations and international standards have been developed for this concept and methods in assessing the level of BCM preparedness have also been established. The need for BCM is currently supported by various drivers and although there are some challenges in implementing the concept, the benefits of BCM are worth mentioning.

Chapter 4
Organizational Culture and Institutional Forces

4.1 Introduction

This chapter reviews culture in organizations and the institutional theory. An overview and definitions on culture is described, followed by review on organizational culture dimensions. Managing organizational culture and its benefits are also discussed. The second section discusses the institutional theory, including its definitions and the three pillars or forces of institutions.

4.2 Culture in Organization

4.2.1 Culture Overview

The roots of cultural studies are from the science of anthropology and have impacts on the knowledge areas of the social and human sciences. In 1872, the British Association for the Advancement of Science (BAAS) produced an inventory of cultural categories in the form of an anthropological field manual that listed 76 unordered culture topics. In 1938, a more comprehensive manual was published which was titled *Outline of Cultural Materials* that listed 79 major and 637 subdivisions of culture topics. Moreover, Sir Edward Burnett Tylor, an anthropologist who worked with the BAAS, published a book named *Anthropology* which is still considered to be relevant in terms of its cultural concepts and theories. Up to today, enthusiastic debate has continued among anthropologists to evolve a universally acceptable definition of culture and its technical attributes (Coffey 2010).

According to Coffey (2010), along with the importance of systems and contingency theory in relation to describing organizations, the following schools of thought also described the effect of these theories on the concept of organizational culture studies:

L.S.R. Supriadi, L. Sui Pheng, *Business Continuity Management in Construction*, Management in the Built Environment, DOI 10.1007/978-981-10-5487-7_4

- Human relations
  This theory described that organizations existed to serve human needs by motivating individuals and developing group dynamics which made organizations more efficient. Research about values, beliefs and attitudes in individuals and groups had a great influence on understanding human relations in organizations, which also have impacts on the growing interest in organizational culture studies.
- Power and politics
  This school of thought is influenced by Pfeffer (1981), Kanter (1983) and Mintzberg (1987) and associated with French and Raven (1959), Porter (1991) and Handy (1993). It proposes that organizations are complex coalitions of groups and subcultures with competing values and preferences. This view is in line with the culture view that organizations are irrational entities that achieve their goals through a mixture of compromise negotiation, conflict dynamics and influence of leaders and subordinates.

The perspective on organizational culture itself has been developed by several scholars and will be described in the following sections.

4.2.2 Definitions of Culture

Hofstede and Hofstede (2005) defined culture as the collective programming of the mind which distinguishes the members of one human group from another. In this sense, culture includes systems of values and values are the building blocks of culture. Furthermore, Mead (2000) implies that:

- A culture is particular to one group and not to others where the values are shared to varying degrees by all members.
- A culture influences the behaviour of group members in uniform and predictable ways.
- A culture is learned and not innate. One is not born with clear understanding of one's culture and it is handed down from generation to generation.
- A culture includes systems of values, where values are defined as an assumption of how people should behave.

According to a study by Coffey (2005), definitions of culture had been defined since the late 1800s until recently. In 1871, Tylor defined culture as a complex whole which includes knowledge, beliefs, art, law, morals, customs, and any other capabilities and habits acquired by man as a member of society. Tylor focused on the individual in the society. In 1945, Linton viewed the culture as the habit of members in a society. Furthermore, Whitten and Hunter (1987) opined that culture consists in the shared patterns of behavior and associated meanings that people learn and participate in within groups to which they belong. A learning process and patterns of behavior are considered as aspects in culture. Harris (2004) also defined culture in a similar way as Whitten and Hunter's (1987). The definition of culture that is viewed to be more representative to a situation of an organization was developed by Cunningham and Gresso (1994) where it is defined as an understanding of "the way we do things around here". Culture is the powerful yet ill-defined conceptual thinking within the organization that expresses organizational values, ideals, attitudes, and beliefs.

In its further studies, culture is categorized into dimensions which provide an important framework not only for analyzing national culture, but also for considering the effects of cultural differences on management and organization. This framework is especially useful for understanding people's conceptions of an organization, the mechanisms that are considered appropriate in controlling and coordinating the activities within it, and the roles and relations of its members.

4.2.3 Organizational Culture Dimensions

Since the 1980s, the concept of organizational culture has been considered and discussed in the field of organizational theory (Smircich 1983). It has been defined as the social or normative glue that holds an organization together, where it expresses the social ideals, values and beliefs that members of an organization come to share. These values or patterns of belief are brought out by symbolic devices such as myths, rituals, stories, legends, and specialized language (Siehl and Martin 1981; Louis 1980; Boje et al. 1982; Deal and Kennedy 1982; Mitroff and Kilmann 1976; Wilkins and Martin 1980; Andrews and Hirsch 1983). These studies also found that culture could have influential consequences in the organization, particularly when they are conducive and enduring (Cheung et al. 2011).

Several cultural studies in organizations proposed that cultures appear to vary in managerial cultural value dimensions. Work goal importance is part of a person's total life situation, where work-related values provide a link between broader cultural values and the behavior of people at the workplace (Hofstede et al. 2010; Littrell 2008). According to Hofstede et al. (2010), the processes of management are carried out in an environment that is man-made. People build organizations according to their values, and societies are composed of institutions and organizations that reflect the dominant values within their culture. The nature of management behavior is such that it is culturally specific, where a management technique or philosophy that is appropriate in one national culture is not necessarily appropriate in another.

Hofstede and Hofstede (2005) also stated that some elements in an organization are influenced by culture, such as:

- Planning and control

  Both of these processes in organizations are strongly influenced by culture. Planning tries to reduce the level of uncertainty, and control is a form of power. These processes depend on how the organization values uncertainties and what type of power is applied. The planning and control systems are often considered as rational tools, but in fact they are partly ritualistic and the implementations mostly depend on the culture in the organization.

- Corporate governance and business goals

  Patterns of corporate governance, the ownership and control of corporations differ between countries. Moreover, the level of capitalism is historically linked to the level of individualism.
- Motivation theories and practices

  Motivation is considered as an assumed force operating inside an individual, inducing the person to choose one action over another. Thus, culture as collective programming of the mind plays a significant role in motivation, where it also influences human behaviors and explains them.
- Leadership

  Leadership and subordinateship in a country are closely related. Vertical relations in organizations are based on the common values of superiors and subordinates. Beliefs about leadership reflect the dominant culture of a country.
- Performance appraisal and management

  Any organization in any culture depends on the performance of its people. Monitoring the performance of subordinates is a theme in most management development programs from the lowest management level upward. A formal performance appraisal program that requires periodic written or oral evaluations are regularly implemented by the superior. Therefore, adopting such programs to organizations in different countries calls for adjustment to their cultures.

In general, organizational culture has been defined as the way things are done and operated within the internal environment of the workplace. It has a number of features that includes: common beliefs; pattern of behavior, norms; values and rules that are exercised among members of the organization. These features are strategically driven and they are usually established through the company founders or the new leaders. One of its characteristics is that the closer the values and beliefs among members of the organization, the stronger the culture will be (Kotler et al. 1992; Schols 1987; Williams et al. 1993). Naoum (2001) found that there are three key elements of a strong culture, which are:

1. A strong organizational culture facilitates goal alignment, because all employees share the same basic assumptions they can agree on.
2. A strong culture leads to high levels of employee motivation.
3. A strong culture is better able to learn from its past.

Regarding the level of adaptiveness, an organization can be considered as having an adaptive culture or vice versa when it is viewed in two main elements, which are (Naoum 2001):

- Core values

  Adaptive organizational cultures strongly value people and processes that can create useful change. On the other hand, unadaptive organizational cultures value the orderly and risk-reducing management process much more than leadership initiatives.

- Common behavior

  In adaptive organizational cultures, the leaders pay close attention to all their constituencies, especially customers, and initiate change when needed to serve their legitimate interests, even if that entails taking some risks. Conversely, unadaptive organizational cultures have their leaders who tend to behave somewhat insularly, politically and bureaucratically. As a result, they do not change their strategies quickly to adjust to or take advantage of changes in their business environments.

In measuring aspects of a culture, dimensions are needed, which is defined as a human construct that aims to represent a cluster of interdependent values bound by some similarities (Hofstede and Hofstede 2005). Hofstede is the founder of the dimension paradigm for cultures, where it was shown that "cultural difference between modern nations could be meaningfully measured and ordered along a discrete set of dimensions, representing different answers to universal problems of human societies" (Hofstede and Hofstede 2005; Maleki 2010). Up to now, there are various studies and findings related to cultural dimensions. This section will describe several findings about cultural dimensions from various scholars.

4.2.3.1 Hofstede's Dimensions

Hofstede has developed seven national cultural dimensions, which are (Hofstede and Hofstede 2005; Low and Shi 2002; Hofstede et al. 2010; Littrell 2008):

- Power distance: the extent of power inequality among members of an organizational society. This refers to the degree to which people accept centralized authority and status differences in society and their organizations. High power distance cultures tend to centralize power more than moderate or low power distance cultures.
- Uncertainty avoidance: the extent to which members of an organizational society feel threatened by and try to avoid future uncertainty or ambiguous situations. Cultures that are high on uncertainty avoidance seek stability and security, whereas weaker uncertainty avoidance cultures can live with more uncertainty.
- Individualism and collectivism: which describes the relationship between the individual and the collectivity that is reflected in the way people live or work together. Cultures that value some collectivism prefer tight social structures in which the group looks after its members; whereas, highly individualistic cultures prefer looser structures in which individuals look after themselves.
- Masculinity and femininity: the extent of roles division between genders to which people in a society put different emphasis on work goals and assertiveness as opposed to personal goals and nurturance. A society is called masculine when emotional gender roles are clearly distinct: men are supposed to be assertive,

tough, and focused on material success, whereas women are supposed to be more modest, tender and concerned with the quality of life. A society is called feminine when emotional gender roles overlaps: both men and women are supposed to be modest, tender and concerned with the quality of life.

- Long-term and short-term orientation: this stands for the fostering of virtues oriented toward future rewards—in particular, perseverance and thrift. Its opposite pole, short-term orientation, stands for the fostering of virtues related to the past and present—in particular, respect for tradition, preservation of "face" and fulfilling social obligations.
- Indulgence and restraint: Indulgence stands for a society that allows relatively free gratification of basic and natural human drives related to enjoying life and having fun. Its opposite pole, restraint, reflects a conviction that such gratification needs to be curbed and regulated by strict social norms.
- Monumentalism and self-effacement: this dimension (developed by Michael Minkov in 2007) had been inspired by the work of Steven Heine (Canadian psychologist) in 2003, who saw a link between self-enhancement (a tendency to seek positive information about oneself) and self-stability or self-consistency (a tendency to believe that one should have unchangeable values, beliefs and behaviors that do not depend on shifting circumstances). The dimension contrasts societies in which the human self is like a proud and stable monolithic monument (monumentalism) versus societies whose cultures promote humility, flexibility, and adaptability to changing circumstances (self-effacement).

Based on these descriptions, the main characteristics of the seven cultural dimensions that were derived from groups in a workplace can be summarized, as illustrated in Table 4.1 These characteristics can be used to analyze which dimension a certain organization or nation possesses.

Along with these seven national cultural dimensions, Hofstede had also found six dimensions for organizational culture, which is based on his cross-organizational studies. The six dimensions are as follows (Hofstede and Hofstede 2005):

1. Process oriented versus results oriented

   In the process oriented culture, people perceived themselves as avoiding risks and spending only a limited effort in their jobs, while each day was pretty much the same. In the results oriented culture, people viewed themselves as comfortable in unfamiliar situations and as putting in a maximal effort, while each day was considered to bring new challenges.
2. Employee oriented versus job oriented

   In the employee oriented culture, people felt their personal problems were taken into account that the organization took a responsibility for employee welfare, and that important decisions were made by groups or committees. In the job oriented units, people experienced a strong pressure to complete the job, they perceived the organization as only interested in the work employees did, not in their personal and family welfare, and they reported that important decisions were made by individuals.

Table 4.1 Characteristics of Hofstede's cultural dimensions

Power distance	Long-term and short-term orientation	Individualism and collectivism	Masculinity and femininity	Indulgence and restraint
• An inequality in the hierarchy of the organizations • Centralization in the organization • The number of supervisory personnel • Manager's reliance either on experience or formal rules • Managing subordinates either consulted or to be told what to do • The ideal boss either resourceful democrat or benevolent autocrat • Subordinate-superior relations • Views on privileges and status • Qualification of lower strata	• The type of main work values • The importance of leisure time • Type of focus, either bottom line or market position • Views of profit importance, either current or future profit • The similarities of aspirations between owner-managers and workers • Type of meritocracy • Level of personal loyalties • Views on economic growth between 1970 and 2000, either slow or fast • The amount of money allocated for investments • Type of investments, either mutual funds or real estate	• Level of occupational mobility • Employees interest in the work environment, either in-group's or self interest • Hiring and promotion decisions • Employer-employee relationship • Management of groups or individuals • Treatment for customer either particularism or universalism • Prioritising task or relationship • Involvement of individuals with organizations • How the organization look after the employee • The basis for policies and practices, either based on duty or allowing individual initiative. • Manager's concern towards modern managerial ideas • The applicability of policies and practices, either to all or varied relations	• The nature of management either based on consensus/intuition or decisive/aggressive • Resolution of conflicts • Basis for rewards • The size of the organization • People's priority between work and living • Preference between leisure time and money • Career priority for either gender, either optional or compulsory • The share level of working women in professional jobs • Type of humanization of work • Type of industry's competitiveness characteristics • The level of organization's interference with people's private lives • Level of job stress • Job restructuring appeal permits either group integration or individual achievement	• The importance of having friends • The importance of thrift • Loose or tight control in organization • The importance of moral discipline • Extroverted or neurotic personalities • The level of optimism • The importance of freedom of speech • The importance of maintaining order

(continued)

Table 4.1 (continued)

Uncertainty avoidance		Monumentalism and self-effacement
• The length of service that relates to the number of changes of employer • The need for rules • The need for working hard • Perspective or views about time • Level of tolerance for ambiguity and chaos • Type of beliefs, either generalist/common sense or expert/technical solutions • Top manager's concern either strategy or daily operations • The number of trademarks • Focus on either decision process or contents • How rules affect intrapreneurs • The number of self-employed people • The capability between invention and implementation	• The level of structuring of activities • The number of written rules • The number between generalist/amateur or specialist/expert • Focus on either pluriform or uniform in the organization • Manager's orientation either impersonal-oriented/flexible or task-oriented/consistent • Willingness to make risky or individual decisions • Level of labour turnover • Level of employee's ambition • The level of satisfaction score • Level of power through control of uncertainty • Level of ritual behavior • Basis for work motivation, either by achievement or security	• The level of seeking self-improvement • The importance of service to others • Pride of own country • The importance of tradition • Family pride or pragmatism • Success attribution whether based on luck or effort • Talent for theoretical or applied science • Appeal of fundamentalism or pragmatism • Appeal of folk wisdom/witchcraft or knowledge and education • The importance of religion • The level of obedience to authority • Level of cooperation and equality • Absolutist or dialectical thinking

Source: Hofstede et al. (2010)

3. Parochial versus professional

   In the parochial culture, people felt the organization's norms cover their behavior at home as well as on the job. They felt that in hiring employees, the company took their social and family background into account as much as their job competence, and they did not look far into the future where they might assumed the organization would do this for them. On the other side, members of professional cultures considered their private lives their own business, and they felt the organization hired on the basis of job competence only.
4. Open system versus closed system

   In the open system units, members considered both the organization and its people open to newcomers and outsiders, and almost anyone would fit into the organization. In the closed system units, the organization and its people were felt to be closed and secretive.

5. Loose control versus tight control

   It refers to the amount of internal structuring in the organization. People in loose control units felt that no one thought of cost, meeting times were only kept approximately, and jokes about the company and the job were frequent. People in tight control units described their work environment as cost-conscious, meeting times were kept punctually, and jokes about the company and/or the job were rare.

6. Normative versus pragmatic

   It deals with the popular notion of customer orientation. In the normative units, the major emphasis was on correctly following organizational procedures, which were more important than results. In the pragmatic units, there was a major emphasis on meeting the customer's needs, results were more important than correct procedures.

Hofstede and Hofstede (2005) argued that the seven national cultural dimensions can be used to understand national and organizational cultures. The difference between them is based on their different mix of values and practices. Hofstede and Hofstede (2005) further explained that national cultures are part of the mental software that are acquired during the first ten years of people's lives, in the family, in the living environment, and at school, and they contain most of the human basic values. Organizational (or corporate) cultures are acquired when entering a work organization, with basic values firmly in place, and they consist mainly of the organization's practices, which are more superficial. It is also mentioned that the seven national cultural dimensions have implications for organization and management processes. Moreover, from the six organizational culture dimensions, dimension number 1, 3, and 4 were to some extent associated with values. For the other three dimensions, number 2, 5, and 6 described practices to which people had been socialized without their basic values being involved.

Regarding these dimensions, Hofstede and Hofstede (2005) also opined that even though the organizational cultures can be meaningfully described by a number of dimensions that can be considered as universal, one should be careful not to claim that the same model applies to any organization anywhere. New studies should choose their own units, compose their own questionnaire covering crucial differences in the practices of these organizations, and use the dimensions as the basic variables to measure.

4.2.3.2 Trompenaars (1993), Schwartz (1994), and House et al.'s (2003) Dimensions

Other scholars that have found cultural dimensions are Trompenaars (1993), Schwartz (1994), and House et al. (2003). Generally, their dimensions have

Table 4.2 Cultural dimensions from other scholars related to Hofstede's dimensions

Hofstede and Hofstede (2005)	Trompenaars (1993)	Schwartz (1994)	House et al. (2003)
Individualism versus collectivism	Universalism versus particularism Individualism versus communitarianism Achievement versus ascription	Intellectual autonomy Affective autonomy Mastery	Performance orientation Assertiveness orientation Institutional collectivism
Power distance		Hierarchy Egalitarian Commitment	Power distance
Uncertainty avoidance	Specific versus diffuse		Uncertainty avoidance
Masculinity versus femininity	Neutral versus emotional	Harmony	Family collectivism Gender egalitarianism Humane orientation
Long-term versus short-term orientation	Attitudes to time Attitudes to the environment	Conservatism	Future orientation

similarities with Hofstede's national cultural dimensions. Table 4.2 illustrates their dimensions compared with Hofstede's dimensions.

From the Table 4.2, it can be seen that there are similarities between these dimensions and Hofstede's. Trompenaars's (1993) seven cultural dimensions have the same meanings with four of Hofstede's dimensions. Three of the dimensions mainly focus on describing the characteristics of individualism and collectivism. As for Schwartz (1994), most of the cultural dimensions focus on the power distance characteristics, which are hierarchy, egalitarian, and commitment. House et al. (2003) has nine cultural dimensions that mostly elaborate the characteristics of individualism and collectivism, and gender (masculinity and femininity). From these studies, it appears that these scholars have their dimensions similar with five of Hofstede's dimensions, where the other two new dimensions (Indulgence vs. Restraint and Monumentalism vs. Self-effacement) were not included by Hofstede until 2005.

4.2.3.3 Naoum (2001), Kotler et al. (1992), Schols (1987) and Williams et al. (1993) Dimensions

Naoum (2001) had described three major dimensions of organizational culture, based on studies by Kotler et al. (1992), Schols (1987), and Williams et al. (1993). These dimensions are as follows:

- Social culture. This refers to the way people behave and the means in which people are motivated at work. It also includes social activities that take place within the organization.
- Technical culture. This means that the organization has a culture related to the techniques used for executing the tasks, equipment, processes and any other facility which transfers inputs into outputs. Naturally, the techniques used depend on the characteristics of the task which in turn affect the organizational structure and culture.
- Managerial culture. This includes the style of management and organizational structure. It refers to ways in which the tasks are divided and the activities are coordinated. These processes will create a pattern of power and authority within the organization which in turn can be developed into a particular culture.

4.2.3.4 Lynch's (2006) Dimensions

Lynch (2006) argued that there are several factors that indicate the type of culture that existed in an organization, which are:

- The age profile of the employees or members. Members with different age groups tend to have different opinions and views.
- The socio-economic group from which members come from. This will have impact on the values that are embedded in people. Members with similar background will often possess similar views and opinions.
- The balance between male and female. Nowadays, this factor is much less significant than in the past, but still has the potential to influence the type of culture the organization possesses.
- Ethnicity, language and religion in a multicultural country are things that should be acknowledged.
- Government policies which affect the training, social welfare, health and pensions, and people development programmes in the organization.

Moreover, from these factors, a number of guidelines were developed for analyzing the organizational culture which includes the following (Lynch 2006):

1. What is the age of the organization and the degree of turbulence that exists?
2. Where is ownership (or stewardship) located and how much power do these people possess?
3. What is the nature of the organizational relationship and how does decision-making take place?
4. What are the metrics that are used to evaluate performance and/or success?
5. How are people (especially leaders) judged?
6. Can changes be made easily to respond to changing environmental conditions?
7. In what way are people controlled (influenced)?
8. Do individuals work cooperatively using teams?

4.2.3.5 Harrison (1972) and Handy's (1993) Organizational Cultural Styles

According to Harrison (1972) and Handy (1993), there are four basic types of organizational style, which are:

- Power

  A power culture in which there is dominance by either one person (often the founder) or a small group, where the beliefs that exist are heavily influenced by those with most power. Any change in these key people will usually have a rapid and insightful impact on people.
- Role

  A role culture relies on the use of rational analysis and formal systems of administration (procedures and processes). People are expected to abide by rules in carrying out the operations that will lead to corporate objectives and deviation is dealt with by sanctions. In such a bureaucratic organization, those who have the ability to write the rules can effectively control behavior.
- Task

  The emphasis in a task culture is on getting things done. The value system is the ability to achieve, where it is likely to be rewarded or celebrated. In this type of culture, the need to abide by rules or procedures will not be considered as important, and there is a risk of failure.
- Personal

  In this type of culture, the strategy is operated to serve the interests of individuals. They may operate competitively, but also cooperatively. Also, work is conducted in teams that are flexible to tackle identified issues.

4.2.3.6 Hansen and Wernerfelt's (1989) Dimensions

Hansen and Wernerfelt (1989) suggested that organizational culture could be analyzed by:

1. The employee's capability to identify and accomplish the company goals
2. The level of awareness of employers in recognizing and rewarding good performance of employees

Based on their case studies on 60 firms from various business fields, they found that the two dimensions are positively correlated with the returns on assets of the firms.

4.2.3.7 Bettinger's (1989) Dimensions

Bettinger (1989) developed eleven artifacts that are regarded as the organizational culture framework. Artifacts are considered as the base level of culture, where they can be observed but are not easy to apprehend the deeper assumptions. Nevertheless, artifacts reflect beliefs and values shared by members in the organization (Cheung et al. 2011). Bettinger (1989) also argued that these eleven artifacts are viewed as constructive artifacts, which are:

1. The level of awareness in establishing company goals and objectives.
2. The sense of pride of the employees on the mission set by the company.
3. The employee's attitude towards the change of goals.
4. The degree of openness in communication, supervision and information sharing among team members.
5. The degree of openness in communication and supervision.
6. The employees' commitment to the organization and teamwork.
7. The atmosphere to reduce conflicts and enhance trust for avoidance of dysfunctional performance outcome.
8. The level of concern of the employees' participation in the decision-making process.
9. The establishment of performance standards and values that contribute towards success.
10. Rituals to support and reinforce values.
11. The presence of a rewarding scheme to recognize good performance.

4.2.3.8 Cameron and Quinn's (1999) Dimensions

A framework named "Organizational Culture Assessment Instrument (OCAI)" was proposed by Cameron and Quinn (1999) for evaluating organizational culture profile based on the core values, assumptions, interpretations and approaches that characterize organizations. They have suggested six artifacts that are included in the framework, which are (Cameron and Quinn 1999; Cheung et al. 2011):

1. Dominant characteristics—assess whether the organization is outcome-based and operating systematically.
2. Organizational leadership—evaluates the competence of the organization's leader in enhancing the organization's efficiency.
3. Management of employees—evaluates the competence of the management team in sustaining the employee's working effectiveness.
4. Organizational glue—evaluates the level of loyalty and mutual trust that are built among members in the organization.
5. Strategic emphasis—assesses an organization in terms of its level of achievement on predetermined strategic goals.
6. Criteria for success—evaluates the competence of the management team in designing a road map for success.

4.2.3.9 Denison's (2000) Dimensions

Denison (2000) developed the Denison Organizational Culture Model to analyze the culture of organizations, and is based on four traits that have a strong influence on organizational performance. The four characteristics that are examined in the model are (Denison 2000; Cheung et al. 2011):

- Involvement—empowerment (measures the authority, initiative and ability that an individual manages his own work); team orientation (measures the degree of cooperation among employees in working toward common goals); capability development (measures the degree that organization continually invests in the development of employee's skills in order to be competitive and able to meet on-going business needs).
- Consistency—core values (measures a set of values that create a sense of identity that members share); agreement (measures the extent that members of the organization are able to reach agreement on critical issues); coordination and integration (measures the extent that different functions and units of the organization are able to work together to achieve common goals.
- Adaptability—creating change (measures the degree of the organization is able to create adaptive ways to meet changing needs); customer focus (reflects the degree to which the organization is driven by the concern to satisfy their customers); organizational learning (measures the degree the organization encourages innovation, gaining knowledge, and developing capabilities).
- Mission—strategic direction and intent (measures the degree that clear strategic intentions are conveyed to make it clear how everyone can contribute); goals and objectives (measures the degree that clear direction is provided for employees in their work; vision (measures the degree that a shared view of a desired future state and core values are conveyed to the employees).

4.2.3.10 Cheung et al.'s (2011) Dimensions

Cheung et al. (2011) had developed organizational culture factors that are significant in construction. These factors had been compiled from various dimensions from various scholars. Table 4.3 describes the organizational culture (OC) factors with their relevant artifacts.

4.2.3.11 GLOBE's Dimensions

The Global Leadership and Organizational Behaviour Effectiveness (GLOBE) study is a large-scale research program involving 160 researchers from many parts of the world. Its objectives are to examine the inter-relationships between societal culture, organizational culture, and organizational leadership (Javidan et al.

Table 4.3 Organizational culture factors

Organizational culture factors	OC artifacts
Goal settings and accomplishment	Clear goals Clear approach to succeed Actions are matched with organization's goals
Team orientation	Emphasize team contributions Amicable opinions and ideas exchange Members' commitment to team
Coordination and integration	Resolve internal problems effectively Encourage inter-departmental collaboration Encourage information sharing
Performance emphasis	Guidance for performance improvement Emphasize good performance Explicit set of performance standards
Innovation orientation	Accept adventurous ideas for sustaining competitiveness Welcome alternative solutions Encourage creative and innovative ideas Allocate resources for implementing innovative ideas
Members' participation	Value employees' ideas Employees' inputs on major decisions Employees' participation in decision-making process
Reward orientation	Emphasize team accountability Emphasize on reward instead of punishment Trust atmosphere Performance-based rewards Accept criticism and negative feedback Recognize and reward members' performance Equitable reward

Source: Cheung et al. (2011)

2006; House et al. 2004). This study identifies nine dimensions of culture as follows (House et al. 2010; Magnusson et al. 2008; Maleki 2010; Chisholm 2010):

1. Uncertainty avoidance—the extent to which members of an organization or society strive to avoid uncertainty by reliance on social norms, rituals and bureaucratic practices to alleviate the unpredictability of future events.
2. Power distance—the degree to which members of an organization or society expect and agree that power should be unequally shared.
3. Societal or Institutional collectivism—the degree to which organizational and societal institutional practices encourage and reward collective distribution of resources and collective action.
4. In-group collectivism—the degree to which individuals express pride, loyalty and cohesiveness in their organizations or families.
5. Gender egalitarianism—the extent to which an organization or a society minimized gender role differences and gender discrimination.
6. Assertiveness—the degree to which individuals in organizations or societies are assertive, confrontational, and aggressive in social relationships.

7. Future orientation—the degree to which individuals in organizations or societies engage in future-oriented behaviors such as planning, investing in the future and delaying gratification.
8. Performance orientation—the extent to which an organization or society encourages and rewards group members for performance improvement and excellence.
9. Humane orientation—the degree to which individuals in organizations or societies encourage and reward individuals for being fair, altruistic, friendly, generous, caring and kind to others.

4.2.3.12 Organizational Behavior Attributes

Organizational culture is defined as the pattern of basic assumptions that a given group has invented, discovered, or developed in learning to cope with its problems of external adaptation and internal integration, and that have worked well enough to be considered valid, and therefore, to be taught to new members as the correct way to perceive, think, and feel in relation to those problems (Osland et al. 2001).

According to Osland et al. (2001), there are seven organizational culture dimensions which are:

- Conformity: The degree to which members feel that there are many rules, procedures, policies, and practices to which they have to conform rather than being able to do their work as they see fit.
- Responsibility: Members of the organization are given personal responsibility to achieve their part of the organization's goals; the degree to which members feel that they can make decisions and solve problems without checking with superiors each step of the way.
- Standards: The emphasis the organization places on quality performance and outstanding production, including the degree to which the member feels the organization is setting challenging goals for itself and communicating these goal commitments to members.
- Rewards: The degree to which members feel that they are being recognized and rewarded for good work rather than being ignored, criticized, or punished when something goes wrong.
- Organizational clarity: The feeling among members that things are well organized and goals are clearly defined rather than being disorderly, confused, or chaotic.
- Warmth and support: The feeling that friendliness is a valued norm in the organization, that members trust one another and offer support to one another. The feeling that good relationships prevail in the work environment.
- Leadership: The willingness of organizational members to accept leadership and direction from qualified others; leadership is based on expertise and the organization is not dominated by, or dependent on, one or two individuals.

As part of organizational behavior, managers can build or change a culture in an organization by influencing other aspects of the organization which is called secondary mechanisms. These mechanisms are (1) the way in which the organization is structured and designed (decision making, coordination, reporting, structure); (2) systems and procedures (performance appraisal, information, control, DSS, etc.); (3) rites and rituals; (4) the design of physical space, facades and buildings; (5) anecdotes, legends, myths, and parables about people and events; and (6) formal statements of philosophy, creeds, and values.

Moreover, in an organization, strong cultures can be seen from the following characteristics (Osland et al. 2001):

- People in the organization can easily identify the dominant values.
- The selection processes target people who are likely to fit into the culture and find it satisfying.
- Socialization and training convey to newcomers the "ropes" they need to learn.
- Employees who do not fit the culture or produce in accordance with its values are sometimes fired.
- People within the company are rewarded for acting in accordance with the dominant values of the organization.
- By their behavior, leaders and managers send clear, consistent signals about desired values and norms.
- Managers measure and control what is important to the culture.

4.2.3.13 Non-technical Attributes from a Quality Assurance System

According to the study by Low (1998), the non-technical framework which affects quality assurance systems is an important element to be considered. Fulfilling the technical requirements is only one aspect of total quality management. The other aspect which focuses on non-technical requirements must not be overlooked. Moreover, the scope of coverage in the non-technical attributes is broader than the technical one. Kanter (1994) stated that the non-technical framework should be reviewed due to its role in helping to promote an integrative environment for the development of change and innovation, with the primary objective of quality improvement.

The attributes identified within the non-technical framework are as follows (Low 1998):

- Authority, power and responsibility

  Authority is linked to responsibility, because in order to accomplish certain results a manager must have the authority to use resources to achieve those results. The authority over resources must be sufficient to enable the manager to meet the output expectations of others.
- Power in organization

  Power refers to a capacity which the manager has over the behavior of an employee, so that the employee will do something he or she would not otherwise do.
- Conflicts in organization

Interpersonal relations in organizations are bound to create occasional conflicts or disagreement between people on substantive or emotional issues. Schermerhorn (1996) observes that managers spend a lot of time dealing with the various conflicts within organizations. People usually respond to conflicts with different emphasis on cooperativeness and assertiveness.

- Adaptability to change

  Whenever changes are introduced into an organization, employees will often either resist of resent such changes. Ways to overcome or break this resistance is by securing top management support; encourage employee participation in the process; foster open communication during the changing phase; and reward contributors who supported the change (Peel 1993).
- Empowerment in organization

  Empowerment is the process of enabling workers to set their own work-related goals, make decisions and solve problems within their sphere of responsibility and authority. Luthans (1992) suggests the following ways in which management can empower employees:

  1. Express confidence in employees' abilities and hold high expectations concerning their performance.
  2. Allow employees to participate in the decision-making process.
  3. Allow employees freedom and autonomy in how they perform their jobs.
  4. Set inspirational or managerial goals for employees.
  5. Use position power in a prudent and positive way and limit the use of coercive power.

A summary of the keywords and indicators from studies of the 13 OC dimensions is shown in Appendix A. From studies of the 13 OC dimensions, 14 main keywords of OC dimensions were found in these dimensions. These 14 keywords have several indicators from various OC studies. It turned out that there are five main studies that describe the 14 main keywords of OC dimensions, which are: Hofstede and Hofstede (2005), Denison (2000), Cheung et al. (2011), Low (1998), as well as the organizational behavior attributes from Osland et al. (2001) and Luthans (2008).

Other descriptions based on Appendix A are as follows:

- It is important to note that Cheung et al. (2011)'s study had compiled studies of other OC dimensions, which were from Bettinger (1989), Denison (2000), and Cameron and Quinn (1999). Also, Cheung et al. (2011) had found that their OCs are significant in the construction industry.
- Hofstede's (2005) national cultural dimensions were related and the benchmark of Trompenaars (1993), Schwartz (1994) and House et al. (2003) was described. Hofstede and Hofstede's (2005) OC dimensions were derived from their cross-organizational studies, but some values were similar with their national dimensions.
- Naoum's (2001) study was based on the studies of Kotler et al. (1992), Schols (1987) and Williams et al. (1993). All of these culture dimensions are similar

with those presented by Hofstede and Hofstede (2005), Cheung et al. (2011) and Low (1998) in their compilation of OC dimensions.
- The 8 OC guidelines from Lynch (2006) are part of the 14 main keywords of OC dimensions. In addition, these 8 OC guidelines are similar with Cheung et al. (2011)'s OC dimensions and the organization behavior dimensions (Low 1998; Osland et al. 2011; Luthans 2008).
- Harrison's (1972) and Handy's (1993) findings were similar with the dimensions presented by Lynch (2006). The attributes from this study have the same indicators as the OC attributes from Cheung et al. (2011) and the organizational behavior studies (Low (1998); Osland et al. (2001); and Luthans (2008)).
- Hansen and Wernerfelt's (1989) OC dimensions were included in the OC dimensions presented by Cheung et al. (2011).

4.2.4 Managing Organizational Culture

According to Hofstede and Hofstede (2005), identifying the organizational culture is part of the crucial element, and the other crucial part is what management does with it. There needs to be a balance in four aspects such as strategy, structure, control and culture. An organization's performance should be measured against its objectives, and it is the top management's role in translating objectives into strategy. Further, strategies are implemented through the existing structure and control system, and their outcome is modified by the organization's culture. Thus, all four of these elements influence each other.

Even though culture is a "soft" characteristic, changing it calls for "hard" measures. Some examples of these changes are (Hofstede and Hofstede 2005):

- Structural changes—which may mean closing departments, opening other departments, merging or splitting activities, or moving employees geographically. When people are moved as individuals or groups, they will adapt to the culture of their new environment.
- Process changes—this means implementing new procedures, eliminating controls or establishing new controls, implementing or discontinuing automation, and introducing new communication links.
- Personnel changes—this means new hiring and promoting policies.

In attempting to change culture in an organization, new symbols usually receive a lot of attention, such as new name, logo, uniforms, slogans, and portraits on the wall, which all belongs to the corporate identity. However, these symbols are only part of the superficial level of culture. These should be supported by more basic changes at the deeper levels of heroes, rituals and values of key leaders in the organization. Culture change needs persistence and sustained attention by the top management.

4.2.5 *Benefits of Identifying Organizational Culture*

In a firm, identifying its organization culture provides practical utility. This process can identify whether the culture fits the strategies set out for the future. Cultural constraints determine which strategies are feasible for an organization and which are not. In addition, in the case of mergers and acquisitions, organization culture identification can help to seek the potential areas of culture conflicts between the partners (Hofstede and Hofstede 2005; Naoum 2001).

Moreover, some scholars have suggested that organizational culture is the key to success. Culture is important because it contributes to how groups of people respond and behave in relation to issues they face. It also has essential influences on the development and change of organizational strategy. Culture is crucial due to its influence where managers can achieve objectives through people's willingness to believe in what is required, rather than having to be forced to do so through threats and imposition of procedures. Therefore, understanding what motivates people and causes them to act in a coordinated way is certainly beneficial to managers (McCabe 2010; Schein 1985).

There are examples on various areas that use cultural dimensions for identifying their organizational culture. Business and marketing have found it useful to determine appropriate marketing campaigns for various culturally segmented markets. Also, engineering and design firms have both made use of cultural dimensions to better understand their customers. Last but not least, other fields such as technical writing, medicine and human capabilities development have started to make use of the information present in the cultural dimensions schemes for their organization's strategies (Bossuyt 2008).

4.3 Institutional Theory

Thompson (1967) and Parsons (1960) have identified three levels in an organization. The first one is the technical level, where the organization carrying on the production function that transform inputs into outputs. The second level is the managerial level, where the organization is responsible for designing and controlling the production system, for procuring inputs and disposing of outputs, and for securing and allocating personnel to units and functions. The third one is the institutional level, where the organization relates to its wider environment, determines its domain and boundaries, and secure its legitimacy.

Research in management has increasingly paid attention to issues where organizations work to influence and shape their environments. Such diverse perspectives as strategic management (Porter 1991), institutional theory (DiMaggio 1988; Oliver 1991), entrepreneurship (Aldrich and Fiol 1994) and organizational ecology (Baum and Oliver 1991) have argued that organizations actively participate in the social construction of their environments. Therefore, a central issue for management research has become the manner in which organizational environments are constituted, reproduced and transformed through organizational action and relationships (Lawrence 1999).

As stated by Scott (1998), the interdependence of the organization and its environment receives primary attention in the open system perspective. This perspective stresses the reciprocal ties that bind and relate the organization with those elements that surround and penetrate it. The environment is viewed to be the definitive source of materials, energy and information, where all of these are essential for the continuation of the system in the organization.

4.3.1 Defining Institutions

Historically, the most influential concept of institutions pervading mainstream sociology throughout the twentieth century has come from the work of Herbert Spencer. Spencer viewed society as an organic system evolving through time. Adaptation of the system to its context was obtained through the functions of specialized "organs" structured as institutional subsystems. Later on, other influential contributions by DiMaggio and Powell (1983) and Meyer and Scott (1983b) described the environmental perspective. They proposed that all organizations are shaped by both technical and institutional forces (Scott 2008).

Institutions are considered as multifaceted, durable social structures made up of symbolic elements, social activities, and material resources that enable or impose limitations on the scope for human agency by creating legal, moral and cultural boundaries. Institutions are inhabited by people and their interactions. Rules, norms and meanings arise in interaction, and they are preserved and modified by human behavior. Institutions ride on various media, where these institutional carriers vary in the process employed to transmit their messages. Moreover, institutions operate at multiple levels, from the world system to interpersonal interaction. Although institutions function to provide stability and order, they may experience change, both incremental and revolutionary. Thus, institutions should not only be viewed as a "property" or state of an existing social order, but also as a "process" (Scott 2008).

There are two variants of the institutional theory, which are the "old" and the "new". The "old" approach focused on regulative dimension represented by coercive pressure of institutions. On the other hand, the "new" approach focused on examining the direct pressure for change by institutions on organizations along three different dimensions, which are regulative, normative, and cultural-cognitive (Scott 2008). Furthermore, the modern institutional theory examines how organizations are influenced by their institutional environment. This stream of theory addresses the relevant aspects of organizational environments, providing concepts for describing and analyzing institutional elements as well as organizational responses to environmental pressures. The pressures can be categorized into coercive (that imposed by other organizations), normative (that which arise by being

Table 4.4 Strategic responses to environmental pressures

Strategies	Tactics	Examples
Acquiesce	Habit Imitate Comply	Following invisible, taken-for-granted norms Mimicking institutional models Obeying rules and accepting norms
Compromise	Balance Pacify Bargain	Balancing the expectations of multiple constituents Placating and accommodating institutional elements Negotiating with institutional stakeholders
Avoid	Conceal Buffer Escape	Disguising nonconformity Loosening institutional attachments Changing goals, activities, or domains
Defy	Dismiss Change Attack	Ignoring explicit norms and values Contesting rules and requirements Assaulting the sources of institutional pressure
Manipulate	Co-opt Influence Control	Importing influential constituents Shaping values and criteria Dominating institutional constituents and processes

Source: Oliver (1991)

part of social groups), and mimetic (that arising from environmental uncertainties) (DiMaggio and Powell 1983; Abdul-Aziz et al. 2010; Dahl and Nesheim 1998).

Regarding responses to environmental pressures, organizations may enact different behaviors. Oliver (1991, p. 151) suggested that "organizational responses will vary from conforming to resistance, from impotent to influential, and from habitual to opportunistic, dependent on the institutional pressures to conformity that are exerted on organizations." In addition, Oliver (1991) had proposed a typology of strategic responses to environmental pressure. Table 4.4 shows the strategy alternatives in responding to pressure to conform with the institutional environment.

The institutional environment influences organizational strategies and practices through a number of mechanisms, ranging from constitutive rules and categorical conformity, to conscious strategies on the part of the actors to dismiss or influence aspects of the environment. The source of environmental influence or pressure may be located at several levels such as in the organizational field, in the societal sector, at the national level and even at the level of the world system (Scott 1998).

4.3.2 Three Pillars of Institutions

Scott (2008) noted that the essence of the institutional perspective in general resides in the three pillars—the regulative, normative, and cultural-cognitive—which provide meaning and stability for social behavior. These pillars have been identified by social theorist as a vital ingredient of institutions. The three elements form a continuum moving "from the conscious to the unconscious, from the legally

Table 4.5 The three pillars of institutions

	Regulative	Normative	Cultural-cognitive
Basis of compliance	Expedience	Social obligation	Taken-for-grantedness Shared understanding
Basis of order	Regulative rules	Binding expectations	Constitutive schema
Mechanisms	Coercive	Normative	Mimetic
Logic	Instrumentality	Appropriateness	Orthodoxy
Indicators	Rules Laws Sanctions	Certification Accreditation	Common beliefs Shared logics of action Isomorphism
Affect	Fear Guilt/ Innocence	Shame/honor	Certainty/confusion
Basis of legitimacy	Legally sanctioned	Morally governed	Comprehensible Recognizable Culturally supported

Source: Scott (2008)

enforced to the taken for granted" (Scott 2008, p. 34). Table 4.5 shows the detailed descriptions about the three pillars, with various indicators for each of them.

4.3.2.1 The Regulative Pillar

The regulative pillar is distinguished by the prominence given to explicit regulative processes; rule-setting, monitoring and sanctioning activities. Scholars view the regulative aspects institutions as constrains and regularize behavior. These processes may operate through diffuse, informal mechanisms, involving folkways like shaming or shunning activities, or they could be highly formalized through specialized actors such as the police or courts. Even though the regulative concepts work to constrain, many types of regulations have also enabled social actors and action, and give special powers and benefits. In other words, this pillar works both to constrain and empower social behavior. Institutions supported by one pillar may, as time passes and circumstances change, be sustained by different pillars. A stable system of rules, whether formal or informal, supported by surveillance and sanctioning power that is accompanied by feelings of fear/guilt or innocence/incorruptibility is one prevailing view of institutions (Scott 1998, 2008).

4.3.2.2 The Normative Pillar

In the normative pillar, the emphasis is on values and norms which introduce a prescriptive, evaluative and obligatory dimension into social life. Values are conceptions of the preferred or the desirable, together with the construction of standards where existing structures or behaviors can be compared and assessed. Norms specify how things should be done, where they define legitimate means to pursue valued ends. Moreover, normative systems define goals or objectives (e.g. winning the game, making a profit), but also designate appropriate ways to pursue them (e.g. rules specifying how the game is to be played, conceptions of fair business practices). This pillar is also viewed as imposing constraints and empowers social action. It confers rights as well as responsibilities; privileges as well as duties; licenses as well as mandates. This conception was first examined through institutions such as kinship groups, social classes, religious systems, and voluntary associations, where common beliefs and values are more likely to exist and constitute an important basis for order. Similar with the regulative pillar, norms can also induce strong feelings, but are somewhat different from those that accompany the violation of rules and regulations. Feelings related with the trespassing of norms basically include a sense of shame or disgrace, or, for those who show exemplary behavior, feelings of pride and honor. The conformity to or violation of norms typically involves a large measure of self-evaluation: heightened remorse and/or effects on self-respect (Scott 1998, 2008).

4.3.2.3 The Cultural-Cognitive Pillar

The cultural-cognitive elements include widely held beliefs and taken-for-granted assumptions, the rules that constitute the nature of reality and the frames through which meaning is created. This cognitive dimension views relationship between the external world of stimuli and the response of the individual organism as a collection of internalized symbolic representations of the world. Symbols like words, signs, and gestures shape the meanings that are attributed to objects and activities. Meanings arise in interaction and are maintained and transformed as they are employed to make sense of the ongoing stream of happenings. Regarding cultural conceptions, it is important to recognize that beliefs are held by some, but not by others. People in the same situation can perceive the situation quite differently—both in terms of what is and what should be. Cultural beliefs vary and are frequently contested, particularly in times of social disorganization and change. The feelings that can be raised from this pillar is expressed through different ranges from the positive effect of certitude and confidence, to the negative feelings of confusion or disorientation. Furthermore, it is considered that actors who align themselves with prevailing cultural beliefs are likely to feel competent and connected, whereas those who are at odds are regarded as "clueless" and at worst, "crazy". A cultural-cognitive conception of institutions stresses the central role played by the socially mediated construction of a common framework of meaning (Scott 1998, 2008).

4.3.2.4 The Three Pillars Varying Carriers and Legitimacy

Scott (2008) espoused that institutions, whether regulative, normative, or cultural-cognitive pillars are stressed, are conveyed by various types of vehicles or carriers. Carriers are important in considering the ways in which institutions change. They point to a set of fundamental mechanisms that account for how ideas move through

Table 4.6 Institutional pillars and carriers

	Pillars		
Carriers	Regulative	Normative	Cultural-cognitive
Symbolic systems	Rules Laws	Values Expectations	Categories Typifications Schema
Relational systems	Governance systems Power systems	Regimes Authority systems	Structural isomorphism Identities
Routines	Protocols Standard operating procedures	Jobs Roles Obedience to duty	Scripts
Artifacts	Objects complying with mandated specifications	Objects meeting conventions, standards	Objects possessing symbolic value

Source: Scott (2008)

space and time, and who or what is transporting them. Table 4.6 describes the various carriers from the three institutional pillars.

The carriers can be in the form of symbolic systems, relational systems, routines and artifacts. Symbols can be considered as transportable, versatile, and flexible. Relational systems are carriers that rely on patterned interactions connected to networks of social positions, which is role system. Routines rely on patterned actions that reflect the tacit knowledge of actors, which are deeply ingrained habits and procedures based on unarticulated knowledge and beliefs. Finally, artifacts are discrete material object, consciously produced or transformed by human activity, under the influence of the physical and/or cultural environment (Scott 2008).

According to Suchman (1995), organizations need more than material resources and technical information if they want to survive in their social environments. Social acceptability and credibility are also needed. These conditions relate with the concept of legitimacy, where it is a generalized perception or assumption that the actions of an entity are desirable, proper, or appropriate within some socially constructed system of norms, values, beliefs, and definitions.

Scott (2008) stated that the three pillars are bases of legitimacy. The regulatory emphasis is on conformity to rules, where legitimate organizations are those established by and operating in accordance with relevant legal requirements. A normative conception focuses a deeper, moral base for assessing legitimacy. Controls in the normative pillar are likely to be internalized than are regulative controls. A cultural-cognitive view points to the legitimacy that comes from conforming to a common definition of the situation, frame of reference, or a recognizable role or structural template. This perspective rests on preconscious, taken-for-granted understanding; hence it is the deepest level of legitimacy.

The bases of legitimacy associated with the three pillars are particularly different and may, sometimes, be in conflict. As an example, a regulative view would determine whether the organization is legally established and whether it is acting according with relevant laws and regulations. A normative view stresses on moral obligations. Many professionals that hold on to normative standards make them depart from the rule-based requirements of bureaucratic organizations. Whistle-blowers claim that they are acting on the basis of a "higher authority" when they contest organizational rules or superiors' orders. Furthermore, an organization such as the mafia may be widely recognized, telling that it exhibits a culturally constituted mode of organizing to achieve specified ends, and it is considered as a legitimate way of organizing by its members. However, it is treated as an illegal form by police and other regulative bodies, and it lacks the normative endorsement of most citizens. What is taken as evidence of legitimacy varies by which elements of institutions are privileged as well as which audiences or authorities are consulted (Scott 2008).

It has been observed by Scott (1998, 2008) that in most institutional forms, not one single pillar is at work, but varying combinations of them. In a stable social system, it was observed that practices that persist and are reinforced are due to them being taken for granted, normatively endorsed and backed by authorized powers. Moreover, in some situations, one or another pillar will operate virtually alone in supporting the social order, and in other situations, a combination of pillars will assume dominance.

4.4 Summary

Organizational culture plays an essential role in the organization. It is considered as the social or normative glue that holds an organization together, where it expresses the social ideals, values and beliefs that members of an organization come to share. Cultural studies had also been included in the previous mainstream management school of thought, and have been further developed into more comprehensive studies with definition and dimensions to measure. Advantages of identifying and managing organizational culture have also been determined by various literature where there are practical utilities in order to have an effective organization. The second aspect that is described in this section is the importance of institutional perspective in organizations. Various studies had found that an organization is supported and constrained by institutional forces. The three institutional pillars which are regulative, normative, and cultural-cognitive are the essence of the institutional perspective that provides meaning and stability for social behavior in organizations. In addition, the main concept of an organization as a special-purpose, instrumental entity is a product of the institutional process, where it is also influenced by environments.

Chapter 5
Mainstream Theories: Implementation by Contractors

5.1 Introduction

This chapter reviews the mainstream theories that have been described in the previous chapters (Chaps. 2, 3 and 4) for implementation by contractors. It starts from describing the implementation of organizational management and traditional organizational theory, including systems theory, contingency theory, complexity theory, and change management. Furthermore, the implementations of crisis management and BCM in construction are discussed. The final section elaborates culture in organization and institutional theory's role in construction.

5.2 Mainstream Theories for Implementation by Contractors

Ofori (1990) defined the construction industry as that sector of the economy which plans, designs, constructs, alters, maintains, repairs, and eventually demolishes buildings of all kinds, civil engineering works, mechanical and electrical engineering structures and other similar works. Thus, the industry includes:

a. Persons, enterprises, and agencies, both public and private, involved in physical construction: both those whose main activity is construction and the relevant parts of entities engaged in other fields of activity who retain some construction capacity (such as the maintenance unit of a university or a manufacturing enterprise).
b. Those providing all kinds of planning, design, supervisory and managerial services relating to construction.

The construction process may be subdivided into the stages described below (Ofori 1990):

L.S.R. Supriadi, L. Sui Pheng, *Business Continuity Management in Construction*, Management in the Built Environment, DOI 10.1007/978-981-10-5487-7_5

a. Conceptual stage: the client identifies the need for the item of construction and appoints and briefs consultants who study the client's requirements, propose an outline of the design and assess the feasibility of the project.
b. Design stage: the concept of the project is further developed with production information and contract documentation prepared. Tenders are called for choosing the suitable constructor.
c. Construction stage: production programs are prepared and construction carried out on site.
d. Operation and maintenance stage: the completed building or works is maintained, repaired or altered as required, over the course of its life.

Construction firms, particularly contractors, are also similar to other organizations in other sectors, where literatures have shown that these organizations also implement the mainstream theories which are described in the previous sub-sections of the preceding chapter. However, Hillebrandt (1984) argued that construction is an industry which has characteristics that makes it different from others. It is an industry that is responsible for creating the built environment, where every member of society relies on the output of construction (McCabe 2010). In relation to contracting firms (contractors), the quantity and quality of management is the key to an efficient firm. Management in contractors is crucial due to the following characteristics (Hillebrandt and Cannon 1990):

1. Each project has to be set up as a production unit on a fresh site starting from nothing.
2. Good man-management is vital to the success of a project due to its labour-intensive characteristic.
3. The production process is very complex involving a large number of inputs.
4. High level of uncertainty due to its weather and ground conditions, and limits of resources. Thus, the number of management decisions to be taken is large and the industry at site level is very decision-intensive.
5. At the level of the operating unit, each project is different and has different clients and professionals, so management relations with outside organizations may be complex.
6. Because of the wide geographical dispersion of activities it is often difficult for the site manager and sometimes for the operating unit manager to obtain help or decisions from head office. This is especially so in overseas work where the managers have to be capable of operating and taking decisions without the benefit of prior consultation.

5.3 Organizational Management in Contractors

According to Lavender (1996), the processes of management established by Fayol in 1916 have been widely used in the construction industry. Contractors have implemented the five processes which are:

- Forecasting and planning

  Planning involves looking ahead, which implies that forecasting is an inherent part. In the market-oriented firm in construction, making strategic decisions about the desired future market position of the firm is the key to planning. It is useful to consider types of planning, and the planning process itself. There are three levels of planning that are used in this phase:

  1. Strategic planning is long term, in very broad details and carried out at a senior level in the organization.
  2. Management planning is more medium term, with a greater degree of details, and carried out by middle management.
  3. Operational or tactical planning is short term, more detailed and carried out at the supervisory level.

- Organizing

  This is about implementing plans and should be taken into consideration at the planning stage. It involves putting into place the necessary structures. Many of the factors concerned with organizing include grouping of tasks, delegation, spans of control and communications.
- Commanding

  This process is needed for giving instructions and ensuring that they are carried out.
- Coordinating

  The process of harmonizing activities is essential, particularly when the organization has complex and interrelated tasks as contractors. They have to manage all of the resources that they have in conducting each activity. Moreover, the process of motivating the people in the organization is crucial. Leadership and work groups are important in construction activities.
- Controlling

  Control systems are set up to ensure compliance with plans. The stages of a control system are as follows:

  1. Set targets or standards of performance.
  2. Measure actual performance.
  3. Compare with target.
  4. Take remedial action if necessary.
  5. Revise targets if necessary.

Planning and control are usually linked together, where good planning is the first phase of a control system. The ability to take remedial action is also an important aspect of control, because without this, all that remains is a system of monitoring. For example, in construction projects, a good deal of monitoring of costs takes place at all stages. To qualify as a control system, the information must be in the hands of management in time to take remedial action in the context of the existing project. It should also be noted that when an examination reveals that actual performance has deviated from the plan, it should not automatically be assumed that remedial action is necessary. It could be that the original targets were incorrectly formulated, or that circumstances have since changed, rendering them unrealistic. Therefore, there are certain basic rules about the development of a control system:

1. Targets should be expressed in measurable forms, such as programmes, budgets and cash flow forecasts.
2. Targets should be realistic and achievable.
3. The system must achieve the appropriate balance of complexity and time, that is, it must yield sufficient meaningful information, but in a time span which enables remedial action to be taken.
4. The setting-up and running costs of the system should be commensurate with the likely benefits.

In construction, the differences between project management and management generally should not be overstated. However, there are some clear distinctions between them. General management is considered as ongoing process, whereas a project is unique in character, is of a temporary nature with a beginning and end, has a series of deadlines and targets, and requires a project-level team (Lavender 1996). Lock (1992) considered project management as having the function to foresee or predict as many of the dangers and problems as possible and to plan, organize and control activities so that the project is completed successfully in spite of the risks.

The implementation of project management by contractors is nothing new. In managing their projects, contractors generally use the project management concepts. As an example, Chen and Partington (2006) found that in the UK, project management (PM) competence is essential for managing the construction activities. The conception of PM is used for the construction process and work details, work interfaces, and handling potential risks and problems. The way in which each conception and its key attributes formed a distinctive structure of project management competence can be seen in Table 5.1.

5.4 Traditional Organizational Theories

Lavender (1996) observed that the development of production methods in construction had been dominated by the scientific management school of thought. Approaches such as de-skilling, the division of labour, mechanization and factorization, and subcontracting are used in the construction industry. In general, there are four historical stages regarding the development of production methods in construction, which are:

1. Traditional, Until About the Late 1950s

   The traditional method is characterized by site operations where skilled craft operatives work on relatively unprocessed materials. Skilled workers are supported by non-craft labourers, and a relatively small amount of plant.

Table 5.1 Project management competence for construction in UK

Conception	Main focus	Key attributes of conception and aspects of competence						
		Ability to plan	Knowledge of construction work	Ability to communicate	Ability to manage team	Knowledge of commercial management	Ability to coordinate	Ability to deal with problems
PM as planning and controlling	Construction process and work details	Plan the work; approve subcontractor's programme and methods; control the work process	Understand the construction process and detailed requirements, particularly of the H&S legislation	By regular meetings; control the implementation of the work and keep the plan updated	Allocate work tasks; chase and help people in their work			
PM as organizing and coordinating	Construction work interfaces		Understand the work interfaces and subcontractors' work interactions	By various approaches; gain trust to assist in organizing and coordinating work on site	Get people to work together; build a team and create a spirit of mutual support, trust and openness	Be aware of the contract and price constraints, manage variations and changes to the contract	Coordinate subcontractors' work interfaces to avoid disruption and inefficiency	
PM as predicting and managing potential problems	Potential risks and problems; Future-oriented		Know the construction industry; be aware of the history and future development of the industry	Communicate with all kinds and levels of people; get updated information from all possible sources	Motivate the team and empower them; care about team member's career development	Be aware of potential risks and problems in the contract and the possible financial constraints	Organize and coordinate all and everything; coordinate to avoid the happening of potential risks and problems	Think forward to predict problems; take precautious actions to be able to resolve problems quickly

Source: Chen and Partington (2006)

Therefore, a high proportion of the value of a building is added on-site, rather than off-site in a factory. Moreover, craft workers tend to have a certain amount of autonomy, where they are free from direct management control. From the management point of view, the craft operative's attitude to their work may not fit with the management's desire to increase productivity and cut costs.

2. First Industrial Phase, During the 1960s

   In this phase, traditional craft-based construction was still the norm, although some off-site prefabrication was used.

3. Subcontracting Phase, During the 1970s

   Changes to social and organizational factors had gathered pace in this period. The main change was the shift from direct employment to self-employment and labour-only subcontracting. The principal effect of this was to create a stronger link between pay and productivity. This was because those employed were now paid a lump-sum for carrying out a set amount of work, rather than a regular weekly wage. Although traditional construction methods continued to be used, design tended to become simpler and more rationalized. Furthermore, many of the traditional materials were increasingly replaced by modern materials, which had a similar appearance but were easier to work with. Thus, by the end of the 1970s, a new kind of workforce had emerged which was more flexible and multi-skilled. Subcontracting had become extremely widespread, and payment-by-results was common.

4. Second Industrial Phase, Since the 1980s

   In this phase, the changes were started with the changes to materials to use more of factory-made components. The impact was that a higher proportion of the value of a building was now being added off-site rather than on-site, where prefabricated components are being used structurally. In a technological sense, building production has increasingly come to resemble the manufacturing sector.

Productivity in construction has been developed from Taylor's scientific management school of thought, where it is considered possible to determine the one best way to carry out a work task, and hence achieve maximum productivity. Productivity is defined as (1) measuring efficiency, relating output to resource input; (2) it influences the costs of production, which in turn affect profits; and (3) some improvements in productivity may be achieved without major expenditure on new resources, while sustained improvements in productivity usually require investments. Increasing productivity in construction activities can be achieved by using factory-made components and dividing the work into work packages where subcontracting can be implemented (Lavender 1996).

5.4.1 Systems Theory

Based on Phua's (2007) study, the literature in construction and project management mostly highlights the utility of the open systems approach in construction firms and project organizations. The firms must continually monitor, respond and adapt to the influences of the external environment in order to succeed and survive. The external environment in this context comprises of all the social, economic and physical factors that exist outside the boundary of the firm, that affect the decision-making behavior of the firm. It was also found that different environmental dynamics place different demands on different firms in terms of the way they structure their strategic functions and activities (for example production, finance, sales and marketing, procurement delivery, etc.) in order to cope with the environmental circumstances.

Furthermore, Griffith (2011) observed that a large principal contractor usually adopts a decentralized organization structure to accommodate construction projects that take place in many and often geographically dispersed locations. Even though a large proportion of construction activities and tasks are the same from project to project, these firms usually have multiple projects done simultaneously that may provide a wide range of interdisciplinary elements. Thus, a contractor must be able to develop a strong corporate structure, organization and management. A framework for establishing effective systems and management system application for managing construction work requires:

1. Corporate, or Company-Wide Structure

   This is to establish organization within which the management of the company and its project sites can be configured. This structure also formalizes the approach to delivering the business outputs through arranging functional management disciplines, roles and responsibilities in relation to the business processes.
2. Management Systems Structure

   This is to establish procedures by which the firm's construction projects can be successfully delivered. It also translates functional management disciplines into management systems, establishing management procedures and implementation plans.
3. Project Structure

   This is developed to establish practices where the construction works will be undertaken at the project site. It translates management procedures and implementation plans into project supervision procedures and working instructions for application to the project.

5.4.2 Contingency Theory

Organizational theories have changed over the past century, from scientific management (mechanical) to contingency theory (natural) as environments have changed from being more stable to more turbulent. In the construction industry, its organizations seem to lead to a position where no individual management school of thought can be regarded as completely right to overcome the changing environment of construction projects. This means that these organizations support the implementation of contingency theory. The construction firms have experienced changes in their policies, processes and controls, and those that are unable to adapt inevitably fail. Some traits that are adopted in these firms are flat structured, collaborative, empowered and share information (Coffey 2010).

Considering the dynamics that occurred in the environment of the firms, risk management has traditionally been applied in construction projects. Its application area has also been expanded in such fields as bid-decision making, feasibility studies, marketability studies, performance evaluations and contingency management by reflecting the various factors spanning all phases of the project life cycle. It has become an integral part of project management and plays an essential role in the construction phase. It also supports the decision-making process in the organization and its projects. Decision-making is considered as a process by which a person, group, or organization identifies a choice or judgment to be made, gathers and evaluates information about alternatives, and selects from among the alternatives. Therefore, this process involves risks in selecting one from various courses of actions (Han et al. 2008).

5.4.3 Complexity Theory

In any construction project, the initial specifications generate a great number of project variables. Some are known, such as the proposed form of the structure, or detail issues associated with the finishing of the structures. However, there are others that are as yet unknown, and they have impacts on the project outcomes. Many of these variables display complex interrelationships and dependencies such that they may directly affect each other in a number of different ways. In this context, thoughts and consideration on how to survive as a construction professional in an ever changing culture, how to survive as a construction company in an ever changing environment, and how to survive as an owner/client among infinite construction choices are necessary.

Griffin (1996) found that there are three basic rules that can be implemented by a construction firm in order to survive its complex environment:

- Avoid failure

  The message for constructors is to know what doesn't work and avoid it. The constantly changing environment in construction may never allow for a blueprint for success, but failure can be recognized and different approach can be selected. Contractors will have to look beyond their projects for the problems that need to be avoided. Planning and forecasting things in a systemic approach can be a useful method for this process.

- Communicate

  Communication in the organization should be implemented continually and managed in a systematic way. Methods ranging from an integrated communication and information system to lengthy meetings can keep all members of the organization on the same information pane.

- Remain on the edge of corporate culture

  Contractors should operate on the leading edge, which means the edge of chaos, where they create a balance between rigid and loose structures/procedures. The balance of the two produces the dynamic at the edge of chaos where beneficial changes can emerge and provide a supporting strength to survive.

5.4.4 Change in Organization

According to Lavender (1996), there are several examples of management of change which had been discussed and occurred in construction firms. First, over the periods, the dominant management schools of thought have changed in response to economic, social and political conditions. The theories on scientific management have developed into the human relations approach, and followed by the contingency theory. Second, as one of the five survival objectives from Drucker's study in 1972, a firm must be adaptable to change, but there are some difficulties that construction firms are facing in responding to changes due to constraints from regulations and public policies. Moreover, the construction firms have to develop organizational structures that are more adaptable to change in order to sustain the business. Third, construction firms are regarded as market-orientated firms; therefore they have to adapt their marketing approach when necessary. Fourth, the production management approach in construction had become more flexible in many of its organizations, with the use of the just-in-time approach and similar ideas. Last but not least, aspects of human behavior within organizations in the construction sector have become a focus, where some management and leadership styles, motivation strategies and group structures were shown to be more adaptable to change.

In construction projects, changes are considered as ways where management may seek time savings, with varying implications for resource usage and costs. Approaches such as site management changes, employing additional resources, construction method changes, and design changes are some of the examples that can be done. Moreover, change in a construction project is also often assumed to be a one-way process. For example, it might be supposed that the development of a new plant or equipment would change the construction method permanently. However, some change is one-way, but other kinds of change may actually be circular, leading back in some respects to what has gone before. It all depends on the current situation that the firm is facing, and whether the preferred option is suitable to the situation (Lavender 1996).

5.5 Crisis Management Implementation by Contractors

Studies have shown that all contractors are prone to crisis, but to different extents, depending upon the nature of work they are engaged in. A crisis occurring in a construction firm would probably stimulate a period of reactionary chaos, and social dislocation, at least in the short term. During this time, costs would escalate, people would be guided by unwritten rules and procedures, and information provided through informal communication networks. The challenges for the management during this period would be to reestablish an element of control and coordination in people's activities by attempting to communicate policies that might have been developed in isolation at a senior managerial level. Nevertheless, communication is not easy during this dynamic environment, particularly during the pressures and high stakes of a crisis. Moreover, if the source of a crisis is at the site level, then the problems would be managed in focusing the relationship between the site management and company level, and by the need to coordinate activities with other stakeholders such as subcontractors and suppliers (Mitroff and Pearson 1993; Loosemore and Teo 2006).

As part of a case study, in Australia, large contractors viewed safety, economic conditions, industrial relations disputes, financial management, IT failure and natural disasters as the most probable and serious crises that they can face. They believed that these crises were inevitable regardless of their firm's managerial effectiveness. However, these firms do not take crisis management as necessary, where it is undertaken as an informal, fragmented fashion with few resources and little strategic guidance and support. The concept is seen as insular and non-integrative in its development, being confined only to the senior management level and is limited in scope to issues such as safety, industrial relations and cost control (Loosemore and Teo 2006).

Loosemore and Teo (2006) further opined that even though many contractors may have survived and able to overcome some crises, continued success and survival in an increasingly dynamic and competitive world would demand the elimination of potential resource wastage and the maximization of potential opportunities. A highly successful firm is able to turn problems to their advantage and fully utilize any opportunities available. Crisis management capabilities are regarded as essential in this view, because crises present both threats and opportunities, where mismanagement can impart a heavy impact towards the organization and in extreme cases destroy its viability.

When implementing crisis management by contractors, Loosemore (1998b) described that there are some specific characteristics that distinguish it from crisis management in other contexts. While crisis management in all contexts is about handling problems that arise when what actually happens differs from what was anticipated, construction crisis management is particularly about dealing, in a temporary environment and corporate level, with a major disturbance to a multitude of inter-dependent activities, carried out by distinct profit-making organizations with a diverse range of opposing interests. Based on these characteristics, components such as social adjustment, behavioral instability, information management, and conflict management should be considered as crucial.

5.6 BCM in Construction Firms

BCM is relatively new in the Asian region. Not many firms in the construction sector had implemented this concept (Broughton 2005; Low, Liu and Kumaraswamy 2008). As for construction firms, based on the survey of large construction firms in China, Hong Kong and Singapore, most of the respondents (around 80%) indicated that they have not implemented BCM. In addition, Teo (2011) had found that the usage of BCM was kept to a minimum by the QS consultants in Singapore. The level of implementation of BCM was also found to be at its early stages. In 2011, the BCM Institute in Singapore revealed that only one QS firm has been BCM certified. The key reason for this is a lack of awareness, which was also punctuated by a corresponding lack of adequate personnel in their organizations who can lead in BCM implementation.

Nevertheless, although lacking of awareness toward BCM, a majority of respondents in China and Singapore has interests in implementing BCM (Low, Liu, and Kumaraswamy 2008). This situation also occurred in the United Kingdom. A study based on the findings of the Chartered Management Institute's BCM survey, which covered thousands of companies across different UK industries, reported that although construction firms had identified threats they had done little to prepare for resilience. It reported that 43% of firms in the construction sector had a business continuity plan in place, ranked below the public sector (68%) and the banking sector (76%). Most of these businesses planned for a disaster only when forced by an external agency such as central government, insurers, or auditors (Broughton 2005).

As an example of BCM implementation by contractors, UK-based contractors had adopted BCM, where it was believed that BCM was a logical next step in order to be seen as setting standards within the construction industry rather than following the competitors. Because some tender documents asked construction firms to offer evidence of a BC plan, they believed that the need for BCM would become more prominent. The firm had conducted the BCM certification based on the BS25999 standards, and they viewed the purposes of implementing BCM are (1) to gain resilience to disruption; (2) to gain better understanding of threats and activities at risk and structured process for implementing measures to protect against these, and (3) to enhance customer relations and improvements to the supply chain (BSI Groups 2010).

Based on these reviews, there should be a good strategy for implementing the BCM concept by contractors. Based on the BCM key concepts and principles, the first step that must be taken is the initiation process. Initiation is the stage at which strategic decisions are made to undertake BCM, to set policy, define structures, allocate resources and agree specific projects and monitoring mechanisms. From this, the organization can establish the objectives and scope of continuity activities and begin to shift the mindset of employees away from recovery to one of recovery and prevention (Elliott et al. 2002). In the initiation stage, it is recommended that the top management of the firm starts this process by committing to develop this concept into the organization's system. The role of the management board is to decide how broad or constrained the focus of business continuity provision is to be. This will involve a consideration of the business processes that are to be covered by the continuity provision, and the extent to which external continuity services will be used. Without top down direction, support and ownership, success in both the BCM process and activating the BC Plan will be difficult, if not impossible (Power 1999). Once the top management has made these policy decisions, it should issue a clear statement about the importance of BCM, and make sure that it appears in communication channels such as the annual report, the firm's newsletter, and the firm's intranet. Along this initiation process, it is also suggested that full support be obtained from the government for implementing this concept in the organization. The Indonesian government had developed various crisis responses for its national regions. Having a good collaboration with the government may foster the initiation process, and hence can continue the process into the next development stages of developing the BC Plan and regular maintenance. Moreover, the firm's business continuity plan can also be developed based on the government's existing disaster management programs and their related regulations (UNR/HC 2005; IFRC 2004; Miyamoto INTL 2007; Tambunan 2006; ICG 2002; Elliott et al. 2002).

5.7 Culture in Organization

Based on Coffey's (2010) findings on the construction industry, relatively little research has so far been done to establish and measure what culture actually is and how it relates to better or worse performance of construction firms. Root (1994) also opined that in the study of the construction environment, very little attention has been paid to the question of culture as an environmental factor or influence values and priorities, assumptions and attitudes, expectations and habits of mind that are developed within different occupational and corporate groupings. This gap in the area of culture studies was considered as necessary, that the applicability of this research area (i.e. culture in the construction industry and firms) is not only generic, but more importantly, is a critical arena in its own right (Fellows and Seymour 2002). Nonetheless, up to now, there are some culturally based studies focusing specifically to the construction industry and its firms. Table 5.2 elaborates some research topics about cultures in the construction industry.

Table 5.2 Research on cultures in construction industry

Researchers	Topics
1991 Engineering and Physical Sciences Research Council (EPSRC)—led by Root (2001)	Do cultural values and attitudes among project teams had a significant effect on how the project proceeds?
Newcombe et al. (1990)	Using their own systems model developed for the construction production process, culture was shown not to be a significant environmental factor influencing the process
Bresnen and Marshall (2000)	The effects of partnering within the construction industry on project performance also studied the complex organizational relationships present within construction companies and how these impacted on their organizational culture. They advocated the need to develop not only "specific business excellence measures" within the industry, but also a set of measures to establish "key culture parameters" of construction organizations in order to develop models to integrate business excellence with culture change and business strategies. The application of a "polar plot model" to ascertain generic team culture plots and subsequent benchmarking against other successful project teams and organizations in the field will enable companies to undertake more successful partnering on projects, share organizational learning and still remain competitive
Seymour and Rooke (1995)	The state of both the organizational culture of the construction industry and of research into the concept of culture within that industry
Maloney and Federle (1991)	Utilized the Competing Values Framework (CVF) model to assess and analyze the culture prevalent at the management levels of US engineering companies
Gale (1992)	Examined the question of the effects of "deculturalising" (i.e. the replacement of male cultural characteristics and stereotyped attitudes with feminine thinking processes) of male construction operatives, as an intervention technique in conflict resolution in building projects
Lingard and Rowlinson (1998)	The establishment and operation of safety culture in the construction industry, particularly in Australia and Hong Kong
Hall and Jaggar (1997), Abeysekera (2003)	The effects of national cultural differences in international construction projects
Brochner et al. (2002)	A study of the Swedish construction industry, who concluded that many of the dimensions first discovered by Hofstede (1991) were present and of significance, and that any improvement in the quality outcomes of construction

(continued)

Table 5.2 (continued)

Researchers	Topics
	projects in Sweden would require a major cultural shift towards greater trust in partnering techniques and less reliance on non-cooperative contract relationships
Liu (1999)	The relationship between cultural traits and job satisfaction in the real estate profession in Hong Kong and noted that "organizational culture appears to be gaining support as a predictive and explanatory construct in organizational science". Liu supported the proposition that, once identified, organizational culture dimensions can be shaped to impact positively on the job satisfaction of individuals. Using a nine dimensional representation of organizational culture, Liu verified that "team oriented ... communicative/supportive cultures" have a positive role in enhancing real-estate professionals' job satisfaction
Noyes (1992)	Concluded that the culture of the real estate industry was somewhat weak due to the finding that estate agents appeared lacking in any strong sense of integration within teams (i.e. no group identity)
Rowlinson (2001)	In-depth study of the structure and culture of a major Hong Kong Government Public Sector development body and described its change to a matrix-based organization
Rowlinson and Root (1996)	Investigated the impact of culture on project management outcomes and success factors
Liu (1999), Liu and Fellows (1999a, 1999b)	Examined the impact of culture on project goals; the impact of culture in the Hong Kong real estate profession; the generic and specific issues of culture related to the construction project procurement systems
Barthforth et al. (1999)	Reviewed the extant literature of both the construction industry and organizational culture and performance studies
Liu (2002)	Explored the Eastern cultural trait of "harmony" as it pertains to project management and is critical of the lack of clear theoretical constructs for examining the phenomenon in the construction literature. Liu opined that the concept of "harmony" is an important attribute to successful project partnering and advocated the use of triangulated methodologies including a more ethnographic (i.e. cultural) approach to future studies of such phenomena in construction scenarios

Source: compiled from Coffey (2010)

According to Cheung et al. (2011), several major industrial reviews (Latham 1994; Egan 1998; CIRC 2001) have highlighted that the construction industry needs to improve its efficiency. The reviews believed that the main foundation for efficiency is a conducive, progressive and enduring culture. Related to this, developing organizational culture is considered as necessary where: it conveys a sense of identity for organization members; it facilitates the generation of commitment; culture enhances the stability of the organization; and culture serves a sense-making device that can guide and shape behavior.

Furthermore, Cheung et al. (2011) had also studied the factors of organizational culture in construction firms, particularly contractors. Based on the study, there are seven ranked significant identifiers of organizational culture in construction firms, which are (from the most significant to the least significant rank): the coordination and integration; goal settings and accomplishment; member's participation in the organization; innovation orientation; performance emphasis; team orientation; and reward orientation.

5.8 Institutional Theory

The institutional compliance framework, which is based on the institutional theory, had been used in a study for construction firms adopting a management concept. Low et al. (2010b) highlighted that Business Continuity Management (BCM) implementation has already been introduced as a set of technical reference and standard, particularly in Singapore. However, the extent to which these have been implemented as well as the drivers behind the implementation in construction firms have remained unclear. For this reason, the present study investigates the motivation and current situations of implementing BCM from an institutional perspective. Different components of the institutional theory explain how these elements are created, diffused, adopted and adapted over space and time, and how they fall into decline and disuse. Collectively, this theory provides a framework to interpret the corresponding implementation issues.

The institutional framework proposed by Scott (2001, 2004) appears to offer a useful platform to elucidate why construction firms do or do not wish to implement BCM (Low et al. 2010b). The study found that lack of awareness is the main reason why large construction firms in China, Hong Kong and Singapore have not implemented BCM in their organizations. From the cultural-cognitive theory, the lack of awareness of the potential risks appears to be the explanation for this observation. Furthermore, because BCM is not presently mandated by law, there is no issue of violation and sanction in these countries. This reason complies with the rational choice theory (regarding rules, laws and sanctions) and shared understanding of no penalty arising from non-compliance in the cultural-cognitive theory. However, it was also found that the normative theory relating to personal morality, social influence and legitimacy are likely reasons for the firms to want to implement BCM in the future. The study concluded that in encouraging construction firms to view BCM for adoption and implementation, attention should firstly be on garnering an understanding of why and what would need to be done for this to take place. Decision makers in China, Hong Kong and Singapore might find it easier to start the BCM implementation process through a good understanding of what the rational choice theory, the normative theory and the cultural-cognitive theory in Scott's (2001, 2004) Institutional Compliance Framework entail and would prescribe, in line with the findings from this study (Low et al. 2010b).

5.9 Summary

This chapter elaborates how organizations in the construction industry implement the theories and concepts described in Chaps. 2, 3, and 4. Even though firms in the construction industry have specific characteristics that differ from other firms, it has been found that these theories are considered as applicable to them. The application of these theories can be seen from the implementation of Fayol's general management functions, the adoption of scientific management in production methods, the use of open systems and contingency theory in organizations, lessons gained from complexity theory regarding its complex and changing environment, and last but not least, the approach in using culture and institutional perspective for management development in construction firms. Regarding crisis management, contractors have understood that they are prone to crises, and this concept is regarded as important for their survival. Crucial elements have also been suggested when implementing crisis management in their organizations. Moreover, although BCM is viewed as relatively new, the concept has been adopted by some construction firms in the region.

Chapter 6
The Indonesian Construction Industry

6.1 Introduction

This chapter describes an overview of the Indonesian construction industry. It starts with introducing a general profile of the country, its location, demography, form of state and legal system, and its economic performance. This section is then followed by an overview of the Indonesian construction industry. It describes the industry's activities, contribution to the national's economic growth, its stakeholders and support infrastructure, regulations and management systems, latest developments and lessons learned. The third section is a description about Indonesian construction firms, which focuses on the contractors and their business activities. The last section of this chapter links the Indonesian contractors with the background of this study, which is a review of the crises faced by the Indonesian contractors.

6.2 Profile of Indonesia

6.2.1 Geography

Indonesia, located in Southeast Asia, is well-known as an archipelago with 13,677 islands, which is the largest in the world. The length of its area extends about 5150 km (3200 miles) from Sumatra in the west to Irian Jaya, the western half of New Guinea in the east. Indonesia's neighboring countries are Singapore, Malaysia, Brunei, Papua New Guinea, and Australia. Indonesia's main islands are Sumatra, Java, Bali, Sulawesi (Celebes), and Kalimantan (part of Borneo island shared with Malaysia and Brunei), which forms a major part of Indonesian territories. Indonesia is also part of the so-called volcanic "ring of fire" on the Pacific Rim, where it has hundreds of volcanoes, with 70 of them still active with several recent eruptions that occurred in these past years. Besides that, earthquakes are also frequent in

L.S.R. Supriadi, L. Sui Pheng, *Business Continuity Management in Construction*, Management in the Built Environment, DOI 10.1007/978-981-10-5487-7_6

Indonesia, especially in the area of the subduction zones on some of Indonesia's main islands. Furthermore, Indonesia has the world's second largest area of primary rainforest, right after Brazil, with various species of plant and animal life (IBI 2000).

6.2.2 Demography and Resources

Based on the Economist Intelligence Unit's report (EIU 2008), Indonesia has a population of around 235 million people, which is the fourth most populous country in the world after China, India and the United States. The population growth has been reduced since the introduction of the family planning programme in the 1970s. In 2003–2007, the population growth was 1.3% per year, which was lower compared with 1.7% in the 1990s, 1.9% in the 1980s, and an average of 2.3% in the 1960s and 1970s. Regarding ethnicity, 95% of the population is of Malay origin, with the other 5% consisting of over 300 minority groupings, including Melanesian, Polynesian, and Micronesian. There are also ethnic Chinese living in Indonesia, estimated to be around four million people. In terms of religion, the population is 87% Muslim, 10% Christian, 2% Hindu (mainly residing in Bali) and 1% Buddhist. There are a number of ethnic and religious conflicts that occurred since 1998, mostly due to large-scale migration conflicts between the ethnic groups in a certain area. Nonetheless, although these isolated conflicts occurred, the traditions of ethnic and religious diversity and tolerance remained significant in the Indonesian people's lives.

The population distribution in Indonesia is highly uneven. Despite implementing the transmigration programme, which attempts to ease congestion in Java, Bali and Madura, 60% of Indonesians still lived on these three islands in 2005, which make up only 7% of Indonesia's land surface area. The most densely populated area is on West Java province with an estimated 1003 per sq km, followed by Bali at 553 people per sq km. Outside Java and Bali, the population density averages are less than 100 people per sq km, with Papua having only 6 people per sq km. Moreover, industrial development has resulted in large-scale migration to urban areas, with 42% of the population having lived in cities in 2000. This is a significant increase compared with 31% in 1990 and 22% in 1980 (EIU 2008).

It has been found that natural resources are the backbone of Indonesia's subsistence and formal economies. Farming, fishing, tree-crop and cash-crop cultivation have been the major activities for millions of Indonesian people. The country also has vast but heavily exploited oceanic resources. Along with that, Indonesia's vast forests have been in the interests of large industrial concerns, which have been depleted by commercial logging since 1970. Furthermore, deposits of tin, coal, copper, nickel, bauxite, gold, silver and iron sands, kaolin, marble, granite, limestone and pumice, which are also found in Indonesia, are the basis of an important mining and quarrying sector. Indonesia also has oil reserves in parts of its area, and was once Asia's only member of OPEC. However, in recent years, oil production has been declining which reduced Indonesia's status as a net crude oil importer on a number of occasions (EIU 2008).

6.2.3 Form of State

Indonesia's system of government is based on the 1945 Constitution, where the state is formed as a unitary republic. The constitution provides for five branches of government, which are the President, the House of Representatives, the Supreme Audit Board, the Supreme Court, and the Supreme Advisory Council. There are 33 provinces which all are lead by a governor, who is responsible to the president through the minister of home affairs and represents the central government in his province. Aside from that, there are three provinces, which are Aceh, the territory of Jogjakarta and the capital Jakarta, that have special status in their provincial government. In leading the country, the president has the executive power, who is elected to a 5-year term by the People's Consultative Assembly, which meets to decide general policy and calls the presidential elections. In supporting the government, Indonesia's legal system is exercised by the Supreme Court, which has the judicial powers of the state (IBI 2000).

6.2.4 Economy

6.2.4.1 Economic Performance

According to the Economist Intelligence Unit (EIU) (2008), domestic consumption has become an increasingly important driver of Indonesia's economic growth since the 1990s. It has helped in easing the economy out of recession in the late 1990s, and has been one of the main engines of growth, compensating for sluggish investments and a slump in the export-oriented manufacturing sector.

Indonesia's economic performance can generally be viewed in some of the main economic indicators, which are its Gross Domestic Product (GDP) growth, annual average exchange rate, population, and annual current account balance. Figure 6.1 illustrates the GDP growth of Indonesia from 2004 to 2011. The GDP growth started from 5.03% in 2004, which had fluctuated around 5% for the next 2 years. After that, it has reached 6.35% in 2007. Although there was a significant decrease from 6.01% in 2008 to 4.63% in 2009, the GDP growth of Indonesia has increased to 6.46% in 2011. Looking back at Indonesia's economic history, it appears that prudent economic management had enabled Indonesia to record consistently high rates of economic growth, well in excess of the expansion in population, for more than two decades. The past economic growth that averaged more than 6% per year between 1970 and 1996 was achieved despite a number of external issues, including oil prices and international exchange rates fluctuations which affected the trade and

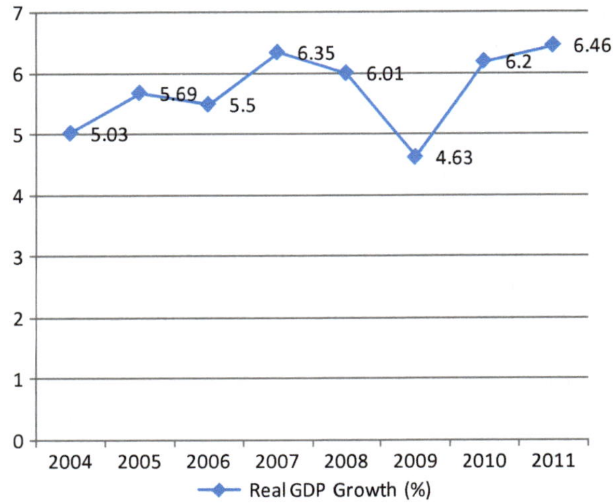

Fig. 6.1 GDP growth of Indonesia. Source: SCA (2012)

value of the country's external debt. Indonesia has transformed from a low-income country in the mid 1960s into a middle-income country in 1996. The 1997–1998 financial crisis had slowed the GDP's growth to 4.7% in 1997 and then contracted by 13.1% in 1998, which was the worst performance since records began. The crisis recovery started in 2000, supported by strong household and government consumption which led to a GDP of over 6% in 2007 for the first time since the 1997 financial crisis (EIU 2008; SCA 2012).

In Fig. 6.2, the annual end-period of the Rupiah's exchange rate against the US Dollar is shown. It can be seen that the average rate of Rupiah for $1 USD during 2006–2012 was around Rp 9404. After the Asian financial crisis in the 1997, the Rupiah has gained its strength in its value, increasing from around Rp 12,000—for $1 USD (the value right after the financial crisis) to the range of Rp 8900–Rp 10,950 in the last few years. In 2010–2012, the value has become more stable than the previous years, maintaining in the range of Rp 8900–Rp 9068. A high and stable value of local exchange rate (particularly towards the dominant exchange rate in the international market) provides a positive environment for international trading and business in Indonesia (EIU 2008, 2012).

Figure 6.3 shows the population of Indonesia in selected years. The population growth increase had been quite stable with an average of 29.6 million additional people per 10 year (SCA 2012).

Last but not least, Fig. 6.4 shows Indonesia's annual current-account balance. The current account is considered as the difference between national (both public and private) savings and investments. A current account deficit may therefore reflect a low level of national savings relative to investments or a high rate of investments. Based on a World Bank's (2012) report, fluctuations had occurred in 2002–2010. The highest account balance was reached in 2006, with US$10,859.5 million. However, the account balance had hit very low level in 2005 and 2008, with the amount of US$277.6 million and US$125.2 million respectively. These

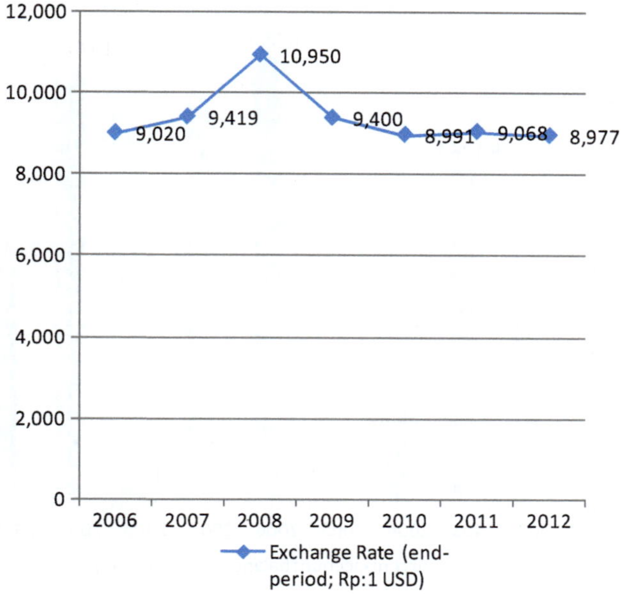

Fig. 6.2 Annual end-period exchange rate Rupiah:$1 (USD). Source: EIU (2012)

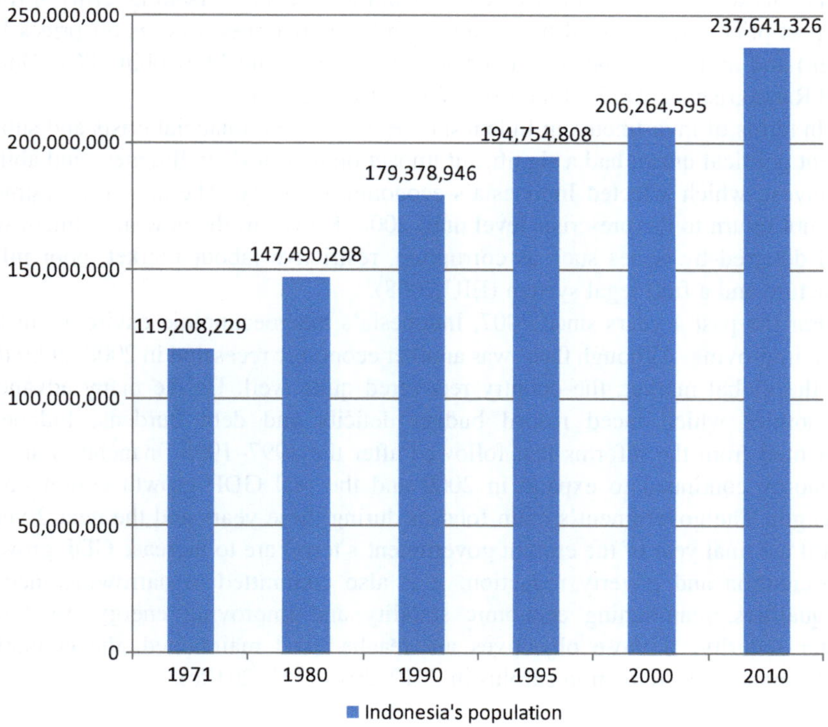

Fig. 6.3 Indonesia's population in selected years. Source: SCA (2012)

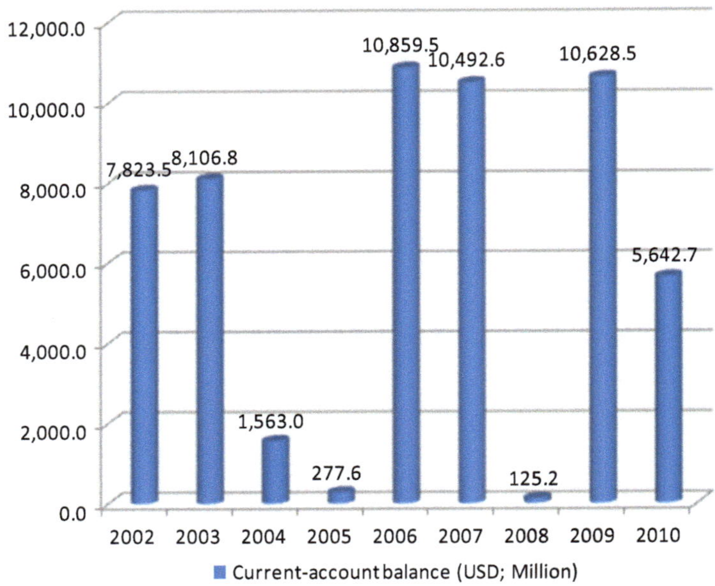

Fig. 6.4 Indonesia's annual current-account balance. Source: World Bank (2012)

situations were due to Rupiah's weakness in those years, worsening terms of trade (export prices such as coal had plunged and import prices such as oil prices had risen) and an unsustainable fuel subsidy policy (Lee and McKibbin 2007; Ghosh and Ramakrishnan 2012; Bofa 2012; World Bank 2012).

In terms of investments in Indonesia, the 1997–1998 financial crisis and subsequent political unrest had a significant impact on investor's willingness and ability to invest, which affected Indonesia's economic recovery. The annual investment did not return to the pre-crisis level until 2006. However, the new investment was still deterred by issues such as corruption, restrictive labour market, poor infrastructure and a frail legal system (EIU 2008).

For the past 4 years since 2007, Indonesia's macroeconomic environment has been improving. Although there was another economic recession in 2008–2009 that hit the global market, the country recovered quite well. Unlike many advanced economies which faced record budget deficits and debt burdens, Indonesia benefited from the reforms that followed after the 1997–1998 financial crisis. Its economy continued to expand in 2009 and the real GDP growth continued in strength. The government's main focuses during these years and the next 2 years (until the final year of the current government's term) are to increase GDP growth, job creation and poverty reduction. It is also committed to narrowing income inequalities, maintaining economic stability and improving energy, food and water security. If these objectives are reached and maintained, the economic performance will remain in surplus in 2010–2014 (EIU 2010c).

6.2.4.2 Infrastructure

As mentioned previously, since the 1997–1998 financial crisis, infrastructure in Indonesia has been poorly neglected, impeding economic growth. Investments have been deterred by the generally low return on infrastructure projects, compounded by legal uncertainties and concerns over security and political stability. All of these issues are now being part of the current government's concerns and who placed infrastructure at the core of its strategy to improve economic growth (EIU 2008). Based on the British Chamber of Commerce reports on Indonesia (BMI 2009), the country needs infrastructure investments of US$70 billion. This will be essential to bring about sustained growth levels about 7%, which is needed to resolve issues in employment and poverty. Below are some descriptions about Indonesia's physical infrastructure, their existing conditions and several issues that are faced by the government (EIU 2008):

- Railways:

  Railways' coverage in Java is around 3100 and 1300 km in Sumatra, and only 10% of this consists of double-track railway. Smaller rail networks that carry cargo and raw materials are also found in parts of Kalimantan. The state-owned company, *Kereta Api Indonesia* (KAI) operates the railways in Indonesia and the network is badly in need of new investments. Recently, a sharp increase of train accidents occurred which are mostly due to the ageing infrastructure. Many railway crossings are found with no security gates, particularly in rural areas, and in 2003, it was reported that 31% of locomotives and 45% of coaches had been operating for more than 30 years. In countering these issues, the House of People's Representative has passed a new rail transport bill in 2007 which will allow the private sector to provide rail transport for the first time. Government regulations are still required before the law takes effect, and the state will continue to set train fares. This new legal framework may promote foreign investments in rail transport, which may support the future long-term planned high-speed 800 km rail link project between Jakarta and Surabaya.
- Roads:

  Land transportation in Indonesia mostly depends on its road network. The road network is measured around 372,929 km in 2004 and 55% of which was asphalt-covered. In terms of coverage and capacity, it is still considered as inadequate. The road's quality also deteriorates substantially in wet weather. There are also a number of payments on the road for road users, which range from tolls, both legal and illegal, to extortion payments demanded by the police or illegal criminal organizations. Hindrance in road network investments includes the risk of rising land acquisition costs. By buying up land earmarked for new toll roads, politically connected land speculators have caused the land acquisition costs of projects to spiral. Problems in land acquisition are also stemming from the poor functioning of the civil service and the judiciary.

  The government's current effort for this matter is to cap such costs at 10% above initial estimates, which may help to ease concerns. Moreover, it has also resolved to outsource the appraisal of land to professional consulting firms, and push more vigorously on reforming the civil service and the judiciary. Resolution of these land-procurement issues may lead to a boom in toll-road development, reducing bottlenecks and create new jobs in the construction industry (EIU 2010a).

- Ports:

  Although Indonesia is well-known as an archipelago, the country has only a small domestic ocean-going fleet, and lacks the port facilities to attract major vessels. Most cargo is transshipped at Singapore and arrives in smaller feeder vessels. Nonetheless, Indonesia's main container port in Tanjung Priok, Jakarta (which managed 28% of exports and 42% of imports in 2005), is partly under private ownership. In addition, future high investments are being planned to upgrade its facilities. Problems that occurred in Indonesia's ports include high insurance costs for Indonesian shippers due to frequent organized crime, piracy, strikes and theft, and recent ferry accidents, including capsize and fire that claimed many lives caused by poor safety.
- Airports:

  Indonesia has 179 commercial airports, with 61 of them that are large enough for wide-bodied jets. Some of the airports has been modernized and extended during the years.
- Energy:

  Annually, Indonesia produces around 100 billion kWh of electricity. In 2004, about 55% of households in Indonesia were able to consume electricity, which stood at 509 kWh per head in 2005. Electricity demand had risen sharply in recent years, with the rate growing at 11% annually in 2002–2006 in the main regions of Java and Bali. The provision of electricity is the responsibility of the state-owned electricity company, Perusahaan Listrik Negara (PLN). However, PLN had inefficiencies in its management, followed by bad business practices and corruption which recently operated in a perpetual situation of near-bankruptcy. PLN's problems have been combined by a heavy reliance on oil-based fuels, which have risen sharply in cost in recent years. Rising oil prices and serious financial problems have forced the company to take vital measures to reduce dependence on oil-based fuels. Based on EIU's current report (EIU 2010a), some progress have been made in order to raise electricity-generating capacity. The efforts were initialized on an US$8 billion programme to develop 10 GW of new coal-fired generating capacity. The programme is planned to be completed by 2012, although delays are likely to occur. The new capacity will come from ten large power stations in Java and 30 smaller plants on other islands, and these projects are mostly funded by Chinese investors. In addition to this programme, the government is also planning a second phase of investments in the power sector, with another 10 GW of capacity which is expected to be generated by 2014.
- Telecommunications:

  The state-owned telephone system covers almost the entire country, and has been greatly extended and more efficient since the mid-1970s by the deployment of telecommunications satellites. There were a total of 8.7 million fixed-line users, which is equivalent to a ratio of 3.7 fixed lines per people in 2005. This low density does not fully reflect the actual degree of coverage provided by the existing network, which is expanded by about 220,000 telephone kiosks located from cities to even the most remote areas in Indonesia. Recently, the mobile telephone system has exceeded the fixed-line technology, where it does not need the same heavy investments in infrastructure. In 2007, the number of mobile subscribers has risen to around 86 million users.

6.2.4.3 Overview of Business Environment

In promoting the investment climate in Indonesia, the government has raised five main issues that must be solved, which are tax reforms to raise the level of competitiveness, labour law revision to give both investors and employees a win-win solution, development with a focus on regulation in the competitor countries, realization of infrastructure projects, and finalizing the Investment Law (BMI 2009). All of these approaches will mostly depend on the political environment in Indonesia, followed by the country's economic performance. Some of the latest socio-politic situations in Indonesia (BMI 2009) include:

- Up to now, central Sulawesi and part of Maluku have been considered as areas of instability due to several ethnic and religion clashes in that area. Nonetheless, ongoing efforts by the government have been conducted to restore peace and build stability, which have started to bear fruit.
- Unrest situation in Papua, due to an independence struggle waged by *Organisasi Papua Merdeka* (OPM), has transformed into something more complex. It has been further complicated by some immigrants from other parts of Indonesia who are supporting for the national government to maintain control. This situation is currently part of the focus of the central government.
- In Aceh, things have remained calm, particularly since the treaty between the central government and the Aceh rebels in 2005. However, there are some concerns about corruption on the part of some former insurgents who now hold political office, and the fragmentation between the former rebel group [*Gerakan Aceh Merdeka* (GAM)] and the activities of the Indonesian military [*Tentara Nasional Indonesia* (TNI)]. This is considered as a risk that may trigger new violence and needs to be monitored by the government.

Furthermore, in supporting the five main issues mentioned above, the Indonesian government has developed policies based on some of their forecasts (EIU 2011):

- Policy toward private enterprise and competition:

  Initial public offerings of shares in several state-owned enterprises will be conducted in 2011, after being postponed due to stock market falls in 2008–2009. However, in 2013–2015, it is forecasted that liberalization slows further ahead of the 2014 elections. Government regulation and price-fixing may deter private investments in important areas of the economy.

- Policy towards foreign investments:
  Starting from 2011, the procedure and process for obtaining investment license will become easier, owing to reforms implemented by the Investment Coordinating Board, but a general suspicion of the motives of foreign investors may continue to deter inward investments. Nonetheless, in 2013–2015, the economic nationalism will strengthen in some quarters as the elections approach.
- Foreign trade and exchange controls:
  Restrictions on foreign exchange and trade which were implemented during the 2008–2009 global recession were rolled back in 2011. The efficiency of port and customs are also improving. Furthermore, the trade liberalization effort will be driven by Indonesia's membership of the ASEAN.
- Taxes:
  Since 2010, tax office modernization and the electronic filing of tax returns had improved with new and efficient collection methods. However, a lack of civil service reforms holds back efforts to bring medium-sized companies into the tax net. In the following years, the government will gradually repeal miscellaneous levies in order to reduce the tax burden.
- Financing:
  Foreign participation in the banking sector continues to encourage competition, modernization, and a more transparent operating regime. This will most likely be followed by improved corporate creditworthiness that encourages loan growth and investments.
- The labour market:
  Indonesia has a relatively young labour force, but most of them are unskilled and semi-skilled labour. People with vocational and managerial training are in short supply, and demand often has to be met by using more foreign workers. In 2010–2014, it is unlikely that the quality of the Indonesian labour force will improve significantly, given that the government has begun to increase investments in education only recently, where the problems start at the elementary-education level. Besides that, the average wage level for the labour force is still among the lowest in Southeast Asia, with only 10–15% of the labour force unionized. Strikes have also happened in recent years, with the vocal trade unions actively opposing privatization and liberalization, which forestalled reforms to labour laws. In future, demand for jobs may prompt the government to confront the trade unions and to make amendments to employment laws. Although the unemployment rate in the overall labour market continued to decline during the global recession, most of the jobs created were low-paid and relatively in the informal economy such as agriculture, informal trade and services. Therefore, job creation where a large proportion of the labour force is in the formal economy should be considered (EIU 2010b).
- Infrastructure:
  In 2010–2014, it is hoped that there will be an improvement in the quality of Indonesia's infrastructure. Previously, the poor quality of the infrastructure was

attributable not just to the government's inability to finance projects but also to the generally poor business environment, which has deterred potential private investors. In an effort to encourage investments in infrastructure, a financing company called Indonesia Infrastructure Finance, was established by the government to provide long-term financing for infrastructure schemes (EIU 2010a). Moreover, more investor-friendly regulations and laws are being created, which will open its sector for foreign participation. One of the measures that had been adopted since 2007 was creating a more robust public-private-partnership regulatory and legal framework (BMI 2009).

6.3 The Indonesian Construction Industry

The important role played by the Indonesian construction industry in the national economy can be seen from the major usage of domestic goods and services such as architects, consultant engineers, economists, building materials, financial services, legal specialists, construction equipment, transport and communications. These goods and services make the total value of building and construction works contribute some significant amount in the total GDP. Moreover, the role of the industry is supported by a broad spectrum of legislation and agents. Under Indonesia's democratic system, the central government and local government (which are the provinces, regencies and cities) have the same with very slight variations in their legislative and administrative systems, relating to licensing and registration of construction related activities. The construction law and Indonesia's national building code also ensure a degree of technical uniformity across the country, setting out the level of performance for building elements (NBCSD 2004).

6.3.1 Construction Activities and National Contribution

The Indonesian construction industry delivers various types of construction projects, which are the residential and non-residential buildings, agricultural structures/irrigation, infrastructures such as roads, bridges, harbours, electricity, gas, water supply, and communication, and others (Raftery et al. 2004). Currently, there are large government projects underway, as well as office and residential building projects. Infrastructure development, residential and commercial construction may provide a major boost to the construction industry (BMI 2009).

In Indonesia, the construction industry is widely spread, where there are some 30 associations accredited and registered with membership firm employers ranging from 14 members to approximately 60,000 members. They compose of general contractors, specialist contractors, material and labour suppliers to equipment leasing firms. Altogether, they employ around 269,000 full time employees and a part time labour force of four to five million skilled and unskilled workers. The larger private and state-owned firms form an association of 65 members which are entrusted with the nation's larger, longer term and more complex projects (NBCSD 2004).

The national contribution that the industry had given to Indonesia can be viewed through its value, growth percentage, percentage of GDP, and total workforce supplied to the industry. A report from Indonesia's Statistical Central Agency (2012) noted that throughout 2004–2009, the value of the Indonesian construction industry had increased steadily. Figure 6.5 shows the increase from around Rp 56 trillion to Rp 110.8 trillion within 5 years.

Regarding its annual average growth, there had been some fluctuations of growth in 2005–2011. The average growth had peaked at 8.51% in 2007, but had steadily decreased to 6.69% in 2011, as seen in Fig. 6.6. This was mostly due to several changes in the infrastructure regulations and political factors (the national election in 2009, central and local government changes) (SCA 2012; BMI 2009). Nevertheless, many projects were still constructed within these years, supported by positive economic performance in the country.

In terms of its contribution to the nation's GDP, in 2004–2009 the industry contributed to around 7.87% of GDP. It is forecasted to contribute around 10% of GDP in 2010–2011 (Fig. 6.7).

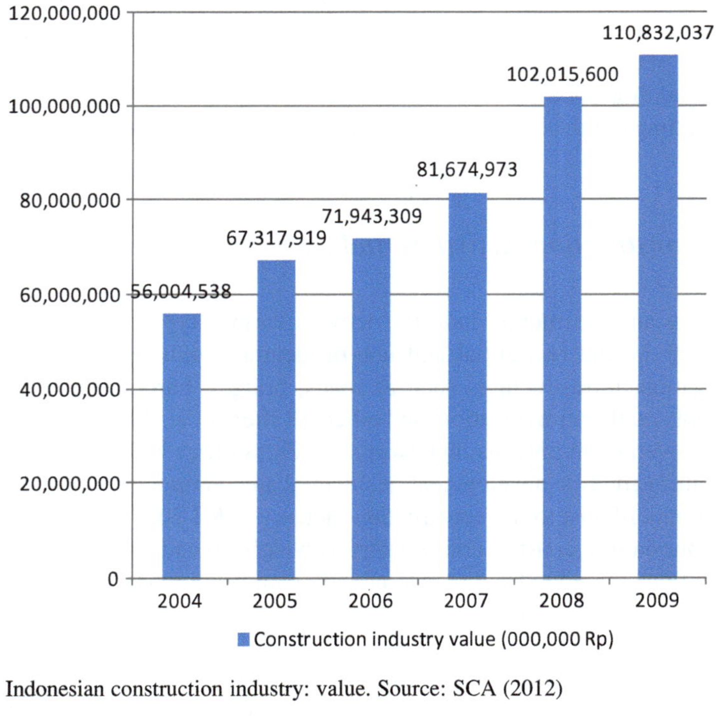

Fig. 6.5 Indonesian construction industry: value. Source: SCA (2012)

Fig. 6.6 Indonesian construction industry: real growth. Source: SCA (2012)

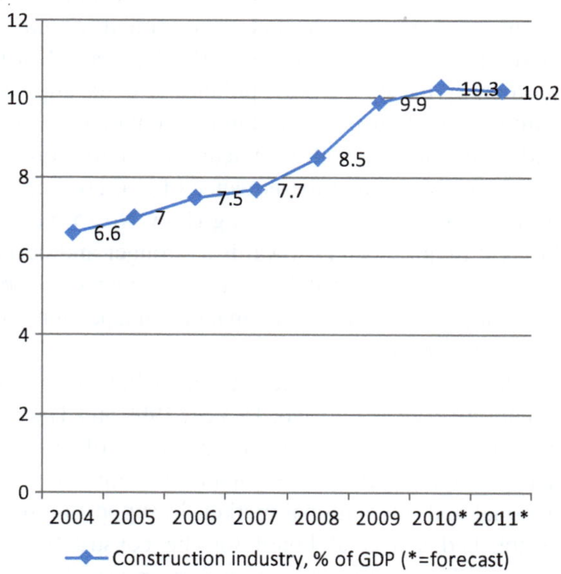

Fig. 6.7 Indonesian construction industry: percentage of GDP. Source: SCA (2012)

The significant value and growth of the construction industry supported the increasing employment. In 2008–2009, the total permanent workforce employed in the Indonesian construction industry had an increase of 7%, followed by a 1.1% increase in the period of 2009–2010 as shown in Fig. 6.8 (BMI 2009; SCA 2012).

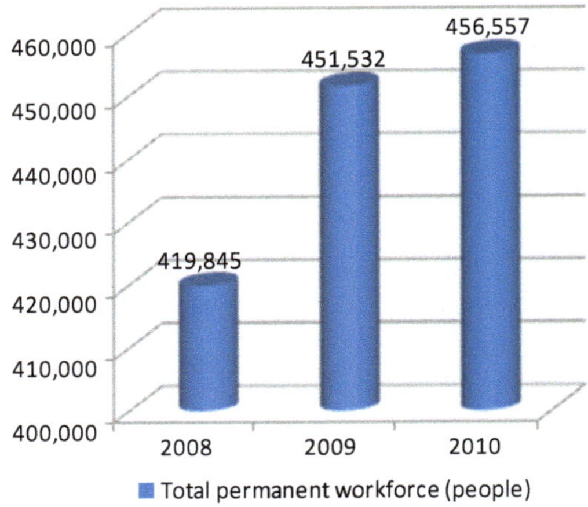

Fig. 6.8 Total permanent workforce in the Indonesian construction industry. Source: SCA (2012)

6.3.2 Role of Construction Administration by Government

Up to now, the Ministry of Public Works is in charge of the central administration of the Indonesian construction industry. One of the main agenda of the ministry is to adopt good governance and management systems both at the central and local government within the framework of decentralization and regional autonomy. In promoting these, the key policies are to implement governance through law enforcement, transparency of implementation, fair treatment to all, professionalism, and public accountability. A leap in progress on the decentralization process has been marked through the issuance of two laws on regional autonomy (Law No.22 Year 1999) and fiscal balancing (Law No.25 Year 1999). These regulations have brought major changes towards a stronger and decentralized local authority in the Indonesian governmental system. These have also become the basis for promoting political, administrative and financial independence of local authorities in the future (NBCSD 2004).

The construction service sector in Indonesia has been regulated by The Construction Services Law No. 18 Year 1999 and The Government Procurement Policy No. 80 Year 2003. The main objectives of these regulations and policies are to achieve a high corporate capability with the competence to compete in both national and international market. In supporting these, the Indonesian government established a special board for the construction industry, which is named the National Board of Construction Service Development (NBCSD). The coordination of all construction administration responsibilities between the central and local governments will be carried out by NBCSD which currently has offices in all provinces in Indonesia. Based on NBCSD's (2004) report, along with the central and local government, NBCSD also coordinates with other construction industry services and professional associations under other ministries for specific construction industry development in mining, energy, transportation, and communication. The role of the central and local governments in regulating and administering national construction development, together with the statutory bodies such as NBCSD and construction services associations, are essential for integrating the whole complex process of construction development and delivery.

6.3.3 Role of Construction Industry Associations

According to NBCSD's (2004) report, its tasks are to conduct research and development in the construction industry, to organize education and training, to register construction workers which include classification, qualification and certification of professional and skilled workers, and to register construction firms. In implementing these tasks, NBCSD has the function as the authority of a construction service society, as an organization for coordinating and communicating the construction process and development, and as the government's partner in developing and enhancing the role of national construction service in contributing to the national economic growth. Furthermore, NBCSD also has the following roles to play:

- To accredit firms and professional associations in carrying out the certification process. Construction designers/planners, constructors, and construction supervisory firms must possess qualification and classification certificates. Also, individual construction designer, supervisor, and construction workers must have professional certificates as set out by NBCSD in association with the related construction or professional associations.
- To issue an equal status of foreign workers certificate and registration of foreign firms.
- To develop construction service information system.
- To disseminate national, regional and international standards.

These functions are mandated by The Construction Service Law No.18 Year 1999 (RI 1999), which stipulates the conditions and standards to perform construction service works in Indonesia.

The NBCSD performs and coordinates its activities with several bodies, which are an important part of the regulatory framework in the construction industry. These bodies represent the various functions, networks for access of information in the construction industry, methods, knowledge, technology, expertise, human resources, management, materials and equipment supply, finance, and other inputs that are needed to carry out the activities of the industry. These bodies are also called the construction industry support infrastructure, which consists of construction and professional associations, government and non-government bodies, construction firms, and universities and educational institutions (Fig. 6.9).

Currently, there are 25 construction services associations that are listed with NBCSD. Generally, these associations are established for firms that provide services in the constructing process (contractors), engineering and designs, pre-construction consultants, and material and equipment supplies. The

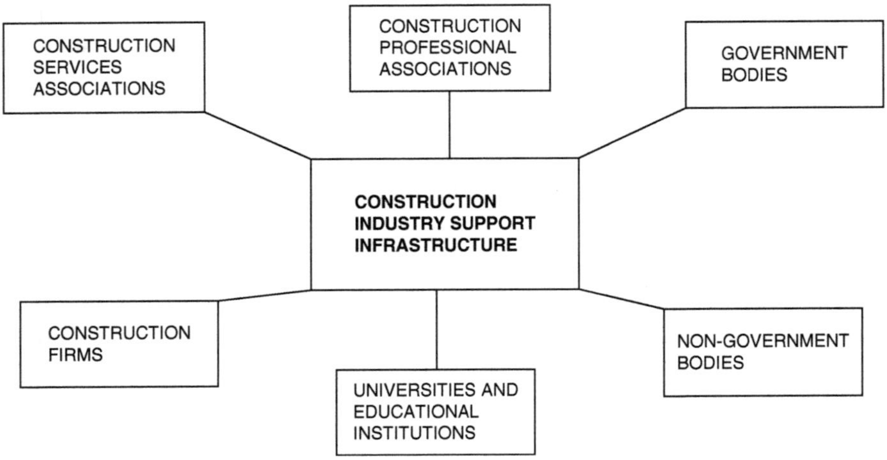

Fig. 6.9 Indonesian construction industry support infrastructure. Source: Adapted from NBCSD (2004)

construction professional associations are created for professionals who work in construction services, such as engineers, construction managers, interior designers, architects, project managers, quantity surveyors, and construction planning experts. There are also 25 professional associations which are currently listed with the board. In terms of the government bodies, all of the bodies in the government that are related to the industry's process and development are listed, which in general consist of the national ministries, banking associations, agencies for statistics and land surveying, and local/regional authorities. The non-government bodies are organizations that support or assist the government in terms of delivering the construction process and development. The higher educational institutions consist of both state-owned and private-owned institutions, which focus more on the research and development process for the industry. The key functions of these construction support industry infrastructure are described in Table 6.1 (NBCSD 2004).

By and large, the construction industry associations have fundamental roles within the regulatory framework of the industry. They provide the technical expertise to advise the government and NBCSD on the conditions of the construction industry for formulating the need for policies or changes. They also perform additional monitoring and control for the industry with regard to its level of competitiveness in order to meet future global challenges. Last but not least, they also provide a platform for communications on a wide range of administrative, technical and social economic issues that occur in and relate to the construction industry (NBCSD 2004).

Table 6.1 The key functions of Indonesian construction industry support infrastructure

Construction industry support infrastructure key functions		
Construction services associations	*Construction professional associations*	*Government bodies*
• Registration of company and firms • Qualification of company and firms • Accrediting of company and firms • Certification of company and firms • Standards of performance • Standards of quality • Monitor and control construction performance • Consultation and information • Training • Communication and networking • Assist government enforcement of policies and standard procedures	• Registration of company and firms • Qualification of company and firms • Accrediting of company and firms • Certification of company and firms • Standards of performance and competency • Standards of quality • Monitor and control professional performance • Consultation and information • Training • Communication and networking • Assist government enforcement of policies and standard procedures	• Formulate regulatory policies and procedures • Enforce policies and standard procedures • Evaluate performance and economy of construction industry • Provide technical support and assistance • Provide construction development and training facilities • Provide access to finance • Monitor and control construction industry • Carry out research and development support • Publish and communicate results of research and development
Construction firms	*Universities and institute of educational institutions*	*Non-government bodies*
• Carry out construction business services, which are planning (consultancy service), constructing (contractors), and supervision services (consultancy) • Each of the business services provides works in areas such as architectural, civil, mechanical, electrical, and environmental works	• Provide education and training • Provide technical support and assistance • Provide technology development assistance • Carry out research and development support • Arrange seminars and workshops • Publish and communicate results of research and development • Benchmarking of international standards and research results	• Carry out survey and analysis for the needed statistics • Carry out research support • Establish national and local standards • Benchmarking of international standards • Assist government in construction industry management and relations

Source: Adapted from NBCSD (2004)

6.3.4 Regulation and Management Systems of Construction Projects

In delivering construction projects, a vast network of relationship between many parties is involved in the process. Both public and private projects are first informed through tender announcements which will be followed by the tendering/bidding process. According to the procurement law, all tenders are to be advertised and given fair treatment and communication to the parties who will register to compete for the projects. In the project execution phase, the standards of the construction industry are maintained and the relevant party in the project will carry out performance based inspection on compliance with planning standards, building regulations standards, safety standards, and other statutory specifications (NBCSD 2004).

6.3.4.1 System of Checks and Acceptance for Projects

Projects awarded are generally based on the competitive bidding process with a single fixed price or traditional lump sum contract. In private projects, the contracts awarded are mostly based on lump sum or unit price. Other types that are quite recognizable in the industry are contracts given based on a negotiated cost plus fixed fee, and guaranteed maximum price agreement. The contractors are usually allowed to deliver the specified works themselves or by sub-contracting part of the work packages to other individual trade contractors. The contractors are then inspected and checked for their past performance regarding time, cost, quality and other specified requirements as stated in the agreed and signed contract (NBCSD 2004; RI 1999).

6.3.4.2 Sureties for Construction Projects

In terms of project payments, the construction law in Indonesia also considers providing protection to the contractors. Many of them have experienced difficulties in project payments, particularly in private projects. The contractor is required to provide a performance guarantee from a recognizable bank or surety bond from a reputable insurance company before they receive their final project payment. Also, the owner usually holds a certain amount of money (retention monies) that will be paid to the contractor after the latter satisfactorily finishes the whole works. However, there are situations where the contractor does not receive the full payment from the owner as stipulated in the contract, and many complaints and claims have been received from the contractors regarding these abuses. These issues are currently being studied by the NBCSD and the government, with a view to developing measures to respond to the interests of the contractor (NBCSD 2004).

6.3.4.3 Insurance for Construction Projects

Based on the Construction Service Law No.18 Year 1999 (RI 1999), the employers of construction projects are legally required to take out insurance policies to cover their liabilities under the law for work injuries of their employees. The major reasons for such insurance policies are (NBCSD 2004):

- Anticipation of the still potentially high accident rates in Indonesia.
- Increasing number of high rise building projects with hazardous transportation of labour, materials and other activities of construction.
- More complex site layout in dense construction area, storage, and other auxiliary support infrastructure.
- Labour intensive projects are still commonly applied.

Generally, the main insurance policies that cover construction projects in Indonesia are as follows (NBCSD 2004; RI 1999):

- Workmen's compensation insurance policy (compliance with the national labour law)

  This policy is for compensating any work-related accidents experienced by the workmen.
- Comprehensive workmen's compensation insurance policy (compliance with the national labour law)

  In some cases, depending on the type of accident, the compensation granted under the workmen's compensation insurance system is inadequate. This policy covers a more detailed provision.
- Contractor's all risks insurance policy

  This insurance is for contractors to cover additional costs involved in recovering losses sustained due to the effects of fire, burglary, lightning or storms on buildings under construction, construction materials, and temporary buildings for use during the construction phase (on-site office and accommodations). Moreover, effects due to design, construction or material defects or work-related errors may be included in the provision.
- Third-party liability insurance

  This insurance is also for contractors that cover the costs of compensating a third party for damages caused by possibility of incidental work-related accidents.
- Product liability insurance

  Another type of insurance for contractors that cover the costs of compensating third parties for incidental accidents resulting from their products or work after the product/building/structure is transferred to another party.

6.3.4.4 Construction Quality Standards

The NBCSD (2004) acknowledged that there has been an increase of concerns and awareness for quality assurance in Indonesia. The ISO 9000 series in Indonesia had been introduced since 1992, when it became adopted as a national standard (SNI-19-9000), which is similar to what have been established in the US (ANSI/ASQ Q9000), UK (BS 5750), and Australia (AS 3900). In 1997, the interest in adopting the standard was growing, but the implementation was still restricted to large construction firms and other services sector. Moreover, in view of a highly competitive industry/sector, many Indonesian firms (including construction firms) have diversified their businesses to higher value-added products and services that meet international standards through the use of international standards and specifications.

As an example, the ISO 9002 standard for building contractors sets out how firms can establish, document and maintain an effective quality management system. In implementing this, the minimum requirements that need to be planned, monitored, and controlled are:

1. Tender and contract
2. Planning and documentation
3. Control of measurement and equipment tests
4. Procurement
5. Sampling, inspection and testing
6. Incoming inspection
7. In-process inspection
8. Final inspection
9. Inspection and test status
10. Material identification and traceability
11. Handling, storage, packaging and delivery
12. Control of production/construction
13. Quality records
14. Control of non-conformity
15. Corrective actions
16. Use of statistical techniques
17. Auditing the quality system

According to Andi and Chandra (2007), the NBCSD, in cooperation with the Department of Public Works, released Guidelines for Quality Management Implementation (ISO 9001:2000) of Construction and Consultant Services. The guidelines consist of key sections explaining the planning and implementation of QM systems and quality documentation, and also provide samples of works instructions.

6.3.4.5 Environment Conservation Approach

Based on the Indonesia Law for Environmental Management No.23 Year 1997, construction project activities that have significant impact on the environment should include the Environmental Impact Assessment (EIA) document. The EIA study, which is conducted at the project feasibility phase, should consider spatial planning, protected areas, historical buildings, sensitive areas, rare and protected biological species, natural resources potentials and the socio-economic cultural society around the project site. This study recommends significant impact mitigation, and changes to project location or cancellation of the project activities if the significant impacts cannot be mitigated by technological, economical and institutional approaches. Due to the complexity of construction works, the EIA should ideally be supported by the Environmental Management System (EMS), which is based on ISO 14000. The EMS provides environmental documentation, recording and auditing process to assure environmental conservation. Recently, some large contractors have been certified to meet the ISO 14000 requirements by the international registrar. It is expected that there will be an increase in the number of firms that will be certified, due to the need to comply with the government regulations on environmental management (NBCSD 2004).

6.3.5 Technology

Some of the large construction firms are starting to adopt the latest information technology in their business. For example, material procurement is conducted through the internet, the bidding processes are managed electronically, and online data sharing within the company (sharing information on CAD, project management information, and network planning control). The online data sharing application is particularly beneficial for firms that deliver construction project in remote areas. Other technologies that have been used include construction automation and mechanization in construction methods. Moreover, there are also several prefabrication technologies that are being used, particularly for high-rise building projects. These technologies reduce the time needed for the projects (NBCSD 2004; Raftery et al. 2004).

6.3.6 Research and Development

NBCSD (2004) found that there were many on-going Research and Development (R&D) works in the construction sector since the past 10 years. Most of the research was developed by academics in universities that focus on the project and construction management knowledge area, and other government funded research institutes collaborating with the Ministry of Public Works. These institutes focus on studies related to water resources, highways and bridges, and urban settlements. Until now, the NBCSD is still currently extending the R&D process by coordinating and collaborating with the academics, professional practitioners, and government institutions. Although the outputs of research and development in the past years have not been very substantial, there is a good example of a technology innovation called the "Hydraulic non Friction Rotating Device", which can be used to rotate heavier load for pier construction of flyovers. This technology has been used by some of the large contractors for constructing highways and roads for over 10 years.

6.3.7 Lessons Learned and Future Developments

For future developments, the construction industry in Indonesia also needs to focus on its existing conditions which may provide some lessons learned to support the industry. Larasati and Tsunemi (2009) highlighted some existing problems which occurred in the industry:

- The production process in the construction industry in Indonesia does not run smoothly. This is indicated by problems in project delivery systems such as the lack of appropriate materials and the necessity to waste valuable resources on reworks, etc.
- Low skill index and experience of construction workers cause the business to face difficulties in undertaking new concepts and technologies.
- Fragmentation in the project delivery process is still large, and this is indicated by a disintegrated relationship between the various parties involved.
- Lack of clear lines of communication and good working relationships in the processes. This situation results in project delays, wastes and disputes caused during construction, and led to an increase in antagonistic relationships. Most of these are due to an overall lack of trust between parties.
- Since the construction industry is in the "change environment" and some changes cannot be well-anticipated by the Indonesian construction sector, the risks that correspondingly arise are also high. Some events like financial crises, natural disasters, political crises and others that have triggered changes in the industry will be further discussed in Sect. 6.5.

In anticipating the current problems in the construction industry, some lessons learned can be useful for further developing more desirable services in the industry. The NBCSD (2004) had summarized these as follows:

- Construction business services need to be carried out based upon standards of competence, registration, certification and licensing of domestic and international knowledge and skills. Furthermore, skills and competence of the professionals should also be standardized and certified. These efforts may help in

delivering higher quality construction products, with better production processes.
- The respective administration and bodies responsible for enhancing the sector's development should continue to expand and intensify the research and development process to focus on the key inputs of the industry.
- The respective administration and bodies in the construction industry should also develop and utilize ICT (Information Communication Technology) capabilities for better coordination and control with other construction industry support infrastructure.
- Construction firms should establish, adopt and apply ICT capabilities and other supporting technology or systems that support the firm's general business performance and drive the necessary competitive edge to contribute more in the industry in a sustainable and productive manner.

6.4 Indonesian Construction Firms

Based on the Construction Service Law No.18 Year 1999 (RI 1999), the services carried out by construction firms in Indonesia consist of construction planning consultancy service, construction execution service, and construction supervision service. The construction work itself is defined as the whole or part of the activities of planning and/or executing (constructing) including supervising which cover architectural, civil, mechanical, electrical and environmental works with their attributes, to produce a building or other physical forms. The work will be bound by the construction work contract, which regulates the legal relationships between the service user and the service provider undertaking the construction works.

6.4.1 Types of Firms

Derived from the general description above (RI 1999, 2000), there are three types of construction firms that are generally acknowledged in Indonesia. These are construction planner (planning consultants), construction executor or constructor (contractor), and construction supervisor (supervision consultants). These firms respectively carry out their specific tasks as part of the whole construction service. There will be firms that will plan the construction projects, which will be followed by organizations which will be constructing them, and supported by consultants that supervise and manage the project.

A construction planner (planning consultant) is defined as a firm that is certified as a professional in construction planning service which is able to produce works in the form of building or other physical forms of planning documents. Its business service provides planning services in construction works which consist of the whole or part of activities that start from feasibility studies until construction work contract documents development. The Indonesia's government regulations (RI 2000) described the scope of construction planning services to consist of:

1. Surveys
2. General planning, macro studies and micro studies
3. Project feasibility studies, industry and production
4. Technical, operational and maintenance planning
5. Research

The second type of firm is a constructor, or more widely known as contractor, which is defined as a firm that is certified as a professional in constructing service which is able to deliver activities in producing a plan into a form of building or other physical forms. The service that it provides are construction works which consist of the whole or part of activities that start from site preparation until final hand-over of the construction work product.

The third type is a construction supervisor (supervision consultant), which is a firm that is certified as a professional in construction supervision service which is able to conduct supervision works since the beginning of construction activities until finishing and the hand-over phase. Its tasks are to provide a whole or part of supervision services in construction works that start from site preparation until final hand-over of the construction work product. The scope of construction supervision services consists of providing construction works supervision, and supervision in quality assurance, time accuracy in the constructing process and results (RI 2000).

Along with these types of firms, there are also some types of firms that have a combination of tasks in the construction project. There are firms that have the scope of integrating the planning, executing (constructing) and supervision services, which also depends on their respective procurement route in the project. The work scope can consists of:

- Design and build (planning and constructing).
- Planning, procurement and construction [similar to Engineering, Procurement, Construction (EPC)].
- Planning and/or supervision services throughout the project phase. This includes services like Project Management, Construction Management, Quality Assurance, and Quantity/cost surveyors.

6.4.2 Firms' Characteristics

The Indonesian Government Regulation No.28 Year 2000 (RI 2000) states that the business areas in construction services consist of:

- Architectural work area: simple/low technology building architecture, middle technology building architecture, high technology building architecture, interior architecture, and landscape architecture including their maintenance.

- Civil work area: roads and bridges, railways, airports, tunnels, underground roads, drainage and flood controls, ports, dams, irrigation and water resources infrastructures, building structures, geotechnical, plants and mining construction, including their maintenance and building demolition works.
- Mechanical work area: HVAC, air installation/air conditioning, oil/gas/geothermal installation, industry installation, thermal and sound isolation, lift and escalator construction, piping and their maintenance.
- Electrical work area: generator installation, transmission and distribution networks, electricity installation, train signal and telecommunication, radio transmitter building, air and water navigation telecommunication and facilities, telecommunication networks, telecommunication central, instrumentation, thunder shields, and their maintenance.
- Environmental work area: urban development/planning, spatial planning, environment impact analysis, environmental engineering, other environment management, regional development, clean water treatment and waste treatment plants, clean water and waste piping, and their maintenance.

In 2006, the number of certified consulting companies was 4118 firms registered by the National Board of Construction Services Development (NBCSD). In the same year, the number of certified contracting companies was 123,676 firms. NBCSD also reported that the total number of registered engineers is around 29,417 professionals (Suraji et al. 2006).

According to Indonesia's Statistical Central Agency (2012), there are three main types of construction projects delivered by the construction firms, which are:

- Building construction: building construction for residential sites, office, industrial buildings, shopping centres, health infrastructure, education infrastructures, accommodations, entertainment buildings, and others.
- Civil construction: road, bridge, railway, tunnel subway, watering building, processing, distribution and reception for the oil and gas, processing building, quay building, communication and electrical and others.
- Specific construction: the fitting of foundation and pillar, producing or drilling of ground water well, steiger fitting, roof covering, prefabrication fitting, fitting of steel framework, dredging and others, building installation and civil building, site preparation, building completion and renting of construction or demolition equipments with operator.

Figure 6.10 illustrates the type of construction projects with the project values completed in 2004–2009. It was found that building and civil construction projects had the highest value during these 5 years. In 2004–2007, the building construction was ranked first and the next 2 years was dominated by the civil construction projects. It was clear that the government's main focus is still in providing better housing and infrastructure for the people (EIU 2008).

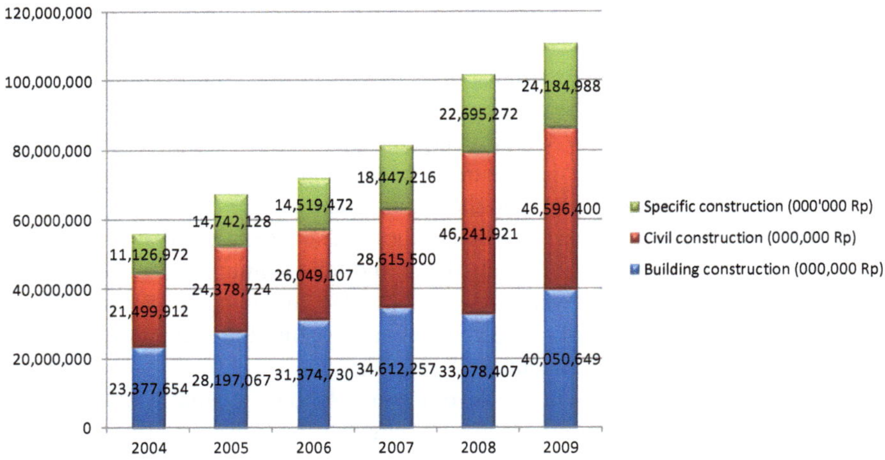

Fig. 6.10 Value of construction projects. Source: SCA (2012)

6.4.3 Indonesian Contractors

In Indonesia, the national contractors are categorized into Large, Medium, and Small firms, which are based on their working capital (net assets and transferred capital) as follows (NBCSD 2002):

- Small firms have working capital of maximum Rp. 1,000,000,000 [US$1 = Rp 9438—at the time of writing (Bloomberg 2012)].
- Medium firms have working capital of more than Rp 1,000,000,000 up to Rp 10,000,000,000.
- Large firms have working capital of more than Rp 10,000,000,000.

Based on the related regulation, Medium and Large firms must be in the form of Corporation [*Perseroan Terbatas* (*PT*)] as approved by the related Ministry.

Until 2009, the total number of Indonesian contractors is 139,964 firms. Based on the categories, most of the firms (approximately 89% of them) are categorized as Small firms. Medium firms are around 10% of the total firms, and the percentage of Large firms is less than 1%. Table 6.2 describes the number of firms for each category. This proportion of contractors is quite interesting, because although Large contractors are still very few, most of the construction projects in the country (particularly large-scale and complex projects with high values) are delivered by these firms. Moreover, Large contractors mostly dominate in constructing projects in the large cities of Indonesia's provinces. Small and Medium firms contribute in delivering small-scale projects within the region, mostly in the smaller towns (BCI 2006; Raftery et al. 2004).

Table 6.2 Number of contractors in 33 provinces of Indonesia

Category	Total
Small	125,222
Medium	13,828
Large	914
Total	139,964

Source: SCA (2009)

6.4.4 Business Activities of Contractors

In Indonesia, the contractors are owned either by the government (state-owned) or private parties (private firms). Until now, the state-owned contractors still dominate the construction market in Indonesia (Sutjipto 1991). The system of contracting in Indonesia consists essentially of three types, which are (Raftery et al. 2004):

1. Construction only (the traditional procurement system): the contractor's task is only to construct the product, based on the given design by the owner.
2. Design and build: The task of the contractor is more than construction only. Here, the contractor has the job to design and also construct the product. The buildability factor had been enhanced by slow but wider recognition of the design and build approach. Until now, the concept had rarely been applied due to the fact that contractors require a proper design network. However, the construction of various government offices, private commercial buildings and mixed buildings have started to use the design and build approach, where engineering detailed design is carried out by the contractor, and basic design by the designer and planner. For process plant investments, the engineering, procurement and construction (EPC) system has been more practiced widely.
3. Build—Operate—Transfer: The tasks defined in the contract are to construct and operate the product, and in a defined time will transfer the product to the owner for future operations.

According to Schaufelberger (2009), there are four primary business activities of a contractor, which are business development, procurement, construction operations, and post-construction services. In business development, the essential activity is to create relationships with existing and prospective customers. This is conducted in order to capture market and obtain projects. A comprehensive market research related to its business area is also important in this phase. Secondly, procurement activities consist of creating and maintaining relationships with subcontractors and suppliers. These relationships are needed for obtaining resources and delivering the work packages. In supporting these, efficient and effective procedures for material purchase and management should be developed. Following or in parallel with the procurement activities, construction operations consist of activities related to constructing the product through the chosen construction methods. This activity is also supported by cost and schedule estimation and control, project management system, quality management and safety management system. Last but not least, post-construction services focus on the after-construction

Table 6.3 Value chain of contractor

Primary activities			
Business development	Procurement	Construction operations	Post-construction services
Support activities	Firm infrastructure		
	Human resource management		

Source: Schaufelberger (2009)

phase. The warranty of the product that was delivered should be well-managed (particularly in handling defects after the hand-over phase) and coordination with the customer or user is essential.

Supporting activities in a contractor's business include the basic organizational structure, its employees and equipment, as well as human resource management programs and policies to attract, develop, and maintain a motivated, skilled team of employees. Elements that must be considered in the firm's infrastructure are the adequacy and location of facilities and equipment, the efficiency and effectiveness of its finance and accounting system, and its information management system. These are necessary because in managing construction projects, the firm will mostly create a project-based (site) office which is near or in the project location. The head office of the firm may not be located in the same area as the site office. Therefore, good coordination between these offices should be managed and maintained during the project phase. Regarding the firm's HRM, establishing procedures for employee recruitment and development, creating an encouraging work environment, creating and maintaining good relationships with unions, and focusing on employee's level of motivation and job satisfaction are needed to achieve the business goals. The value chain for contractor in doing its business is illustrated in Table 6.3, describing the elements in its primary and support activities.

In delivering a project, a contractor needs to manage the materials, people, and equipment in a project site and assembling the materials in the proper sequence to construct a project that meets the customer's requirements. In meeting these requirements, the contractor also has to consider their stakeholders, which can vary from the customers, suppliers, creditors, investors, employees, subcontractors, governments and the public, to existing and new competitors. Moreover, the business management challenges for contractors are to ensure that (Schaufelberger 2009):

- The revenue generated by the construction activity exceeds the cost of doing the work.
- The firm has adequate demand for its services.
- The firm has adequate financial resources to finance construction projects until reimbursed by its customers.
- The firm has a skilled, motivated workforce of sufficient size to meet anticipated requirements.
- The cost of the firm overheads is affordable based on the projected workload.

Schaufelberger (2009) also mentioned that each type of construction projects has its unique set of technical challenges, but the following business responsibilities for contractors are similar, which are acquisition of the work, performance of the work, and management of the financial, capital, and human resources of the firm.

The existing conditions of the Indonesian contractors have been analyzed. Suraji (2003) had completed an analysis showing the strengths, weaknesses, opportunities, and threats in general. Based on these findings, the Indonesian contractors have the strengths of delivering more projects due to their many labour resources and the current needs of infrastructures in many cities in Indonesia. There are many opportunities for collaboration with other foreign contractors arising from their recent businesses in Indonesian construction projects. Thus, this can enhance the capability of Indonesian contractors in delivering better projects through technology and knowledge transfer. Some weaknesses identified are the lack of funds and technologies, lack of skilled workers, the high level of competition in the national construction industry, and management inefficiencies. Factors such as the competencies of human resources, research and development, certification and support from other sectors may threaten the firm's growth and sustainability in the future if not considered comprehensively.

6.5 Review of Crises Faced by Indonesian Contractors

A crisis may give various consequences to an organization, whether financial, legal, or operational consequences. It may disrupt the business process from a few minutes to up to several months or years in extreme cases. These consequences can impact the business process, and hence might threaten the firm's sustainability.

6.5.1 Crisis Overview

Based on the Oxford Dictionary (2006), the word crisis means a time of great danger, difficulty, or confusion when problems must be solved or important decisions must be made. A crisis can be a threat because it has the possibility of trouble, danger, or disaster. Moreover, Barton (1993) described crisis as an abnormal situation or perception which threatens operations, staff, customers, or the reputation of the organization. There are various examples of crises that may occur in a firm. Elliott et al. (2002) had listed examples of crises as illustrated in Table 6.4. These events may result into a disaster for a firm if not handled or responded well and quickly.

Various studies had also grouped crises into several types and categories. Shrivastava and Mitroff (1988) had developed a crises typology, which is based on its root causes, whether from the internal, external, technical/economic, or human/organizational/social aspects. Moreover, Karakasidis (1997) had grouped

Table 6.4 Unstructured list of crises

Unstructured list of crises (Elliott et al. 2002)	
– Adverse weather	– Loss of important staff
– Computer breakdown	– Major industrial accidents
– Computer bugs	– Media crisis
– Computer failure	– Natural disasters
– Currency fluctuations	– Product injuries
– Disease/epidemics	– Product tampering
– Fire	– Sabotage by outsiders
– Floods	– Sabotage by staff
– Hostile takeover	– Societal crisis
– Illegal activities	– Supplier crisis
– Industrial action	– Telecommunications failure
– Kidnapping	– Terrorism

crises based on the intentional or unintentional cause. Low et al. (2008a) had likewise compiled the categories into four types of crises, which are:

- Acts of nature such as floods, snowstorms, earthquakes, etc.
- External, man-made events such as terrorist attacks, bomb attacks, riots, etc.
- Internal, unintentional events such as an accidental loss of files, a computer crash, etc.
- Internal, intentional events such as sabotage, data deletion, etc.

By grouping crises based on their characteristics, it is beneficial for the firm to detect and respond to the crisis more accurately.

6.5.2 Possible Crises Faced by Contractors

Like any other firms, contractors may also experience various crises within their businesses. The types of crises can be grouped into categories, where each of them has various levels of impacts towards the business. Compiled from various references, the types of crises that contractors may encounter can be categorized based on the firm's primary and supporting business activities as follows (Edwards 1995; Low et al. 2008b; Schaufelberger 2009):

- Business development

  During the phase of creating relationships with existing and prospective customers for capturing market and obtaining projects, there are some events that may threaten or interrupt these activities. Scandals such as corruption in the firm and other malicious contamination within the organization may affect the firm's reputation and image. Actions or protests by environmentalist/pressure groups toward the contractor due to its actions that may not be beneficial to the public or the environment could also disrupt the firm's business development activities. Other crises such as financial crisis (either national or internal financial crisis), and political instability (that may lead to changes of project scope or project cancellations, sanctions and embargoes, and others) could affect the contractor's current and future business development strategies. Moreover, issues such as regulation or legislation changes, client insolvencies, and delays due to resolving disputes with clients or other stakeholders may also disrupt the business development process.

- Procurement

  In creating and maintaining relationships with suppliers and subcontractors for purchasing resources for the projects, there are some threats that can interrupt the process such as material shortages (particularly key materials), material damages or faulty goods from the manufacturers, unexpected price escalation for the resources, and subcontractor's insolvencies.

- Construction operations

  During the construction operation phase, activities that relate to and support the methods for constructing the product is monitored and managed. The time, cost, quality and safety measures during this process are essential, supported by a comprehensive project management system. Crises such as serious accidents, construction plant breakdown, fire, explosion, theft, sabotage, and limitation or restriction to some vital access may create significant interruption for this phase, and may threaten the final delivery of the product.

- Post-construction services

  In this phase, the contractor is mainly focusing on managing the warranty period and further developing relationship with the client after handing over the product. If any serious defects or failures on the product or its components occurred, this could give rise to adverse impacts that may lead to liabilities on the contractor's part.

- Firm's infrastructure

  The process of managing the firm's infrastructure, which includes its offices (head and site offices), finance and accounting system, and information system, can be disrupted by crises such as natural disasters, loss of confidential information, war, riot, and terrorism. These crises can damage the firm's infrastructure and the continuity of its business.

- Human resources management

  This supporting activity, which focuses on the firm's employee recruitment and development, maintaining relationships with various human resources unions, and managing the firm's working environment, can be disrupted by several events. Crises such as virus pandemic (flu pandemic, SARS, etc.), labour strikes or disputes, lack of competent workforce, loss of management personnel or key staff, and kidnap of or ransom for employees may threaten the firm in continuing its business activities, particularly its human resources activities. Considering that most of the activities of a contractor are led and operated by people, these events should not be overlooked as these can cause significant impacts toward the firm's business.

Table 6.5 compiles these crises, which are grouped into the contractor's business activities categories, and followed by their respective classifications into the various main types of crises, including the acts of nature, external man-made events, and internal, intentional or unintentional events.

6.5.3 Crises Experienced by Indonesian Contractors

Indonesian contractors have also experienced various threats or crises that have significant impacts on their business activities. Some factual crises that occurred and documented in the past years, followed by their impacts toward contractors and the firms' responses toward those events are described below:

- 1997 financial crisis (Agustinus and Luhur 2008; Firdausy 2002; Kartasasmita 2000; Saparini 2009)

  The 1997 financial crisis started when the Indonesian currency depreciated sharply more than three times. The impacts were that the investment level of property, real estate and construction diminished significantly, with a sharp drop in construction business which led to bankruptcy, where the growth rate became −37.5% and companies laid off more than 1.4 million workers, and with high unemployment faced by the contractors. In responding to this crisis, the government provided public work programs for the unemployed construction workers. But this was a temporary program and there were problems related to the program such as the mismatch in job tasks, lack of program dissemination, and ineffectiveness in managing the program. Moreover, currency and inflation control by the government provided some slow recovery for the construction sector.
- 2008 financial crisis (Suryadharma 2008)

  This crisis started from the global economic downturn, which led to increased interest rates (up to 9%), and skyrocketing costs of raw materials, including steel and cement, which accounted for approximately 20% of an average project's cost structure. This had led the contractors to be more selective in managing their costs of contracts because hedging raw materials prices was by no means a sure way to counteract the impacts of high building material prices. Furthermore, a rejection of escalation clauses may in fact have darkened the future of the construction sector. The National Planning and Development Agency (*Bappenas*) has rejected a project value escalation proposal by the contractors' associations, electrical and mechanical associations and other construction-related associations. The sector was then also likely to suffer from postponed infrastructure projects in 2009 as the Public Works Ministry was likely to see its original expenditure plan slashed to Rp 35.6 trillion from Rp 58.7 trillion due to budget constraints amidst higher oil-related subsidies. Diversification into other businesses may help the contractors reduce risks in this context. Some firms have

Table 6.5 Possible crises faced by contractors

Contractor's business activities (Schaufelberger 2009)	Possible crises (Low et al. 2008b)	Type of crises (Low et al. 2008a)
Business development: • Creating relationships with existing and prospective customers • Obtaining the work • Market research	Loss of public goodwill, reputation, image due to malicious acts	Internal, intentional event
	Action by environmentalist/pressure groups (protests)	External, man-made event
	Financial crisis	External, man-made event
	Corruption scandal	Internal, intentional event
	Political instability i.e. those leading to changes of project scope or cancellations, sanctions and embargoes, tighter exchange controls, repatriation of funds	External, man-made event
	Changes in regulations and statutory legislation	External, man-made event
	Client insolvency leaves outstanding debts for work done	External, man-made event
	Delays or uncertainty in resolving disputes	Internal, unintentional event
Procurement: • Relationship with subcontractors • Relationship with suppliers • Efficiency and effectiveness of material purchasing procedures and management procedures	Increase in price of raw materials (unexpected price escalation)	External, man-made event
	Shortage of key materials	External, man-made event
	Material damages (during deliveries or faulty products from manufacturer)	Internal, unintentional event
	Subcontractor insolvency	External, man-made event
Construction operations: • Cost and schedule estimation and control • Project management system • Quality management system • Safety management system	Breakdown of key construction plant	Internal, unintentional event
	Serious accidents in a project	Internal, unintentional event

(continued)

Table 6.5 (continued)

Contractor's business activities (Schaufelberger 2009)	Possible crises (Low et al. 2008b)	Type of crises (Low et al. 2008a)
	Fire	Internal, unintentional event
	Explosion	Internal, unintentional event
	Theft	Internal, intentional event
	Sabotage	Internal, intentional event
	Access/approval restriction or limitation	External, man-made event
Post-construction services: • Warranty management system • Customer relationship development program	Serious product defects or component failures	Internal, unintentional event
Firm infrastructure: • Adequacy and location of facilities and equipment • Efficiency and effectiveness of finance and accounting system • Information management system	Loss of confidential information	Internal, unintentional event
	Natural disasters (earthquake, floods, tsunami, etc.)	Acts of nature
	War	External, man-made event
	Riot	External, man-made event
	Terrorism	External, man-made event
Human resources management: • Procedures for recruiting and developing employees • Working environment • Relationship with unions • Levels of employee motivation and job satisfaction	Lack of component workforce	Internal, unintentional event
	Loss of management personnel or key staff	Internal, unintentional event
	Kidnap and ransom (effect on production and share price of loss of key personnel)	Internal, intentional event
	Strikes, labour disputes	Internal, intentional event
	Health issues (flu pandemic, SARS, etc.)	Acts of nature

already begun engaging in the LPG-tank business or have acquired stakes in toll road or power plant firms to ensure continued growth and reduced risks.

- Terrorism (CMfEA 2009; PTX 2008[1]; UNR/HC 2005)

  Some serious terrorism events that occurred during the past 10 years in Indonesia were the Bali bombing in 2002, Jakarta bombings (Australian Embassy, JW Marriott Hotel, and Ritz Carlton Hotel) in 2003 and 2009. There were tens to hundreds of casualties due to these bombings. These events had resulted in a 1-day business disruption in the Jakarta area which led to building evacuation and higher security within that week. For this type of crisis, most of the contractors have emergency responses for bomb threats (as applied to their head and site offices).
- Riots (Herlijanto 2004; ICG 2002; Lee 2009; PTX 2008; Tirtosudarmo 2005; UNR/HC 2005)

  Various riots and conflicts have occurred in Indonesia, such as: The Poso conflict in 2005, Jakarta riots in 1998, Aceh conflict (during the 1980s up to 2005), and Papua conflict in 2002. These conflicts were mainly due to political issues and instability with various diverse root causes. These caused various business disruptions, from 1 day to 1 week in the conflict area. As an example in the Jakarta riots, the capital city had a level 1 alert (highest level of national security alert) within the week. People were trapped in their offices, there were no transportation and commercial activities, and there was mass departure from certain ethnic groups going abroad. The government responded to the crises with high level of security and military responses in the area. In addition, after the 1998 Jakarta riots, several contractors had subsequently developed emergency responses to riots/conflicts.
- Fire (PTX 2008; Tadie 2008)

  Forest fires and short circuits in slums area and buildings mostly occurred in Indonesia's dense cities and villages. In populated cities like Jakarta and Surabaya, any fires due to short circuits in the buildings or houses had led to building evacuation. Considering this, in protecting the firm's head or site office, most of the contractors have emergency responses for fires in their facilities and sites.
- Accidents/safety issues (Permana 2007; PTX 2008)

  Recently, it has been found that accidents in construction projects are increasing, particularly involving the small contractors. The accident severity rates are increasing from year to year. Impacts from these events included small injuries to fatalities, loss of workforce due to injuries, and loss of productivity. Responding to these, most of the large contractors have developed detailed safety plans and certification. However, only a few small contractors have safety plans in their firms.
- Earthquakes and tsunami (MiyamotoINTL 2007; PTX 2008; Tambunan 2006; UNR/HC 2005)

[1]PTX (2008)—This document was provided by an Indonesian contractor who requested for anonymity.

Indonesia had suffered from many earthquakes and a huge tsunami disaster within these past years. Some recent events included earthquakes in Aceh and Nias (2004), the Yogyakarta area (2006), and West Sumatra (2007 and 2009). Impacts from such natural disasters are severe, such as building and infrastructure damages, injuries and fatalities. The largest fatalities were in the Aceh and Nias earthquake caused by the tsunami disaster that followed after the earthquake. Emergency responses were provided by local authorities, followed by local and international aids in the following days after the disaster. The local and national government also provided temporary shelters, health aid facilities, and surveys of damage around the area. Due to the significant impact towards the business sector, including construction, an earthquake site evacuation procedure had been planned by most of the contractors. These procedures are particularly vital to the firms that have projects or offices in earthquake-prone area.

- Floods (IFRC 2004; PTX 2008; Sutardi 2006)

  Generally, floods were due to torrential rains, such as those seen in Sumatra and Sulawesi islands in December 2004, the Jakarta floods in 2002, and North Sumatra flash floods in 2003. The impacts included loss of home and building damages, business disruptions from 1 day to 3 days, and infrastructure damages. Most of the contractors have emergency responses for floods in their facilities and sites, supported by emergency evacuation and responses by the local authorities. In addition, the government had also developed Integrated Flood Management Policies, including corresponding emergency measures and responses.

6.5.4 Current Findings

Based on the literature review presented earlier, the contractors appeared to have reacted differently to different types of crises. In financial crises, some contractors postponed their current projects, sought assistance from the government such as proposing project value escalation, while others diversified their businesses to ensure continued growth. These reactions gave rise to different impacts in their businesses, with some firms that could not save their businesses due to high debts, while others lay off their workers temporarily until they can resume their projects. In other cases, because of currency intervention and inflation control by the government, some firms have also survived the crisis to recover within several years.

In crises related to external man-made events (such as terrorism and riots) and internal unintentional events (such as fires), these firms experienced business disruptions which can vary from 1 day to 1 week, depending on their office and project location. Most of the contractors applied emergency responses (evacuation process) to give priority to people's safety. These responses had been written in the firm's emergency manual. For further business recovery, the contractors will need to create a recovery team who will assess all the loss and develop steps that are needed to resume the business. From this procedure, it can be seen that the firms are mostly focusing on the evacuation response process. The process and steps for continuing the business after the crisis have not been planned in details yet, thus resulting in an even longer disruption period.

Crises related to natural disasters (such as earthquakes, tsunami, and floods) caused the contractors to apply emergency responses (site or office evacuation procedures), and further coordination with the local government. The processes after the disaster appeared to be mostly dependent on the government's aid. These included temporary shelters, health aid facilities, and surveys of damage around the area provided for by the government. Similar to the previous types of crises, the contractors do not have a detailed back-up plan and continuity procedure to recover and resume business. They will only create the recovery team after the situation had evolved to be conducive enough to resume business.

In addition, in crises such as accidents in construction projects, most of the larger contractors have detailed safety plans. The emergency responses are therefore well managed with detailed procedures. However, smaller contractors do not have detailed safety plans and only a few are certified for emergency preparedness. This may cause the disruption period to last longer and affect the productivity of projects.

From these cases, it can be seen that most of the contractors have provided relevant emergency responses for evacuating people during the crises (in external man-made events and natural disasters). However, a detailed recovery procedure for their businesses to resume after the crisis had not been planned in advance. The firms will create the recovery team and develop steps to resume based on the management's decision. As for financial crises, the reactions were varied, which were again dependent on the management's decision to sustain the business. Most of the contractors faced difficulties during these events. They were significantly reliant on government's actions in currency intervention and in managing the spiraling inflation rate.

Some extremely damaging crises have major impact on the local or provincial economy, and some to the national economy. For example, the 1997 financial crisis resulted in the collapse of the banking sector which led to projects and businesses being postponed in various sectors. This caused negative growth in the Indonesian economy (Agustinus and Luhur 2008). As for natural disasters that have occurred in various regions in Indonesia, the impact of these crises caused severe damages to areas and infrastructure in the location, high death tolls, and disrupted business activities which led to associated economic problems in the location. These situations required aids and assistance from the central government.

From these findings, it appears that the Indonesian contractors have not developed their crisis responses into a holistic management approach in their organizations, and there is a lack of detailed responses for their business stakeholders. There remain patches of responses that have not yet been sewn together for the Indonesian contractors to survive and continue their businesses. Therefore, these firms should start to adopt a management concept within their organizations to recover and sustain their businesses in an effective manner. Considering the types of crises and the severe impacts that have occurred, the existing responses made for these crises were not fully effective for assuring the firm's business continuity.

6.6 Summary

As the largest archipelago in the world, Indonesia has various and abundant natural resources, with a large population distributed across the islands. Its economic performance has been improving throughout the years, after facing financial and political crises that occurred in the late 1990s. Regarding its infrastructure, Indonesia still needs to develop this sector in a systematic and practical manner. This sector also provides attractive opportunities for investments and the construction industry can play an important role in developing this.

The Indonesian construction industry is one of the important sectors in Indonesia. Its role can be seen from the major usage of domestic goods and services that contribute significantly to the country's total GDP. There are some problems faced by the industry, particularly due to its position in a volatile environment. In anticipating this matter, the construction industry has gained some lessons learned and engaged in an effort to develop the industry into a more valuable sector.

The Indonesian contractors cannot overlook any events that may cause various consequences to them, whether financial, legal, or operational consequences. A crisis may disrupt the firm's business processes from a few minutes to up to several months or years in extreme cases. Based on the literature review, a contractor's business activities may be vulnerable to various types of threats and crises. These were proven by some factual crises that occurred and documented in the past years, which had been experienced by the Indonesian contractors. From these findings, the contractors appeared to have reacted differently to different types of crises. The impacts from these crises also differ, from several days of disruption to bankruptcy. In order to be resilient and able to response to such threats, the Indonesian contractors should start to adopt a holistic management concept, such as BCM in their organizations.

Chapter 7
Knowledge Based Decision Support System (KBDSS)

7.1 Introduction

This chapter reviews the Knowledge Based Decision Support System (KBDSS) as part of the decision making tools. Decision making process will be described in general, followed by reviews on various decision making tools. The second part will elaborate on an overview of DSS, including its development and implementation. KBDSS will be described afterwards with a discussion of its general concept and its implementation, particularly in the construction industry and for emergency management (aspects that are focused in this study). The final part will discuss the development of KBDSS, its formulation, system development, and validation process.

7.2 Decision Making Process

Most businesses are facing challenging situations with respect to the changes in their environment (both internal and external to the organization). This is because the environment has become more complex and uncertain. Human decision making in complex environments is not a well understood process. Decision making can be defined as an outcome of evaluation processes leading to determine the most appropriate choice from among several alternatives. Every decision making process produces a final choice. The output can be an action or an opinion of choice. Making a decision implies that there are alternative choices to be considered, and in such a case it is not only to identify as many of these alternatives as possible but to choose the one that has the highest probability of success or effectiveness and best fits with the goals, desires, values, and other criteria (Kaya and Kahraman, 2010; Srinivas and Shekar, 1997).

L.S.R. Supriadi, L. Sui Pheng, *Business Continuity Management in Construction*, Management in the Built Environment, DOI 10.1007/978-981-10-5487-7_7

According to Asemi et al. (2011), in the 1950s Herbert Simon and James March for the first time introduced a different decision making framework for understanding organizational behavior. The model suggested that when an individual makes decision, he examines a limited set of possible alternatives rather than all available options. Satisfactory or good enough choices will be accepted, rather than an insistence on optimal choices. Choices that are good enough are made because the individual does not search until a perfect solution is found to a problem (Gordon, 1993).

Based on Simon's (1997) study, decisions are grouped into two basic types: programmed and non-programmed decisions. Programmed decisions are routine and repetitive decisions, and the organization usually develops specific ways to handle them. For this kind of routine and repetitive decisions, standard arrangement decisions are normally made according to the management guidelines established. In contrast, non-programmed decisions are typically one-off decisions that are usually less structured than programmed decisions (Certo, 1997).

After Simon's study in the 1950s, Gorry and Morton (1971) continued to classify decisions into three levels: (1) structured decisions, which the ingredients, or variables, that comprise a decision are known and they can be measured quantitatively; (2) unstructured decision, which is one that the ingredients, or variables, that comprise a decision cannot be measured quantitatively; and (3) semi-structured decision, which is in between structured and unstructured decisions. Most business decisions are usually in the semi-structured form (Simon, 1997).

7.3 Decision Making Tools

When a complex problem has an increase in the number of alternatives, goals and criteria, the decision making process will be more difficult. Due to this difficulty, there have been many studies on how to solve such problems. Making decisions in the real world would take place in an environment with unknown or imprecise goals and constraints that need tools to assist with the process (Kaya and Kahraman, 2010).

Many tools have emerged for understanding, analyzing, and aiding the decision making process. Some of these tools are models of the decision situation (such as Markov decision problems; decision trees; influence diagrams, and Bayes nets), and other tools can be used to analyze these situations to come to a (sometimes optimal) decision (examples: expected and multi-attribute utility analyses; game theory; Bayesian inference; stochastic optimal control theory; partially observable Markov decision processes; reinforcement learning models; and rule-based cognitive architecture) (Busemeyer and Pleskac, 2009; Keeney and Raiffa, 1993; Luce, 2000; Myerson, 1991; DeGroot, 1970; Pearl, 1988; Clemens, 1996; Stengel, 1986; Puterman, 1994; Haykin, 1999; Sutton and Barto, 1998; Newell, 1990).

Moreover, Buyukozkan and Feyzioglu (2004) observed that different decision methods have been developed to overcome the uncertainty related problems, which are summarized as follows:

- Probabilistic modes: These include Monte Carlo simulation and decision trees. Monte Carlo analysis uses the process of simulation to achieve a range of solutions to a problem. Decision tree is a diagram that provides a structured approach to decision making that incorporates uncertainty of outcome.
- Scoring models and checklist: Various alternatives are rated and scored on a variety of qualitative questions.
- Behavioral approaches: These are tools designed to bring managers to a consensus in terms of which alternative to choose, and include methods such as the Delphi method which is a qualitative forecasting method that uses a panel of experts. These are particularly useful for studies at the early stages, where only qualitative information is available.
- Analytical hierarchy process (AHP): These are decision tools based on paired comparisons of criteria.
- Fuzzy logic: It deals with problems where a source of vagueness is involved (Zadeh, 1965). In general, the probability concept is related to the frequency of occurrence of events, captured by repeated experiments whose outcomes are recorded, while the fuzzy sets provide the appropriate framework to evaluate the possibility of events rather than their probability.
- Sensitivity analysis: It examines how the optimal solution and the optimal objective value are affected from the changes of the uncertainty parameters (values and probabilities) that are considered to be important.
- Scenario analysis: This technique has been widely preferred and used by many decision makers. Here, a combination of possible values of the uncertainty parameters are assumed pertinent to different points of views (e.g. pessimistic, neutral and optimistic), and the resulting scenario is solved. By solving the problem repeatedly for different scenarios and studying the solutions obtained, the decision maker observes sensitivities and heuristically decides on an appropriate solution.

All of these techniques can be used exclusively or in a hybrid way for the decision making process.

7.3.1 Decision Making Tools Using Fuzzy Approach

According to Bashiri and Badri (2011), due to incomplete or unavailable information in the real-world situation, data (attributes) are often not so deterministic and the majority of these attributes can be assessed by human perception and human judgment. Hence, these attributes are typically fuzzy imprecise.

During the last decades, decision making methods have been developed and along this growth, a variety of hybrid algorithms has also been developed. A large number of these hybrid methods have been created in a fuzzy environment. Generally, vague decision making problems in the real world are expressed by the human's thinking and subjective perception in words, rather than probability and statistics. This is the reason why fuzzy numbers and fuzzy sets have been utilized to make this process more realistic. Moreover, multi-criteria decision analysis models have been used in fuzzy environment.

Among these methods, AHP (Saaty, 1998) has been widely used and viewed as a weight estimation technique in many areas such as selection, evaluation, planning and development, forecasting, and others (Vaidya and Kumar, 2006). A combined fuzzy AHP with multi-dimensional scale (MDS) have been established to identify the preference similarity of alternatives (Chen, Tzeng and Ding, 2008). Moreover, Chen (2000) developed a fuzzy TOPSIS method for group decision making where decision matrix and criteria weights in fuzzy environment are considered.

Another most widely used method is the simple additive weight (SAW) or the weighted sum method. The basic principle of SAW is to obtain a weighted sum of the performance ratings of each alternative under all attributes (Chen et al., 2008; Chen and Hwang, 1992). This method consists of two basic steps: (1) scale the values of all attributes to make them comparable; (2) sum up the values of all the attributes for each alternative. This method, similar with AHP, can also be combined with the fuzzy approach (Hwang and Yoon, 1981; Chang and Yeh, 2001).

Other examples of the combined fuzzy and multi-criteria decision analysis models are found in the manufacturing sector. Borenstein (1998) developed a visual interactive multi-criteria decision analysis model for the evaluation of flexible manufacturing systems' design alternatives. This method tries to take into account the designer's preferences and wishes, in order to customize the manufacturing system for the user's particular situation. Chan et al. (2006) designed a fuzzy multi-criteria decision making procedure for the selection of manufacturing technologies. Their method considers different justification parameters such as strategic, economic and analytic evaluations. These models are not fixed for manufacturing processes only. Various adjustment and modification can be done in order to render them applicable to other sectors.

7.4 Decision Support System (DSS) Overview

As part of the decision making process, it has been recognized that computer based information systems play important roles, particularly for semi-structured or unstructured decision making activities. The term Decision Support System (DSS), which is a computer-based system, has been known since the early 1970s.

7.4.1 Definition and Development of DSS

DSS is defined as an interactive computer-based system which supports a decision maker in utilizing data and models to solve unstructured or semi-structured decisions. It produces output in the form of periodic or special report or the results of mathematical simulations (Asemi et al., 2011; Raymond, 1990; Toda et al., 1991; Raman and Phoon, 1990). Raman and Phoon (1990) added that a DSS is mostly used to support managers in their decision making in activities such as planning, coordination, control, organizing, forecasting, budgeting, administration and general management.

Sen and Biswas (1985) found that semi-structured and unstructured decisions are typically solved by examining different scenarios and asking "what if" type of questions. DSS are designed for solving this type of domain. Moreover, it is recommended that DSS is designed to be user-friendly, so that it can be used by users who are knowledgeable in their fields of expertise, but may have very limited experience in computer usage. The DSS should have the knowledge about the user's domain, the type of problems encountered, and the tools needed to solve them.

The main objective of DSS is to help the decision maker make effective decisions by identifying what should be done and ensure that the chosen criterion is relevant. Within its development, DSS has evolved to give more effective function as a decision support tool. Factors such as the discovery of structure in some judgmental tasks and improvement in technology which allows the computer to do more tasks have influenced its evolution. Examples of recent technologies that are incorporated in a DSS are modern database technology, graphical user interface, hypermedia, multimedia, expert systems, neural networks, fuzzy logic, generic algorithms, distribute systems, client-server, and the object-oriented approach. These technologies were not feasible in the 1970s (Fazlollahi et al., 1997).

There are three conceptual components of a DSS, which are as follows (Sen and Biswas, 1985):

- Database management

  A database is a collection of related data which is organized so that useful information may be extracted for decision making and duplicate data collection is minimized. The effectiveness of a database is derived from the fact that much of the information that is relevant to a variety of organizational purposes may be obtained from one single, comprehensive data set.
- Modelbase management

  The modelbase concept considers a model as a data abstraction consisting of equations, elements and solution procedures. The ideal model management facility should provide: (1) a modelbase management system (MBMS) to generate, retrieve and update parameters, to restructure models, and to include a model directory for maintaining information about available models; (2) model execution to control the actual running of the model and to link models together when integration is needed; (3) a modeling command processor to accept and interpret modeling instruction as they flow out of the dialogue component and to route them to the MBMS or the model-execution function; and (4) a database interface to retrieve data items from the database for running models, and eventually, to store model outputs in the database for further processing, perusal, or as inputs to other models.

- Dialogue management

  DSS usage involves a dialogue between the user and the DSS. The most complicated mathematical models, the most comprehensive databases, or the most detailed instructions are all ineffective as decision support devices unless their methods, information, and results can be effectively communicated to the user who must make the decisions. Even if a DSS provides extremely powerful functions, it may not be used if the dialogue is unacceptable. Dialogue style—the nature of the interface between the system and the user—includes question/answer (Q/A) style, command mode, input/output (I/O) form, and menu-driven dialogue. Moreover, some combinations of these are also possible.

Asemi et al. (2011) had compiled the general characteristics of a DSS. As a support for the decision maker mainly in semi structured and unstructured situations, DSS is found to improve the effectiveness of decision-making, particularly the accuracy, timeliness, and quality (Davis and Olson, 1985). Moreover, when less structured decisions frequently require the involvement of several individuals, which lead the need for group decision making, DSS can provide this support.

A DSS is also able to handle large amount of data and can be developed using a modular approach. With this approach, separate functions of the DSS are placed in separate modules—program or subroutines—allowing efficient testing and implementation of systems. It also allows various modules to be used for multiple purposes in different systems (Asemi et al., 2011).

Furthermore, a DSS has a graphical orientation. It can provide a more attractive and informative graphical presentations on screens or can be printed. A line drawing, pie chart, trend line and other graphs can be incorporated in the system to help the user to better understand the situation of the problems. This characteristic can become part of the basic approach that a DSS has, which is performing "what-if" and goal seeking analysis. "What-if" analysis is the process of making hypothetical changes to the problem data and observing impact of the results (Asemi et al., 2011; Stair, 1992). Figure 7.1 illustrates the main characteristic of a DSS, which is focusing on semi-structured problems and incorporating the analysis based on computer and the user (or manager).

7.4.2 DSS Implementation

DSS has been a major area of IT practice and the decisions made using IT-based decision support can provide a significant effect on the nature and performance of an organization (Arnott and Pervan, 2008). Its functions might be for supporting

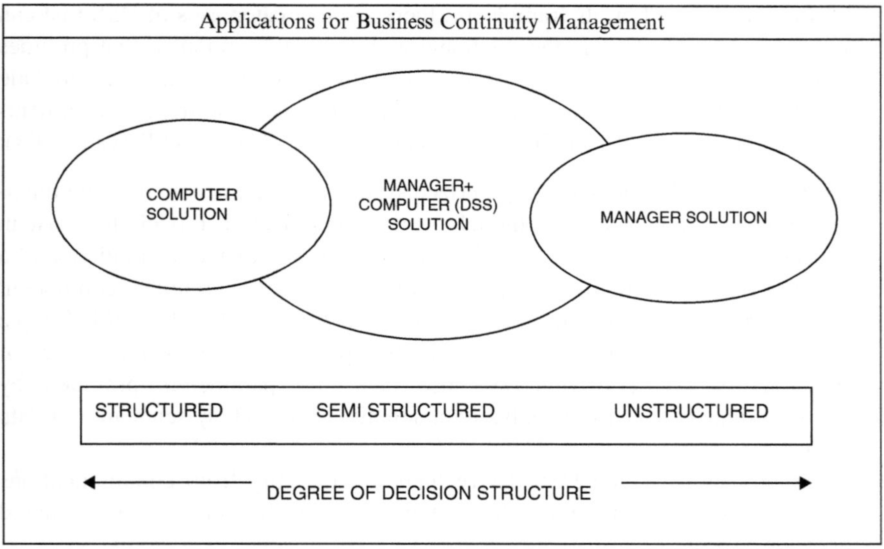

Fig. 7.1 DSS characteristics. Source: Adapted from Raymond (1990)

operational, tactical, or strategic decision making. Moreover, a DSS can provide summaries of data; forecast future developments, which take account of uncertainties, and help the decision makers explore their own perceptions and values. DSS can be used for individuals or groups, who may work in the same time and place or different locations (French and Turoff, 2007).

According to the studies of Arnott and Pervan (2008), the major DSS sub-fields are:

- Personal DSS (PDSS): Usually small-scale systems that are developed for one manager, or a small number of independent managers, to support a decision task.
- Group support systems (GSS): The use of a combination of communication and DSS technologies to facilitate the effective working of groups.
- Negotiation support systems (NSS): DSS where the primary focus of the group is negotiation between opposing parties.
- Knowledge Based DSS (KBDSS): Systems that support decision making by aiding knowledge storage, retrieval, transfer and application by supporting individual and organizational memory and inter-group knowledge access.
- Data warehousing (DW): Systems that provide the large-scale data infrastructure for decision support.
- Enterprise reporting and analysis systems: An enterprise focused DSS including executive information system (EIS), business intelligence (BI), and more recently, corporate performance management (CPM) system. BI tools access and analyze data warehouse information using predefined reporting software, query tools and analysis tools.

Furthermore, based on its framework, there are five categories of DSS that can be recognized by identifying the dominant architectural component that provides the functionality for supporting decision-making. The five categories include model-driven DSS, as well as communication-driven, data-driven, document-driven, and knowledge-driven DSS (Asemi et al., 2011; Arnott and Pervan, 2008):

- Model-driven DSS include computerized systems that use accounting and financial models, representational models, and/or optimization models to assist in decision-making. Model-driven DSS emphasize access to and manipulation of a quantitative model and hence the model or models are the dominant component in the DSS architecture that provides the functionality for the DSS. Simple analytical tools based on algebraic models provide an elementary level of functionality. Model-driven DSS use data and parameters provided by decision-makers to help in analyzing a situation, but such systems are not data intensive.
- Communications-driven DSS derive their functionality from communications and information technologies that are used in the system to support shared decision-making.
- Data-driven DSS include file drawer and management reporting systems, data warehousing and analysis systems, Executive Information Systems (EIS) and data-driven Spatial DSS. Business intelligence systems are also examples of data-driven DSS.
- Document-driven DSS integrate a variety of storage and processing technologies to provide sophisticated document retrieval and analysis to support decision-makers.
- Finally, knowledge-driven DSS suggest or recommend actions based upon knowledge that has been stored using AI or statistical tools like case-based reasoning, rules, frames, and Bayesian networks. The knowledge component provides the primary functionality for the DSS or subsystem.

Table 7.1 shows the categories of DSS framework subsystems based on its subfields. A DSS can be developed from more than one functional component. As an example, a DSS may include both a data-driven and a model-driven subsystem.

Implementing DSS provides various benefits. The benefits are the interactive user interface, and non-procedural data analysis language of DSS that provide direct access and control over information to decision makers. It is also considered to help decision makers reduce the time required in the intelligence, design, and choice phases of decision making, where it can improve managerial productivity. It can achieve a competitive advantage for the organization that uses it, and is also considered as a part of competitive strategy among organizations that offer the same range of products and services, that refers to their qualitative differentiation (Raman and Phoon, 1990; Porter and Millar, 1985). In addition, Singh et al. (2008) found that the broad benefits from this system are that it provides a quick access for the user to all relevant information, the process is direct and personalized and the

Table 7.1 DSS framework subsystems

DSS sub-fields	DSS subsystems
Personal DSS (PDSS)	Model-driven; Communications-driven; Document-driven
Group Support Systems (GSS)	Model-driven; Communications-driven; Document-driven
Negotiation Support Systems (NSS)	Model-driven; Communications-driven
Knowledge Based Decision Support System (KBDSS)	Knowledge-driven; Model-driven; Document-driven; Communications-driven
Data Warehousing (DW)	Data-driven; Document-driven
Enterprise reporting and analysis systems (EIS or BI)	Data-driven; Model-driven; Document-driven; Communications-driven

Sources: Arnott and Pervan (2008), Asemi et al. (2011)

problem models that are evaluated in the system can be integrated into a logical framework.

In utilizing a DSS, there are several issues that need to be considered. Firstly, identifying the user's information needs may take time. The problems and their related elements need to be identified in order to support the user for making decisions. Secondly, selling the idea of the system to senior management is considered essential. This is for the purposes of committing adequate resources for setting up DSS facilities and creating a conducive organizational environment for managers to use the DSS facilities. Thirdly, the issue of standards, application and database development standards, personal computer and networking of computers should also be considered due to its process of managing data from a variety of internal and external sources. Last but not least, the effectiveness of the system's planning, design and implementation is essential to achieve good standards and integration (Raman and Phoon, 1990).

7.5 KBDSS Overview

Making strategic decisions is viewed as a difficult task that often takes long periods of time and the coordinated effort of many people throughout the organization. It is well known that as issues become more strategic, they also become more complex. Furthermore, they are often unstructured and filled with uncertain relationships.

Generally, a successful top executive will be a person endowed with many skills for making strategic decisions. Even accepting this fact it will be difficult for any individual to perform well in having many skills, where the range of necessary knowledge will become too broad. It is here that technology can be of benefit. If the organization is to be a knowledge-based organization, then it is essential that top management gain the potential provided by information technology. The computer holds the key to storing the knowledge needed by top management to improve their ability to operate in the business.

Knowledge is data endowed with relevance and purpose. It has been shown that knowledge is an essential prerequisite for strategic decision makers. The decision support system that can provide this element is a knowledge-based system (Jackson and Browne, 1992).

7.5.1 KBDSS Concept

According to Mockler (1989), a knowledge base (KB) is a collection of knowledge from experts that consists of information related to a certain topic/issue, which are collected to gain a specific decision. A KB system is a system that is designed to form a function that is similar to how an expert decides. This computer system can be used as a consultation process with the user that provides questions related to the topic discussed. This system can also provide solution recommendations for the topic. Arinze et al. (1992) were of the same views, whereby this includes eliciting knowledge from domain experts, and representing it in the form of facts and rules, frames, semantic nets, or other knowledge representations, KB system may "reason with" both stored and interactively-entered knowledge to diagnose problems, offer solutions, create plans, and monitor and control processes.

Incorporating knowledge, in the form of rules, into DSS has been recognized for some time as a means of gaining competitive advantage, formulating better problem solving processes, improving decision quality, and refining business operations. The phrase KBDSS is often used to describe the efforts to integrate DSS architectures with knowledge based system technologies. Specific approaches to incorporate rule derivation, using induction, into DSS have developed over the past years. Such approaches provide both new modeling and analysis capabilities to the decision maker (Owens and Philippakis, 1995).

Schuwer and Kusters (1993) noted that when building a knowledge based system (in this case a KBDSS), in a similar way to a conventional system, the important relations of interest in the domain must be input. The first process is to define a KBDSS as a program to compute the range of knowledge of a knowledge base. In order to determine if a problem can be adequately solved using a KBDSS, the knowledge must be analyzed into a set of data and a set of rules. Based on the properties of these sets (size, complexity, completeness, robustness and the order of use of rules), the appropriateness of the implementation mechanism can be determined.

The architecture of the main components of the KBDSS is shown in Fig. 7.2. The model contains two main components, i.e. a knowledge-base and a decision support shell. The knowledge-base was developed through initial sieving and organization of data from the database. The decision support shell provides decision support through a structured process consisting of building the hierarchy among the main criterions and the suggested controls, rating the controls, and analyzing the controls for selection through multiple analytical techniques. From the various analyses gained from its interaction with the knowledge base (that has been processed with

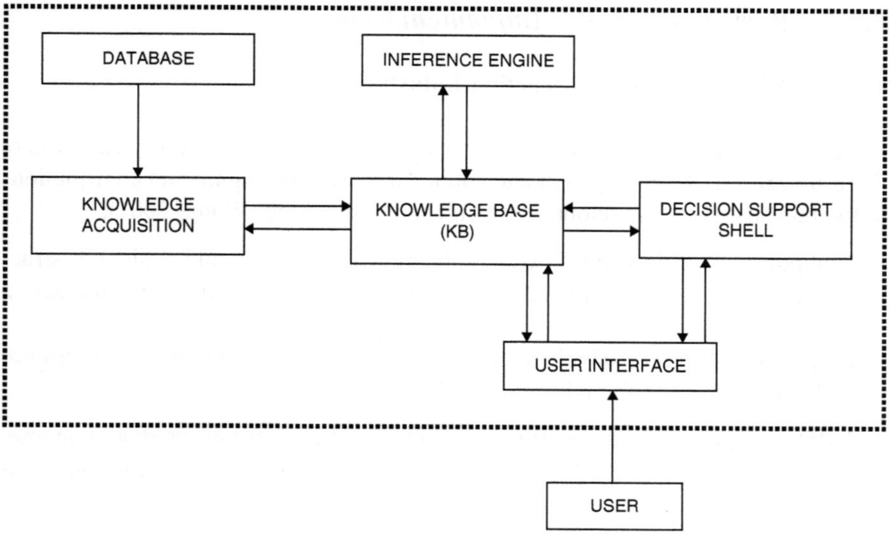

Fig. 7.2 Main components of KBDSS. Source: Adapted from Arain and Low (2006)

the inference engine), the decision support shell will provide the recommended output to the user (Arain and Low, 2006).

According to Levitt (1987), there are three categories of KBDSS, which are:

1. Operational KBDSS. A validated system that has been used regularly by the client's personnel.
2. Operational-prototype KBDSS: A prototype that has been successfully applied, but still needs validation and testing.
3. Developing KBDSS: A system that has passed the conceptual phase and its first prototype has been formed.

Owen and Philippakis (1995) stated that in utilizing a KBDSS, it is necessary to understand that significant risks may occur if the system is unable to cope or adapt to new information. Specifically, the use of decision rules that are not adapted to new information may result in poor decisions. Over time, KBDSS must adapt to changes from the environment. The lack of change processes may result in brittle systems. Moreover, it should be noted that KBDSS is not designed to make decisions for users, but rather it provides pertinent information in an efficient and easy-to-access format that allows users to make more informed decisions.

7.5.2 *Review of KBDSS Implementation*

7.5.2.1 KBDSS in the Construction Industry

Development of IT in construction has been increasing throughout the years, where there are several IT development and application that have significant contributions to the industry. Some developments that can be found are (Wang, 2001):

- Computer aided drawing (CAD) application for drawing and designing structures. CAD data are interpreted as drawings that are also needed for construction management.
- Artificial intelligence development that stores construction expert's experience and knowledge into the KB of a KBDSS.

KBDSS applications in construction can be used to interpret, predict, diagnose, monitor and plan. Several KBDSS that had been developed for construction are as follows:

- KBDSS for construction project planning, which is a material procurement planning system, where the system can recommend the proper material procurement model that had considered various factors such as the company's conditions, market conditions, and supplier's needs and conditions. The system is named PLANEX (Mockler, 1989).
- KBDSS for determining the optimum activity duration from various considerations. The system is named MASON (Hendrickson, 1989).
- KBDSS that is called CALLISTO, which is used for recommending resource allocation and configuration (Mohan, 1990).
- KBDSS for recommending project budgets, resources needs, optimum duration and project's profitability. This system is named ELSIE (Mohan, 1990).
- KBDSS for choosing construction project equipments that was developed by Amirkhanian and Baker (1992), and also Alkas and Aronian (1993).
- KBDSS for construction work safety, developed by Elzarka et al. (1995).
- KBDSS for determining risks in construction projects, developed by Mohan (1990) with KNOW-HOW system, and Risk Management Expert System named IRIS (Mohan, 1990).
- KBDSS for recommending construction project planning and execution strategy, followed by the success probability by ASHLEY (Mohan, 1990).

7.5.2.2 KBDSS in Emergency Management

Emergency situations, both man-made and natural, need coherent and effective emergency management involving complex decisions. Many conflicting objectives must be resolved and priorities must be set while the various perspectives of many stakeholder groups must be brought into some form of consensus. Multi-criteria decision analysis can help to ensure transparency during the decision making process. Moreover, the role of a KBDSS has been known for supporting the decision making process in this phase. A KBDSS for emergency response was developed to assess the state of preparedness of a transportation agency to respond to emergencies, enable the development of new SOPs, and to better train and empower employees in the decision making process (Yoon et al., 2008).

Geldermann et al. (2009) created a KBDSS named RODOS for emergency management that provides descriptive reports of the situation such as maps of the predicted, possible and, later, actual impact patterns and distributions, and detailed evaluation of the benefits and drawbacks of various countermeasure strategies and their ranking according to the societal preferences.

An integrated KBDSS aimed at centralizing the information in an operational emergency unit and monitoring and coordinating the activities of all the groups involved has also been developed. One of the examples is FLOODSS, which was developed in Italy, with the objective of implementing a DSS for inundation risk evaluation and emergency management aimed at analyzing and anticipating catastrophic flood events through various coordination processes and at preventing and mitigating their effects on the economical, social, environmental and cultural heritage (Todini, 1999).

Moreover, Yoon et al. (2008) found that a similar system but for a different context, named FALCON has been designed to assist emergency organizations with environmental management decisions. It was developed by integrating chemical inventories information, security, health readiness and population demographics into one information system that allows for assessment of response readiness, training and security. This type of KBDSS typically provides the following set of information needs; synthesized information, planning/plan testing/exercise support, resource tracking/capacity management and geographic data.

Based on these examples, access to a KBDSS provides a decision maker with the ability to (1) identify, secure and deploy the types of resources; (2) determine the current inventory of available resources, personnel and their location; (3) review historical records of similar or related events; and (4) store on-time decisions for further review and the creation of organizational memory system as a lesson-learned process (Yoon et al., 2008).

7.6 KBDSS Development

7.6.1 Formulation

Bajcsy et al. (2010) stated that there are six questions that should be answered during the development of a KBDSS system, which are:

1. What information should be preserved?
2. How to gather information?
3. How to store gathered information?

4. How to retrieve gathered information?
5. How to reconstruct decision process activities?
6. What questions can be answered?

This set of archival questions related to computer-assisted decision processes may be used as the basic guideline in formulating the system.

KBDSS has a specific characteristic in its approach for the decision making process, which is using the *if-then* rule. In this rule, a situation is obtained from an input of a user or a result from another *if-then* rule. An activity can provide a result or summary from other *if-then* rules. With this approach, KBDSS is able to handle specific problems in a structured manner. The major components in a KBDSS consist of (Mockler, 1989; Turban, 1995; Angehrn, 1991):

- Knowledge base. The knowledge representation from experts. The basic knowledge is formed from facts and information about objects and rules.
- Inference engine. A component that functions as the think-tank and inferencing pattern in the system. This mechanism analyzes problems and recommends the proper answer/solution. The capability to dynamically analyze and interpret the models described by the decision-makers and to identify suitable problem-solving methods are applied in this engine.
- User interface. It is a linkage between the KBDSS and the user. This section consists of procedures and rules in the form of dialogues between the programme and the user.
- Computer hardware. This part consists of hardware and supporting software.

To develop a KBDSS, a computer programming language is needed. The specific programming languages usually used are the object oriented programming language, such as LISP, Prolog, and C++ (Azis, 1994). In the beginning of KBDSS development, LISP (List Processing) was popular, where it is a symbolic programming language that represents knowledge in the form of a list. PROLOG (Programming in Logic) is another option that can be used, which uses the calculus approach. Finally, C++ is the latest option that has been developed into a more user friendly programming process.

7.6.2 System Development

KBDSS is developed from phases as follows (Mockler, 1989):

1. Analyzing and developing the decision situation. In this phase, scenarios for making a decision are created and grouped into decision situations. The decision making process is developed per component, therefore the analytical thinking process in the decision making can be viewed in details.
2. Formulating the decision situation. In this phase, the decision situation being analyzed is formulated into a structured situation diagram. This diagram will

further be completed with decision rules and thus will be formed into a more complete diagram called dependency diagram.

3. Synthesizing the dependency diagram into the computer programming language. This is the last phase, where the decision situation, completed with diagrams, rules and logics will be transformed into a computer programming language format for further application.

Mockler (1989) also stated that as the knowledge representation, the knowledge collected from the data analysis will be represented in a systematic form, which will consist of:

- Rules—Rules will be given for every decision situation;
- Framework—The decision framework is developed based on the knowledge provided in the system. In the framework, components such as who the users are, inputs from users, processes after receiving inputs from users, knowledge to be used in the processes, and the outputs are considered; and
- Logic representation—Logical sequence is used for every structured situation diagram and dependency diagram.

In developing a KBDSS, a large amount of time and costs are needed; therefore, the process is usually started by developing its prototype before using substantial amount of resources for its real operations. The benefits of developing a prototype are for the ease of the system's evaluation and testings, and providing sufficient time for the system's developer to further understand the KBDSS technology before creating a more complex system.

7.6.2.1 The Use of Fuzzy Logic in the Inference Engine

One of the methods that can be used in the inference engine of a KBDSS is fuzzy logic. Fuzzy logic can be treated as a tool having the ability to compute with words for modeling qualitative human thought processes in the analysis of complex systems and decisions. In fuzzy logic, qualitative perception-based reasoning is represented by *if-then* fuzzy rules (Liu and Lai, 2009; Zadeh, 1996).

Dweiri and Kablan (2006) found that fuzzy logic is a problem-solving methodology that provides a simple way to extrapolate definite conclusions from vague and imprecise information. Fuzzy set theory was first introduced by Zadeh in 1965, who was motivated after observing that human reasoning can utilize concepts and knowledge that do not have well-defined boundaries. Fuzzy set theory is a generalization of the ordinary set theory.

Technically, a fuzzy set is a set whose elements belong to the set with some degree of membership μ. In general, a fuzzy set F in a universe of discourse U is characterized by a membership function μ_F that takes values in the interval [0,1]. i.e., $\mu_F{:}U \to [0,1]$. Hence, the fuzzy set F in U can be represented as a set of ordered pairs of a generic element u and its degree of membership function as the following (Lee, 1990; Schmucker, 1984):

$$F = \{(u, \mu_F(u)), u \text{€U}\}$$

The probability that u belongs to F is the membership function μ_F (u).

Although fuzzy logic covers a wide range of theories and techniques, it is mainly based on four concepts: fuzzy sets, linguistic variables, possibility distributions (membership functions), and fuzzy if-then rules (Yen and Langari, 1999). The values of a linguistic variable are both quantitatively and qualitatively described by as fuzzy set. Possibility distributions or membership functions are constraints on the value of a linguistic variable imposed by assigning it a fuzzy set. Fuzzy *if-then* rules reflect a knowledge representation scheme for describing a functional mapping between antecedents and consequences. Fuzzy if-then rules are important for most industrial applications of fuzzy logic including many fuzzy logic control systems.

Fuzzy control was first used by Mamadani (1974). Fuzzy controllers are among the important applications of fuzzy set theory. Expert knowledge in terms of linguistic variables and fuzzy if-then rules is used to describe the system. A fuzzy rule usually has a form similar to the following:

If *a* is Low and *b* is Medium, then *c* is High. Where Low, Medium and High are linguistic values of the linguistic variables *a, b,* and *c* in the universe of discourse *U, V* and *W*, respectively.

Furthermore, a fuzzy decision making system is comprised of four main components which are a fuzzification interface, a knowledge base, decision making logic, and a defuzzification interface as described below (Lee, 1990):

- The fuzzification interface: It measures the values of the input variables on their membership functions to determine the degree of truth for each rule premise.
- The knowledge base: It comprises experts' knowledge of the application domain and the decision rules that govern the relationships between inputs and outputs. The membership functions of inputs and outputs are designed by experts based on their knowledge of the system and experience.
- The decision making logic (DML): It is similar to simulating human decision making in inferring fuzzy control actions based on the rules of interference in fuzzy logic. The evaluation of a rule is based on computing the true value of its premise part and applying it to its conclusion part. This results in assigning one fuzzy subset to each output variable of the rule. In minimum inferencing, the entire strength of the rule is considered as the minimum membership value of the input variables' membership values (Mamadani and Assilian, 1975).

$$\mu_{\text{Output}} = \min\left\{\mu_{\text{Input1}}, \mu_{\text{Input2},\ldots,} \mu_{\text{InputN}}\right\}$$

A rule is said to "fire", if the degree of truth of the premise part of the rule is not zero.

- The defuzzification interface: It converts a fuzzy control action (a fuzzy output) into a nonfuzzy control action (a crisp output). The most commonly used method in defuzzification is the center of area (COA) method. The COA method computes the crisp value as the weighted average of a fuzzy set. The result of applying the COA defuzzification method to a fuzzy conclusion "*u is F*" can be expressed according to the following formula (Yen and Langari, 1999):

$$U_0 = \frac{\Sigma i\, \mu_F(u_i) \times u_i}{\Sigma i\, \mu_F(u_i)}$$

Where u_i is the representative value of the fuzzy subset member i of the output, and $\mu_F\ (u_i)$ is the confidence in that member (membership value) and u_O is the crisp value of the output.

7.6.3 Validation

Validation is an essential phase in developing a KBDSS. There is a general agreement in the literature about the need to validate complex model-based systems in order to ascertain what a system knows, knows incorrectly, or does not know. Validation can be considered as a fundamental step for more scientific and effective computer based systems (Borenstein, 1998).

According to Finlay (1989), a DSS validation is defined as the process of testing the agreements between behavior of the DSS and that of the real world system being modeled. DSS validation is not concerned with proving that a DSS is a truthful representation of the real world—since this is impossible—but with demonstrating that the DSS has appropriate underlying relationships to permit an acceptable representation.

It is necessary to distinguish between the concepts validation and evaluation. Validation is the process of defining whether the model behavior represents the real world system in a particular problem domain. Validation has two dimensions, which are verification and substantiation (O'Leary et al., 1990). Verification is defined as the process of testing the extent to which a model has been faithful to its conception, whether or not it and its conception are valid (Miser and Quade, 1988). Substantiation is defined as the demonstration that a computer model, within its domain of applicability, possesses a satisfactory range of accuracy consistent with the intended application of the model (Balci and Sargent, 1981).

Evaluation is defined as the process of assessing a software system's overall value. It includes (1) verification, validation and quality control of the usability of the model and its readiness for use; and (2) investigations into the assumptions and limitations of the model, its appropriate uses and why it produces the results it does. The focus of evaluation is on the software and the real-world.

O'Leary et al. (1990) and Preece (1990) asserted that the use of qualitative methods is more effective during a prototype development of a system, where time and costs are more important factors than detailed data collection and analysis. A

KBDSS can be evaluated through a two-stage procedure, which are laboratory testing and field tests.

The validation process follows the cyclic strategy of prototyping methodology development. The two-stage validation occurs iteratively throughout the system development. The results from any stage (or substage) may require changes in the prototype. Also, whenever the prototype is modified or expanded the system must be re-evaluated.

The validation methodology can be further described as follows (Borenstein, 1998):

A. Laboratory testing

These testing take place in settings constructed by the development team for evaluating the KBDSS. Some of these tests may involve potential users that will contribute to the validation through questionnaires and interviews. The following laboratory tests are specified in the validation approach.

- Face validation. The main objective of a face validation is to achieve consistency between the system designer's view and the potential user's view of the problem in a timely and cost-effective way. More specifically, face validation ensures that the formulated problem contains the entire actual problem and is sufficiently well structured that a credible solution can be derived before extensive and detailed software development proceeds (O'Leary et al., 1990). The face validation acts in the approach as a feedback mechanism for prototype refinement, reformulation, and revision.
- Subsystem Verification and Validation. It consists of testing, verifying, and/or validating the DSS modules one at a time as they are developed. The main objective is to guarantee the quality of each model in the sub-model component of the KBDSS. It is much more focused on prototype's details and specifics than face validation, and therefore has as its main function the identification of areas where the prototype needs further detailed development and/or revision.
- Predictive validation. It consists of validating systems using test cases in which the results are known. A KBDSS is driven by past input data from the test cases, and its results are compared with corresponding known results.
- User assessment. It can be defined as the process by which interested parties (who were not involved in a model's origins, development, and implementation) can determine, with some level of confidence, whether or not the model's results can be used in decision-making (Gass, 1983). The main objectives of the user assessment testing are as follows:

  1. To obtain a statement of the applicability of the system by possible users.
  2. To assess the impact of the computational system's assumptions, simplifications, methods and generic structure from an independent source.

B. Field tests validation

Field tests place a KBDSS in the field, and then seek to identify those performance errors that occur. Miser and Quade (1988) stated that field experiments are the most effective of all validity tests if a situation makes it possible. Field test validation of a computer based system is a very desirable step to take before full implementation.

According to Borenstein (1998), attributes to assess during face validation and field test are as follows:

- Graphical modeling
- Integration of modules
- Presentation of results
- Manufacturing terminology
- Logical description
- Global efficiency
- Program consistency

These attributes can be assessed based on the assessor's level of satisfaction with the system. Measurement such as a range from poor to very good can be used as the scale.

Additionally, a KBDSS prototype developed by Sudarto (2007) has similar validation attributes that can be used to rate the system's overall performance. The assessment used a 5-point rating scale (from not good to very good), and the attributes that were evaluated are:

- The level of completeness of the systems' knowledge base.
- The level of completeness of the system's glossary/terms regarding its knowledge base.
- The processing speed of the system.
- The system's user-friendliness.
- User's comprehension towards the system.
- The level of accuracy from the system's results or output.
- The level of the system's applicability.
- The computer program's overall performance and layout.

7.7 Summary

As part of the essential activities in an organization, human decision making process is a complex process, with various characteristics. In business, the decisions are mostly in the form of unstructured or semi-structured decisions. These types of decisions can be assisted by using decision making tools. Nowadays, many tools has emerged for understanding, analyzing, and aiding the decision making process.

One of the tools that has played an important roles, particularly for semi-structured or unstructured decision making activities is a computer-based system known as DSS. DSS had evolved since the 1970s, with its basic functions to support managers in their decision making in activities such as planning, coordination, control, organizing, forecasting, budgeting, administration and general management.

DSS has been developed into various types of subfields. KBDSS is one of the subfields with its main characteristic as a system that supports decision making by aiding knowledge storage, retrieval, transfer and application by supporting individual and organizational memory and inter-group knowledge access. Its implementation has been found in various sectors, particularly in the construction industry and in emergency management, and it is found to be beneficial in those areas.

The development of a KBDSS is similar to other computer-based system development and mainly consists of system formulation of the data needed, system development (which is more focused on the technical aspects of rules and logic development) and system validation.

Chapter 8
Conceptual Framework

8.1 Introduction

This chapter describes the development of the conceptual framework. The framework is developed based on the literature reviews from previous chapters and the relationships between the theories and concepts involved. It starts with descriptions about the development of factors needed to understand the Indonesian contractor's knowledge on BCM. Furthermore, relationships between organizational culture, institutional forces, and BCM will be discussed, complete with the relevant attributes needed for the framework. The development of BCM guidelines for Indonesian contractors is explained in the following section and the next section discusses automating BCM through KBDSS. The final section shows the conceptual framework that compiles the findings from previous sections. The framework is the basis for the study in developing its research design and methodology.

8.2 Indonesian Contractor's Knowledge About BCM

Based on the literatures, there are findings that showed that Indonesian contractors have not developed their crisis responses into a holistic management approach in their organizations, and there is a lack of detailed responses for their business stakeholders. Even though emergency responses have been developed in their firms and they have experiences in managing such situations, it is not viewed as sufficient enough. There are still patches of responses that have not yet been integrated for their effectiveness in overcoming crises. Moreover, in some cases, responses to recovery and restoration have not been planned in advance. The firms will create the recovery team and develop steps to resume based on the management's decision. Based on this, it is viewed that there should be some efforts that could be mounted in order to minimize the impacts from the crises. Adopting BCM

L.S.R. Supriadi, L. Sui Pheng, *Business Continuity Management in Construction*, Management in the Built Environment, DOI 10.1007/978-981-10-5487-7_8

within their organizations appeared to be necessary for these firms (Agustinus and Luhur 2008; Firdausy 2002; Herlijanto 2004; ICG 2002; Kartasasmita 2000; Lee 2009; MiyamotoINTL 2007; PTX 2008; Tambunan 2006; UNR/HC 2005).

In addition, the BCM concept is found to be relatively new for construction firms, particularly in Asia (e.g. Singapore, Hong Kong and China). Many large construction firms in these countries have yet implemented BCM due to their lack of awareness (Low et al.. 2008). The receptiveness of BCM among the construction firms is far from ideal, even though its importance is clear. Nonetheless, a majority of the firms in China and Singapore had interests in implementing BCM (Low et al. 2008b, 2010a). Therefore, as the first step before implementation, it is important to determine whether Indonesian contractors have understood and are aware of the BCM concept.

Table 8.1 shows the constructs that were used for identifying the Indonesian contractor's knowledge about BCM. First, it is important to understand how the firms currently respond to crises. This can be viewed from their knowledge on the types of crises; standard operating procedures on handling crises; their views on crisis management, emergency plan and BC plan; and their coordination with the government and stakeholders during a crisis. The second construct identifies the firm's knowledge about BCM, its definition, and whether they have implemented BCM and have been certified for it. Last but not least, their knowledge about BCM principles would be identified and whether these principles have been implemented in their organizations.

8.3 Relationships Between BCM, Organizational Culture and Institutional Forces

8.3.1 Defining Organizational Culture Dimensions

Various scholars defined organizational culture as the way things are done and operated within the internal environment of the workplace. The features in organizational culture are common beliefs; pattern of behavior, norms; values and rules that are exercised among members of the organization. They viewed that the closer the values and beliefs among members of the organization are, the stronger the culture will be (Kotler and Heskett 1992; Scholz 1987; Williams et al. 1993).

Based on the various dimensions that were established by scholars, there are five organizational culture studies that have shown essential values within their dimensions. The five studies are as follows:

1. Hofstede six dimension of organizational culture (Hofstede and Hofstede 2005)
   Hofstede provided six dimensions of organizational culture. The difference between Hofstede's national culture dimensions is based on their different mix of values and practices. Regarding studies for organization level, although the seven national cultural dimensions have implications for organizations and

Table 8.1 BCM knowledge variables

Constructs	Variables	Sub-variables
(1) The contractor's current response towards crises	Types of crises acknowledged in the firm	Crises related to the contractor's business activities
	SOP (Standard Operating Procedure) on crises	Does the firm have SOP for every identified crisis?
		The contents of SOP: Evacuation—Communication—Recovery—Restoration
	Crisis management (CM plan)	Does the firm have CM Plan in place for every identified crisis?
		The contents of CM Plan: Signal detection—preparation and prevention—damage containment—recovery—learning
	Emergency plan	Does the firm have an Emergency Plan in place for every identified crisis?
		The contents of emergency plan: evacuation procedure—communication
	BCM (BC Plan)	Does the firm have BC Plan in place for every identified crisis?
		The contents of BC plan: Emergency response plan—Setting up Emergency Operation Center—Recovery and resumption of Critical Business Function within their Recovery Time Objective and Recovery Point Objective
	Coordination with government/stakeholders	Communication with government / local authority during crisis
		Communication with stakeholders during crisis (owner—client—community—supplier)
(2) BCM knowledge (general)	Do they know about BCM?	The definition of BCM
		Source of knowledge of BCM: Regulator; Other firms (competitor); Higher level management; Media; Seminar/workshop; others
	Do they implement BCM?	
	Are they certified to BCM?	Certification by: Government/regulator; BCM organization

(continued)

Table 8.1 (continued)

Constructs	Variables	Sub-variables
		(local/regional/international); others
(3) The knowledge about BCM principles: Are these being implemented in the organization? Note: an organization may not know the formal concept of BCM, but it may have implemented part of BCM principles in its organization.	Risk analysis on threats in organization (risk identification—risk analysis—risk treatment)	
	BIA for responding to threats	
	Determining strategies for maintaining the operation of CBFs (especially during threats)	
	Developing plans guiding the response and recovery actions when disaster occurs	
	Conducting tests and exercises for verifying the developed plans	
	Programme management on the developed plans	

Source: Adapted from Spring Singapore (2008), BCIGPG (2010)

management processes, it is recommended to use the six organizational culture dimensions. Hofstede and Hofstede (2005) further explained that national cultures are part of the mental software that are acquired during the first 10 years of people's lives, in the family, in the living environment, and at school, and they contain most of the human basic values. Organizational (or corporate) cultures are acquired when entering a work organization, with basic values firmly in place, and they consist mainly of the organization's practices.

2. Denison organizational culture dimensions (Denison 2000)

   Denison had developed the Denison organizational culture model with four traits that have a strong influence on organizational performance. One of the traits that is important and is relevant to BCM is adaptability. It relates to how the organization views change, the customer, and its innovation and learning capabilities.

3. Cheung et al. organizational culture dimensions (Cheung et al. 2011)

   This study had developed organizational culture factors that are found to be significant in the construction sector. Cheung et al. (2011) had compiled these factors from various dimensions from various scholars, which mostly come from Bettinger (1989), Denison (2000) and Cameron and Quinn (1999).

4. Non-technical attributes from a Quality Assurance System (Low 1998)

   Low (1998) found that the non-technical or human behavioural attributes encountered in the process of implementing and maintaining a quality system to deliver total service quality is essential. These attributes pave the way for successful documentation, implementation and improvement that have an adverse influence on a quality assurance system. Furthermore, the purpose of the non-technical approach is to help promote an integrative environment for the development of change and innovation, with the primary objective of quality improvement. Therefore, based on these considerations, the attributes in this study can be useful for BCM implementation as a management system.

5. Organizational behavior attributes (Osland et al. 2001; Luthans 2008)

   These studies have focused on the sources of an organizational culture, which are the values of the founders or strong leaders of the organization and the solutions to problems that other members have learned over time. Other than that, industry, environment and national culture also influence the culture of an organization. The studies also concluded that in leadership, the willingness of organizational members to accept leadership and direction from qualified others is important. As needs for leadership arise, members should feel free to take on leadership roles and are rewarded for successful leadership. Leadership is recommended to be based on expertise, where the organization is not dominated by, or dependent on, one or two individuals.

Hence, based on the five studies, 14 dimensions have been identified and used as the organizational culture (OC) attributes in this study. The 14 dimensions represent the values from the five studies and are described in Table 8.2.

From the reference column in Table 8.2, it can be seen that each dimension may have be mentioned in all of the five studies, or it may only be mentioned in one or two studies.

8.3.2 Relationships Between OC Dimensions and BCM Principles

Several literatures had shown that OC dimensions are required in BCM. These attributes support the BCM implementation and can be applied to the six main BCM principles:

- Empowerment

The degree of empowerment through to lower levels in an organization is increasingly seen as a critical facet of an organization's culture. It has been identified that empowerment is a key part in business continuity planning and response. Furthermore, BC planning and response can be effective only if developed cooperatively, in which a wide range of individuals at various levels in the organization is involved (Drennan and McConnell 2007; BCI 2010; McManus et al. 2008). As an example, Sheffi (2006b) found that empowerment has been an important approach for the successful response by the US Coast Guard during the Hurricane Katrina disaster which saved over 24,000 lives. In business continuity

Table 8.2 Organizational Culture (OC) dimensions

No	Organizational culture dimensions	References
1	Empowerment: Empowering individual to manage their own work; Decision making by individual; Value employee's ideas; Employee's input in major decisions; Employee participation in decision making (members' participation); Confidence in employee members	Denison (2000) Osland et al. (2001) Luthans (2008) Cheung et al. (2011) Low (1998) Hofstede and Hofstede (2005)
2	Team orientation: Cooperation among employees; Team contributions; Amicable opinions; Commitment to team	Denison (2000) Low (1998) Hofstede and Hofstede (2005) Cheung et al. (2011)
3	Developing employee's skills: Performance improvement; High expectations of performance	Denison (2000) Cheung et al. (2011) Low (1998) Osland et al. (2001) Luthans (2008)
4	A set of values: Clear goals-direction-approach-strategic intentions; Action = Goals; Shared visions	Denison (2000) Cheung et al. (2011) Low (1998) Osland et al. (2001) Luthans (2008)
5	Coordination and integration: Resolve problems effectively; Inter-department collaboration; Information sharing; Agreement on critical issues; Different functions work together; Decisions made by groups/individuals; Trust atmosphere; Managing conflict; Good communication	Cheung et al. (2011) Denison (2000) Hofstede and Hofstede (2005) Low (1998)
6	Employee/job oriented: Consider employee's welfare; Level of job pressures; Friendly workplace (warmth and support)	Hofstede and Hofstede (2005) Low (1998) Osland et al. (2001) Luthans (2008)
7	Adaptability to change: Creating change; Developing innovation-knowledge-capabilities; Innovation orientation; Employee's resistance to change	Denison (2000) Cheung et al. (2011) Low (1998)
8	Setting standards and good performance: Organizational structuring-meeting times-cost conscious (tight or loose control organization); Emphasize good performance; Set of performance standards	Osland et al. (2001) Luthans (2008) Hofstede and Hofstede (2005) Cheung et al. (2011)
9	Process/results oriented: Conformity to procedures and rules	Hofstede and Hofstede (2005) Osland et al. (2001) Luthans (2008)
10	Customer orientation: Focus on customer	Hofstede and Hofstede (2005) Denison (2000)

(continued)

Table 8.2 (continued)

No	Organizational culture dimensions	References
11	Reward orientation: Team and members accountability; Emphasize on reward; Performance-based rewards; Equitable reward	Cheung et al. (2011) Osland et al. (2001) Luthans (2008)
12	Power in organizations: Sources of power; expert (knowledge-based); leadership	Low (1998) Osland et al. (2001) Luthans (2008)
13	Parochial / professional; Cover either social/family background plus job competence or job competence only	Hofstede and Hofstede (2005)
14	Open / closed system: Organization open to newcomers or not; How people fit into the organization	Hofstede and Hofstede (2005)

planning, Light (2008) observed that empowering the employees can be in the form of:

1. Sharing authority throughout the organization by empowering subordinates while still maintaining a strong chain of command.
2. Creating flexible and efficient decision making processes.
3. Designating responsibility for crisis-readiness to one individual or a specific team.

- Team orientation

As mentioned earlier, planning for BCM can be effective when it is developed cooperatively. Cooperation among employees is essential and this means that BCM principles support team orientation in its process. In defining the scope of the BCM processes, dependencies between individuals and parties should be clearly defined. Therefore, every individual is aware of the dependencies involved, whether it is external or internal (such as key suppliers, personnel, operating system, etc) to successfully mitigate the specified crisis. Commitment and cooperation within a team can be fostered by providing bonding activities in the organization. These activities can be a platform to encourage the employees in getting to know and understand each other's beliefs, cultures and practices (Drennan and McConnell 2007; Low et al. 2008a; Ministry of Manpower 2010; BCI 2010; Light 2008).

- Developing employee's skills

BCM principles also focus on building employee's capabilities to handle crisis situations. A BC plan or programme will only be effective if the business can continue and recover its critical business functions and processes. The staff and the recovery team members should understand their roles and responsibility when disaster strikes. The need for training and awareness programme are necessary in developing the staff's skills and hence improving their performance (Ministry of Manpower 2010; BCMI 2011; BCI 2010; Light 2008). Along with providing training for staff, Elliott et al. (1999) and Smith (2003) also suggested that each staff in the recovery team should be determined based on the relevant skills needed to support the organizational recovery strategy. The skills can be benchmarked against the competencies provided by standards of business continuity.

- A set of values

According to Smith (2003), BCM activities must match, focus upon and directly support the organization's business strategy and goals. Moreover, BCM should focus on the business priorities of the organization by relating the campaign message to corporate and individual factors (BCI 2010). There must be policies to guide BCM efforts by the staff. This strategic intention or set of values should set out the organization's aims, principles and approach specifying what is to be achieved or delivered, and will serve as the rationale and support for all BCM areas. Also, policies provide the rationale for establishing the processes, people, and infrastructure (as BCM components) to support BCM on an ongoing basis (Spring Singapore 2008; Denison 2000; Cheung et al. 2011). As part of the strategic efforts in implementing BCM, the BC plan can be incorporated into the organization's strategic management process and its overall business planning process (Light 2008).

- Coordination and integration

Upon the occurrence of a crisis, many parties could be affected. Not only the employees of the company, but other stakeholders such as investors, suppliers and clients may also be affected. Coordination with the stakeholders is important during the planning and execution phase of BC. Moreover, an integrated approach within the business functions should be conducted in order to execute a coordinated response to a crisis situation. Lack of integration will result in a segmented response that leads to gaps in continuity or quality of response (Elliott et al.. 2002; McCrackan 2005). Developing and publicizing the BCM policy throughout the organization can be reinforced with suitable communications. Everyone involved in the BCM should be informed and consulted. Consultation in itself helps raise awareness and may help prepare the way for commitment to new working practices in BCM. Managing relationships with important stakeholders inside and outside of the organization should be supported with fast and efficient communication flows that also have clear chains of command (BCI 2010; Light 2008; Robinson et al. 2009; Low et al. 2008a; Singapore Business Federation 2003; Elliott et al. 1999). Pauchant et al. (1991) also provided several communicational efforts that can be adopted in developing BCM, which are:

1. Media training for BCM team (training to handle the media).
2. Major efforts in public relations.
3. Increased information to local communities.
4. Increased relationships with intervening groups (police, media, etc).
5. Increased collaboration or lobbying among stakeholders.
6. Use of new communication technologies.

During crises, employees will most likely want to be kept informed of the latest updates. Thus, it is necessary that the organization has proper plans and procedures to disseminate updates and information. Good communication may also help reduce any fears and anxieties which occurred during that critical time. Otherwise, fears and anxieties, which are likely to exacerbate rumours, would affect harmonious working relations and morale of the employees. As a consequence, these would adversely affect the productivity and operations of the business (Ministry of Manpower 2010).

- Employee or job oriented

According to Drennan and McConnell (2007), business continuity must protect business assets, which are staff, equipment, facilities, IT systems, reputation, market-share, liquidity, etc. Relating to that, people are the principal asset of any business, because without them, the business does not function (McCrackan 2005; Low et al. 2008a; Singapore Business Federation 2003). Therefore, effective BCM needs the people element to be involved. The BCP should consider that recovery planning is not just about a technical solution, but about people. Furthermore, understanding the people/staff's issues and needs during the relocation plan development is essential. By adopting the people/employee focused approach to BCM, the confidence of the employee, customers, shareholders and regulators can be maintained. Thus, the organization's reputation can be protected (BCI 2008).

- Adaptability to change

The implication of BCM must be considered as an essential part of the business change management process. Its best practice is about adding value and creating an attitudinal change throughout the organization and considering its associated stakeholders (Smith 2003; Elliott et al. 1999). Drennan and McConnell (2007) opined that in building resilience to business interruptions, there is a need to manage the risks involved. A risk management process is conducted in BCM in order to assess, analyze, and treat risks (Hiles 2007). Light (2008) added more approaches that can support in addressing and managing the firm's key risks, which are:

1. Monitoring trends in the external environment.
2. Engaging in scenario planning.
3. Encouraging innovative and creative solution development.

Moreover, McManus et al. (2008) noted that the concept of adaptive capacity is at the core of current organizational resilience methodology. Adaptive capacity is defined as the ability of an enterprise to alter its "strategy, operations, management systems, governance structure and decision-support capabilities" to withstand perturbations and disruptions (Starr et al. 2004). Organizations that focus on their resilience in the face of disruption generally adopt adaptive qualities and proactive responses. They emphasize positive behavior within the enterprise and within employees, and look at disruptions as being opportunities for advancement (Mallak 1998; Folke et al. 2002).

Nonetheless, facing employee's resistance to change in this type of process is not easy. However good and creative the BC plans and approach may be, it is almost inevitable that opposition towards the programme, claims of information overload or plain ignorance may occur. Not everyone in an organization will want to read the BC plan, understand or take any actions regarding BCM. There will be resistance in implementing BCM unless there are significant reasons why it should be implemented (Price Waterhouse Coopers 2009).

- Setting standards and good performance

BCM is a management process that must optimize cost efficiencies. Financial impact has been one of the major impacts that need to be considered in BCM, because it has long term effects toward the organization. Therefore, standards for BC procedures and processes should be set in order to develop a pragmatic, cost effective, and operable recovery plan, hence to enable the firm to run its critical business processes during disruptions. As a support, implementing and maintaining a robust exercise, rehearsal and testing programme could ensure that the business continuity capability is effective, up-to-date and fit-for-purpose (Smith 2003; Hiles 2007; Health 1999; O'Hehir 1999). To be cost conscious, there are factors that need to be considered during BC plan development (Low et al. 2008a; Singapore Business Federation 2003; Pauchant et al. 1991):

1. Managers
   The managers who control the situation should provide cost effective solutions in handling the impact of the crisis and the effects on business resumption and transparency of operations.
2. Accountability
   Individual accountability for managing the risk and ensuring that the nominated person has the appropriate technical expertise and authority to manage the risk.
3. Being realistic
   The management should be ready to accept certain risks and should be prepared to spend the necessary funds to mitigate the risks involved.
4. Performance measures
   Establishing measurement indicators to enable assessment and monitor the effectiveness of the process.
5. Creating budget for BCM
   The BCM programme should be planned with a dedicated budget.

- Process or results oriented

Part of the essential BCM components is processes. It is the set of activities with defined outcomes, deliverables and evaluation criteria to attain the objectives of the BCM policies. They also include formal change control and documentation processes. Processes are available in all of the six BCM principles. Therefore, conformity to procedures and rules is considered as necessary in BCM implementation (Spring Singapore 2008; BCI 2010). Health (1999) and O'Hehir (1999) added that along with developing detailed procedures, a BC plan should be using a well structured and comprehensive methodology to assure the quality and accuracy of mandatory processes and documentation.

- Customer orientation

According to EIU (2007), customers are the stakeholders that are viewed as most important in driving decisions about business continuity, with 59% citing them as a significant influence. McCrackan (2005) also supported this statement describing that customer service is one of the main factors to take into consideration when assessing impact of a disastrous event. Moreover, Hiles (2007) opined that the focus of continuity in customer service should be on defining what level of service must be maintained throughout a disaster, and what is required to achieve that level of service. In defining time lines for the resumption of support and services, and transparency of operations in a crisis, the impact on customers should also be considered (Singapore Business Federation 2003). By demonstrating the importance of customers in BCM implementation, it can be used as a competitive advantage to gain new customers and to improve margins by using it as a "customer care" approach (BCI 2007a).

- Reward orientation

BCI (2002) stated that performance-based rewards can ensure the active involvement of managers and staff at all levels of the organization, especially the operational middle management who has to implement and maintain BCM. Performance management and rewards are one of the mechanisms that can exert influence upon what is seen as important and how it is done. When performance and its measurement are aligned to rewards and recognition, it provides a strong incentive. Creating performance management and reward systems in BCM should be consistent to improve the accountability of the individuals and teams involved in the process (Elliott et al. 1999).

- Power in organization

The quality of leadership and top management's commitment are essential in BCM. The support from top management is important as BCM should be driven from a business perspective. The development of BCM will be implemented successfully and continue to be successful as a programme when there is central control and coordination (McManus et al. 2008; BCI 2010; BCMI 2011). According to EIU (2007), an ineffective BCP is often the fault of poor leadership. The factor that distinguishes organizations that do well from those that do not is leadership. The leader needs to show an interest in the process, ask questions, and then put in place the appropriate governance and controls to make sure it happens. Moreover, the leader should have a shared vision of mission and encourages the BCM team (Light 2008). Therefore, the fundamental activity required prior to the establishment of a BC plan is to obtain senior/top management approval, support and commitment (Elliott et al. 1999, 2002; Pitt and Goyal 2004).

- Parochial or professional

In appointing staffs for BCM, participation from various business units in the firm should be established to oversee BCM efforts. The skill sets and competence of participants are essential to the success of BCM. Moreover, the roles and responsibilities of staff involved should be clearly defined (Spring Singapore 2008). An independent expert or professional can be appointed as a supporting member in BCM team, provided that the expert has complied with certain competencies needed in developing BCM. The expert will have a role in determining the adequacy of the response to the crisis through regular meeting, and reporting to higher management to signify the importance of BCM (Low et al. 2008a).

- Open or closed system

BCI (2008) opined that BCM should be using a people-focused approach. How the process of appointing staff with necessary skills for BCM development that involves HR department, considering the impacts toward the staff during disruption or relocation process and managing how the people will fit into the BC process are needed in order to have an effective BCM. Prioritizing the people element in BCM will also help in gaining satisfaction from the staffs and shareholders. According to Hofstede and Hofstede (2005), in the open system units, employees considered the organization is open to newcomers and the process of how the employees would fit into the organization and its division is transparent. This approach would support BCM implementation, particularly in managing the people who will be in charge of the BC process.

8.3.3 Defining Institutional Forces Attributes

According to Scott (2008) institutions, whether regulative, normative, or cultural-cognitive pillars are stressed, are conveyed by various types of vehicles or carriers. The carriers for each pillar are described as follows:

- Regulative pillar: Consider the principle as a rule that has legal sanctions, carried through coercive manner, comply by expedience and stem from regulatory forces.
- Normative pillar: Consider the principle as a value and expectations (that is morally governed), that provide certification/accreditation status, carried through normative manner, comply by social obligation and stem from validating forces.
- Cultural-cognitive pillar: Consider the principle as a typification/schema (that is culturally supported-recognizable), that provide common beliefs, comply by taken for granted/shared understanding and stem from habitualizing forces.

Table 8.3 shows the constructs of the institutional forces for identifying which forces drive or hinder BCM implementation for Indonesian contractors. The table adopts the institutional compliance framework from Low et al. (2010b)'s study, which were used to understand why construction firms in China, Hong Kong and Singapore do or do not wish to implement BCM.

Table 8.3 Institutional Forces (IF) attributes

Constructs	Variables	Sub-variables
Regulative forces	Rules, laws and sanctions	Probability of violation detection and being sanctioned
		Non-compliance cost is not small
	Gains, losses and consequences	Improve company procedures
		Easy to integrate with other management systems
Normative forces	Personal morality	Improve worker health, safety and welfare
	Social influence	Insisted upon by stakeholders and parent company
		Concern for social reputation
		Increase competitiveness of company
		Company and peer groups are compliant
	Legitimacy	Procedures are fair
		Appropriateness and effectiveness of the law
Cultural-cognitive forces	Shared understanding of compliance	Law compliance as taken for granted activities and belief in abiding the law
		Awareness of potential risks
		BCM is already part of company culture

Sources: Adopted from Scott (2008), Low et al. (2010b)

8.3.4 Relationships Between IF Attributes and BCM Principles

Explanations of how institutional pillars may force the organization to adopt BCM are described as follows:

- Regulative forces

According to BSI Groups (2010), BCM was developed and implemented by some construction firms in the UK due to the need to meet legislative and corporate governance requirements. Regulations requiring construction firms to offer evidence of a BC plan in the tender documents have been found in many tendering processes. EIU (2007) also reported that pressure from regulators in developing BCM have become more pronounced, not only in the construction sector, but also in other sectors mostly in the financial services. In ranks, the influence that the following external organizations have over the firm's decision about BCP are: Regulators; Customers; Shareholders; Governments; Auditors; Insurance Companies; Risk Consultants; The Media; BC/Security vendors; and Emergency Services.

- Normative forces

In relation to its relevance towards normative forces, customers are the stakeholders that is viewed as important in driving decisions about business continuity, with 59% citing them as a significant influence, derived from a survey of companies in Europe by EIU (2007). Social influence has been part of the drivers for initiating

BCM such as concerns over the firm's reputation, the need to increase the competitiveness of the firm, and compliance to peer groups in the same sector have supported the decision to adopt BCM. These forces push the firm to obtain BCM certification from the relevant association (BCI 2010).

- Cultural-cognitive forces

The implementation of BCM by a firm can also be pushed by cultural-cognitive forces. In this context, the firm views BCM implementation as a logical step where its principles are already part of the firm's culture. The firm has high awareness of the risks involved in the organization and shared an understanding of BCM's benefits for the organization (Low et al. 2010a; BSI Groups 2010; Scott 2008).

8.3.5 The Importance of Organizational Culture and Institutional Forces in Adopting BCM

A. The importance of Organizational Culture in adopting a concept
Hofstede et al. (2010) stated that modern management theory identifies cultural differences as having a critical impact on management theories, and managers need to adapt their tools and practices to suit the environment in which they are operating. This is supported by Peter Drucker's opinion where he quoted "What managers do is the same around the world. How they do it is determined by tradition and culture."(Brake et al. 2002, p. 247). Moreover, several scholars found that there is a US centric focus to management literature, thus the majority of management research and methodologies need to be evaluated before they are applied in a cultural environment outside North America (Trompenaars 1993; Chisholm 2010).

Regarding culture, the importance of organizational culture towards adopting a concept or system in the organization can be seen from studies which increasingly push organizational culture as the guide for organizational strategies. It is suggested that organizational culture can impact manager's ability to process information, rationalize and exercise discretion in their decision-making processes (Hofstede and Hofstede 2005). Liu et al. (2010) found that identifying organizational culture helps its members to understand organizational functioning. It affects how the firm responds to external events and makes strategic choices. Moreover, organizational culture is a key to many change initiatives, where success in implementing a concept depends on the organizational culture (Leidner and Kayworth 2006). In line with this, Ogbonna and Harris (2000) also found that when a change is initiated due to external conditions, competitive and innovative cultures in an organization are considered as important factors. An adaptable organizational culture to external contingencies may provide a source of sustainable competitive advantage.

B. The importance of Institutional Forces toward organizational behaviors
Various studies had found that an organization is supported and constrained by institutional forces. Also, it incorporates many institutionalized features in the form

of symbolic systems, relational systems, routines and artifacts within their own boundaries. The main concept of an organization as a special-purpose, instrumental entity is a product of institutional process, where it is also influenced by the environment (Scott 2008).

Moreover, Scott (2008) and DiMaggio and Powell (1983) described that there are various ways which institutions are diffused. The three contrasting mechanisms are through coercive, normative, and mimetic ways that form varying forces or motives for adopting new structures and behaviors in organizations.

The use of coercion relates with the regulative process, where it requires clear demands, effective surveillance and significant sanctions. In addition, it also matters whether the mechanisms employed are primarily those of power, involving imposition of authority (Scott 1987; Scott 2008). For example, Cole (1989) found that in the adoption and retention of innovative small-group activities, such as quality circles, there are differences among firms in Japan, Sweden and the US. Cole's (1989) studies emphasized the role played by national government agencies, trade associations and union organizations, in legitimating, informing and supporting the innovations. Regarding the process of spreading and maintaining the innovation process, Japan is more stable than Sweden and the US, respectively. This shows that these three countries varied in the relative strength of regulatory authority.

Normative processes focus on the importance of network ties and commitments. Normative standards may be established by self-appointed arbiters employing representative bodies and deliberative procedures. As an example, although accreditation is not legally mandated in areas such as health care, organizations lacking accreditation may not be eligible for reimbursement from certain funding sources. Therefore, this suggests that organizations that are accredited by appropriate professional bodies are more likely to survive than those lacking such normative support (Scott 2008).

According to Strang and Meyer (1993), mimetic ways focus on the centrality of cultural-cognitive elements in institutional diffusion. For diffusion to occur, the actors involved need to regard themselves as similar in some important respect. An example of this type of diffusion is described by Cole (1999), regarding the efforts of American firms in exploring new concept in their firms. In the mid 1970s, fierce competition had been proclaimed by Japanese automobile and electronic manufacturers. This led to the US firms in approaching a range of practices that came to be labeled as "total quality management" (TQM). The US firms were found to be slow to respond, unsure of the nature of the challenge they faced or what to do about it. Even though various experts and consulting companies have given advice and professional associations have offered normative support, little consensus had been developed. The movement was not supported by adequate normative and regulative forces to diffuse widely in the country. Moreover, the firms felt the need to change, but the directions did not provide clear guidelines. The most important change associated with TQM adoption was in the cognitive framing of quality, shifting attention from the concerns of internal engineers to external customers, and from a "detect-and-repair" to a "prevent-and-improve" mentality.

Liu et al. (2010) also described that scholars have increasingly referred to the institutional theory as an important perspective for studies on organizational innovation or concept adoption. They argued that institutional pressures emanating from the environment and transmitted through operational channels can strongly affect firm's predisposition towards concept adoption.

C. Determining drivers and hindrances of implementing BCM

Based on the previous discussion, it appears that institutional forces and organizational culture may work together and interact with each other to affect concept or system adoption. Firms may react differently to the same levels of perceived institutional forces to adopt BCM due to the differences in their organizational cultures. Complementing these, studies also had suggested that the immediate motivation for concept or system adoption comes from institutional forces. Furthermore, organizational culture, as a stable element of the organization, moderates the effects of institutional forces. These relationships may also provide an understanding of the different effects of organizational culture between different dimensions of institutional forces and the intent of organizational concept adoption (Liu, et al.. 2010).

Based on this understanding, IF and OC attributes can be synthesized with BCM principles in developing a framework for determining the drivers and hindrances of

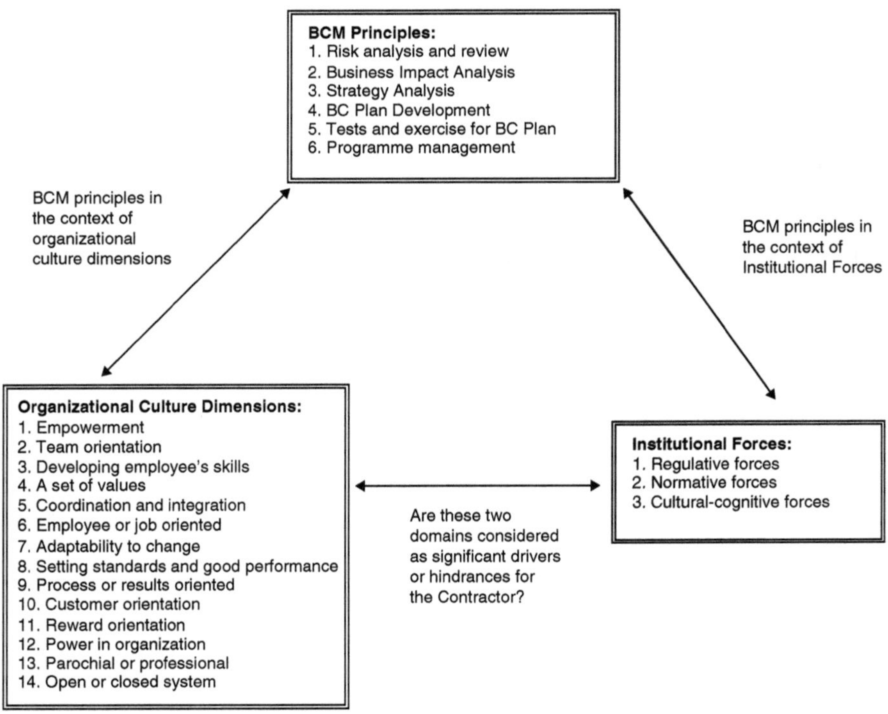

Fig. 8.1 Framework of IF–BCM–OC relationships

implementing BCM by Indonesian contractors. Figure 8.1 illustrates the framework for this purpose.

The analysis from the framework may also explain in more details about the critical success factors (CSFs) in BCM implementation. These can be derived from the significant drivers from the two domains. These factors can be a good feedback for the Indonesian contractors in developing strategies for initiating BCM. Based on the framework, it can be analyzed whether BCM implementation (and all its principles) is supported by the contractor's organization culture and institutional forces. If it is supported, then the process of initiating BCM in the firm and developing the BC plan can be conducted. On the other hand, if the contractor's organization culture and institutional forces do not support BCM implementation, then finding strategies to embed BCM principles in the firm through highlighting the CSFs is recommended. This means that there may be changes or adjustments needed in the firm for initiating BCM.

Furthermore, the detailed variables for determining these relationships can be seen in Figs. 8.2 and 8.3. Figure 8.2 shows the variables used in BCM–IF

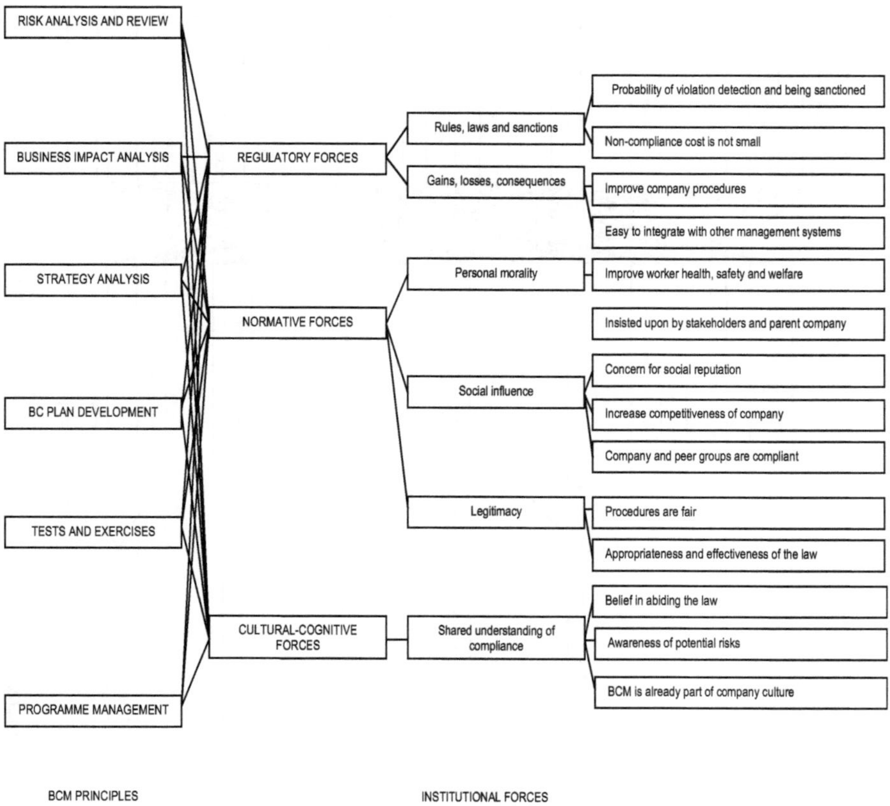

Fig. 8.2 BCM–IF variables

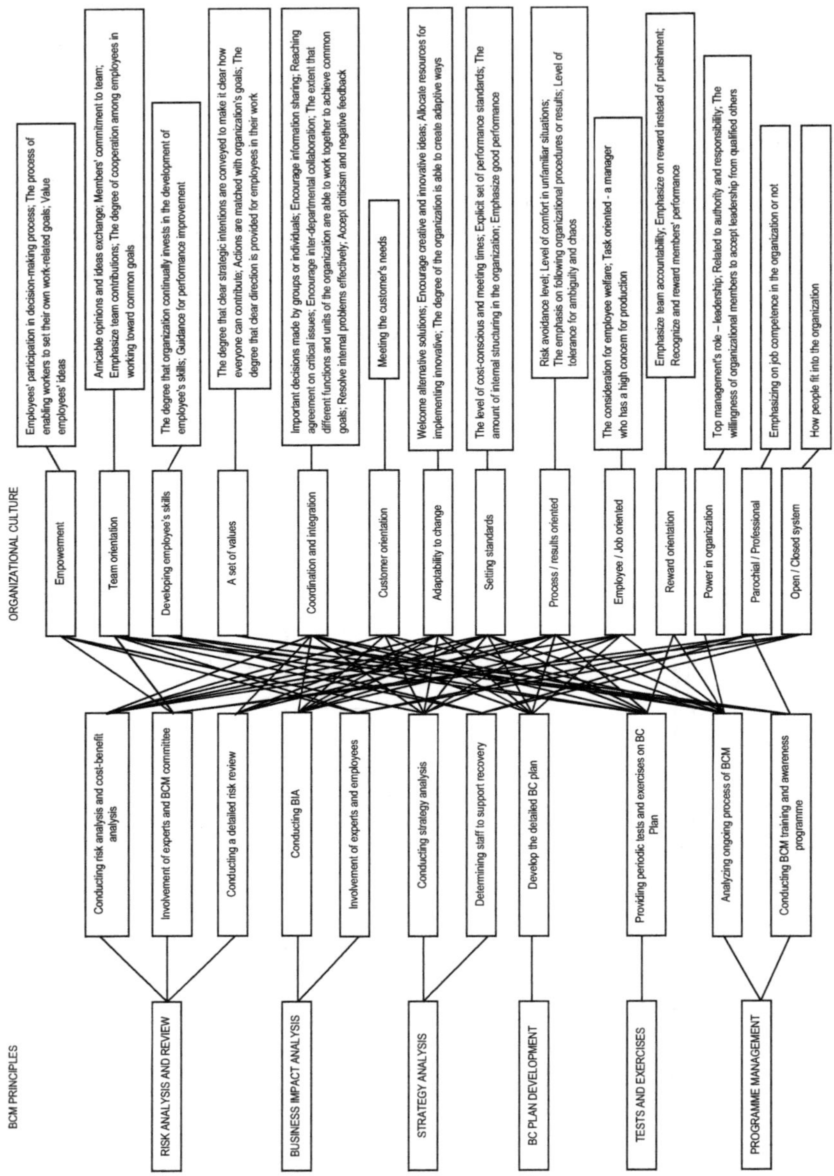

Fig. 8.3 BCM–OC variables

relationships and Fig. 8.3 illustrates the variables in BCM–OC relationships. The framework is supported by the variables compiled from IF, OC and BCM attributes.

8.4 Developing BCM for Indonesian Contractors

Various BCM standards have been developed by several countries. The interrelationships between these standards and the BCM planning process have shown some similarities and differences. Their similarities are the seven main methodologies for BCM planning that are all included in the standards, which are:

- Project management
- Risk analysis and review
- Business impact analysis
- Recovery strategy
- Plan development
- Testing and exercising
- Programme management

The differences are on the way each standard compiles or expands the methodologies into various details with different terms but in a way provides the same approach and meanings.

In Singapore, the SS540:2008 standard has been formally used as the standard for implementing BCM in a firm. This Singapore Standard is applicable to all organisations regardless of their size. This standard emphasizes resilience and protection of critical assets, in the human, environmental, intangible and physical domains. It focuses on continuity management and recovery of critical business functions (Spring Singapore 2008). Up to now, Singapore is the only country in Asia that has established a BCM standard, whereas other BCM standards came from Europe, North America and Australia (Elliott 2010). Therefore, due to the same geographical location and regional similarities in socio-culture-economic background, SS540:2008 can be a good benchmark for Asian countries in adopting BCM, particularly for the construction industry. Along with its higher economic growth, the construction industry in Singapore is one of the most well regulated when compared to other countries in terms of quality and skill enhancement, research and development. The introduction of benchmarking and buildability scores is seen only in Singapore while other Asian countries have been far slower to move in this direction. Harmonized standards will be the key to a harmonized construction market in the region (Raftery 2004). Due to this reason, the BCM guidelines for this study will refer mostly to the SS540:2008 standard. Nonetheless, some significant principles in the guidelines are also compiled from other main standards such as BS25999:2006, NFPA1600:2007, ANZ5050:2009, and ISO22301:2012.

In developing the BCM implementation guidelines for Indonesian contractors, the main steps that will be involved are based on BCM's main principles, which are

Table 8.4 BCM steps for developing implementation guidelines

Constructs	Variables (for construction business variables)	Sub-variables
Risk analysis	Risk analysis on threats in organization (risk identification—risk analysis—risk treatment)	Threats identification in the firm; Business units (BU) in organization (BU—BF—CBF); Risk identification and treatment for each BU; Disaster identification (for key disaster scenario)—Disaster may be compiled from one or more identified risks; Risk review (per BU)
BIA	BIA for responding to threats	Business function; MBCO (minimum business continuity objective); Establish priority for analyzing impact (per disaster); Establish CBF (for each BU); Dependencies of each CBF (each BF can span across one or more business operations); CBF requirements (this construct relates to previous CBF dependencies construct—as above); Resource requirements and capabilities (inventory for each BU)
Strategy	Determining strategies for maintaining the operations of CBFs (especially during threats)	Recovery strategy selection (based on selected disaster—per disaster)
BC Plan	Developing plans to guide the response and recovery actions when disaster occurs	Identify triggers and response (per disaster); Establish the command and control structure to respond to incident; emergency; disaster situations (per disaster) Prioritize activities; Time sequence of a BC plan for a selected disaster Activities and tasks should be prioritized based on the time sequence; Coordinate and finalize commitment; Gather requirements (list of pre-incident measures); Gather detailed requirements for each CBF; Checklists for writing the BC Plan (based on the tables and procedures/ lists); Confirm the BC plan; Distribute BC Plan; Not all BU require the entire BC Plan content; Based on need to know and need to hold basis

(continued)

Table 8.4 (continued)

Constructs	Variables (for construction business variables)	Sub-variables
Tests and exercises	Conducting tests and exercises for verifying the developed plans	Establish practice to operate BC Plan; Prepare for tests and exercises ; Conduct tests and exercises; Assess the results; Infrastructure to support tests and exercises; Identify and implement corrective actions;
Programme management	Programme management on the developed plans	Align BCM with organization operations; BCM can be aligned with the organization's operations (checklists); Review key BCM elements by BCM SC (checklists); Review BC plan (minumum once a year by BCM SC) (Checklists); Provide continuous training and awareness (checklists); Perform BCM audit; Track BCM trends and practices (checklists)

Source: Adopted from Spring Singapore (2008), Goh (2010), BCIGPG (2010)

risk analysis, business impact analysis (BIA), strategy development, BC plan development, tests and exercises and programme management. Table 8.4 describes the constructs and variables needed to develop the guidelines.

BCIGPG (2002) had provided some criteria for understanding the preparedness of the firm towards BCM. These criteria can be used as a benchmark in line with the BCM principles. In this study, some of these criteria were observed in order to identify the preparedness of Indonesian contractors. Table 8.5 shows the variables used as the benchmark to better understand aspects for BCM preparedness.

Indonesian contractors have their specific organization strategies that are related to their business value chain. According to BCIGPG (2010), factors that can be utilized for identifying the organization's objectives and strategies are as follows:

- Mission and vision.
- Organization objectives.
- How to achieve objectives.
- Organization's products and services.
- Direction and focus.
- Short and long term plans for growth, downsizing, restructuring, acquisition or even disposal.
- Timescales for new products or services.
- Operational geographic scale.
- The geographic extent of a disruption.
- The extent of resource loss.

Table 8.5 BCM preparedness criteria

Variables to be used as a benchmark—BCM good practice criteria
Write in statement BCM vision and policy statement
Communicating vision and policy throughout the organization
BCM committee from senior management
Senior management support and strong commitment
Middle management support and strong commitment
Monitoring and evaluating BCM implementation and maintenance
Defined BCM roles and responsibilities at all levels within job descriptions
Integration with organization's reward and recognition system
Integration with organization's performance management and appraisal system
Defined BCM roles and responsibilities at all levels within personal annual performance contracts
Defined KPI for BCM
BCM KPI linkage to personal annual performance contracts
Integral part of organization's project change management process
Integral part of organization's project management process
Using defined MIS to monitor and evaluate BCM competency of the staff and managers
BCM awareness and training programme
Strong commitment from senior management on BCM awareness and training programme
Assurance by middle managers on allocating BCM roles and responsibilities to their staffs
Defined BCM training programme
BCM training budget
Defined BC Plan
Awareness of the significance and importance of legal privilege in all communication and documentation regarding a crisis or BCM event
Awareness of the communication and call-out tree
Employing BCM professionals
Attending external BCM seminars and courses
Provide formal training and professional development plans for BCM personnel
Promoting BCM as an issue for continuous professional development for its staffs
Developed and distributed BCM awareness information (aid-memoire)
Defined BCM exercising programme
Defined BCM maintenance programme
Defined BCM audit programme
Achieving the outcomes set out from BCM standards regarding its awareness and training programme
Using BCM standards as a process for embedding a BCM culture
Providing deliverables set out from BCM standards regarding its awareness and training programme

Source: BCI (2002)

- Current and expected market conditions of its business.
- Competitors and the competition.
- Likely reaction of customers and competitors to its operations being disrupted.

- Competitors' reactions: taking advantage during difficulties or supporting one another (for protecting reputation of the sector).
- The regulatory environment of the business.
- Types and total number of suppliers.
- The timescale for finding alternative suppliers.
- Total number of customers.

Table 8.6 Firm Y—State owned: Relationships between business units and value chain

	Business Development			Procurement			Construction Operations				Post-Construction Services		Firm Infrastructure			Human Resources Management			
INDONESIAN CONTRACTOR'S ORGANIZATION STRUCTURE	Creating relationships with existing and prospective customers	Obtaining the work	Market research	Relationship with subcontractors	Relationship with suppliers	Efficiency and effectiveness of material purchasing procedures and management procedures	Cost and schedule estimation and control	Project management system	Quality management system	Safety management system	Warranty management system	Customer relationship development program	Adequacy and location of facilities and equipment	Efficiency and effectiveness of finance and accounting system	Information management system	Procedures for recruiting and developing employees	Working environment	Relationship with unions	Levels of employee motivation and job satisfaction
STATE OWNED CONTRACTOR																			
Main Director																			
HR and Development Director:																X	X	X	X
Finance Director							X	X						X					
Partnership and Environment Support	X	X	X	X	X							X	X		X				
Dept. of Human Capital																X	X	X	X
Dept. of Finance							X	X						X					
Corporate Secretary																			
Dept. of Legal	X			X	X			X			X	X						X	
Operational I Director:	X	X	X	X	X	X	X	X	X	X	X	X	X	X	X	X	X	X	X
Procurement Division				X	X	X													
General Civil Dept							X	X	X	X	X	X							
Business Development Division	X	X	X									X							
General Civil Division							X	X	X	X	X								
Overseas Dept	X	X	X	X	X	X	X	X	X	X	X	X	X	X	X	X	X	X	X
Overseas /regional Division	X	X	X	X	X	X	X	X	X	X	X	X	X	X	X	X	X	X	X
Building Dept	X	X	X	X	X	X	X	X	X	X	X	X							
Business Development Division	X	X	X									X							
Building Division				X	X	X	X	X	X	X	X								
Operational II Director:	X	X	X	X	X	X	X	X	X	X	X	X							
Procurement Division				X	X	X													
Industrial Plant Dept							X	X	X	X	X	X							
Business Development Division	X	X	X									X							
M/E Division							X	X	X	X	X								
Energy Dept	X	X	X	X	X	X	X	X	X	X	X	X							
Energy Division				X	X	X	X	X	X	X	X								
Business Development Division	X	X	X									X							
Investment Division	X	X	X																

Source: Firm Y website (2011)

- Customers who pay insurance for improved reliance on delivery.

As mentioned before, the Indonesian construction firms are owned either by the government (state owned) or private parties (private firms) (Raftery et al. 2004). The type of works to be delivered and their business functions are generally similar, but there can be various types of organizational structures adopted in the firm.

Table 8.7 Firm Z—Private owned: Relationships between business units and value chain

VALUE CHAIN FOR CONTRACTOR

	Business Development			Procurement			Construction Operations				Post-Construction Services		Firm Infrastructure			Human Resources Management			
INDONESIAN CONTRACTOR'S ORGANIZATION STRUCTURE	Creating relationships with existing and prospective customers	Obtaining the work	Market research	Relationship with subcontractors	Relationship with suppliers	Efficiency and effectiveness of material purchasing procedures and management procedures	Cost and schedule estimation and control	Project management system	Quality management system	Safety management system	Warranty management system	Customer relationship development program	Adequacy and location of facilities and equipment	Efficiency and effectiveness of finance and accounting system	Information management system	Procedures for recruiting and developing employees	Working environment	Relationship with unions	Levels of employee motivation and job satisfaction
PRIVATE OWNED CONTRACTOR																			
President Director																			
Management Representative													X	X	X				
Internal audit													X	X	X				
Corporate secretary																			
Director I	X	X	X	X	X	X	X	X	X	X	X	X							
Marketing	X	X	X																
Estimation							X												
Project Development	X	X	X	X	X	X		X				X							
Constructing Engineering & R/D								X	X	X	X								
Director II	X			X	X			X	X		X	X						X	
Legal	X			X	X			X	X		X	X						X	
Customer care												X							
Product quality									X		X	X							
Director III	X	X		X	X				X	X	X	X	X				X	X	X
Equipment				X	X								X						
SHE									X	X	X		X				X		X
General Affairs	X	X										X						X	
Director IV				X	X	X					X	X	X	X	X	X	X	X	X
Logistic				X	X	X													
HRD																X	X	X	X
Personnel Administration																X	X	X	X
Property & Building Management											X	X	X	X	X				
Director V	X	X	X				X	X	X	X		X		X	X				
Accounting														X					
Cash Operation														X					
Project Control							X	X	X	X									
IT															X				
Investor relations	X	X	X									X							

Source: Firm Z Website (2011)

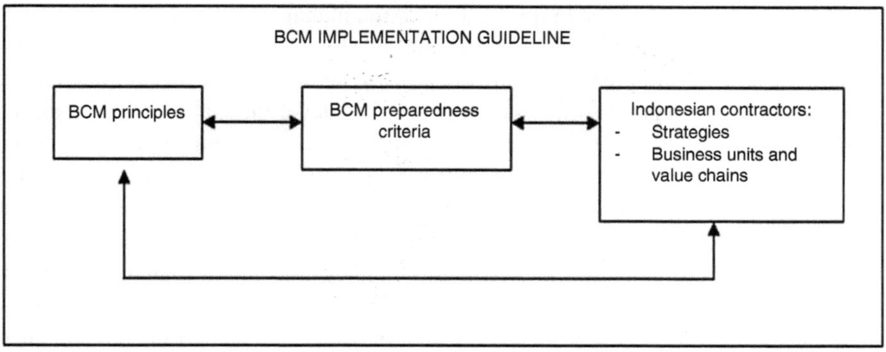

Fig. 8.4 BCM implementation guidelines framework

Tables 8.6 and 8.7 describe the relationships between the firms' business units (based on their organizational structures) and the business' value chains. The two samples were derived from two large firms in Indonesia, one a state-owned firm, and the other a private-owned firm. These relationships can be utilized for developing the BCM guidelines, in identifying the firm's business units and functions (BU and BF) that will further be used for determining the firm's critical business functions (CBF), MBCO, Recovery time objective–Recovery point objective (RTO–RPO) and their dependencies.

In summary, the BCM implementation guidelines for Indonesian contractors are developed based on the framework shown in Fig. 8.4.

8.5 Development of KBDSS for BCM

8.5.1 Benefits of KBDSS for Management Process and Construction

According to Angehrn and Jelassi (1994), there are three issues that can be used as a guidelines for developing a KBDSS in an organization. These issues are as follows:

1. Conceptually, KBDSS focuses on addressing the nature of individual and organizational decision-making processes.
2. Methodologically, KBDSS focuses on integrating the evolving computer-based tools, techniques and systems into the human decision-making context.
3. KBDSS is applied for addressing the real organizational needs by extending decision support to business teams

Furthermore, as a computer-based system that support decision making by aiding knowledge storage, retrieval, transfer and application by supporting knowledge access (Arnott and Pervan 2008), KBDSS provides benefits for the management process.

Most scholars viewed the major benefit for the management process as improvement in managerial productivity. Managerial productivity is considered as a function of the time spent in retrieving information, generating value added information and finding problems in the intelligence phase, and developing alternative solutions in the design phase of decision making. A good KBDSS may reduce the time required in this process and thereby improves managerial productivity. Moreover, the broad benefits from this system are that it provides quick access for the user to all relevant information, the process is direct and personalized and the problem models that are evaluated in the system can be integrated into a logical framework (Raman and Phoon 1990; Singh et al. 2008). Nonetheless, before using KBDSS for the management process, it is necessary to understand that KBDSS is not designed to make decisions for users, but rather it provides relevant information in an efficient and easy-to-access format that allows users to make more informed decisions (Arain and Low 2006).

In the construction industry, there are several KBDSSs that have been developed for various functions such as for material and procurement planning in construction projects (Mockler 1989), time management in projects (Hendrickson 1989), integration of AUTOCAD and construction scheduling (Wang 2001), resource allocation (Mohan 1990), resource or equipment selection (Amirkhanian and Baker 1992; Alkas and Aronian 1993), risk management process (Mohan 1990) and project success forecasting (Mohan 1990). It can be seen that a DSS or KBDSS is quite applicable in the construction industry, and there could be other processes or concepts that can be applied into this tool.

Yoon et al. (2008) also found that KBDSS have been developed for emergency response and management. A KBDSS for emergency response was developed to assess the state of preparation of an agency to respond to emergencies, enable the development of new SOPs, and to better train and empower employees in the decision making process. It is utilized to reduce the time to make critical decisions such as task assignment and resource allocation and to guide long-term decisions, training and the control capabilities of the organization. Considering the function and benefit of this tool, developing a KBDSS can be proposed in order to improve the effectiveness and efficiency of the decision making process in BCM.

Although there are many benefits in developing a KBDSS for management process and construction, the challenges in developing this system should not be ignored. Raman and Phoon (1990) discussed the issues in developing a KBDSS:

- Selling the idea to senior management

This is considered essential for purposes of committing adequate resources for setting up KBDSS facilities and creating an organizational environment conducive for users to use the system.

- Standards and integration

KBDSS draw data from a variety of internal and external sources. The issue of standards and integration includes hardware standards, operation system standards, application and database development standards, personal computer, and networking of computers that must be considered in advance.

Moreover, Kerr (2004) observed that in order to develop a KBDSS that is used by the target industry or organization, it is essential that effective and enthusiastic development committees be formed. This can only be done if the industry/organization is in favor of KBDSS development. Furthermore, effective evaluation techniques must be established in advance for the successful implementation of a KBDSS. Lastly, the system of KBDSS required regular updates on its knowledge base, which need regular monitoring process by the system's controller (Arain and Low 2006).

8.5.2 The Application of KBDSS for BCM

According to Hiles (2007), KBDSS can be utilized for the management's decision making process when developing the response plan (BC Plan), when responding to crises (for resource allocation) and during the recovery phase (for resource allocation and managing restoration). The use of this tool depends on the need of the management at that period, and which phase of BCM that needs shorter time for better decision making. Moreover, it is always important to understand that this tool does not replace the BCM analysis phase, the BC planner, the selected methodology or the ongoing management commitment to the process. It is developed purely to assist with developing and maintaining business continuity programmes.

Until now, there are various automated tools that can be used for BCM implementation. Originally, the automated tools or software were designed to support the actual development and documentation of BC plans. Tools that automate planning functions such as information updates and action plans were mostly favoured. It was in this area that a majority of the software products was targeted and still forms a high percentage of the market for BCM tools. The consideration of removing as many of the manual processes as possible to increase the amount of automation and routing of data is viewed to be necessary (Hiles 2007; Continuity Insights 2009).

Other software tools that also exist are tools that support the risk analysis, business impact analysis, and emergency response processes (Hiles 2007). COOP Systems (2010) had proposed developing an automated business impact analysis (BIA) process. The main function of this tool is in providing the BIA manual details to be tracked, maintained, and updated by the user. Furthermore, a web-based tool for organizations to measure and compare their resilience has also been developed. Stephenson Resilience, which collaborated with the Resilient Organizations Research Programme at the University of Canterbury in New Zealand, designed the tool for providing resilience scores, establishing a baseline of data, and enabling repeat measurements to track changes in resilience over time. This tool is considered to be useful because organizations can sometimes find it hard to invest in resilience due to the difficulty in identifying their level of resilience, and demonstrating progress or success (Continuity Central 2011).

Based on these findings, automated tools for BCM appear to be useful, where it provides a central repository of the information necessary to manage various aspects of BCM. Aspects of the BCM lifecycle can be accommodated, and changes to information covering all parts of the organization—staff, IT, suppliers, business processes and so forth, can be captured and propagated across the BCM structure. Automation delivers greater efficiency to the organization by managing the day to day details of BCM (Brooks 2010).

For this study, KBDSS was proposed as a supporting tool for the management team in developing a BCM. Considering that the scope in this study is on the phase of identifying and understanding BCM implementation for Indonesian contractors, the automated tool can be beneficial for fast and effective decision making process. Furthermore, the system can provide the knowledge needed by the management team in developing BCP, where the knowledge base can be updated and upgraded regularly.

According to Crabb (2011), there are parts in BCM that need consistency of approach, consistency of information and structure and also consistency of process. This type of phase can be supported by a KBDSS. The content for the proposed KBDSS is to develop a BCM preparedness assessment and recommendation (in the form of action plans). It provides technical guidelines or steps on each BCM principle based on the standards. The non-technical attributes that consist of the organizational culture and institutional forces that relate with BCM principles are included in the system. The answers from the user (the firm) were assessed to determine the level of BCM preparedness, and followed by recommendations on that specific level. In the proposed system, the level of BCM preparedness is grouped into four levels, which starts from an undeveloped BCM, beginner level, moderate level and comprehensive level. These levels are adapted from various BCM level of preparedness studies (MOF-BC 2007; Lansley and McAtee 2009; Marsh Risk Consulting 2010; Smit 2005).

Automating this process will provide the BCM information and knowledge in one place that allows the management to take a more rounded view of issues, resources and situations. This allows organizations to develop a much more strategic-based approach for planning its business resilience. Its content is not in providing the exact solution, but supports the decision maker (the user) with information for obtaining the proper solution in that phase. Knowledge is provided for the decision maker in assisting him to select the decision for developing BCM through the recommended action plans. Figure 8.5 illustrates the general process in the proposed KBDSS.

In developing the proposed KBDSS, all aspects from the previous phases of the study were needed, where the data were compiled and synthesized. Furthermore, the data were processed as the knowledge base for the system. Inferencing patterns were developed based on relevant rules for each decision situation. A structured situation diagram that includes the decision rules was formed into a more complete diagram called the dependency diagram. In the last phase, a computer programming language was used for synthesizing the dependency diagrams. In this phase, knowledge base, diagrams, rules and logics were transformed into a computer programming language format for further application as a KBDSS. Application using the computer programming language is supported by the user interface application, where this section consists of procedures and rules in the form of dialogues between the programme and the user (Mockler 1989; Turban 1995).

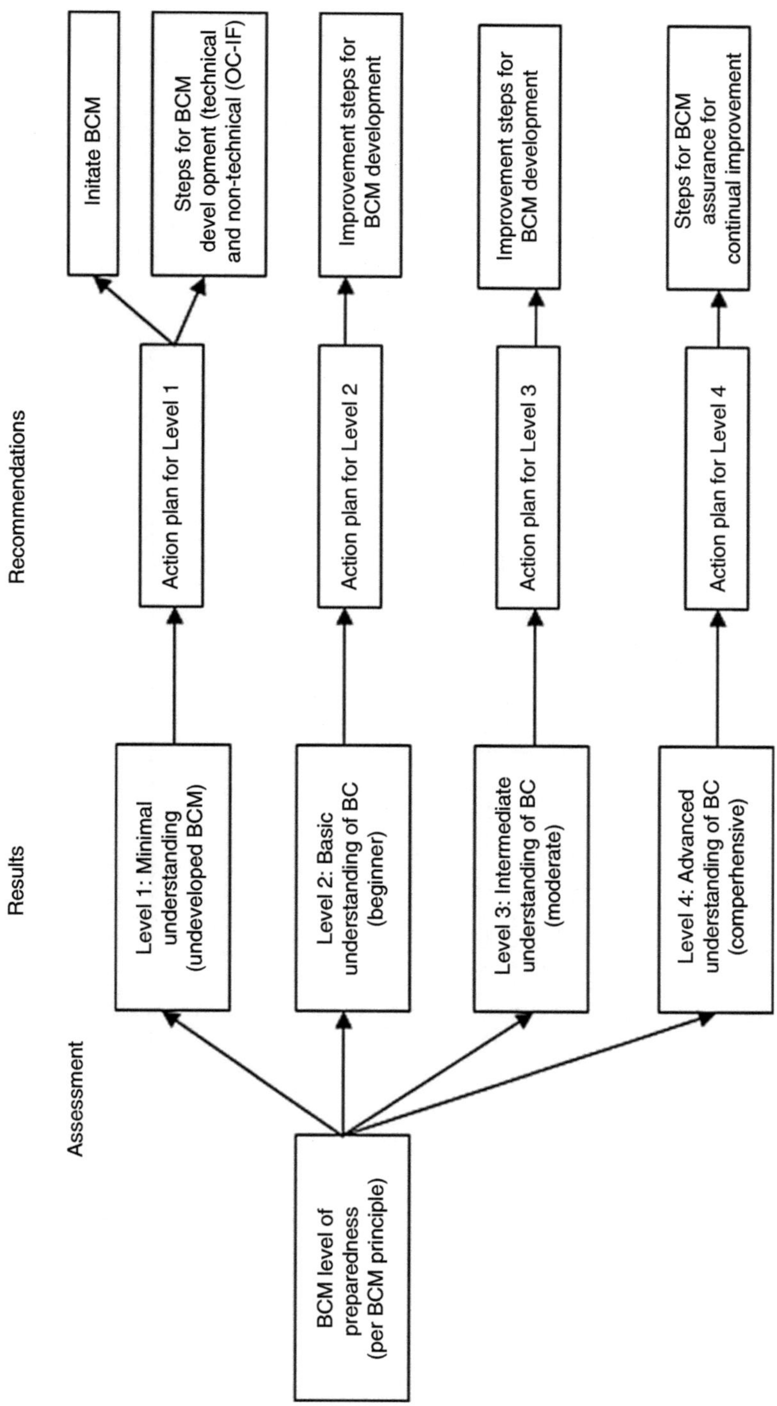

Fig. 8.5 KBDSS process: BCM level of preparedness assessment

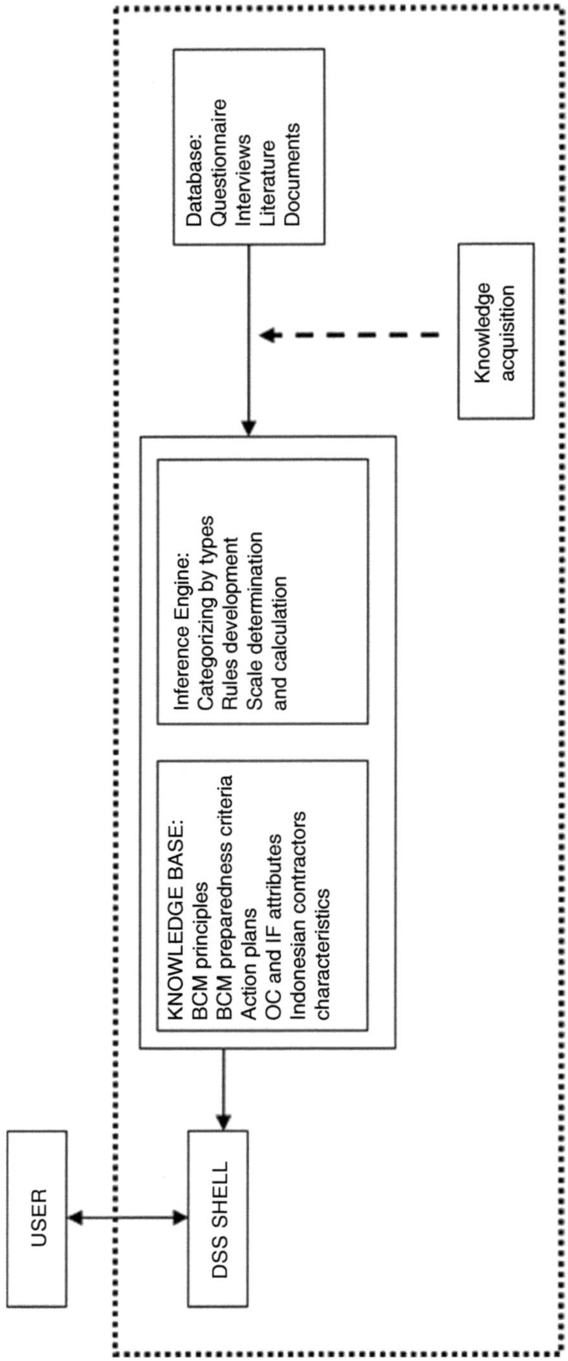

Fig. 8.6 KBDSS components

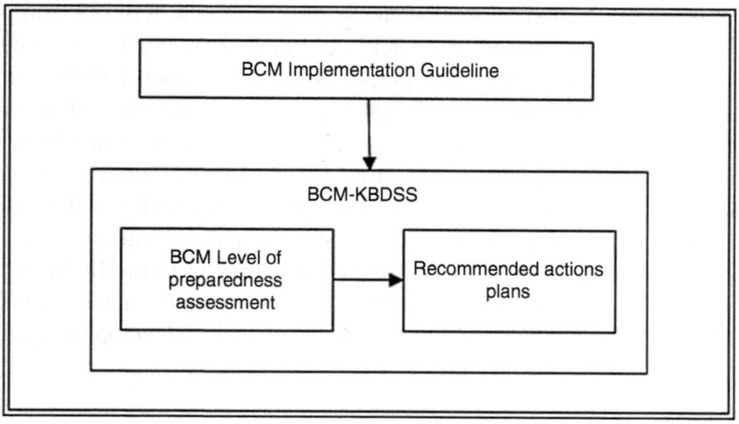

Fig. 8.7 BCM–KBDSS framework

Table 8.8 Variables for BCM–KBDSS model

Constructs	Variables	Sub-variables
Knowledge base (KB) (groups/or layers)	BCM technical and non-technical guidelines	BCM principles
		BCM preparedness criteria
		Compiled IF–BCM–OC attributes
		Organization strategy and objective
		Indonesian contractor business value chain
Rules development (per layer/groups)	BCM guidelines	Providing steps in BCM development (from six principles)—will be supported by a template for each step
		Providing non-technical recommendation in each of the six BCM principles
Frame (per section)	User	
	Input from user	
	Process (what KB will be used)	
	Output	
Logical representation (per section)	Structured situation diagram	
	Logic sequence (in details)	
	Dependency diagram (with rules)	
Program algorithm (per section)	Coding	
	Rules and logic	
	Program	
Computing language	Borland C++; DSS Shell	
Validation	Laboratory testing	Face validation
		Subsystem verification and validation
		Predictive validation
		User assessment
	Field testing	Field test validation

For further illustration, Fig. 8.6 shows the proposed KBDSS components, Fig. 8.7 describes the proposed BCM–KBDSS framework, and Table 8.8 describes the variables needed to develop the proposed KBDSS model. KBDSS comprises components such as DSS shell, knowledge base, inference engine, and database. All the data from the database were analyzed to acquire the relevant knowledge needed in the system. Inference engine was used to develop the knowledge base that assessed inputs from the user and provide the outputs. DSS shell was the component that structured these processes into a user-friendly interface with the user.

Figure 8.7 shows that all of the knowledge obtained from the BCM implementation guidelines were placed in the KBDSS for developing the assessment process.

Furthermore, Table 8.8 shows the constructs needed to develop the system.

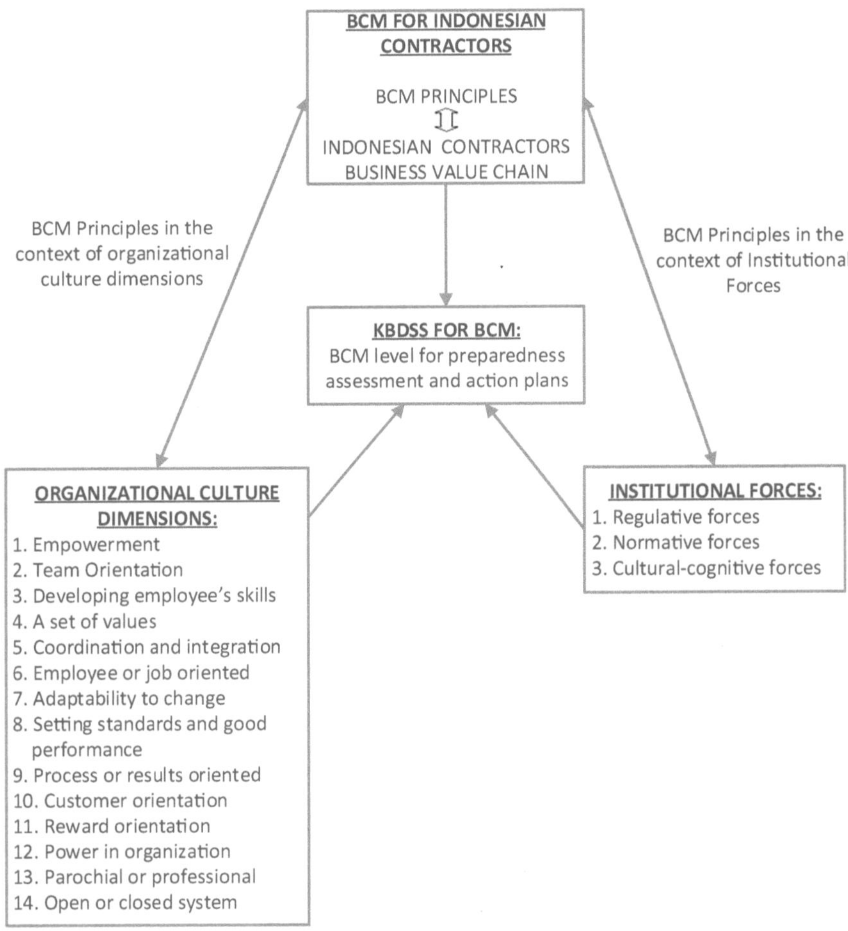

Fig. 8.8 Conceptual framework

8.6 Conceptual Framework

The main conceptual framework for this study is shown in Fig. 8.8. This framework was utilized for answering the research's objectives. The framework had synthesized the BCM–Cultural-institutional relationships, BCM implementation guidelines, and BCM–KBDSS aspects.

8.7 Summary

This chapter discusses the development of the study's conceptual framework. The constructs needed to understand the Indonesian contractor's knowledge of BCM were developed, followed by defining the relationships between BCM, OC, and IF. These relationships were considered as the outer layer of the conceptual framework. Furthermore, in developing BCM for Indonesian contractors, various constructs were identified which were developed into a guideline for BCM implementation. BCM principles, BCM preparedness criteria and characteristics of Indonesian contractors were the aspects used in this phase. The role of KBDSS is discussed following that, with the description of the proposed KBDSS framework, complete with its general process and components descriptions. The development of BCM implementation guidelines into a KBDSS was considered as the inner layer of the conceptual framework.

Chapter 9
Research Design and Methodology

9.1 Introduction

This chapter discusses the research design and methodology for the study. Descriptions about the research framework that are derived from the research questions and hypotheses are described in the first section. The middle section of the chapter elaborates on the research design and methods for collecting data. The final section describes the methods of analyses that were used in the study.

9.2 Research Framework

An essential outcome from the literature review is the development of the conceptual framework. Theories from the review provide the framework for the research. It will indicate the data which should be collected. The bodies of theory that were examined and evaluated will provide the basic structural framework to identify and explain facts and the relationships between them. The process of developing the conceptual framework model is called conceptualization, where there are three tasks that will be conducted, which are: (1) identifying the variables and constructs, (2) specifying hypotheses and relationships and (3) developing the conceptual model that visually represents the theoretical basis of the relationships that will be examined (Fellows and Liu 2003; Hair et al. 2011). Furthermore, the conceptualization process will be conducted based on the research questions which are identified from the research problems.

L.S.R. Supriadi, L. Sui Pheng, *Business Continuity Management in Construction*, Management in the Built Environment, DOI 10.1007/978-981-10-5487-7_9

9.2.1 Research Questions and Hypotheses

Based on the background, motivation, research problems and knowledge gap, the research questions for this study are:

1. What are the BCM implementation guidelines for different level of preparedness for Indonesian contractors? (RQ1)
2. How can the BCM implementation guidelines for Indonesian contractors be automated as a KBDSS? (RQ2)

Moreover, the objectives of the study are to:

1. identify Indonesian contractors' knowledge about BCM;
2. identify the significant drivers and hindrances from institutional forces for implementing BCM;
3. identify the significant drivers and hindrances from organizational culture dimensions for implementing BCM;
4. develop BCM implementation guidelines for different levels of preparedness for Indonesian contractors; and
5. automate BCM implementation guidelines for Indonesian contractors into a KBDSS.

A conceptual framework has been developed to achieve these objectives. The conceptual framework in Fig. 8.8 shows the main constructs of each aspect that were analyzed. The BCM principles were grouped into six constructs, based on its principles. The organizational culture (OC) has 14 dimensions that were analyzed on their relationships with BCM. Furthermore, the three main pillars of institutional forces were used to better understand the BCM principles within its context. All of these constructs were synthesized into a knowledge base, a form that was developed into a decision support system.

9.2.1.1 Hypotheses

According to Tan (2008), a hypothesis is the researcher's guess of the factors (causes) affecting the outcome. It is a preliminary or tentative explanation by the researcher of what the researcher considers the outcome of an investigation will be. This involves reviewing the literature to determine the possible causes and mechanisms linking causes and effects. As a tentative answer to the research question, the hypothesis may be rejected, refined, or supported after empirical testing.

For RQ1, which is related to Research Objective 1, 2, 3, and 4 (RO1, RO2, RO3, RO4), three hypotheses are formulated. These hypotheses are tentative answers based on literature reviews related to BCM, OC and IF. The three hypotheses are:

Hypothesis 1. *(H1): Indonesian contractors have not heard of BCM, but have emergency plans in place in their organizations.*

Hypothesis 2. *(H2): The institutional forces that consist of Regulative, Normative and Cultural-cognitive forces are the significant drivers for Indonesian contractors to implement BCM.*

- *Sub-Hypothesis 2.1 (H2.1): The regulative forces that consist of rules, laws, sanctions, gains, losses and consequences are significant drivers for Indonesian contractors to implement BCM.*
- *Sub-Hypothesis 2.2 (H2.2): The normative forces that consist of personal morality, social influence and legitimacy are significant drivers for Indonesian contractors to implement BCM.*
- *Sub-Hypothesis 2.3 (H2.3): The cultural-cognitive force that consists of shared understanding of compliance is a significant driver for Indonesian contractors to implement BCM.*

Hypothesis 3. *(H3): The organizational culture dimensions that consist of 14 attributes are the significant drivers for Indonesian contractors to implement BCM.*

- *Sub-Hypothesis 3.1 (H3.1): Empowerment is a significant driver for Indonesian contractors to implement BCM.*
- *Sub-Hypothesis 3.2 (H3.2): Team orientation is a significant driver for Indonesian contractors to implement BCM.*
- *Sub-Hypothesis 3.3 (H3.3): Developing employee's skills is a significant driver for Indonesian contractors to implement BCM.*
- *Sub-Hypothesis 3.4 (H3.4): A set of values is a significant driver for Indonesian contractors to implement BCM.*
- *Sub-Hypothesis 3.5 (H3.5): Coordination and integration is a significant driver for Indonesian contractors to implement BCM.*
- *Sub-Hypothesis 3.6 (H3.6): Employee oriented is a significant driver for Indonesian contractors to implement BCM.*
- *Sub-Hypothesis 3.7 (H3.7): Adaptability to change is a significant driver for Indonesian contractors to implement BCM.*
- *Sub-Hypothesis 3.8 (H3.8): Setting standards and good performance is a significant driver for Indonesian contractors to implement BCM.*
- *Sub-Hypothesis 3.9 (H3.9): Process oriented is a significant driver for Indonesian contractors to implement BCM.*
- *Sub-Hypothesis 3.10 (H3.10): Customer orientation is a significant driver for Indonesian contractors to implement BCM.*
- *Sub-Hypothesis 3.11 (H3.11): Reward orientation is a significant driver for Indonesian contractors to implement BCM.*
- *Sub-Hypothesis 3.12 (H3.12): Power in organization is a significant driver for Indonesian contractors to implement BCM.*
- *Sub-Hypothesis 3.13 (H3.13): Professional attribute is a significant driver for Indonesian contractors to implement BCM.*
- *Sub-Hypothesis 3.14 (H3.14): Open system is a significant driver for Indonesian contractors to implement BCM.*

Before developing a hypothesis, it is necessary to recognize whether it is proper to include a hypothesis in the research or not. A hypothesis is appropriate to be included in research when it is based on theory and previous work that sets out to test the existence of certain variables and/or any relationship between them. Studies that have a hypothesis to test are mostly quantitative studies. As for qualitative study, it is viewed that a hypothesis may not be needed due to the nature of the study. A qualitative study seeks to carry out a fundamental investigation to identify what is occurring, such as to observe behavior in a highly novel environment or in a new community where established values or theories may not apply (Fellows and Liu 2003).

Therefore, understanding that in achieving Research Objective 4 (RO4), a qualitative approach was used (data collected to generate ideas or framework), the proposition for this objective is: A BCM implementation guidelines model for Indonesian contractors can be developed for different levels of preparedness based on the BCM principles and Indonesian contractor's characteristics.

Moreover, for Research Question 2 (RQ2) (related to Research Objective 5 (RO5)), the approach used in this part is also a qualitative one. It is a discovery oriented approach, with analysis using the data collected to generate ideas or theories or framework. This study is based on inductive reasoning, where it is a type of thinking that involves identifying patterns in a data set to reach conclusions and build a system from the data collected. This approach is used because it is to propose a new idea, which in this case is a guidelines model automated by a decision support system that can eventually be tested with quantitative research (Hair et al. 2011). *Thus, the proposition for RQ2 is: A BCM implementation guidelines model for Indonesian contractors can be automated into a KBDSS.*

9.3 Research Design

A research design provides the basic directions for carrying out the study. It is a plan for testing the hypothesis or for interpreting events. The purpose of a research design is to rule out alternative explanations or false conclusions that will be vulnerable to attacks from critics. Therefore, the research design should be chosen based on its relevance to the research questions and hypotheses and its ability to complete the study in the most efficient manner (Hair et al. 2011; Tan 2008).

9.3.1 Research Design for the Study

In testing the hypotheses (H1, H2, H3), as part of answering RQ1, the survey method was used as the research design. Survey is a systematic method of collecting data based on a sample. It is used to explore particular issues; describe phenomenon; determine preferences; and ascertain reasons (Tan 2008). In this case, it was used for understanding the Indonesian contractors' knowledge of BCM and gathering information on the significant drivers and hindrances in implementing BCM. The type of surveys that were conducted involves a cross-sectional study, which gathers information about a population at a point in time. The details of the samples will be discussed in the sampling section.

For interpreting the proposition based on RO4 and RQ2, surveys and case studies were used as the research design. The surveys were utilized for obtaining knowledge regarding BCM implementation for Indonesian contractors from construction experts, and case studies were used for obtaining a more in-depth knowledge about it, particularly from the view of the management level in Indonesian construction firms.

A case study is defined as an in-depth investigation of a particular unit (country or organization) or phenomenon to test theories, explore the ground, offer new insights or suggest new variables. The case study should be holistic and aim at thick description using multiple sources of evidence (Tan 2008).

This study used a descriptive case study, which presents a complete description of a phenomenon within its context, which in this case is about BCM implementation by Indonesian contractors. The theories used cover the scope and depth of the object (case) being described. Moreover, this type of case study is part of an exploratory research, where it is meant to discover new relationships, patterns, themes, or ideas. It is not intended to test specific research hypothesis. It relies more on qualitative techniques, although quantitative approaches may also be used (Yin 2003; Hair et al. 2011). This study used multiple case studies to compare and contrast different cases. Furthermore, by using the purposeful sampling strategy, this study selected two cases, which are from a large private-owned firm and a large state-owned firm respectively (Tan 2008; Patton 2002).

9.3.2 Sampling

Sampling is defined as ways of selecting subsets (samples) from a population (set of all elements) (Tan 2008). In collecting data, sampling is an important phase, where samples that are chosen will provide the information needed. The results of a survey depend on the data collected from the samples.

According to Hair et al. (2011), during sampling design, there are things that need to be considered: (1) determining whether to use a sample or a census; (2) determining which sampling approach is best; and (3) determining the size of a sample. The procedures to obtain representative samples are as follows:

- Defining the target population, determining which of the complete group of objects or elements relevant to the research will be used.
- Choosing the sampling frame.
- Selecting the sampling method.

- Determining the sample size. The challenge in this phase is to obtain an acceptable balance among the many factors that need to be considered. These factors are the variability of elements in the target population, the type of sample required, time available, budget, required estimation precision, and whether the findings are to be generalized with what degree of confidence.

  In computing the sample size, formulas based on statistical theory can be used. However, for pragmatic reasons such as budget and time constraints, alternative ad-hoc methods are mostly used. Sample sizes based on rules of thumb, previous similar studies, one's own experience, or simply what is affordable are some of the alternative methods. Rules of thumb are usually derived from some common statistical principles, aimed at obtaining balance between resources and ideal sample sizes. Another approach is to use the same sample size as those of studies similar to the current study (Israel 2013). Moreover, Tan (2008) described that sample size is largely based on judgment rather than precise statistical computation. Formulas based on statistical theory are seldom used in actual surveys because the variables often do not have means or standard deviations. In such cases, a sample size of 10% of population is sometimes recommended but it has to be used with care. If the population is ten million, 10% of the population is one million, which is too large. If the population is only 50, and it is not costly, the study could sample the entire population. Whichever method that will be used in determining the sample size, the important thing that needs to be kept in mind is that a sample size should be of a sufficient size and quality to yield results that are seen to be credible in terms of their accuracy and consistency (Tan 2008; Hair et al. 2011).

A. Survey (for RO1, RO2, RO3)

The population for conducting the survey in achieving Research Objective 1–3 was the Indonesian contractors. Based on the scope of the study, the type of contractors that was surveyed was the large firms. A sampling frame was developed for this survey, where a list of registered large contractors in ICA was used as the frame. The type of samples that was chosen was the stratified sample, a part of probability sampling in which elements are selected by chance using a sampling frame. For the samples, the respondents that were chosen to answer the surveys were from the management level (Directors/General Managers) and the staffs/executives (Coordinators/Supervisors/Engineers) from both state-owned and private-owned contractors. This study identified 127 large Indonesian contractors, which are registered in ICA as the sampling frame. For the questionnaire survey, the respondents were representing their own organization (i.e. the Indonesian contractor). Hence, one respondent represented one firm. The respondents were contacted directly based on their job positions. The person who completed the questionnaire was the incumbent who had many experiences in the relevant field and had communication flow directly with their senior managers. The respondents' main offices are mainly located in Jakarta and some firms are in Surabaya. Their project-based offices are in various cities in Indonesia.

B. Survey (for RO4 and RO5)
In achieving Research Objective 4 and 5, the sampling method for the survey was a non-probability sample type, which was purposive sampling. In this sampling, judgment is preferred rather than random sampling. The respondents for this survey were the management level staff from contractors (Directors/General Managers) and construction experts from construction development institutions, academics and the government. They were chosen due to their expertise in their knowledge area that relates to construction, and in order to gain more understanding of BCM for contractors. This is in line with the criterion sampling method, which is applied for investigating an in-depth aspect derived from various knowledgeable sources (Brikci and Green 2007; Patton 2002). Five respondents were interviewed, all of whom met each criteria for the selection of respondents. The five respondents provided in-depth description about BCM implementation by Indonesian contractors. As stated by Patton (2002), fewer samples could be chosen for greater depth of discussion towards the topic of the study.

9.4 Methods of Data Collection

The following step after determining the research design is to select the methods of collecting data. The methods for this phase can be in the form of questionnaires and interviews, observation techniques, analysis of past documents and simulation (Tan 2008; Hair et al. 2011). In this study, questionnaires, interviews and analysis of past documents were used.

In developing a questionnaire, components that need to be considered are details of sections in the questionnaire, which generally consist of the respondent's information and sections of questions related to the variables to be measured, and scales used in the questionnaire. The scales can be in the form of nominal, ordinal, interval, or ratio. As for conducting interviews, the process can be categorized as structured or non-structured interviews. A structured interview means that the interviewing process uses a defined list of questions. A non-structured interview does not focus on a set of defined questions, and the topic discussed can be an open-ended one. In relation to the past document analysis method, the most important aspect that must be kept in mind is the details of the published sources, the period of the publications and the names of the authors.

In the process of data collection, the criteria of the interviewers/surveyors/observers and the types of equipment used should be determined. A consideration for providing training to the surveyors may be necessary if they are new to the process. Moreover, supervision and field checks during the data collection stage should be conducted. After data collection, data processing in the form of data editing for errors or omissions is necessary. Data should be checked for computer input errors, coded, classified, and processed for subsequent data analysis (Hair et al. 2011).

9.4.1 Methods of Collecting Data

A. Data collection for RO1, RO2, RO3

A questionnaire was used for collecting data that are related to Research Objective 1–3. The methods of the questionnaire distribution were in the forms of postal mail, online mail, and the self-administered process. In general, the variables that were measured and included in the questionnaire were the BCM knowledge variables and the BCM–IF and BCM–OC variables. The scales used for this questionnaire were nominal (for yes or no questions); ordinal (rating and ranking); Likert scales; and categorical scales (choosing from a group of given attributes).

The sections of the questionnaire consist of three main sections, which were:

- Section A: Firm's information. This section used nominal and categorical scales.
- Section B: Crises response and general BCM knowledge. The nominal, categorical and Likert scales were used.
- Section C: Significant drivers and hindrances for implementing BCM for Indonesian contractors in the context of cultural and institutional perspectives. Nominal and Likert scales were used in this section.

Before the data collection from the questionnaire was undertaken, a pilot study was conducted. The purposes of this study were to identify the questions in the questionnaire that respondents might found difficult to answer and to test the appropriateness of the questionnaire as an instrument to achieve the research objectives. During the pilot study, the respondents completed the questionnaire and provided feedback. The fieldwork for the questionnaire survey was conducted based on the amended questionnaire from the pilot study (Tan 2008).

The data collected were processed by editing and coding them for data analysis. The questionnaire for this part of the study is attached in Appendix B.

After analyzing the data, structured interviews were conducted with respondents from the management level of the firms. Purposive sampling was used and the interviews were meant to gain a better understanding of the results from the survey, asking the respondents reasons or descriptions relating to the results. The interview consists of sections discussing BCM knowledge aspects from the contractors' point of view and the significant drivers and hindrances for implementing BCM. The data were processed and analyzed qualitatively.

B. Data collection for RO4 and RO5

In developing a BCM guideline model for Indonesian contractors which is automated in a form of a KBDSS, structured interviews were conducted. These interviews will consist of questions relating to detailed knowledge about BCM in construction and recommendation for BCM implementation by Indonesian contractors.

Furthermore, data collection in the case study phase was in the form of checklists, interviews to top and middle management level staff in the firm, and analysis of related internal documents. The data collected include the firm's characteristics (its vision, mission, business value chain and its views on crises), BCM preparedness description, and the firm's practices relating to how crises are handled. The questions for the structured interview and case study are attached in Appendix C and Appendix D.

9.5 Methods of Data Analysis

Fellows and Liu (2003) stated that data can be analyzed quantitatively and qualitatively. Qualitative data emphasizes on determining the meaning of the data, where what is analyzed is not numbers, but texts. These data must be handled systematically, where the process consists of (1) organization and categorization of data into concepts, (2) establishing links between concepts and (3) refinement and development. A content analysis can be used for this type of data, where it is determining the main facets of a set of data, by simply counting the number of times an activity occurred, a topic is mentioned, etc. According to Tan (2008), the validity and reliability of a qualitative analysis are the use of good hypotheses, the use of experienced observers, triangulation of observations such as by using different observers, establishing trust so that respondents tell the truth and conducting independent checks by using simple surveys to validate key findings.

Analysis of quantitative data can utilize the approach of the qualitative form to yield numerical values of the categorized data such as ratings, frequencies, rankings and others that may be subjected to statistical analyses. Analyses such as descriptive statistics (summarizing data), population parameters (assuming data come from a particular distribution), and statistical inference (using sample statistics (e.g. sample mean) to make inferences about corresponding population parameters) are often used. Before analyzing the data with these methods, it is also suggested that an exploratory data analysis (EDA) is conducted. EDA is used to examine data patterns such as their relations among variables, outliers, trends and turning points and distributional assumptions.

The mixed-method combination of qualitative and quantitative methods may offset the weakness of each method, thus, increasing the validity and reliability of research findings (Miles and Huberman 1994). In this study, the sequential mixed-method is used where the researcher elaborates the findings of one method with another method. The research begins with a quantitative method and further explore in details with the qualitative method (Creswell 2009).

A. Methods of data analysis for RO1, RO2, RO3

(1) Quantitative analysis

The data collected from the questionnaire were analyzed quantitatively. For section A and B in the questionnaire, descriptive statistics of the means, modes and frequency of distribution were used. Reviews of respondents from different type of firms were also analyzed. Moreover, the univariate tests of significance (non-parametric test and *t*-test) were used to test the hypothesis.

For section C, Exploratory Factor Analysis (EFA) was utilized. EFA techniques were used because they can meet the objectives of data analysis, which are to identify the structure of relationships among the variables and to identify representative variables from a much larger set of variables for use in subsequent analyses. EFA is a statistical procedure commonly used for extracting underlying constructs out of a set of observed variables. It is typically used to condense a large set of variables into a few meaningful "factors". In EFA, the focus of investigation is on uncovering the underlying latent variable structure, since the researcher does not have existing knowledge about it (Tjandra 2004; Raykov and Marcoulides 2008; Tan 2008).

In determining the number of factors to be retained, the Kaiser criterion was used in the study. This is the most common method and the rule is to drop the least important factors from the analysis, by dropping all components with eigenvalues under 1.0. The eigenvalue denotes the relative importance of each factor, where it measures the variance in all the variables that is accounted for by that factor. If a factor has a low eigenvalue, then it is contributing little to the explanation of variances in the variables and may be ignored as redundant with more important factors (Tjandra 2004; Dunteman 1989).

Furthermore, to obtain more interpretable results, varimax rotation was used. It is also the most common rotation option that minimizes the number of variables that have high loadings on any given factor. Each factor will tend to have either large or small loadings of particular variables on it. A varimax solution yields results that make it as easy as possible to identify each variable with a single factor (Tjandra 2004; Hair et al. 1995).

According to Thompson (2004), understanding EFA may facilitate deeper insight into other statistical analyses. An important idea in statistics is the notion of the general linear model (GLM). Central to the GLM are realizations that all statistical analyses are correlational and can yield effect sizes analogous to r^2, and apply weights to measured variables to obtain scores on the composite variables that are actually the focus of all these analyses. In fact, because all parametric methods (e.g. analysis of variance, regression, multivariate analysis of variance) are special cases of canonical correlation analysis, and one kind of EFA is always implicitly invoked in canonical analysis, it can be shown that an EFA is an implicit part of all parametric statistical tests.

(2) Qualitative analysis

The structured interviews following the completion of the questionnaire survey were analyzed using content analysis.

B. Methods of data analysis for RO4 and RO5

As with the previous process, the data collected from checklists and internal documents were analyzed quantitatively using descriptive statistics. Other data collected from interviews and reviews of internal documents were analyzed qualitatively using content analysis.

The results from these analyses were compiled and synthesized into a knowledge base for KBDSS development. In developing the proposed KBDSS, the knowledge acquired from data analysis was represented in a systematic form which consists of rules, decision framework, and logic representation. Rules were given for every decision situation gained from the knowledge. The decision framework was developed based on the knowledge, where components such as who the users are, inputs from the user, processes after receiving inputs, knowledge to be used in the processes and the outputs were determined. Logic representation was incorporated in the decision framework. These processes were structured, calculated, and compiled in the inference engine of the system. Furthermore, in developing the DSS shell, an appropriate computing language was used with its program algorithm that represents all the processes inside the KB and inference engine (Mockler 1989). The tools used for developing the KBDSS were a DSS shell software (e.g. Visual Basic and MySQL) and mathematical function software (e.g. Microsoft Excel).

C. Validation

In validating the results of the analyses, particularly for the significant drivers and hindrances for implementing BCM by Indonesian contractors, structured interviews were conducted for this purpose. The interviews were conducted to gain feedback regarding the results of the survey. They were conducted with construction experts and directors of Indonesian construction contracting firms.

Furthermore, the KBDSS validation was evaluated through two stages, which were laboratory testing and field testing (Borenstein 1998). Laboratory testing involves the testing of the system's prototype development, starting from testing its consistency between the system designer's view and the potential user's view of the problem, testing the KBDSS modules, and validating the system using test cases in which the results are known. Following this, field testing was conducted to identify the system's general performance. Attributes such as graphical modeling, integration of modules, presentation of results, logical description, program consistency, the processing speed of the system, the system's user-friendliness, the level of system's applicability and accuracy of the output were used for evaluation. The validation process utilized the scoring or rating approach to evaluate the attributes. These scores were analyzed quantitatively using descriptive statistics. The respondents for this process were construction experts, management and staff working for Indonesian construction contracting firms. Statistical software (SPSS) and DSS shell software (MySQL) were used for the validation process.

9.6 Summary

Based on the study's background, research problems, and knowledge gap, the research questions that lead to the study's objectives were developed. The literature reviews relating to the objectives aided the process of developing the conceptual framework and hypotheses of the study. Following that, the appropriate research designs were developed to achieve the study's objectives. Surveys and case studies were used in the research designs, where the data were collected in the forms of questionnaire, interviews, or analysis of past documents. The methods for analyzing data include both quantitative and qualitative analysis. Statistical analysis and content analysis were mostly used in the process. Moreover, a system development process using computing language was conducted for developing the KBDSS. Last but not least, validation was conducted for the results derived from the analyses.

Chapter 10
Data Analysis: Surveys

10.1 Introduction

This chapter elaborates upon the data analysis and results from the surveys. A description of the pilot study conducted before the survey is provided in the first section. Following that, details of the questionnaire survey are provided, including the results of each section from the questionnaire. Lastly, the survey validation is explained thoroughly.

10.2 Pilot Study

Before the fieldwork was undertaken, a pilot study was conducted. The purposes of this study were to identify the questions in the questionnaire that respondents might found difficult to answer and to test the appropriateness of the questionnaire as an instrument to achieve the research objective. The pilot study was conducted among five respondents with different job levels, work experience, and education levels. The details of the respondents can be seen in Table 10.1.

During the study, the respondents filled out the questionnaire and provided feedback. Seven questions needed to be answered by the respondents. Table 10.2 shows the details of the feedback.

Four general cases of feedback to the survey were received, which were used to amend the questionnaire. The feedback is summarized as follows:

1. In the introduction section, it was described that the time for filling in the questionnaire would be around 30–45 min.
2. Most of the suggestions resulted in revising the sentences into shorter statements and using practical terms.
3. The BCM terms in Section C were written in bold (to make the respondents easier to read).

L.S.R. Supriadi, L. Sui Pheng, *Business Continuity Management in Construction*, Management in the Built Environment, DOI 10.1007/978-981-10-5487-7_10

Table 10.1 Respondent details

Respondent	Working experience (years)	Job level	Education level
1	15	Manager	Master degree
2	15	Senior Lecturer	Master degree
3	27	Senior Manager	Master degree
4	23	Senior Manager	Master degree
5	16	Senior Manager	Bachelor degree

4. For the survey, the methods used were the face-to-face approach (the researcher meets with the respondents to help them fill in the questionnaire form) and the workshop approach (deliver a short presentation about BCM in a class and follow by distributing the questionnaire which will be discussed when the survey is completed).

The fieldwork was conducted using the amended questionnaire with the recommended survey methods.

10.3 Questionnaire Survey

The sampling method for the questionnaire survey (a sample of the questionnaire can be seen in Appendix B) adopted the probability sampling of stratified samples, which uses a sampling frame (Tan 2008). As described in Sect. 9.3.2, the sampling frame used in this phase was the large contractors which have been registered as members of the Indonesian Contractors Association (ICA). The members of the ICA were chosen as the sampling frame because they are members of a national contractors association that has facilitated the work of large-scale contractors in Indonesia. ICA members are the dominant players in the Indonesian construction market. They have taken active roles as consortium members as well as consortium leaders. The association has also played active roles in the development of the construction industry in Indonesia among others as one of the National Board of Construction Service Development (NBCSD) founders (NBCSD 2002b; ICA 2012). In order to achieve Research Objectives 1–3, questionnaires were distributed to potential respondents in the sampling frame. The sampling frame has a total of 127 samples, with disproportional sampling of 13 state-owned firms and 114 private firms. Of the 127 firms, 65 firms responded to the survey, and 56 samples were found to be suitable for the data analysis. This yielded a 44.09% total response rate. The surveys were conducted from January to April 2013.

Before analyzing the data collected, the reliability of the questionnaire was examined to determine its internal consistency (whether all items in the questionnaire measured the same thing). Cronbach alpha (α) test was used as the measure of reliability. The alpha value typically varies between 0 and 1. The closer the alpha is to 1, the greater the internal consistency is. A coefficient scale of an alpha value

Table 10.2 Respondent feedbacks

Questions	Respondent 1	Respondent 2	Respondent 3	Respondent 4	Respondent 5
How long does it take to fill out the questionnaire?	30 min	35 min	35 min	30 min	45 min
Does it take too long to fill out the questionnaire?	Yes, particularly for section C, need to think thoroughly on the factors in the table	Yes, need some time to think on the BCM terms (new terms for me)	Yes, need time to read through and understand the BCM terms in the questions	Yes, need time to understand the BCM terms	Yes, only for section C. Sections A and B are quite fast. Section C needs to understand the BCM terms (the descriptions of BCM principles are well described)
Which questions (per section) in the questionnaire are difficult to answer?	None	Already commented in the questionnaire	Section C = needs more time to read and think thoroughly	Already commented in the questionnaire	None
Is there any mistake or ambiguity in the wording or terms used?	None	There are some ambiguous terms. Already commented in the questionnaire	There are some ambiguous terms. Already commented in the questionnaire	None	None
Are any additional questions needed for each section of the questionnaire?	None	None	For companies that already have good crisis management in place, how can one compare the existing system with BCM?	None	None

(continued)

Table 10.2 (continued)

Questions	Respondent 1	Respondent 2	Respondent 3	Respondent 4	Respondent 5
Is there any suggestion for each section of the questionnaire?	In section C, some statements should be described in shorter sentences	Already commented in the questionnaire	Already commented in the questionnaire	None	None
Is there any suggestion of methods for conducting the questionnaire survey (e.g. using mail/online or face-to-face or workshop approach)?	Face to face approach, online and workshop	Face to face approach, workshop, and online/mail	Face to face approach and workshop	Face to face approach and workshop, for more interaction between the surveyor and the respondents	Face to face approach, online/mail and workshop

Table 10.3 Respondents' job levels

Job level	Frequency	Percentage
Director/CEO	1	1.8
Senior Management/General Manager	1	1.8
Manager/Assistant Manager	13	23.2
Senior Executive/Senior Staff	8	14.3
Executive/Staff	24	42.8
Junior Executive/Junior Staff	9	16.1
Total	56	100

Table 10.4 Years working in the company and construction sector

Years working in the company			Years working in the construction sector		
Years	Frequency	Percentage	Years	Frequency	Percentage
<5	8	14.3	<5	4	7.1
5–10	47	83.9	5–10	31	55.4
11–15	0	0	11–15	19	33.9
16–20	0	0	16–20	1	1.8
>20	1	1.8	>20	1	1.8
Total	56	100	Total	56	100

above 0.7 is considered reliable (Tan 2008; Pallant 2001). In this study, the Cronbach's alpha coefficient for the section of BCM principles, Institutional Forces, and Organizational Culture attributes perspectives were 0.938, 0.814, and 0.967, respectively. This shows that the questionnaire has good internal consistency.

Among the 56 valid responses, 46 samples were from private firms, and 10 samples were state-owned firms. 75% of the respondents had a Bachelor degree as their highest level of education, and 25% of them had obtained a Master degree. Regarding their job level, most of them are at the executive/staff position (42.8%) and the manager/assistant manager level (23.2%). Table 10.3 describes the job levels of the respondents.

Furthermore, Table 10.4 shows the respondents' years of experience working in the current firm and in the construction sector. The majority of the respondents have 5–10 years of experience working in the current firm (83.9%), and 33.9% of them have worked for 11–15 years in the construction sector. This shows that they have been exposed to the business environment within their firm and sector for quite a while.

10.3.1 Crises Response

Based on Section B of the questionnaire, descriptive statistics of the respondents' crises response were obtained. Table 10.5 describes the number of crises that the

Table 10.5 Crises that occurred in the firm within 5 years

	Frequency	Percentage
No crisis	4	7.1
1 time	9	16.1
2 times	15	26.8
3 times	11	19.6
>3 times	12	21.4
Do not know	5	8.9
Total	56	100

Table 10.6 Types of crises that occurred in the firm within the last 5 years

No	Types of crises that have occurred in the last 5 years	Frequency	Percentage
1	Access/approval restriction or limitation	13 out of 56	23
2	Delays or uncertainty in resolving disputes	13 out of 56	23
3	Increase in price of raw materials (unexpected price escalation)	12 out of 56	21
4	Changes in regulations and statutory legislation	11 out of 56	20
5	Natural disasters (earthquake, floods, tsunami, etc)	10 out of 56	18
6	Loss of management personnel or key staff	10 out of 56	18
7	Lack of component workforce	9 out of 56	16

respondents claimed to have occurred in their firms within the last 5 years. Most of them responded that their firms have experienced 2 crises. Only 7% of the respondents claimed to not have experienced any crisis, and 8.9% of them were not aware of any.

Regarding the types of crises that occurred in the respondents' firms within the last 5 years, Table 10.6 explains them in detail. This table shows the top 7 crises (from 32 types of crises suggested in the questionnaire), and the gaps between the percentages of respondents who had chosen the crises were small. It turned out that crises such as access/approval restriction or limitation, delays or uncertainty in resolving disputes, and increase in price of raw materials were chosen the most by the respondents.

The respondents also responded about the impacts that have occurred due to crises. The results show that delays in work/dissatisfied customers was the most chosen crises impact, followed by revenue impact and loss of productivity. The least chosen crises impact was huge data loss and client impact. The details about the results can be seen in Table 10.7.

Moreover, among the 56 samples, 50% of the respondents confirmed to have a crisis SOP in their firms. 25% of them confirmed that their firms do not have a crisis SOP in place, and the other 25% of the samples do not know whether they have a crisis SOP. Table 10.8 explains in greater detail the input from the respondents that have a crisis SOP in place. The table shows that the majority of the respondents (71.4%) confirmed to have in place the four main elements of a crisis SOP, which are emergency response, communication procedures, EOC, and restoration and

Table 10.7 Impacts from crises

Impacts from crises (Top 3 impacts)	Frequency	Percentage
Delay in work/dissatisfied customers	27 out of 56	48
Revenue impact	26 out of 56	46
Loss of productivity	25 out of 56	45
Impact from crises (bottom 3 impacts)	Frequency	Percentage
Huge data loss and client impact	1 out of 56	2
Building had to be evacuated	4 out of 56	7
Failure of few systems	5 out of 56	9

Table 10.8 Crisis SOP elements

	Yes (%)	No (%)	Do not know (%)
Emergency response in place	75	10.7	14.3
Communication procedure in place	89.3	0	10.7
Emergency Operation Center (EOC) in place	53.6	25	21.4
Restoration and recovery in place	71.4	10.7	17.9

recovery procedures. Regarding the name of the SOP, 34% of the respondents claimed that their firms named the SOP as a Crisis SOP, 28% of them referred to it as a Crisis Management Plan, 21% called it a Business Continuity Plan, and 17% of them knew it as an Emergency Plan.

The respondents who have confirmed the existence of a crisis SOP also stated that most of them have prepared external coordination with external agencies such as the police, the fire department, and the hospital. Other parties such as the government were also contacted for coordination, but not as often as the previous parties. Last but not least, it had been found that, during crises, the respondents viewed the clients and employees as the top priorities as points of communication, followed by the firms' partners, regulator/government, and vendor/supplier.

10.3.2 BCM General Knowledge

Section B of the questionnaire also obtained the respondents' understanding about BCM. It had been found that, of the 56 samples, 87.5% responded that they did not know about BCM and 12.5% did. Among the samples that knew BCM, 27% percent were familiar with this concept from their mentor or colleague. Also, most of them agreed that the main reason their firm implemented BCM was to protect the firm and ensure long-term survival (26%). The details of these results can be seen in Tables 10.9 and 10.10, respectively.

The one-sample t-test was performed to measure the effectiveness of BCM implemented in the respondents' firms. This parametric statistic test was used to

Table 10.9 Sources of BCM knowledge

Sources of BCM knowledge	Frequency	Percentage
1. Mentor/colleague	4 out of 7	57
2. Information from company	3 out of 7	43
3. Media (newspaper, magazine, TV, internet)	3 out of 7	43
4. Academic reference (books, papers)	3 out of 7	43
5. Workshop/seminar/lecture class	2 out of 7	29

Table 10.10 Reasons for implementing BCM

	Frequency	Percentage
1. To protect the firm and ensure long-term survival	6 out of 7	86
2. Part of risk management	5 out of 7	71
3. Company requirements	3 out of 7	43
4. Response to past disruptions or crises	3 out of 7	43
5. Market practice within the industry	3 out of 7	43
6. To protect revenues and minimize potential penalties	2 out of 7	29
7. Regulations within Indonesia	1 out of 7	14

Table 10.11 Effectiveness of BCM implemented in a firm [based on their agree-disagree perspectives; Disagree-Agree (1–5)]

No.	Description	Mean	Rank	P-value
1	It can effectively reduce the impact of the disruption	4.43	2	0.030
2	It enables continued delivery of key products and services without interruption to clients	4.00	3	0.042
3	It enables the organization to return to normal operations more quickly than otherwise would have been possible	4.57	1	0.030
4	It helps to cope with the immediate effects of an incident on employees	4.43	2	0.030
5	It supports employees after recovery	4.43	2	0.030

test the mean value of a distribution. It will show its significance if the p-value is less than the significance level (Hair et al. 2011). Table 10.11 shows that the respondents significantly agreed that BCM provides effectiveness to their firms (mean scores > 4.00; p-value < 0.05). The view of BCM as enabling the organization to return to normal operations more quickly than otherwise would have been possible was the highest rank, with a mean score 4.57.

In addition, 57.1% of the respondents that had implemented BCM responded that their firms did not use any BCM standards. The other 42.9% of them either use NFPA1600:2007 or ANZ5050:2009 standards. Last but not least, the 56 respondents were interested in obtaining more knowledge about BCM if relevant assistances are given to them. From the data, the average of their interest was 3.68 (P-value: 0.000) (from the scale of 1–5; Not interested-Very interested).

10.3.3 BCM Principles

Section C of the questionnaire obtained the respondents' view of the BCM principles and whether this principle has been implemented or not in their firms. This aspect is analyzed because an organization may not know the formal concept of BCM, but it may have implemented part of the BCM principles in its organization. Table 10.12 describes the results of the survey based on the one-sample t-test. The results showed that the top five ranks of the significantly implemented BCM principles were from the Risk analysis (RA1, RA2, RA3) and Strategy analysis principles (S1 and S2), with the mean scores ranging from 3.52 to 4.04 (p-value < 0.05). The other principles (BIA1, BIA2, BCP1, TE1, PM1, and PM2) were found to be the lower ranks (p-value > 0.05).

Furthermore, in obtaining the correlations of the BCM principles implementation between private and state-owned firms, the Spearman rank correlation was performed. This test was applied to investigate whether each party shares the same perspectives regarding its ranking of the implemented BCM principles. As shown in Table 10.13, the results suggested that all correlations between the rankings by these two parties were significant.

10.3.4 Institutional Forces (IF) that Support BCM Principles Implementation

In obtaining the meaningful factors from the institutional forces (IF) that support the BCM principles implementation, Exploratory Factor Analysis (EFA) was performed. This analysis was performed by using SPSS. There were 11 tests for each BCM principle. Measurement of the Kaiser–Meyer–Olkin (KMO) measure and Bartlett's Test of Sphericity was firstly conducted to examine the sampling adequacy of the responses, ensuring that EFA was appropriate for the study. The perceived agreement from the respondents based on the given institutional forces variables were entered as the input for EFA.

The results showed that, for every BCM principle, the KMO measure of sampling adequacy was >0.5, indicating that the sample was acceptable for the analysis. Also, the high number of the Bartlett test of Sphericity and its significance level (0.000) suggested that the population correlation matrix was suitable for performing EFA. This means that the data obtained could be grouped into a smaller set of the underlying institutional forces factors (Raykov and Marcoulides 2008).

Table 10.12 BCM principle implementation in the firm (whether this principle has been implemented or not; Scale: 1 = strongly disagree, 2 = disagree, 3 = neutral, 4 = agree, 5 = strongly agree)

Code	BCM principles	Mean	Rank	P-value
RA1	Conducting risk analysis and cost benefit analysis	3.82	2	0.000
RA2	The involvement of experts and BCM committee in risk review	3.62	4	0.002
RA3	Conducting a detailed risk review that examines and assesses the availability of critical equipment, technology, and facilities for BU/CBF (including location of facilities, essential utilities and telecommunications, transportation to premises and physical security of premises)	4.04	1	0.000
BIA1	Conducting business impact analysis (BIA) that focuses on assessing the impact of losses if the corresponding business operations and processes are disrupted	3.50	6	0.009
BIA2	The involvement of experts, employees from related BUs and key staff members in BIA. BIA participants selected from related functional area/BU, including experts from the business, technology, financial, facility, and legal domains. BIA will be conducted by the BCM coordinator, committee, and experts	3.21	9	0.261
S1	Conducting strategy analysis for maintaining the operations of CBFs that cover pre-incident preparedness, response and recovery	3.71	3	0.000
S2	Determining staff members to support the recovery strategy and providing training and awareness programmes	3.52	5	0.012
BCP1	Developing the detailed BC plan (compiled from BCM principles 1, 2, 3), including its emergency response, Emergency Operations Center (EOC) plan, and identified Critical Business Functions (CBFs) with their Recovery Time Objectives (RTOs) and Recovery Point Objectives (RPOs). BC plan caters to 4 sets of activities (pre-incident preparation, response to incident/emergency/disaster, recovery and resumption of CBFs, restoration and return of all business operations from temporary measures adopted during recovery to supporting normal business requirements after disaster)	3.50	6	0.013
TE1	Providing periodic tests (to verify the capability of BC plan) and exercises (to train and condition BC team members to highlight any weaknesses in the operation and effectiveness of BC plan in following corrective actions) to ensure that the BC plan is viable and workable	3.30	7	0.139
PM1	Analyzing the ongoing efforts and activities to maintain the effectiveness of its BCM, including providing systematic training and awareness programmes to staff members	3.25	8	0.163
PM2	Conducting BCM training and awareness programmes for all staff and related external parties	2.89	10	0.549

The EFA results for each BCM principles, based on their rotated factor matrix can be seen in Appendix E—Analysis for Chap. 10. As the factors were extracted in order of decreasing eigenvalues, which denote the importance of the factors, Factor

Table 10.13 Correlations between private and state-owned firms

Firm type	State-owned	Private
State-owned	1.000	0.725*
Private	0.725*	1.000

*Correlation is significant at the 0.05 level (2-tailed)

Table 10.14 Most important institutional forces factors (IF) that support risk analysis and review (variable orders by size of loadings)

RA1	RA2	RA3
Awareness of potential risks	Appropriate and effective for better planning	Awareness of potential risks
Fair procedures for better planning	Awareness of potential risks	Appropriate and effective for better planning
Cost impact	Management full support	Comply with the regulation that is taken for granted
Management full support	Company's reputation	Non-compliance impact
Concern for reputation	Comply with the regulation that is taken for granted	Concern for reputation
Part of the company culture	Improve the company's procedure	Management full support
	A fair procedure	Part of the company culture
	Non-compliance impact	
	Part of the company culture	

Table 10.15 Most important institutional forces factors (IF) that support Business Impact Analysis (variable orders by size of loadings)

BIA1	BIA2
Fair procedures for better planning	Appropriate and effective for better planning
Management full support	Fair procedures for better planning
Improve the organization's procedures	Awareness of potential risks
Improve the employee's safety and welfare	Company's reputation
Easily integrated with other management systems	Management full support
Comply with the regulation that is taken for granted	Improve the company's procedure
Not implementing BIA can result in sanctions	Non-compliance impact
Company culture	Comply with the regulation that is taken for granted
	Company culture

1 (for each BCM principle) was the most important factor. Its importance was also showed by the variance explained by this factor (Dunteman 1989; Tjandra 2004). For further explanation of the results, the tables below (Tables 10.14, 10.15, 10.16,

Table 10.16 Most important institutional forces factors (IF) that support strategy analysis (variable orders by size of loadings)

S1	S2
Fair procedures for better planning	Fair procedures for better planning
Appropriate and effective for better planning	Management full support
Part of the awareness of potential impacts	Company's reputation
Management full support	Helps to improve the organization's procedures for facing crisis
Company's reputation	Part of the awareness of potential impacts
Easily integrated with other management systems	Easily integrated with other management systems
Non-compliance impact	
Company culture	
Improve the employee's health, safety and welfare	

Table 10.17 Most important institutional forces factors (IF) that support BC plan development and tests and exercises (variable orders by size of loadings)

BCP1	TE1
Appropriate and effective for better planning	Appropriate and effective for better planning
Part of the awareness of potential impacts	Company's reputation
Company's reputation	Improve the organization's procedures for facing crisis
Comply with the regulation that is taken for granted	Management full support
Company culture	Comply with the regulation that is taken for granted
Non-compliance impact	Not providing periodic tests and exercises to ensure that the BC plan is viable and workable may lead to negative impact
Not implementing them can result in sanctions	Easily integrated with other management systems

10.17 and 10.18) illustrate the most important IF factors for each BCM principle (Factor 1 for each BCM principle).

The first factor for these principles showed the variety of reasons and motivations that can support the risk analysis and review process in the firm. Table 10.14 explains the most important IF for implementing risk analysis and review. The respondents agree that, for conducting risk analysis and cost-benefit analysis (RA1), the awareness of potential risks can mostly drive employees to conduct this process. Also, in conducting risk reviews (RA3), this awareness was viewed to be the most agreeable IF that can support this process. In implementing this process, the involvement of experts and the BCM committee (RA2) was viewed as appropriate and effective for better planning. From these three aspects, IF, such as awareness of potential risks, management full support, concern for reputation, and part of the

Table 10.18 Most important institutional forces factors (IF) that support programme management (variable orders by size of loadings)

PM1	PM2
Fair procedures for better planning	Fair procedures for better planning
Management full support	As appropriate and effective for better planning
Part of the awareness of potential impacts	Management full support
Improve the organization's procedures for facing crisis.	Comply with the regulation that is taken for granted
Comply with the regulation that is taken for granted	Not implementing them can result in receiving sanctions
Easily integrated with other management systems	Company culture
Improve employees' health, safety, and welfare	Improve employees' health, safety, and welfare
Company culture	Non-compliance impact

company culture were the most agreeable factors that can support the risk analysis and review process.

In implementing the Business Impact Analysis (BIA), the respondents opined mostly regarding the better planning aspects of this process. They agreed that BIA is an important part of fair procedures and effective for better planning. It can also enhance or improve the firm's procedures, particularly for resiliency. Other forces such as the need to comply with the regulation, awareness of its sanction when not complied with, and already part of the firm's culture were viewed to be important reasons to conduct BIA. In conducting a comprehensive BIA (BIA1), the motivation to improve employees' safety and welfare and its easy integration with the firm's current management system were confirmed to be beneficial. Also, management's full support and the concern for the company's reputation were opined to be important when this process involves experts and various BU employees (BIA2). Table 10.15 describes the factors in greater detail.

Similar to BIA, the respondents also asserted that strategy analysis should be implemented for better planning. This can be seen in Table 10.16, where the management's full support, concern for the company's reputation, the awareness of the potential impacts of not conducting this process, and its easy integration with the firm's current management system were also chosen as important drivers to its implementation. In addition, the motivation to improve employees' health, safety, and welfare and acknowledging that this process was already part of the firm's culture were supporting factors in conducting a thorough strategy analysis (S1).

Table 10.17 describes the most important IF factors that support BCM's principles four and five (BC Plan development; Tests and Exercises). The view of their effectiveness for better planning was also the most important factors that can support implementation. Other normative forces, such as concern for the firms' reputation and regulative force such as non-compliance impact, were also considered to be important and able to support these principles' implementation.

Regarding the implementation of BCM's programme management, Table 10.18 shows the IF factors that can support it. Similar to other principles, normative forces were the dominant factors, where the motivation for better planning, full support from the management, and improvement of employees' health, safety, and welfare were viewed as important. In addition, cultural-cognitive forces such as complying with the current regulation and the consideration as part of the company's culture can be important drivers of this principle.

Lastly, from the analyses, it had been found that there were two normative forces considered to be the least supporting factor for BCM implementation. In implementing all of the six BCM principles, the firms' competitiveness and stakeholders/clients' requirements were viewed as less important than other factors mentioned above.

10.3.5 Organizational Culture Attributes (OC) Implemented for BCM Principles

Similarly, EFA was also performed to obtain the meaningful factors from the organizational culture (OC) attributes that had been implemented. There were also 11 tests for each BCM principle. The KMO measure of sampling adequacy and the Bartlett test of Sphericity were also >0.5, which was acceptable for the analyses. The input for EFA was also the perceived agreement from the respondents based on the given OC variables. The EFA results for each BCM principle, based on their rotated factor matrix, can also be seen in Appendix E—Analysis for Chap. 10. For further explanation about the results, the tables below (Tables 10.19, 10.20, 10.21, 10.22, 10.23, 10.24, 10.25, 10.26, 10.27 and 10.28) illustrate the most implemented OC factors for each BCM principle (Factor 1 for each BCM principle) and the least implemented factors (the last factor for each principle's rotated factor matrix).

Tables 10.19 and 10.20 describe the OC attributes that were viewed as the most and the least implemented factors that can support risk analysis and review in the respondents' firms.

In supporting the RA1 process, OC attributes which were grouped into adaptability to changes, coordination and integration, and employee-oriented were opined to have been implemented. The involvement of experts and the BCM committee (RA2) was supported by the firm's team orientation, open system, and empowerment to their employees. In terms of implementing risk review, OC attributes such as process-oriented nature, coordination, and adaptability to changes were viewed to be beneficial. The results also showed OC attributes that were less implemented and may need to be considered to support this process (Table 10.20), such as setting standards and increasing the level of risk avoidance. In supporting the BCM team and experts of these analyses, professionalism, team orientation, and

Table 10.19 Most implemented organizational culture attributes for risk analysis and review (variable orders by size of loadings)

RA1	RA2	RA3
Adaptability to change • Open to alternative solutions • Encourage creative and innovative ideas • Allocate sufficient resources for implementing innovative ideas	*Team orientation* • The experts and the BCM committee's commitment to the risk review team are highly valued	*Process oriented* • Organizational procedures is essential
Coordination and integration • Good communication • Able to reach agreement on critical issues	*Open system* • The experts (from an external party) and the BCM committee (from an internal party) fit very well into the risk review team, and their opinions are appreciated	*Coordination and integration* • Good communication is essential
Employee oriented • Responsibility for its employees' welfare	*Empowerment* • The member of the risk review team is enabled to decide or solve problems within his/her sphere of responsibility or authority	*Process oriented* • Unfamiliar situations are managed and identified
		Adaptability to change • Allocate sufficient resources for implementing innovative ideas
		Coordination and integration • Inter-departmental collaboration is encouraged

Table 10.20 Least implemented organizational culture attributes for risk analysis and review (variable orders by size of loadings)

RA1	RA2	RA3
Setting standards • High level of cost-consciousness • Detailed set of performance standards	*Professional* • Emphasizes job competence	*Setting standards* • High level of cost-consciousness • Provides a detailed set of performance standards
Customer orientation • Meeting the client's needs and satisfaction	*Team orientation* • Emphasizes team contributions • Amicable exchange of opinions and ideas between members is facilitated	*Adaptability to change* • Able to create adaptive ways to meet changing needs
Process oriented • High level of risk avoidance	*Empowerment* • Member of the risk review team to participate in the decision-making process	*Process oriented* • High level of risk avoidance

Table 10.21 Most implemented organizational culture attributes for business impact analysis (variable orders by size of loadings)

BIA1	BIA2
Adaptability to change • Encourages creative and innovative ideas • Open to alternative solutions • Able to allocate sufficient resources for implementing innovative ideas	*Empowerment* • Supports the member of the BIA team to participate in the decision-making process
	Team orientation • Emphasize on team contributions • The experts, employees from related BUs and key staff's commitment to the BIA team are highly valued by the company

Table 10.22 Least implemented organizational culture attributes for business impact analysis (variable orders by size of loadings)

BIA1	BIA2
Setting standards • The company has many internal structuring (procedures) in the organization and considers that meeting times is essential	*Open system* • The experts (from an external party), employees from related Bus, and key staff (from an internal party) fit very well into the BIA team, and their opinions are appreciated
Customer orientation • Main focus on meeting the client's needs and fulfilling satisfaction	*Team orientation* • Amicable exchange of opinions and ideas between members is facilitated
Setting standards • High level of cost-consciousness	*Professional* • Emphasis on job competence
	Empowerment • The member of the BIA team is enabled to decide on or solve the problems

Table 10.23 Most implemented organizational culture attributes for strategy analysis (variable orders by size of loadings)

S1	S2
Adaptability to change • Encouragement of creative and innovative ideas • Open to alternative solutions • Able to allocate sufficient resources for implementing innovative ideas	*Team orientation* • High degree of cooperation among employees
Coordination and integration • Good communication	*Developing employee's skills* • Invests in the development of employees' skills • Guidance for employees' performance improvement is provided
Empowerment • Values employees' ideas	*Team orientation* • Amicable exchange of opinions and ideas between members is facilitated
Coordination and integration • The members of the committee or business units are able to reach agreement on critical issues	

Table 10.24 Least implemented organizational culture attributes for strategy analysis (variable orders by size of loadings)

S1	S2
Empowerment • Employees' participation in the decision-making process	*Team orientation* • Support for the recovery strategy is highly valued • Emphasis on team contributions
	Professional • Emphasis on job competence
	Open system • The staff members support the recovery strategy and fit very well into the BCM team, and their opinions are appreciated

Table 10.25 Most and least implemented organizational culture attributes for BC plan development (variable orders by size of loadings)

Most implemented	Least implemented
Customer orientation • Has a main focus on meeting the client's needs and satisfaction	*Employee oriented* • The manager in the company has a high concern for the procedures in the operation of the BUs
Setting standards • Provides a detailed set of performance standards • Has many internal procedures in the organization	
Adaptability to change • Able to create adaptive ways to meet changing needs	
Setting standards • High level of cost-consciousness	
Adaptability to change • Able to allocate sufficient resources for implementing innovative ideas	
Coordination and integration • The members of the committee or business units are able to reach agreements on critical issues	

empowerment of the team for a better decision-making process should also be considered.

In supporting BIA, attributes related to adaptability to change were the most important factors which had been implemented by the firms. The involvement of experts and employees in this process should also be supported by the firm's empowerment and team orientation. These descriptions can be seen in Table 10.21. Similar to the risk analysis and review process, attributes such as setting standards and professionalism were not implemented as much. However,

Table 10.26 Most and least implemented organizational culture attributes for tests and exercises (variable orders by size of loadings)

Most implemented	Least implemented
Coordination and integration • Different functions and units of the organization have a high capability to work together • The members and BUs accept criticism and negative feedback	*Process oriented* • Unfamiliar situations are managed • Following organizational procedures
Team orientation • High degree of cooperation among employees	
Coordination and integration • The company can resolve internal problems effectively	
Adaptability to change • Allocate sufficient resources for implementing innovative ideas	

Table 10.27 Most implemented organizational culture attributes for programme management (variable orders by size of loadings)

PM1	PM2
Setting standards • Provides a detailed set of performance standards • High level of cost-consciousness • Many internal structuring (procedures) in the organization	*Developing employees' skills* • Continually invests in the development of employees' skills • Guidance for employees' performance improvement
Customer orientation • Main focus on meeting the clients' needs and fulfilling their satisfaction	*Team orientation* • Amicable exchange of opinions and ideas between members is facilitated • Emphasis on team contributions • The staff members' commitment to be involved BCM training and awareness programmes are highly valued by the company
A set of values • High degree of clear strategic intentions	
Reward orientation • Emphasis on team accountability	
Leadership—power in organization • The leadership of the company's top management has a high influence	
Coordination and integration • Different functions and units of the organization have a high capability to work together to achieve common goals	

there were different views for the involvement of experts and BCM committee in BIA (BIA2). Implementing an open system, exchanging ideas between members, and enabling them to solve problems within their responsibility were not as highly considered. More details were shown in Table 10.22.

Table 10.28 Least implemented organizational culture attributes for programme management (variable orders by size of loadings)

PM1	PM2
Coordination and integration • Important decisions are made by groups as a consensus	*A set of values* • Actions are matched with the company's goals
	Professional • Emphasis on job competence
	A set of values • Provides a clear direction for employees in their work

The most implemented OC attributes for strategy analysis were also similar with the previous principles, which are adaptability to changes, coordination and integration, and empowerment. In implementing the staff members' support for recovery strategy (S2), the firm's culture of encouraging good cooperation among employees, developing their employees' skills, and facilitating opinions and ideas between staff members were viewed to be beneficial. Table 10.23 explains the detailed attributes of this process. Furthermore, Table 10.24 describes the least implemented OC attributes for this process, which were other attributes of empowerment, team orientation, professionalism, and open system.

Regarding developing a BC plan, focusing on meeting the client's needs and setting standards were considered as the most important attributes, followed by the firm's adaptability to changes and coordination and integration. On the other hand, the attribute that was least implemented was the manager's concern for the procedures in their business units. Table 10.25 provides the details of these attributes.

Moreover, Table 10.26 shows the OC attributes that were most and least implemented for BCM principle 5; Tests and Exercises. In conducting these attributes of coordination and integration, team orientation, and adaptability to changes were mostly needed. Attributes such as managing unfamiliar situations and following procedures were found to be the least implemented ones.

Lastly, in conducting programme management for BCM which maintains the whole processes from the BCM principles, Tables 10.27 and 10.28 explain the OC attributes that should be considered.

In addition to coordination and team orientation, implementing attributes such as setting standards, customer orientation, a set of values, reward orientation, leadership, and development of employees' skills were viewed to be essential. Attributes such as group decision making, emphasis on job competence, clear directions, and actions that are suited with the firm's goals were found to be less often implemented.

Table 10.29 P-value range of OC attributes (between the perceived importance and implemented) for each BCM principle

Code	P-value range
RA1	0.274–0.986
RA2	0.435–0.855
RA3	0.264–0.980
BIA1	0.439–0.971
BIA2	0.259–0.921
S1	0.308–0.995
S2	0.165–0.891
BCP1	0.284–0.988
TE1	0.190–0.979
PM1	0.322–0.985
PM2	0.259–0.977

10.3.6 Perceived Importance and Implemented for OC Attributes

In order to ascertain whether there was any significant difference between the perceived importance and perceived implementation of the OC attributes towards BCM principles, the Wilcoxon Signed-Rank two-tailed test was used. This non-parametric test for significance between attributes is the most appropriate, as the attributes were measured on an ordinal scale. One hypothesis was formulated, and a 5% level of significance was used. Based on the calculated p-value, if the significance value is <0.05, there is a significant difference. If the significance value is >0.05, there is no significant difference (Hair et al. 2011; Tan 2008).

Based on the analyses, the results show that there were no significant differences between the respondents' perception of the OC attributes' importance (as supporting factors to implement BCM) and their implementation. The p-value for each OC attributes for each BCM principle was >0.05. This means that their determination of what they have perceived as important (or not) is in line with what they have implemented in their firms. Table 10.29 shows the overview of the p-value range of the tests for each BCM principle. To view in a more micro result of the analyses, figures of each BCM principle's OC attributes with its mean scores were provided in Appendix E—Analysis for Chap. 10.

10.3.7 Survey Validation

After all responses were received and analyzed, a set of face-to-face interviews was carried out to validate the findings of the data analysis. These interviews were conducted with five respondents who all had more than 15 years of experience in the construction sector. Three respondents (Respondents 2, 3, and 5) were from private contractor companies, and two respondents (Respondents 1 and 4) were from state-owned contractors. These respondents were chosen due to their willingness to participate in further discussion about the results from the survey. The results of the interview were as follows:

10.3.7.1 The Knowledge of BCM

The survey results showed that the majority of the survey respondents did not know about BCM. It turned out that four respondents who were interviewed also did not know about BCM, and only one respondent had heard about it. The respondents' comments were as follows:

Respondent 1 I do not know the BCM term. This is new for me. I have been in this sector for about 30 years and have not known about this. So yes, this is the first time I heard about BCM.

Respondent 2 I have heard about BCM, but I do not know what the abbreviation stands for. I heard this is related to managing crisis.

Respondent 3 I have not heard about BCM. I think the Indonesian construction industry has not been familiar of this concept. Probably other sectors may have adopted this, but I do not think this has been adopted in construction.

Respondent 4 No, I do not know about BCM. And since you've brought this topic up, I would like to know more about it.

Respondent 5 No, I have not heard about BCM. This is the first time for me to hear this term.

10.3.7.2 Crises Occurred Within the Firm in the Last 5 Years

Based on the result of Table 10.6, the five respondents agreed on the top seven crises (from 32 types of crises given in the questionnaire). All of them briefly described their experiences based on these crises. Below were their descriptions about those crises:

Respondent 1 I agree with the results. These top crises have occurred more frequently in the construction sector. I had many experiences regarding the lack of component workforce. Due to this problem, there were so many project delays. Also, the unexpected price of raw materials had impacts on the project's costs. The project office had to propose to the main office to increase the budget for the project and this needs some time to be approved. Regarding natural disasters, I had an experience with overcoming floods in the project area. When this occurred, the project stopped until the water receded after several days. There was nothing that we could do about it. For issues regarding disputes, this had occurred when the locals wanted to work in our project as our laborers, and they were doing demonstrations that lead to riots near our project site. We cannot hire them directly because they need to have skills. We had delays in our project due to this problem, and we needed to resolve it by communicating with the locals and their leaders.

Respondent 2 Access restriction or limitation is the top issue here, mainly due to the lack of good coordination between stakeholders. For example, for road/toll projects, the first thing that needs to be done is land acquisition. This single process can take a long duration of time, and sometimes it is not resolved for months. Therefore, the project is delayed or even stopped until the acquisition process is finished. This is mostly due to the land owners who disagree with the settlement that they received. There are many problems regarding land acquisition for road projects. Regarding natural disasters, Indonesia is a country that experiences natural disasters every year. So I agree this is included in the top crises for construction firms. As a construction firm that may have projects anywhere and may be located in earthquake-prone areas, it needs to be prepared for this type of crisis. I think price escalation may be related to changes in regulations. When this occurred, we usually negotiate with the regulator or work together with the constructor association (ICA) to handle this matter effectively.

Respondent 3 I think changes in regulations may also be related with the project financing. If something changes in the financing process, this can lead to higher impact for the project, and the project can be delayed/stopped. Among all of these crises, I think natural disasters are the crises that need to be prepared for. I have so many experiences, particularly in overcoming floods. When floods occur near your project site, or even in your head office, you need to save a lot of things. Documentations, equipments, materials, and even the workers' belongings in their project lodge. I think a comprehensive and socialized flood forecast is needed for providing information to us so we can prepare to overcome it.

Respondent 4 I think these results are portraying the crises that we have been experiencing in this sector. For natural disasters, I have experience handling floods in our projects. Even though we have a force majeure clause listed in the contract, but handling it during the flood is quite a tough time. There were so many clean-ups in the site and the project had to stop for several days. Another crisis that also has a high impact is the lack of component workforce. When your skilled workforce is not available, you need to have a good resource allocation. If not, the project will be delayed.

Respondent 5 For me, uncertainty in resolving disputes, natural disasters, and loss of management personnel are the top 3 crises that I have experienced in 5 years. My projects are mostly located in Sumatra, where land acquisition issues occurred frequently. I need to coordinate and communicate with the locals and their governments. These issues could take several months, mostly due to lack of good law enforcement. For natural disasters, haze and forest fires are among the top crises that we have experienced. These events occurred annually, and there were 5–7 days of disruptions due to these. The impacts were high, where it affects the cost of equipment leases and labor salary. The loss of management personnel was also viewed as a crisis, because there were some periods of adjustment needed in the company. The new person who undertook the position needed to adjust to the work environment, job assignment, skills needed, coordination with other units, etc. Conflicts could occur due to this matter.

10.3.7.3 Impacts from Crises

As seen in Table 10.7, the top three impacts and the bottom three impacts of crises were chosen as significant to the surveyed respondents. All of the five interviewees also agreed on the results, with comments as follows:

Respondent 1 Yes, delays in work are the highest impact for construction. Any crisis will lead to this. Other than this, I think loss of productivity is also the top impact from crises. Due to a disruption, the worker/employee cannot do their usual job in a normal time, there will be productivity loss.

Respondent 2 I think the three top impacts have their correlations. If a crisis occurs, there will be disruptions that will lead to delays. Delays will lead to loss of productivity of the worker, and due to this, there will be more budget needed to finish the job, which has an impact toward the firm's financial aspect, including its revenue.

Respondent 3 Delays, revenue, and loss of productivity are the major impacts that need to be considered for people working in the construction sector. Regarding the impact of buildings that need to be evacuated, I think this may occur if the crises are natural disasters, fire, or terrorism. For building evacuation, all of the company's documentation should be put in a safe place.

Respondent 4 I agree that delays are the top impact from crises. Delays from crises may not be avoided, but the most important thing is to minimize the period of delays for the project to continue.

Respondent 5 I could not agree more. Delays and loss of productivity are impacts that need to be considered by the firms. I think, in overcoming these, non-technical skills such as management skills are needed here. Also, the quality of human resources is essential.

10.3.7.4 Crisis SOP in Place

The survey results showed that the majority of the respondents had a crisis SOP in place. Respondents 1, 2, 4, and 5 had also agreed to have a crisis SOP in their firms, as mentioned below:

Respondent 1 My firm has provided a crisis SOP in place, but I think it is not for every crisis. Mostly it is used for accidents in projects and force majeure. The details in the SOP include the emergency response procedures, communication, and recovery.

Respondent 2 Yes, we have a crisis SOP in place. This is part of our Safety, Health and Environment (SHE) program. But I am not sure whether this SOP is applicable to all of the possible crises in construction.

Respondent 4 There is an SOP for overcoming force majeure and accidents in projects and main office. This document includes the evacuation process, emergency response, communication process with the internal and external parties during the incident, temporary operation center, and recovery process after the incident.

Respondent 5 A crisis SOP is part of our SHE program. We have defined the possible crises and documented the responses needed to overcome those. Every site office has one document about this, but I am not sure whether they have done any evacuation drills at the site. For our main office, we have fire drill once a year, but that is it.

However, Respondent 3 was not sure whether it was in place:

Respondent 3 I am part of the samples that do not know whether my firm has a crisis SOP in place. I am aware that we have procedures to overcome accidents in projects or force majeure, but it is not documented as an SOP yet.

10.3.7.5 External Coordination During Crises

All of the interviewees confirmed the survey result regarding external coordination during crises. They all agreed about the need for coordinating the external agencies and government. Also, Respondents 2 and 5 respectively mentioned the military and the contractors association as other external agencies with which to coordinate for crisis preparation. The feedback was as follows:

Respondent 1 Communication during a crisis is essential. I think keeping the contact numbers of the external agencies and the related government official is necessary.

Respondent 2 If you have projects in a rural area, you need to have good external coordination, particularly with the police, hospital, fire department, and the military.

Respondent 3 Yes, external agencies such as the police, hospital, fire department, etc., are important in case anything happens in the project or the main office. You need to update their contact numbers regularly and maintain a good coordination.

Respondent 4 Once you are in the project's location, you should know about the external agencies' particulars: their location, contact number, person in charge, etc. You need to have good coordination with them, especially during a crisis. The government's role is mostly for proposing approval to some regulations or access approvals.

Respondent 5 I agree with the results. This external coordination is important, especially during a crisis. I would like to add one more party, which is the contractor's association. Good coordination with this party is beneficial, because the association can assist you in solving problems related to regulations. They are helping you voice your concern to the regulator.

10.3.7.6 Communication Coordination During Crises

The five parties that were considered important to communicate with during crises were found to be significant. All five interviewed respondents supported this assertion. Their responses were as follows:

Respondent 1 If anything happens with your project, you need to communicate this with your internal parties first, then coordinate it with the partners, clients, and other external parties.

Respondent 2 Every project will have its coordination meetings in place. All of the listed parties can be involved in these meetings. But during crises, the internal parties such as the employees/workers should be communicated with first, then the coordination with other parties can continue.

Respondent 3 I think all of the five parties are essential to communicate with during crises. The clients and partners should know what really has happened in the project/firm, the employees should know this in advance of other external parties, the suppliers should be informed, especially when they also have impacts from the crises, and the government should be aware of the situation for a larger-scale response.

Respondent 4 During crises, particularly the high-impact ones, clients and partners should be contacted. They have a big influence on the projects/firm, and they may have an important role in overcoming the crises.

Respondent 5 A higher percentage for clients and employees regarding communication coordination during crises means that the respondents are aware that these parties are considered vital to the firm. Any impact that rises from the crisis will have an effect mostly on the employees and clients.

10.3.7.7 Reasons for Implementing BCM

Based on Table 10.10, Respondents 1, 3, and 5 expressed views about the regulations within Indonesia. They recommended that BCM be regulated by the government. Their comments were as follows:

Respondent 1 Based on your description about BCM, I agree that this concept can help the firm to protect its assets and prepare for long-term survival. I am not sure whether BCM implementation has been regulated by our government, particularly for construction. But it should be considered.

Respondent 3 *If BCM implementation is regulated by the government, I think every construction firm will comply with this. Some may do this just for following the rules, but others, the ones that are financially stable, may do this to increase their competitiveness and reputation. In Indonesia, we already have the Disaster Management Agency as the national scale. It would be very beneficial if the construction firms, regulated by the government, integrate their BCM with this agency for better resilience.*

Respondent 5 *I do not think that the construction industry has implemented BCM; therefore, more socialization about this concept should be conducted. This concept should be known to the regulators so they can inform this to the construction firms and provide a sound regulation.*

Moreover, Respondents 2 and 4 agreed on the view of BCM as a long-term plan:

Respondent 2 *To firms that focus on short-term plans, BCM may be viewed as a high-cost burden. But to firms that focus on long-term plans, this concept is beneficial and will reduce their future risks. These firms will view this implementation as an investment.*

Respondent 4 *I agree that BCM is part of risk management, but also this concept is more than that. Ensuring long-term survival of the firm is mostly what BCM will provide.*

10.3.7.8 Effectiveness of BCM Implemented in Firms

Table 10.11 showed the results of BCM's effectiveness in firms. The respondents all agreed on the survey results, with comments below:

Respondent 1 *Based on the description of the concept, BCM has a comprehensive approach for overcoming crises and resuming activities back to normal. I agree that this statement has shown the effectiveness of BCM. For a firm to implement BCM into this kind of effectiveness, I believe that it has trained and prepared its human resources to comply with the BCM procedures.*

Respondent 2 *The nature of the concept of continuing business during crises should make BCM principles and techniques focus on the resuming business quickly and effectively.*

Respondent 3 *This effectiveness can be achieved due to its thorough planning concept. I agree that, by identifying the CBF and MBCO, the firm can coordinate its units more effectively.*

Respondent 4 *An organization able to return to normal operations more quickly should be supported not only by BCM's techniques, but also its employees/workers. Responsive and prepared human resources are needed for this.*

Respondent 5 This means that BCM provides ways to handle crises effectively. It provides plans that are needed to coordinate the firm's resources in order to resume to normal operations after any disruption.

10.3.7.9 BCM Standards Used in Firms

The survey results showed that more than 50% of the respondents who had BCM in place do not use any BCM standards in their firms. There were several recommendations provided by the five respondents regarding this matter:

Respondent 1 I think if the ISO provides the BCM standards, many contractors would be able to develop their BCM from this standard. ISO standards are mostly used by Indonesian contractors.

Respondent 2 There should be more information regarding BCM and its standards. I just heard now that Singapore has its own BCM standard. As part of an ASEAN country, Indonesian firms could adopt this standard.

Respondent 3 The Indonesian contractors should do more benchmarking with other countries, especially in understanding more about BCM and its standards.

Respondent 4 In informing the construction firms about the latest standards, etc., I think the Construction National Board (NBCSD) and the ICA have this role. They can introduce the BCM standards to the firms and provide assistance for the implementation. But I'm not sure whether these boards have been aware of BCM and its standards.

Respondent 5 If the construction firms are involved in international projects or having joint ventures with international firms, they may have been aware of BCM and its standards.

10.3.7.10 Interests in BCM

From the result of a good degree of interests from the respondents in terms of learning more about BCM, the five interviewees supported this and provided feedback on which parties can provide BCM workshop/training, as follows:

Respondent 1 It is a good notion that the respondents are interested in understanding more about BCM. The top management of the firms should think about what kind of relevant assistance to be provided for their employees in learning about BCM. Once this is managed from the top level, it can be implemented.

Respondent 2 This response will be more effective when the interest comes from the top management level. A system or a concept can be implemented once the top management fully supports it. If the idea comes from the lower level of the firm, time may be needed to propose and persuade those at the higher level.

Respondent 3 This interest in BCM should also be supported by the respondents' superiors/bosses.

Respondent 4 Regarding this level of interest, I think the firm, supported by the ICA and NBCSD, is able to carry out some kind of BCM workshop for the contractors employees. Or maybe it can hire a certified BCM trainer to inform and socialize BCM more thoroughly. Also, the Ministry of Public Works could support this type of program. This could be a good start for the Indonesian construction industry to understand more about BCM.

Respondent 5 If the interest in BCM is high, then the relevant parties can provide the BCM training/awareness program. I think the ICA, Public Works Ministry, and NBCSD can support this event.

10.3.7.11 BCM Principles Implementation in the Firm

The results of the top five ranked implemented BCM principles from Table 10.12 were viewed to be significant by the five respondents. They supported these results with comments below:

Respondent 1 I am aware that risk management has been adopted in construction, especially for managing project risks. Maybe that is why most of the respondents have agreed to have implemented risk analysis.

Respondent 2 Risk analysis and strategy analysis are necessary during project planning. I could not agree more with this result. The firms should have implemented these analyses in their daily operation. BIA, BC plan, tests and exercises, and programme management are the next phases that they need to understand in regard to BCM implementation.

Respondent 3 In managing construction projects, risk and strategy analyses have been implemented by most of the firms. Now they should learn about BIA and BC plan development in order to adopt BCM.

Respondent 4 Usually large contractors have implemented sound risk management, particularly for their projects. Strategy analysis is included in the risk management.

Respondent 5 Part of the BCM principles are risk and strategy analysis, where these analyses have been implemented by the contractors. Therefore, understanding BCM would not be difficult for them, because they just need to understand and apply the latter principles for a holistic BCM implementation. I know this sounds simple, it will not be that simple, but this result provides positive feedback to introducing BCM to the contractors.

10.3.7.12 Correlations Between Private and State-Owned Firms

Table 10.13 showed that there were correlations between the rankings of private firms' BCM implemented principles and those of the state-owned firms. The respondents agreed on these results, with reasons provided as follows:

Respondent 1 In Indonesia, private and state-owned firms have similar traits. They delivered similar projects, in similar locations. Therefore, there should be no significant difference between them.

Respondent 2 From my point of view, the way the private and state-owned firms deliver their projects are similar. Regarding their knowledge/skills, I think both have the same sets of skills, especially that they are involved in the same business environment.

Respondent 3 I have some experience working together with the state-owned contractors in the form of joint ventures. We all have similar management and technical skills.

Respondent 4 The differences between private and state-owned firms are mostly related to the regulations. There are some government projects that can only be delivered by state-owned firms. So you can say that there are some differences in the procurement system between the private and state-owned firms. Other than that, I think the technical and non-technical skills provided are similar.

Respondent 5 Both of these types of firms are under the same ministry, construction board, and association. They both have learned and developed in the same environment. So that is why they have similar ranks toward BCM implementation.

10.3.7.13 The IF Drivers for Implementing BCM in Contractors

Regarding the drivers related to the institutional forces, the five respondents provided their views and supported the survey results. The responses were as follows:

Respondent 1 I think regulative forces still dominate the way firms implement a system or a concept. If a regulation or a rule has been given, like it or not, they will obey it in order to avoid any sanction or penalty. So if there is an official regulation to implement BCM for contractors, I think they will implement it.

Respondent 2 It depends on the type of the firms. For a large contractor, normative forces may dominate the way it behaves. Large firms tend to view that increasing their level of competitiveness and legitimacy are important. Implementing a concept like BCM that can increase their market value would be beneficial to them. On the other hand, small contractors may not experience the same thing. These firms would still focus on their financial stability. Implementing BCM may be viewed as a burden to them.

Respondent 3 I don't think our contractors have the cultural-cognitive forces to implement BCM. Regulative forces such as regulation and sanctions are still the factors that help them implement such a concept. Normative forces may support BCM implementation, but only for firms that have long-term plans.

Respondent 4 I think regulations are important for driving the contractors to implement BCM. For example, to implement quality and environmental management, there are official regulations from the government. Therefore, the initiative to implement BCM in construction should be a top-down approach, particularly from the regulator to the firm's top management.

Respondent 5 Both regulative and normative forces are the drivers for implementing BCM in contractors. Once BCM is the official requirements for contractors to comply with, they will start to adopt this concept. Maybe if part of the tender requirements is having a BCM certification, the firms will obtain this certification.

10.3.7.14 The IF Hindrances for Implementing BCM in Contractors

Furthermore, the respondents provided their views on the hindrances for implementing BCM in contractors, within the context of institutional forces. Only Respondent 4 was in line with the survey results, whereas the other respondents provided opinions regarding cultural-cognitive forces. Brief descriptions from each of the respondents were provided below:

Respondent 1 I would say the quality of the firm's human resources has the greatest effect on BCM implementation. Would this be included in cultural-cognitive forces? If the human resources are still lacking in their attitude and their awareness towards risks, the firm may have a hard time implementing BCM.

Respondent 2 Cultural cognitive forces may be the hindrance for implementing BCM. Most contractors still view things in profit/non-profit ways. If something is not profitable to them, they may not take or implement it.

Respondent 3 Most of the workers for contractors have low risk avoidance. This is due to their educational background that may not be the same as others. Therefore, cultural-cognitive forces can hinder the BCM implementation. BCM may be planned very well by the firm's management, but when a crisis really happens, I'm not sure that the workers are prepared mentally or culturally.

Respondent 4 I think the mindset of the Indonesian contractors is still focused on profitability. Aspects like competitiveness and stakeholders' requirements are not their main focus yet. So I think normative forces will not drive them to implement BCM.

Respondent 5 *Cultural-cognitive forces may hinder the BCM implementation. The culture of discipline and preparedness attitude toward risks/crisis is not yet dominating, particularly for the firm's low-level employees.*

10.3.7.15 The Significant OC Attributes Among Contractors

When discussing the significant OC attributes that may drive or hinder BCM implementation in contractors, the five interviewees supported the survey results. They also provided their views on some of the OC attributes as follows:

Respondent 1 *I think the habit of communicating and coordinating during executing projects is essential and may be a good driver for implementing BCM. Also the employees/workers in contractors are used to follow procedures for constructing their projects. Regarding the hindrance for BCM implementation, I might say the attitude of the human resources plays an important role. I'm not sure they have a high risk avoidance, particularly in terms of awareness of threats/crises. They may need to be trained for higher awareness towards risks.*

Respondent 2 *In projects, the top management gives authority to the project officers and their project staffs to conduct their own risk analysis regarding the project that they are managing. This is implemented because the management understands that the one who is on the project site is the one who fully understands the situation. So empowerment has been implemented in the firm and can support the BCM implementation. As for the hindrance, the employees are still not tidy enough to compile and document reports. Documentation is still a problem for our firm. If not instructed directly and monitored regularly, documentation can be messy.*

Respondent 3 *I think the adaptation level of the employees in construction firms is quite high, because they need to move from one location to another when delivering the projects. In a crisis situation, adaptive thinking may be beneficial. In my opinion, the skill of documenting reports for construction firm employees is still low. They are not used to writing and documenting events or lessons learned from the project activities. If they have documented the report, the storage facility is not sufficient. This can hinder the BCM implementation.*

Respondent 4 *The level of coordination in construction projects is quite complex. There are a lot of parties involved, including the client, supplier, worker, and the community around the project site. This shows that the employees are used to coordinating with other parties, and if something occurs unexpectedly, the habit of coordinating can support the effort to overcome the unexpected event. What the employees are lacking is their discipline in keeping up reports and documentation. They will be busy preparing the reports if only there were audits coming up to their units.*

Respondent 5 *I think leadership supports a good BCM implementation. Most of the construction firms are using the top-down approach, so when the top level initiates the implementation, the lower level will follow through. This would be very beneficial if the leader of the firm also has good persuasion and managing skills toward his/her employees. The subordinates would implement the procedures willfully. Regarding hindrance, documentation has been an issue in my firm. Although we have provided forms for any progress reports or activity reports, it all depends on who will be disciplined enough to write the reports. This task seems to be easy, but not doing it may lead to problems. For example, the firm does not have a complete documentation for conducting claims, so they fail to claim what may be beneficial to them. BCM may need many documented reports for its plan development, so the firm needs competent employees to conduct this documentation process.*

From this process, it can be seen that the five interviewed respondents mostly agreed and supported the survey results. They also have provided some examples, reasons for the results, and recommendation regarding the results.

10.4 Summary

This chapter presented the results of the pilot study and the questionnaire survey. The results of the pilot study provided feedback for questionnaire amendments. After the amendments, the questionnaire survey was conducted and the data was analyzed. The results of each section from the questionnaire were provided with a brief description about the analyses used. Some results that provided more details were shown in the Appendix E—Analysis for Chap. 10. Following the results descriptions, the survey validation in the form of interviews were explained. These results will be described more extensively in Chap. 13 with further findings and discussion.

Chapter 11
Data Analysis: Case Studies

11.1 Introduction

This chapter is the second part of the study's data analysis. Firstly, it describes the case studies results, followed by the survey results in the form of interviews to experts discussing BCM implementation for Indonesian contractors. The last section will elaborate the development of the BCM implementation guidelines.

11.2 Case Studies

To interpret the proposition based on Research Objective 4 (RO4) and Research Question 2 (RQ2), descriptive case studies were conducted with two Indonesian contractors (state and private-owned firms). The case studies explored the firms' characteristics and BCM implementation in the two firms.

11.2.1 Case Study 1: Firm A

Firm A is a state-owned firm, where its head office is located in Jakarta, Indonesia. It is 70% owned by the Government of Republic of Indonesia and 30% owned by the public (which included 2% share owned by the employees and management of Firm A). For this study, there were four respondents who participated and provided the data about Firm A. A brief description about the respondents: The first respondent (Respondent 1A) was a Project Manager with 23 years of experience; the second respondent (Respondent 2A) was also a Project Manager with 20 years of experience; the third respondent (Respondent 3A) was a Senior Manager with 30 years of experience; and the fourth respondent (Respondent 4A) was a Project

L.S.R. Supriadi, L. Sui Pheng, *Business Continuity Management in Construction*, Management in the Built Environment, DOI 10.1007/978-981-10-5487-7_11

Manager with 13 years of experience. Interviews were conducted with them and they also provided related documents for the study. The study was conducted for 3 weeks in March 2013.

11.2.1.1 Organization Characteristics

Organization's Objectives

Firm A's vision is to become one of the best construction and investment firms in South East Asia. Its mission is to make breakthroughs by evolving into an integrated infrastructure firm through the development of a number of subsidiaries which focus on pre-cast manufacturing and also property.

Firm A's objectives are:

1. To provide total solution services in construction, operations and maintenance, including financial through a synergy in Firm A subsidiary groups.
2. To provide assurance that the products/deliverables to the customers have met the performance and requirements as agreed; where it also focuses on satisfying the expectation of key stakeholders.
3. To apply appropriate business ethics and norms of good corporate governance principles, which are fairness, transparency and accountability.
4. To apply reliable management systems such as Safety and Health Management System, Risk Management System, Quality Management System, Security Management System and Integrated Environmental Management System.
5. To develop overseas market which is conducted in stages, starting with a strategic market that provides comparative advantage compared to other countries.

Strategies developed to achieve Firm A's objectives:

- Marketing strategy. This is based on its specific market. The strategy for the domestic market is to select projects coming from the government with adjusted budget and private firms which offer better profit. While for overseas market, Firm A stays focused on its existing market.
- Financial strategy. This is focused on the optimization of liquidity. This strategy was to be achieved through centralized financial strategy and self financing project policy. These strategies are expected to keep the company's liquidity healthy.
- Operational strategy. Firm A believes that centralized procurement is one of the substantial ways to reduce the costs especially inventory costs and to gain higher income. Also, risk mitigation is necessary by reducing the potential risks faced by the company.
- Investment strategy. Firm A has chosen the value added and competitive advantage investment strategy. Value added strategy is conducted by making the best investment decision which can be expected to give additional value to the company in the form of Return on Investments. In addition to this, the

competitive advantage investment strategy refers to Firm A's investment capability to select the type of investments based on its core competence.
- Human capital development strategy. Firm A supports a leadership program to anticipate the firm's growth. Also, talent management is determined by putting the right people in the right jobs. Hence, the talent management strategy can be specifically developed through developing the specified talent for each individual in the company.

Organization's Products and Plans
Firm A's products and services consist of:

1. Construction. There are two types of construction services provided by the firm, which are civil construction (road and bridges, seaport, airport, jetty, dam, barrage and water supply) and building construction (high rise building, commercial building and utility building).
2. Energy. This service consists of investments in oil and gas utility and steel fabrication.

Firm A's operational geographic scale consists of cities in Indonesia and abroad. In Indonesia, 45% of the projects were in west Indonesian islands (Sumatra, Kalimantan, Java and Bali), and 55% of them were in east Indonesian islands (Sulawesi, NTT, NTB, Papua). Firm A also had overseas projects in Algeria, Iraq, Libya, Myanmar, Brunei Darussalam and Timor Leste. Until the end of December 2012, Firm A had obtained new contracts valued at Rp17.13 trillion that has surpassed the 2012 targeted new contracts which has been predicted at Rp16.52 trillion.

Currently, Firm A's direction and focus are on revitalization and integration inside the firm. The focuses are:

- Revitalization of business architecture holding company and subsidiary
- Business process reengineering and control
- Investment, forward-backward integration
- State owned enterprise's synergy.

For the short and long term plans, Firm A intends to focus on its internal growth and collaborate with other multi-national companies. In obtaining internal growth, Respondent 3A described that they need to have a strong financial capacity to conduct market penetration of domestic and overseas markets, network with licensed technologies and internationalize human capital. As for the long term plan, these collaborations are expected to create new business expansion for Firm A.

In the short term, Firm A will perform more intensive development of its market segmentation into the domestic and global market and specific market of oil and gas, as well as investment projects. To be in line with such strategies, Firm A will continue to implement a management system of Quality–Safety–Health–Environment (QSHE) as one of its differentiation strategies. Integrated QSHE implementation has been carried out consistently with continuous and relevant improvements to cope with the requirements set by the project owner or other stakeholders. Moreover, this shall be implemented through Firm A's integrated management system.

Respondent 3A also mentioned that in implementing and building the QSHE culture, Firm A has put in place the following:

- Established QSHE policy as the direction and guidance in implementing QSHE in every work unit.
- Established QSHE responsibility, starting from the Director level up to the project level through the Site Implementation Team (SIT) establishment.
- Acquired ISO 9001:2008 for Quality Management System, OHSAS 18001:2007 for Safety Management System, and ISO 14001:2004 for Environmental Management System.
- All managers and the entire work units to comply with QSHE competency requirements through relevant training and or certification.
- Provided feedback and guidance on QSHE implementation.

Organization's Market and Competitors

According to Respondent 3A, due to the stable economic situation in Indonesia, with a stable rate of inflation, a stable gasoline price and a stable national economic growth (6.3% growth derived from the investment sector, export and public consumption), Firm A will remain focus in the business. There will be more infrastructure projects in Indonesia that is in line with the government's short term development plan. In addition, Firm A will also increase its market share in the private sector. Since 2007, Firm A had expanded its projects to other countries and will continue to capture the global market in the future. The latest project won was in Brunei Darussalam (infrastructure projects).

According to Respondent 4A, the construction industry is a tough and very competitive industry. Firm A had to compete with many companies that produce similar products and provide similar services. No single company has a dominant market share in Indonesia's construction industry. Most of the market opportunities that Firm A pursued were subject to a competitive tendering process. However, Firm A is quite confident that its strategies, combined with excellence in execution, safety, cost containment and experience in the construction industry, provide the opportunities for convincing and competitive business offerings to their clients.

Respondents 3A and 4A both viewed that competition in the business process occurs in the procurement or tendering process. This is where Firm A should really prepare and promote its competence. There are six criteria that Firm A focused for competing during the procurement process, which are the highly competence team, excellent project execution, financial strength, safety, excellence in working with international companies, and risk management.

Organization's Suppliers and Clients

In general, Firm A does not have a policy to place the local suppliers as the principal suppliers. However, it considers using the local potential suppliers for the basic materials needed in a construction project. These needs are then met by the local suppliers. Firm A sets the standard of quality and the sustainability of supplies to be the main requirements in determining the choice of local suppliers. In addition, the firm also considers their qualifications based on four criteria:

1. Provide expertise, experiences, technical and managerial abilities in the business that can be proven with the testimonies issued by the related associations.
2. Provide the needed resources to execute the tasks.
3. Provide an adequate track record.
4. Never having made false statements about their qualifications.

Table 11.1 shows the total number of suppliers used by Firm A, grouped into local, national and international suppliers.

Furthermore, in Firm A, finding alternative suppliers had already been included in the project plan, which is before project execution. The alternative suppliers could be local, national or international suppliers.

Regarding clients, Firm A delivers projects to local (Indonesia) and international clients. The projects can be funded by the government and private sector. According to all of the respondents, Firm A had arranged for insurance coverage between the clients and the firm for improved reliance on delivery. They have insurance policies for contractor all risk (CAR), erection all risk and machinery breakdown. These are all provided for in the contract.

Organization's Business Value Chains

In relation to the terms used in BCM, the four respondents of Firm A have described the firm's BUs, CBFs and MBCOs in general. Table 11.2 describes the types of business units (BUs) involved in the business value chain of construction service in Firm A. It also describes the critical business functions (CBFs) for the business value chain in general. This description is an overview and can further be expanded into more details (up to 3–4 task position level per business unit). However, Table 11.2 only shows the main business unit and its general functions. It shows that for every business value chain, the BUs will have intertwining functions and interdependencies. Table 11.2 was completed by the four respondents through discussions and by referring to their organizational structures and functions.

Table 11.1 Documented suppliers in 2012 (Firm A, 2012)

	Total number of suppliers			
Description	Local	National	International	Total
Procurement of goods	35	357	15	407
Procurement of services	19	295	3	317
Total	54	652	18	724

Table 11.2 Firm A's BUs and CBFs

Business value chain	Business Unit(s) (BU) (Division is under Department)	Critical Business Functions (CBF) in Business Unit(s) (BU)
Business development • Creating relationships with existing and prospective customers • Obtaining the work • Market research	• Human Capital Department • System and Business Development Department • Corporate Social Responsibility • Corporate Secretary • Finance Department • Legal Department • Business Development Division (per Operational Department) • Investment Division (Energy Department)	• System and Business Development Department (all business developments report to this unit) • Finance Department (financial management regarding Business Development) • Legal Department (Contracts administration and management)
Procurement • Relationship with subcontractors • Relationship with suppliers • Efficiency and effectiveness of material purchasing procedures and management procedures	• Legal Department • Procurement Division (per Operational Department) • Projects Division (per Operational Department) • Finance Department • Human Capital Department	• Legal Department (Contracts administration and management) • Procurement Division (per Operational Department) (Procurement management) • Finance Department (financial management regarding Procurement)
Construction operations • Cost and schedule estimation and control • Project management system • Quality management system • Safety management system	• Finance Department • Human Capital Department • Legal Department • Operational Departments (General Civil Engineering Department; Regional and Overseas Department; Building Construction Department; Industrial Plant Department; Energy Department) • Projects Division (per Operational Department)	• Operational Departments (Project Management) • Projects Division (per Operational Department) (Site Management) • Finance Department (Financial management regarding construction operations) • Legal Department (Contracts administration and management)
Post-construction services • Warranty management system • Customer relationship development program	• Legal Department • Finance Department • System and Business Development Department • Operational Departments (General Civil Engineering Department; Regional and Overseas Department; Building Construction Department; Industrial Plant Department; Energy Department) • Projects Division (per Operational Department)	• Legal Department (Contracts administration and management) • Finance Department (Financial management regarding Post-construction) • Operational Departments (Project Management) • Projects Division (per Operational Department) (Operation and Maintenance)

(continued)

Table 11.2 (continued)

Business value chain	Business Unit(s) (BU) (Division is under Department)	Critical Business Functions (CBF) in Business Unit(s) (BU)
Firm infrastructure • Adequacy and location of facilities and equipment • Efficiency and effectiveness of finance and accounting system • Information management system	• Finance Department • System and Business Development Department • Operational Departments (General Civil Engineering Department; Regional and Overseas Department; Building Construction Department; Industrial Plant Department; Energy Department) • Projects Division (per Operational Department) • Internal Control Department	• Finance Department (Financial management regarding Firm Infrastructure) • System and Business Development Department (Integrated Management System) • Internal Control Department (Information Management System) • Operational Departments (Project Management) • Projects Division (per Operational Department) (Site Management)
Human resources management • Procedures for recruiting and developing employees • Working environment • Relationship with unions • Levels of employee motivation and job satisfaction	• Human Capital Department • Finance Department • Legal Department • Operational Departments (General Civil Engineering Department; Regional and Overseas Department; Building Construction Department; Industrial Plant Department; Energy Department) • Projects Division (per Operational Department)	• Human Capital Department (HRM for head office and projects) • Finance Department (Financial management regarding HRM) • Legal Department (Contract administration and management) • Operational Departments (Project Management) • Projects Division (per Operational Department) (Labor Management)

Moreover, the process of defining Minimum Business Continuity Objective (MBCO) had only been conducted by the management and only provided for the core business units. This process is within the top management's (BOC and BOD) authority. Based on Table 11.2, where CBFs had been determined for each business value chain, the overview of the MBCO for the most influential CBFs is shown in Table 11.3.

Part of BCM is developing coordination between the firm's internal structures and external stakeholders. Regular coordination and communication between them are essential for business continuity planning and in executing the BC plans when needed. Table 11.4 describes the overview of coordination between Firm A's head office, project-base office and external agencies.

Table 11.3 Firm A's MBCO for each CBF

CBF	Minimum Business Continuity Objective (MBCO)
• Finance Department (financial management)	Managing and protecting all financial matters in all of the business value chains
• Legal Department (contract administration and management)	Managing and protecting all of contracts and legal administration
• Procurement Division (procurement management)	Managing procurement for all projects (materials; men; machines)
• Operational Departments (site and project management)	Managing projects; documentation for completed projects and ongoing projects
• Projects Division	Managing project sites (materials; equipment; facilities) and labor
• System and Business Development Department	Maintaining the company's integrated management system; Managing and protecting the company's business development documentation
• Human Capital Department (human resources management)	Managing the human resources in head office and projects (employee's wage and welfare)

11.2.1.2 Crises Response

The four respondents from Firm A viewed the following crises as important that need to be prepared for:

- Changes in regulations and statutory legislation. Any changes in regulations or statutory legislation may lead to changes in the firm's budget plans, resource allocations and project financing.
- Increase in price of raw materials (unexpected price escalation). This would certainly has high impact for the ongoing projects. For the upcoming projects, the firm needs to review the cost and schedule plans regarding this.
- Shortage of key materials. The firm needs more time to coordinate for overcoming the materials supply shortage.
- Subcontractor insolvency and conflicts at the project site. This may lead to delays in completing the projects (conducted by subcontractors).
- Weather issues/climate or natural disasters.

Crises such as lack of materials, conflicts at the project site and natural disasters have occurred in Firm A. These are some of their views and experiences of those disruptions:

Respondent 1A When a crisis such as lack of materials at the project based office occurred, this could have significant impact towards the continuity of the project. There could be delays due to waiting for the materials to arrive on site. Sometimes, we need to consult with the head office regarding this issue, whether we should buy the materials from alternative suppliers in the nearest location from the project, or get the materials from the firm's warehouse (which may need more time). This situation may be more complicated when the project is on different island from the firm's material warehouse or in a very remote area.

Table 11.4 Firm A's internal and external coordination overview

Coordination	Description
Head office and project-base office	
How do they develop, communicate and distribute the reports on performance progress?	Regular reports on performance progress are conducted daily, weekly, monthly and annually These reports are distributed from project site office [approve by the project manager (PM)] to Head office. Reports distributed via regular meetings and emails (firm's intranet)
What types of communication are there for this relationship? (verbal; non-verbal) Who is in charge for the communication between them?	Meetings; emails; intranet; video conference Persons in charge: PM (site office) and Operational BU (Projects division)
How to allocate resources (from head office to project office; in the project office)?	Permanent employees—assigned to projects Outsourced employees—based on the number needed (outsourced from Head office or locals in the project) The policy on recruiting employees is part of the head office's authority in Jakarta and the number of accepted new employees is adjusted to the company need. In line with the principle of equality, Firm A does not have specific regulations to prioritize local communities around the head office, regional offices or locations of construction projects to become employees. However, the company still seeks to involve the communities around locations of construction projects as freelancers. They are accountable for tasks that do not require specific expertise. They are involved as freelancers within a certain period of time, adjusted along with the process of the occurring construction works. Due to this condition, Firm A has not taken any specific data on the number of people from local communities becoming freelancers in its projects.
What types of infrastructure are there for coordination between head office and project-based office?	• IT infrastructure (hardware and software) • Document storage/documentation management
Head office and governments; public; clients; suppliers (supply chain); sub-contractors	
What types of communication are there for these relationships? (verbal; non-verbal) Who is in charge for the communication between them?	Verbal: meetings Non-verbal: Reports, emails, letters Persons in charge: Operational BUs The external parties' representatives
What types of coordination/reports are there for these relationships?	Coordination between the Operational BUs and external parties Regular meetings are conducted providing weekly/monthly/annual reports Conference call meetings can also be conducted

(continued)

Table 11.4 (continued)

Coordination	Description
What are the resources needed for these relationships? How to allocate resources between these relationships?	Teams from the head office and representatives from the external parties Usually the management representatives of the head office will coordinate with the representatives from the external parties
What types of infrastructure are there for coordination between these relationships?	• IT infrastructure (hardware and software) • Document storage/documentation management
Project-base office and public; suppliers; sub-contractors	
What types of communication are there for these relationships? (verbal; non-verbal) Who is in charge for the communication between them?	Verbal: meetings Non-verbal: Reports, emails, letters Persons in charge: PM and resident engineers The external parties' representatives
What types of coordination/reports are there for these relationships? How do they develop, communicate and distribute the reports?	Coordination between the PM team and external parties Regular meetings are conducted providing weekly/monthly/annual reports Conference call meetings can also be conducted
What are the resources needed for these relationships? How to allocate resources between these relationships?	The PM team (project administrative team and PM or its representative) with the external parties' representative
What types of infrastructure are there for coordination between these relationships?	• IT infrastructure (hardware and software) • Document storage/documentation management

Respondent 2A Crises such as natural disasters always have significant impact towards the project. One major impact is delays and loss of productivity due to the natural disaster. There was a landslide that occurred near our project, and our material and equipment warehouse were damaged. Due to this situation, we needed to coordinate with our head office intensively and it took us a week to finally clean up the damage and on the second week, we were able to resupply the damaged resources. The good thing is that this event occurred in the same province as our head office. If this occurred in a different province or island, it must take more than 2 weeks to recover.

Respondent 3A Incidents such as conflicts with the locals at the project site may disrupt the project. There were times where for certain projects (mostly related to infrastructure projects), the locals around the project area were interested in working at the project as laborers. The project manager had to coordinate and communicate wisely with them, because sometimes we do not need additional labor resources. But due to the educational background and local culture (where the locals believe that they have the right to be involved in the project located near their village), the project manager and team have to address the issue in the most gentle manner and approach. There was this one time where the locals were so upset because they could not work at the project, they conducted demonstrations and burned some tires in front of the project site gate. This crisis lasted for almost one week due to protracted negotiating with the demonstrators and the local village leader. Representatives from the head office had to come to the project, to help overcome the situation.

According to the four respondents, the most significant crisis that had occurred in the firm that resulted in the highest impact is weather changes. Several projects that are disrupted due to weather changes include the following:

- 50 MW diesel power plant in Bali. Due to high tides, unloading of materials in the seaport was disrupted. Works in the project were delayed by 22 days and Firm A had to pay the commercial operator late fines.
- 60 MW Gas Power Plant, South Sumatra. Due to weather anomaly, Firm A had to postpone the foundation works for 1 month. The estimated loss was approximately Rp90 million.

Despite disrupted projects due to the climate change phenomenon, Firm A had successfully completed all its projects.

The four respondents agreed that the significant impact from the crises that had occurred in Firm A were: loss of productivity, waste in materials, delay in work/dissatisfied customers and revenue impact. When crises occur, the four respondents agreed that resources such as materials, equipment and workers can be at stake. The materials are wasted if the disruptions damage the project site; there could be problems with the equipments if they are damaged due to the disruptions (particularly due to floods/landslide/earthquake/volcano) and it may need some time to repair and restore the equipments. In addition, workers may need to secure themselves when natural disasters occur. The workers would most likely lose their productivity.

Moreover, Firm A has carried out line communications aimed at creating customer loyalty. The activities undertaken during the reporting period were regular meetings, visits to Firm A project sites and provision of all information on project execution to the client or prospective end users. If the firm's operations are being disrupted, the likely reactions of customers would be to ask for clarification and they needed to be informed regularly about the disruptions. This situation will be managed directly by the QSHE of Firm A.

As for competitor's reactions if Firm A is experiencing crises, Respondent 3A explained that if the disruption led to project failures, other firms may take over the project. But if the disruption was caused by natural disasters (where other firms may experience it too), this situation would be overcome by firms supporting one another. One example was the 2006 earthquake in Yogyakarta, where several reconstruction processes were conducted by joint venture contractors.

Up to now, Firm A's business regulation environment does not provide any support for overcoming crisis. However, Firm A's internal regulation had provided standard operating procedures (SOPs) for overcoming selected incidents and these

Table 11.5 Firm A crises response

Description	Provided or not provided	Further comments (procedures; resources; infrastructures)
Evacuation procedure	Provided	Procedures particularly for accidents/fires/earthquakes/floods
Communication procedure during crisis	Provided	Procedures particularly for accidents/fires/earthquakes/floods
EOC/alternative facilities	Provided	Using the branch office if needed
Recovery procedure	Provided	Procedures particularly for accidents/fires/earthquakes/floods. Need more coordination with head office
Restoration procedure	Provided	Procedures particularly for accidents/fires/earthquakes/floods; Need more coordination with head office

SOPs provide coordination with external agencies. According to Respondent 3A, every state-owned enterprise is supported financially by the government, where they will be supported to continue business as usual if any financial crisis occurred. This can be viewed as a support for business continuity. Within the reporting period, Firm A has never received financial aid from the government or other facilities in terms of fiscal or financial matters. Furthermore, internally, the management of Firm A has also created a risk management road map. In risk management, the role of supervision executed by the Board's Chairman is specifically assisted by the committee for financial planning and business risk.

Table 11.5 shows the description of Firm A's crises responses. The firm has documented evacuation and communication procedures. However, the further recovery and restoration procedure still need more coordination with the firm's head office. These plans are used for overcoming accidents, fires, earthquakes, floods and weather issues. Procedures for other types of crises have not been developed and documented. According to the respondents, so far, the response plans for accidents in projects and floods have been implemented.

In terms of lessons learned from the crises, Firm A viewed that:

- Each business unit and function needs more coordination holistically. Defining the interdependencies of each BU is necessary for developing a holistic procedure to overcome crises.
- Communication procedures should be provided in details in order to reduce waiting time or delays.

11.2.2 *Case Study 2: Firm B*

Firm B is a private firm, with its head office located in Jakarta, Indonesia. It was established 44 years ago and focuses in materials and construction. Having gone through the professionalism process, the company strived to stake its position in the construction service market by starting out as a building contractor. It is 70% owned by the founder and 30% owned by the public. The case study in Firm B included three respondents. All of them are project managers, with experiences of 22 years (Respondent 1B); 11 years (Respondent 2B); and 15 years (Respondent 3B) in the firm. Document analysis and interviews were conducted with the respondents for two weeks in April 2013.

11.2.2.1 Organizations Characteristics

Organization's Objectives
Firm B's missions are as follows:

1. To become a construction organization known for integrity, respect, fair dealing, quality, pride and excellence.
2. To satisfy customers by producing quality work and to deliver excellence in service.
3. To deliver excellence and pride in construction.

In order to achieve its missions, Firm B focuses on:

1. The construction and development of premium, high-rise residential, commercial and office developments.
2. Increasing the provision of its value-added design and build service.
3. Undertaking building projects primarily as the main contractor; and
4. Customer service excellence.

In achieving these objectives, Firm B developed strategies such as:

- Increasing revenues from existing markets and new markets in Indonesia. Firm B plans to expand sales in its primary market in Java from both the private and public sectors. Penetrating new market outside Java is also another strategy of the firm.
- Increasing growth and development through prudent corporate strategy and maintaining stringent financial controls (by carefully managing and controlling costs and exercising lean construction).
- Improving human resources and hold regular programs on corporate values as well as frequent training programs for employees.

Organization's Products and Plans

The types of products and services offered by Firm B are mostly in building construction. The firm has delivered projects such as:

- Office and retail buildings
- Residential buildings
- Institutional buildings such as schools, hospitals and medical facilities
- Industrial buildings for manufacturers.

Moreover, Firm B's role is to function as the main contractor of the projects with the following tasks:

1. Project administration for projects from commencement to completion.
2. Selection and procurement of building materials.
3. Implementation of an optimal construction method.
4. Field site management.
5. Recruitment and management of all subcontractors; and
6. Joint operation projects for large scale and international standard projects.

Currently, Firm B's operational scale is only in Indonesia. 70% of the projects were delivered in west Indonesia (Sumatra, Kalimantan, Java and Bali) and 30% of projects were in east Indonesia (Sulawesi, NTT, NTB and Papua). Currently, Firm B is focusing on building construction with quality as its main priority, but is still continuously seeking new market opportunities.

Firm B had developed its short and long term plans, which are internal growth and domestic building projects, including green building projects, respectively. Respondent 1B described that for its short term plans, it will establish a more synergistic approach to its repeat customers, yet still maintaining a prime service to the potential customers, by strengthening the company's internal foundation through the improvement of qualified human resources. Firm B will also focus on large scale industrial projects as its subsidiary market such as power plant and factories. Last but not least, Firm B's Corporate Social Responsibility (CSR) thrust led to a disaster resources partnership (DRP) which takes innovative measures in providing access for life-saving efforts in remote areas in Indonesia.

Moreover, for its long term plans, Firm B will still focus on constructing buildings which have a relatively high investment value, such as hotel, apartment, mall and office buildings. This is due to the high demand for construction services in Indonesia and by Firm B's customers who retain their interests in cooperating with the firm. Along with that, there will be more emphasis on constructing green buildings in the future. Firm B plans to conduct a number of tests, scorings, as well as innovations in green standardizations for projects that will be adjusted to the requirements of green building office issued by the Green Building Council Indonesia.

Organization's Market and Competitors

According to Respondent 2B, the current market condition of Firm B's business focuses on domestic building projects. As for the expected market conditions in

5–10 years, Firm B was optimistic about the prospects in the construction service business. Its optimistic attitude towards the accelerated prospects is based on two reasons, namely macro-economic growth conditions and domestic consumption. It also views that the number of building projects in Indonesia will still be rising and supported by an increasing number of infrastructure projects. Therefore, Firm B is most likely to still focus on domestic building projects in the future.

The three respondents agreed that the construction industry is a highly competitive industry. Firm B's competitors were the local state-owned contractors, private firms and international firms working in Indonesia. The competition occurred in the tendering process (procurement phase). Respondent 3B shared that Firm B had so far been able to deliver high-performance in competing for the development of various prestigious projects in Indonesia. Success is reflected in the company's project portfolio that consists of the construction of unique and high-rise buildings. Other than that, Firm B provided its expertise based on its achievements in project deliveries such as (Firm B, 2013):

- Repeat businesses from customers.
- Renowned for providing high quality, customer oriented, premium construction services in Indonesia.
- Attracts new customers who are seeking for the premium contractor.
- Track record of over 40 years in construction projects across various sectors in Indonesia.
- Highly experienced and competent employees; management and employees have been with the company for 5 years in general; all project directors and project managers are engineers.
- Offers a full range of construction services, from the inception stage through to the after-sales service of a project; provides value-added design services for projects utilizing its in-house design team and reputable third party consultants.
- Integration of design services with the company's construction services.
- Adopts a prudent business strategy, in particular relating to financial management.
- Has not assumed any material level of indebtedness throughout the company's operating history. It is positioned to weather downturns, particularly when compared to its highly-geared competitors.

Organization's Suppliers and Clients

Firm B had local and international suppliers involved with its projects. 75% of the suppliers were local and 25% were international suppliers. The raw materials such as sands, gravels, cements and steels were mostly from the local suppliers known for their high quality. As for the finishing materials, this mostly depends on the customer's demands. According to Respondents 1B and 2B, alternative suppliers were decided and planned since the project plan development phase. Usually, finding the nearest located (and up to quality standards) alternative suppliers is the main priority.

Although Firm B's projects were primarily all in Indonesia, overseas companies from Singapore and Malaysia have also become Firm B's customers. These companies have projects delivered in Indonesia. Other than that, Firm B's local clients were mostly the government, state-owned firms and private firms (mostly banks and developers). In addition, insurance coverage is provided by the clients, arranged in the contract agreement. The insurance policies provided by the clients include the contractor all risk (CAR) and equipment all risk (EAR) insurance.

Organization's Business Value Chains

The three respondents of Firm B have described the firm's BUs, CBFs and MBCOs in general BCM terms. This description is an overview and only shows the main business unit and its general functions. Table 11.6 describes the types of business units (BUs) involved in the business value chain of construction service in Firm B. It also describes the critical business functions (CBFs) for the business value chain in general. It shows that for every business value chain, the BUs will have interdependencies. Table 11.6 was completed by the three respondents through discussions and by referring to their organizational structures and functions.

According to the respondents, some detailed core business MBCO of Firm B must be provided by top management in order to develop the corporate risk strategies. Continuing from Table 11.6, where CBFs had been determined for each business value chain, the overview of the MBCO for the most influential CBFs of Firm B is shown in Table 11.7.

As earlier mentioned, developing coordination between the firm's internal structures and external stakeholders is part of BCM. Regular coordination and communication between them are essential for business continuity planning and in executing the BC plans when needed. Table 11.8 describes the overview of coordination between Firm B's head office, project-base office and external agencies.

11.2.2.2 Crises Response

The respondents from Firm B viewed this list of crises as important and need to be prepared for:

- Action by environmentalist/pressure groups (protests)
- Access/approval restriction or limitation
- Delays or uncertainty in resolving disputes
- Materials shortage
- Increase in price of raw materials (unexpected price escalation)
- Subcontractor insolvency
- Serious accidents in a project
- Natural disasters (earthquake, floods, tsunami, etc)
- Loss of management personnel or key staff.

Table 11.6 Firm B's BUs and CBFs

Business value chain	Business Unit(s) (BU)	Critical Business Functions (CBF)
Business development • Creating relationships with existing and prospective customers • Obtaining the work • Market research	• Marketing Unit • Project Development Unit • Legal Unit • General Affairs Unit • Investor relations Unit • Human Resources Unit • Management Representative	• Legal Unit (Contract management) • Investor relations Unit (Client relations) • Management Representative (Representing the company for business development) • Marketing Unit (Marketing Intelligence)
Procurement • Relationship with subcontractors • Relationship with suppliers • Efficiency and effectiveness of material purchasing procedures and management procedures	• Project Development Unit • Legal Unit • Equipment Unit • Logistics Unit • Accounting Unit • Cash Operation Unit • Human Resources Unit • Personnel Administration Unit	• Project Development Unit (Procurement management) • Legal Unit (Contract management) • Equipment Unit (Procurement management) • Logistics Unit (Procurement management) • Accounting Unit (Financial management) • Cash Operation Unit (Financial management) • Human Resources Unit (HRM) • Personnel Administration Unit (HRM)
Construction operations • Cost and schedule estimation and control • Project management system • Quality management system • Safety management system	• Estimation Unit • Project Development Unit • Construction Engineering and R/D Unit • Legal Unit • Customer Care Unit • Product Quality Unit • Equipment Unit • Logistics Unit • Safety, Health, Environment Unit • Human Resources Unit • Personnel Administration Unit • Accounting Unit • Cash Operation Unit • Project Control Unit • IT Unit	• Project Development Unit (Project Management) • Legal Unit (Contract Management) • Customer Care Unit (Customer relationship) • Product Quality Unit (Quality Management) • Equipment Unit (Equipment operations and maintenance) • Logistics Unit (Materials management) • Safety, Health, Environment Unit (Safety management) • Human Resources Unit (HRM) • Accounting Unit (Financial Management) • Cash Operation Unit (Financial Management) • Project Control Unit (Project Management) • IT Unit (Information Management System)

(continued)

Table 11.6 (continued)

Business value chain	Business Unit(s) (BU)	Critical Business Functions (CBF)
Post-construction services • Warranty management system • Customer relationship development program	• Project Development Unit • Construction Engineering and R/D Unit • Legal Unit • Customer Care Unit • Product Quality Unit • Equipment Unit • Logistics Unit • Accounting Unit • General Affairs Unit • Property and Building Management Unit • Investor Relations Unit	• Project Development Unit (Operations and Maintenance) • Legal Unit (Contract Management) • Customer Care Unit (Customer Relationship) • Product Quality Unit (Quality Management System) • Accounting Unit (Financial Management) • General Affairs Unit (Customer Relationships) • Property and Building Management Unit (Warranty and Maintenance) • Investor Relations Unit (Customer Relationship)
Firm infrastructure • Adequacy and location of facilities and equipment • Efficiency and effectiveness of finance and accounting system • Information management system	• Internal Audit Unit • Property and Building Management Unit • IT Unit • Accounting Unit • Management Representative	• Internal Audit Unit (Company's internal control) • Property and Building Management Unit (Internal asset management) • IT Unit (IT maintenance) • Accounting Unit (Financial management)
Human resources management • Procedures for recruiting and developing employees • Working environment • Relationship with unions • Levels of employee motivation and job satisfaction	• Legal Unit • Safety, Health, Environment Unit • General Affairs • Project Development Unit • Human Resource Unit • Personnel Administration Unit • Accounting Unit	• Legal Unit (Contract Management) • Human Resources Unit (HRM) • Personnel Administration Unit (HRM) • Accounting Unit (Financial management) • Safety, Health, Environment Unit (Human resources safety and welfare)

Table 11.7 Firm B's MBCO for each CBF

CBF	Minimum Business Continuity Objective (MBCO)
Legal Unit (Contract Management)	Managing legal and contractual matters
Investors relations Unit	Managing relationships with the company's investors and clients
Marketing Unit (Marketing Intelligence)	Managing the company's market forces and future business development
Project Development Unit (Project management)	Managing the company's existing and current projects (head office—project site management)
Accounting Unit (Financial Management)	Managing financial matters
Human Resources Unit (Human resources management)	Managing the company's human resources (employed and outsourced)
Equipment Unit (Equipment operations and maintenance)	Managing equipment resources used by the company for all projects
Logistics Unit (Procurement management)	Managing logistics for all projects
Safety, Health, Environment Unit (Safety management and environmental management)	Managing safety for the human resources in the company
Project Control Unit (Project management)	Managing the project during execution
Customer Care Unit (Customer relationships)	Managing relationships with customers (before, during and after projects)
General Affairs Unit (Stakeholder management and customer relationships)	Managing all affairs related with the company (representing the company to external parties)
Property and Building Management Unit (Warranty, maintenance and internal asset management)	Managing the company's physical assets (head office, project office, resources)

As to the extent of a disruption, the respondents viewed that disruptions such as lack of materials and natural disasters have occurred in Firm B. These are their views and experiences of those disruptions:

Respondent 1B If natural disasters occur (such as earthquake, floods, volcano, landslide), this may lead to disruption to the firm's business. This can affect not only the project based office, but also the head office. When these events occur at the project based office (which may be located far from the head office), the project site team need to coordinate intensively with the head office for recovering from the event.

Respondent 2B Supplier issues such as lack of materials can affect the firm's business. In overcoming these, regular updates and control about the material availability should be conducted, both by the project based office and head office (the Logistic department).

According to Respondent 2B, increase in material prices (price escalation) in 2008 had the highest impacts for Firm B. It needed assistance from ICA to liaise with the regulators regarding this matter. The positive thing that arose in this

Table 11.8 Firm B's internal and external coordination overview

Coordination	Description
Head office and project-base office	
How do they develop, communicate and distribute the reports on performance progress?	Regular reports (daily, weekly, monthly) as well as progress and performance reports were developed and distributed via regular meetings and emails The reports are from project site office (by the project manager (PM)) to Head office (Project Development Unit)
What types of communication are there for this relationship? (verbal; non-verbal) Who is in charge for the communication between them?	Verbal: meetings Non-verbal: reports, notes, letters, emails, social media Persons in charge: PM and his/her administration team; and Project development and construction division
How to allocate resources (from head office to project office; in the project office)?	Permanent employees are assigned to projects Outsourced employees—based on the number needed, the employees will be outsourced from Head office or locals in the project
What types of infrastructure are there for coordination between head office and project-based office?	• Document storage • IT infrastructures (computers, telephone, facsimile, laptops, smartphones)
Head office and governments; public; clients; suppliers (supply chain); sub-contractors	
What types of communication are there for these relationships? (verbal; non-verbal) Who is in charge for the communication between them?	Verbal: meetings; conference; workshops Non verbal: emails, letters, reports Persons in charge: • External parties' representatives • Management representatives and Managers under Board of Directors
What types of coordination/reports are there for these relationships? How do they develop, communicate and distribute the reports?	• Progress and performance reports to clients • Other reports related to projects that need to be distributed/informed to government/public/supplier/sub-contractors • Financial, technical and administration matters/coordination to suppliers and subcontractors These reports will be addressed by providing meetings or distributing them via mails/emails
What are the resources needed for these relationships? How to allocate resources between these relationships?	Human resources and documentation are needed Scheduled coordination between parties should be developed
What types of infrastructure are there for coordination between these relationships?	• Document storage • IT infrastructures (computers, telephone, facsimile, laptops)
Project-base office and public; suppliers; sub-contractors	
What types of communication are there for these relationships? (verbal; non-verbal)	Verbal: meetings; conference; workshops Non verbal: emails, letters, reports Persons in charge:

(continued)

Table 11.8 (continued)

Coordination	Description
Who is in charge for the communication between them?	External parties representatives PM for the project and the resident engineer and administration teams
What types of coordination/reports are there for these relationships? How do they develop, communicate and distribute the reports?	• Progress and performance reports to suppliers and subcontractors • Other reports related to projects that need to be distributed/informed to government/public/supplier/sub-contractors • Financial, administration, and technical matters/coordination to suppliers and subcontractors These reports will be addressed by providing meetings or distributing them via mails/emails
What are the resources needed for these relationships? How to allocate resources between these relationships?	Human resources and documentation are needed Scheduled coordination between parties should be developed
What types of infrastructure are there for coordination between these relationships?	• Document storage • IT infrastructures (computers, telephone, facsimile, laptops)

situation was that the financial management in this firm was prepared and able to cope with the price adjustments. However, not many projects were delivered within that year due to the situation. In addition, other situations like access limitation or approvals (particularly for projects in Sumatra) also impacted the firm. There were delays to starting the projects. These situations were due to regulations in the specific area, and Firm B needed to negotiate and coordinate intensively with the local regulator and clients.

The respondents of Firm B explained that the significant impacts from crises that occurred in the firm would be: delays in work, dissatisfied customers, employees were not able to reach the office, and revenue reduction. As to the extent of resource loss, the firm viewed productivity losses, material wastes and equipment damages as the main losses. This was elaborated by Respondent 3B:

Respondent 3B When disruptions occur, usually, there would be productivity losses from the workers (human resources) because their activities may be disrupted for some time. Other than that, damages or wastes of materials/equipments due to natural disasters can be viewed as resource loss.

Furthermore, when the firm's operations were disrupted, the customers needed to be informed regularly of the disruption, and where communication and management system of Firm B are essential. All the respondents viewed that Firm B's value regarding customer orientation is very high and have always been reflected in their project deliveries. The firm is known for quality and safety that constantly meets customers' expectations, and where projects are always completed and handed over on schedule. It also has an excellent service record with its customers, from the commencement to project completion. Therefore, maintaining communication with the customers during normal business operations or during disruptions should always be prepared by Firm B.

According to Respondent 1B, generally when the firm is experiencing crises during project deliveries, the competitors may not be aware of the situation. The firm will usually overcome these crises internally and coordinating them only with its direct stakeholders (internal and external). But if the crises were known publicly and may lead to project failure, some competitors might be opportunistic to take over the project. Other than that, competitors that have good and mutual relationships with Firm B (who had previous entered into joint ventures with them in some projects) might support the firm to help overcome the crises.

In terms of regulations for handling crises, the three respondents pointed out that there is presently no business regulation that focuses mainly on overcoming crisis. However, Firm B has its own internal regulations and procedures (SOP) for overcoming some selected incidents. These plans also need coordination with external agencies such as the local government, NBCSD, securities (police and army), hospitals, fire brigade and the local community.

With respect to support for business continuity, there are regulations from the government that encourage the firm to comply with specific standards for safety, quality and environmental management. These standards may not focus specifically on business continuity. Therefore, all of the respondents opined that Firm B needs to prepare internally by complying with specific BCM international standards.

Table 11.9 describes Firm B's crises responses. Firm B has provided procedures for evacuation, communication during crisis, alternative facilities, recovery and restoration. These procedures are used for handling fires, accidents, earthquakes and floods. Procedures for other types of crises however have not been developed and documented.

Up to now, response plans for floods (for the head office) and accidents in projects have been implemented. Respondent 3B shared that during the 2012 floods in Jakarta, Firm B's head office had to be closed for one week due to the flooding of all road accesses to the office. The benefit of the plan was that Firm B had secured its vital physical assets at a remote place away from the floods area. Firm B's head office only needed to clean up the lower and underground level of the building,

Table 11.9 Firm B crises response

Description	Provided or not provided	Further comments (procedures; resources; infrastructures)
Evacuation procedure	Provided	Specifically for fires/accidents/earthquakes/floods
Communication procedure during crisis	Provided	Call out diagrams and detailed contact lists
EOC alternative facilities	Provided	
Recovery procedure	Provided	
Restoration procedure	Provided	

which were parking areas. However, Firm B still needed to improve its communication procedures during floods, where some employees were trapped in the floods and traffic when they did not receive current information about the head office. They were trapped for most of the first day of the floods. Some of them were able to reach the office only to find out that it was closed. This situation was an important lesson for Firm B.

From the crises that occurred in Firm B, all respondents agreed that they have learned the following lessons:

- Effective communication during crisis is essential. Firm B needs to review and improve its communication procedures. Regular communication drills relating to certain crisis should be conducted.
- Not all crises/threats have been analyzed by Firm B. It needs to have a more comprehensive plan for overcoming crises.
- The role of top management is essential during crises, and the management should appoint a specific person for developing the BC plans for various crises.

11.2.3 BCM Preparedness

The case studies also provided descriptions about the firms' practices related to BCM. The following tables illustrate some lists of activities that have been implemented (✓) or not (✗) by the two firms, based on each BCM principle. In addition, the studies had identified some hindrances that have occurred in the two firms when implementing these principles.

11.2.3.1 BCM Start-Up Phase or How to Initiate BCM

Table 11.10 shows that both Firm A and Firm B have not used the term "BCM" and "BC plans" (Point 1). All respondents from both firms confirmed that this is the first time they heard about the term BCM and BC plan. However, their firm's practices on crises responses and continuing their business operations during disruptions had incorporated some aspects of the BCM principles. Both firms had also not focused more intensely on their stakeholder's interests (as in Point 3) and perceptions. Both firms had implemented the meaningful value of business continuity in their products and services (Point 4) and they had also communicated their firms' vision, goals and objectives to their staffs (Point 8). Moreover, Firm B had almost all the start-up phase implemented, whereas Firm A has only implemented Point 4 and Point 8.

As for hindrances in implementing this phase (Table 11.11), both firms have promptly notified the designated client/user representatives during a business disruption. However, Firm A still does not have adequate financial resources to

Table 11.10 Practices that have been implemented in the initiation phase

Descriptions	Firm A	Firm B
1. Using the term "BCM" and "BC plans" in the firm	✗	✗
2. The requirements for business continuity, taking into account of the firm's objectives; obligations—legal, regulatory, contractual	✗	✓
3. Identifying interests of key stakeholders: (a) What are their requirements and perceptions? (b) Who are they? (shareholders; employees; suppliers; regulators; financial investors; insurers; auditors; professional bodies; trade associations; government departments; competitors, the community; media and "vested interest" groups)	✗	✗
4. Scope of business continuity in terms of products and services	✓	✓
5. Accountability and responsibility for key areas were defined at the time that the framework was implemented	✗	✓
6. There are clearly defined and approved management processes to manage business continuity	✗	✓
7. The management and staff adopted an attitude to continuity management planning that ensures a positive control environment	✗	✓
8. The entity regularly communicates its vision, goals and objectives to staff	✓	✓

Table 11.11 Hindrances in implementing the initiation phase (in the context of its current condition)

Descriptions	Firm A	Firm B
1. The entity has inadequate financial resources to implement BCM	✓	✗
2. Designated user representatives are not promptly notified if a business disruption occurs	✗	✗

implement BCM holistically. In the other case, Firm B felt more confident for initiating and implementing BCM.

11.2.3.2 Risk Analysis and Review

Table 11.12 shows that both firms had implemented risk analysis and review within their organization. Only Firm A had not implemented a detailed risk review for each of its BU (Point 5). As for the hindrances (Table 11.13), both firms had not regularly maintained their contact details for interdependent entities (Point 1) and in reviewing the risks identified, they still had minimum capability for implementing possible solutions because of remote access (Point 2).

Table 11.12 Practices that have been implemented for risk analysis and review

Descriptions	Firm A	Firm B
1. Threat identification in the firm	✓	✓
2. Management carefully analyses and assesses risks and opportunities before authorizing new ventures or significant changes	✓	✓
3. Risk identification and treatment for each BU (Business Unit)	✓	✓
4. Disaster identification (for key disaster scenario)—Disaster may be compiled from one or more identified risks	✓	✓
5. Risk review (per BU)	✗	✓

Table 11.13 Hindrances in implementing risk analysis and review (in the context of its current condition)

Descriptions	Firm A	Firm B
1. Contact details for interdependent entities are not regularly maintained	✓	✓
2. Lack of capacity to adequately implement a possible solution due to remote access	✓	✓
3. No alternative contacts have been identified	✗	✗
4. No ongoing effort to minimize exposure to disasters and operations/ systems vulnerabilities	✗	✗

11.2.3.3 Business Impact Analysis (BIA)

As for BIA, Table 11.14 illustrates the practices both firms had implemented. It can be seen that Firm A had not implemented all of the practices, particularly on further analyses of its CBFs (Points 5, 6 and 7). Also, both firms had not identified specific recovery timeframes for each CBFs.

Regarding hindrances in implementing BIA, Firm B did not have any problems in implementing this principle (see Table 11.15). The firm had prepared all the necessary items for overcoming its defined crises. On the other hand, Firm A still needed to have periodical evaluation of its critical systems (Point 1); regular off-site processes for reconstructing critical data (Point 2); and ready access to public network following a disruption (Point 6).

11.2.3.4 BC Strategy

Table 11.16 describes the process for developing a BC strategy. Both firms still need to implement more detailed planning for a specific disaster in its strategy (Point 3). Additionally for Firm A, it should analyze the strategies in more details, as listed in Point 4.

Based on Table 11.17, Firm A still needs to prepare its infrastructures for the recovery strategy to manage their defined crises. As mentioned in Point 3 and

Table 11.14 Practices that have been implemented for BIA

Descriptions	Firm A	Firm B
1. Defining business function	✓	✓
2. Defining minimum business continuity objective (MBCO)	✓	✓
3. Establish priority for analyzing impact (per disaster); the impact if key services and products are disrupted—for whatever reason	✓	✓
4. Establish CBF (critical business function) (for each BU)—what are the critical activities?	✓	✓
5. Defining dependencies of each CBF (each BF can span across one or more business operations)	✗	✓
6. Defining CBF requirements (this construct relates to previous CBF dependencies construct—above)	✗	✓
7. The BIA identifies the recovery timeframes of the critical business functions	✗	✗
8. Defining and reviewing resource requirements and capabilities (inventory for each BU)	✓	✓

Table 11.15 Hindrances in implementing BIA (in the context of its current condition)

Descriptions	Firm A	Firm B
1. Critical systems are not periodically evaluated and their minimum essential features cannot be provided for in a disaster	✓	✗
2. Daily transactions needed to reconstruct critical data are not rotated off-site with adequate frequency	✓	✗
3. Critical operations and systems documentation for each platform are not stored off-site	✗	✗
4. Replacement equipment is not readily available	✗	✗
5. Appropriately skilled IT personnel, or specialist equipment are not available	✗	✗
6. Ready access to public network following a disruption is not available	✓	✗
7. Lack of access to communications hardware (e.g. pager, fax, email)	✗	✗
8. Vital records are stored in a single location	✗	✗

4, Firm A's staff support areas were still not prepared for recovery operations and alternative processing facilities were not prepared (facilities for material workshops, equipment warehouses, data storage and operational center).

11.2.3.5 BC Plan Development

Table 11.18 shows the detailed practices in developing a BC plan. It can be seen that Firm B had implemented most of the practices, except for detailed communication management in the plan (Point 10). On the other hand, Firm A had not implemented half of the practices tabulated. In general, both firms had initiated their response plan development adequately (Point 1–5), but practices after documenting the plan were conducted differently. Firm B had more detailed

Table 11.16 Practices that have been implemented for BC strategy

Descriptions	Firm A	Firm B
1. Recovery strategy selection (based on selected disaster—per disaster)	✓	✓
2. Recovery strategy must cover: people; premises; technology; information; supplies; stakeholders	✓	✓
3. In general you should consider four high level scenarios and what alternative working arrangements could be made if		
(a) Cannot gain access to the building	✗	✗
(b) A high percentage of the staff is unavailable	✗	✗
(c) The ICT systems are unavailable	✓	✓
(d) A key supplier/partner is disrupted	✗	✗
4. Recovery strategies must		
(a) Recognise critical functions, dependencies and single points of failure	✗	✓
(b) Enable organisation to perform critical activities	✓	✓
(c) Allow decisions to be taken by responsible managers	✗	✓
(d) Signed off by senior management	✗	✓
5. The continuity strategies that best meet the entity's needs have been implemented based on a cost-benefit analysis	✓	✓

Table 11.17 Hindrances in implementing BC strategy (in the context of its current condition)

Descriptions	Firm A	Firm B
1. An insufficient number of qualified personnel are available to perform user tasks during the recovery phase	✗	✗
2. Personnel who play a role in recovery are unaware of their responsibilities and may not have been adequately trained to perform the recovery tasks	✗	✗
3. Staff support areas are not prepared to support the recovery operation	✓	✗
4. Lack of alternative processing facilities available as and when required	✓	✗

follow-up practices such as in confirming the plan (approvals to related authorities), distributing the plan systematically, implementing information management for the plan, updating the plan, and formally evaluating the plan. Firm A still needs to develop the response plan more thoroughly.

As for the hindrances (Table 11.19), it turned out that both firms had not developed a response plan that will cover any event which could render both primary and alternative facilities inoperable and inaccessible (Point 1 and 2). Both firms viewed that this type of event will most likely not happen. They are still focused on developing response plans for events that have a high chance of impacting their firms.

Table 11.18 Practices that have been implemented for BC plan development

Descriptions	Firm A	Firm B
1. Identify triggers and response (per disaster)	✓	✓
2. Establish the command and control structure to respond to incident; emergency; disaster situations (per disaster)	✓	✓
3. Prioritize activities; Time sequence of a BC (Business Continuity) plan (or response plan) for a selected disaster; Activities and tasks should be prioritized based on the time sequence	✓	✓
4. Coordinate and finalize commitment	✓	✓
5. Gather requirements (list of pre-incident measures)	✓	✓
6. Gather detailed requirements for each Critical Business Function (CBF)	✗	✓
7. Checklists for writing the response plan (based on the tables and procedures/lists)	✓	✓
8. Confirm the response plan	✗	✓
9. Distribute the response plan; Not all BU (business unit) require the entire response plan content; Based on need to know and need to hold basis	✗	✓
10. Incorporating communication management in the response plan		
(a) Regularly update senior management	✓	✓
(b) Keep the customers informed	✗	✗
(c) Mechanisms to inform employees	✗	✗
(d) Keep other stakeholders informed	✗	✗
(e) Ensure media is briefed	✗	✓
11. Incorporating information management in the response plan		
(a) Collate situation reports	✗	✓
(b) Access to contact details	✓	✓
(c) Access to staff records	✗	✓
(d) Insurance policies, contracts	✗	✓
(e) Monitor the media	✗	✗
(f) Maintain a log of decisions, activities and actions	✓	✓
12. The response plan covers: critical products and services as specified in the scoping document; High level plans; Departmental plans; Unit plans	✗	✓
13. The response plan is documented and endorsed	✓	✓
14. The response plan is up-to-date	✗	✓
15. The response plan is linked to the emergency management and incident management plans for the entity	✓	✓
16. The response plan has been formally evaluated as part of the entity's overall corporate governance arrangements	✗	✓

11.2.3.6 Tests and Exercises

Regarding tests and exercises (Table 11.20), both firms had prepared some practices to test the response plan. However, Firm A had not implemented those practices in the form of tests or exercises. Firm B had conducted tests and exercises, but had not further identified and implemented related corrective actions from these practices (Point 6).

Table 11.19 Hindrances in implementing BC plan development (in the context of its current condition)

Descriptions	Firm A	Firm B
1. The response plan will not cover any event which simultaneously renders both the primary and all alternative data/resource centre facilities inoperable	✓	✓
2. The response plan will not cover any event which simultaneously renders the data/resource centre inoperable and the essential off-site storage inaccessible	✓	✓
3. Critical users do not have response plans developed to be able to proceed at the alternative processing facility	✗	✗

Table 11.20 Practices that have been implemented for tests and exercises

Descriptions	Firm A	Firm B
1. Establish practice to operate the response plan	✓	✓
2. Prepare for tests and exercises	✗	✓
3. Conduct tests and exercises	✗	✓
4. Assess the results	✗	✓
5. Infrastructure to support tests and exercises	✗	✓
6. Identify and implement corrective actions	✗	✗
7. Considerations for implementing the tests and exercises		
(a) Risk, impacts and capabilities	✓	✓
(b) Types of exercise to be used	✗	✓
(c) Involvement of senior management	✗	✓
(d) Process of delivering exercises	✓	✓
(e) Planning exercises which minimise the risk of disruption and the risk of an incident occurring as a direct result of the exercise being minimised	✗	✗

Table 11.21 Hindrances in implementing tests and exercises (in the context of its current condition)

Descriptions	Firm A	Firm B
1. Tests and exercises do not involve interdependent entities	✓	✓
2. Periodic test and exercise of the response plan is not conducted	✓	✗

Table 11.21 shows the hindrances in implementing tests and exercises of the firms' response plans. Both firms had not conducted tests and exercises outside of their organization (Point 1). Other external stakeholders had not been involved directly during the practice.

Table 11.22 Practices that have been implemented for programme management

Descriptions	Firm A	Firm B
1. Align BCM with organization operations	✓	✓
2. Review key BCM elements by BCM Steering Committee	✗	✗
3. Review BC plan/response plan (minimum once a year by BCM Steering Committee)	✗	✗
4. Provide continuous training and awareness	✗	✓
5. Perform BCM audit; An internal audit or external review of the implemented framework has been undertaken	✗	✗
6. Track BCM trends and practices	✗	✓
7. The environment in which the firm operates is constantly changing so BCPs and BCM arrangements need reviewing	✓	✓
8. Where changes are needed this will lead to re-writing, re-issuing and re-training and endorsement by management team	✗	✗

Table 11.23 Hindrances in implementing programme management (in the context of its current condition)

Descriptions	Firm A	Firm B
1. There is an insufficient number of personnel possessing the appropriate skills available to implement Business Continuity (BC) operations	✓	✗
2. There is inadequate maintenance of BC (Business Continuity) plan	✓	✓

11.2.3.7 BCM Maintenance (Programme Management)

Table 11.22 elaborates on the detailed practices for implementing BCM programme management. Both firms had aligned their practices in developing and managing their response plans with their operations (Point 1). They had also made some efforts in reviewing the current plans (Point 7). However, further approaches in managing the plans such as forming an official steering committee (Point 2), regular review on the plan (Point 3), auditing the plan and its development process (Point 5), and formally updating or changing the plan (Point 8) had not been implemented by both firms. Moreover, Firm A had not conducted any continuous training or awareness program relating to their crises response plans (Point 4) and had not tracked any best practices related to their response plans (Point 6). On the other hand, Firm B had implemented those practices (Point 4 and Point 6) for managing their crises response plans.

As for the hindrances in implementing BCM programme management (Table 11.23), both firms did not have adequate maintenance for their crises response plans. Moreover, Firm A needed to provide sufficient personnel with appropriate skills to implement business continuity planning.

11.3 Interviews with Experts on BCM Implementation

The second approach used to interpret the proposition based on RO4 and RQ2 was by interviewing five experts from the Indonesian construction industry. They were asked several questions relating to BCM implementation for Indonesian contractors. Their responses were synthesized into the proposed BCM implementation guidelines for Indonesian contractors. The five experts were as follows:

- Respondent 1C: An ICA board member, who owns a construction firm, with 35 years of experience.
- Respondent 2C: An academic, a member of NBCSD, with 20 years of experience.
- Respondent 3C: A director of a construction firm, with 23 years of experience.
- Respondent 4C: A director of a state-owned construction firm, with 21 years of experience.
- Respondent 5C: A director of a state-owned construction firm, with 33 years of experience.

11.3.1 The Benefits of Implementing BCM for Contractors

On the benefits of implementing BCM for Indonesian contractors, all respondents provided several aspects that relate to the firm's resilience and competitive advantage. Respondents 1C, 2C and 5C also highlighted some technical benefits of BCM implementation such as providing systematic data back-ups, preparedness for economic recession, preparedness for floods, securing the firm's assets, and better coordination with stakeholders. The interviewees provided their respective opinions below.

Respondent 1C From your description about BCM and some examples of implementing the concept, I think this concept can help contractors to be more resilient and able to prepare for possible crises.

It can help them in disaster preparation, which is currently occurring in Indonesia. Also the firms can develop data back-ups using the latest technology (such as using cloud storage). Other than that, I think BCM can also help them to prepare for any future economic recession. I believe the BIA analysis can assist the firm to analyze the impacts from economic recession and which BUs and BFs are affected.

Respondent 2C In my opinion, by implementing BCM, the firms can be more prepared in unexpected situations. For example, they can be more ready to overcome floods in Jakarta, which have high impact to the firm's operational tasks, particularly in the head office. The implementation of this concept can also help the firms to be more resilient and competitive globally, because I think construction firms in other countries may have adopted this concept.

Respondent 3C *Although this is a new concept that I've heard of, I think implementing this may have an additional value to the firm. I think this concept has been implemented by developed countries; therefore contractors in Indonesia (as a developing country) should be aware of this concept and able to adopt this to become more competitive globally.*

Respondent 4C *The benefits of implementing BCM are that the contractors can become more resilient towards threats or crises. Secondly, by adopting this concept, they can exercise a resilient corporate culture, which are always be prepared and aware of any situation; react effectively based on the plans that had been exercised and trained; and able to continue their minimum task objectives during certain conditions.*

Respondent 5C *In my view, by implementing BCM, the firms' assets (tangible and intangible) can be secured properly. Having a thorough plan to overcome defined crises can be more time-effective than not having it/not have thought about it in advance. Also, the BCM principles provide steps for better coordination with the stakeholders particularly in overcoming crises.*

11.3.2 The Drawbacks of Implementing BCM for Contractors

All respondents viewed that there will be drawbacks in implementing BCM. Respondents 1C, 4C and 5C opined that implementing BCM may not be the firm's main priority. Moreover, Respondents 2C and 3C thought that BCM is a holistic and sophisticated concept to be implemented. Therefore the top management of the firms should fully support it. Without their support, the process would not be implemented effectively. The interviewees provided their respective opinions below.

Respondent 1C *The drawbacks? Well, as for now, I'm not sure that Indonesian contractors are ready to implement this concept comprehensively. Mostly, the attitude of these firms is not interested in learning new knowledge or process. And they are still mainly focusing on getting as many projects as they can, to increase their profitability. The thought of being resilient may not be their first priority.*

Respondent 2C *Implementing this concept needs to have full support from the top management. If not supported, it may not be applied effectively.*

Respondent 3C *I think the drawbacks of implementing BCM for contractors is the sophisticated approach from this concept, particularly integrating the six main principles may not be easy to implement. The firm needs strong support from the top management to implement this, financially and non-technically. Also, the firm needs knowledgeable and committed members/coordinators to prepare for the BC plan, providing tests and exercise, and managing the programme regularly. The question that each contractor has to consider and answer is whether they have these human resources in place.*

Respondent 4C Getting all of the stakeholders to participate in the implementation may need more time. More time to convince them the benefits of this concept and more time to make them organized. Why do I point out the process to convince them? Because mostly, they have the thought that "if those crises have never happened to our firm, then we should not focus more on preparing something that are less likely to happen."

Respondent 5C Most of the contractors are focusing on how to be profitable every year. I'm not sure they will prioritize and invest their time and money to implement such concept. I think, BCM is a long-term investment. For a more sustained firm, they may have the time to prepare for this, but for others, they may see this as their second goals. So, the goals of these firms may not support this concept. Contractors have been securing their assets by having insurances. Although BCM is not fully about insurance, and it's all about how to prepare, react and restore effectively during crises, they may view this as the same thing. They may think that by having insurances, it will secure them effectively. So, there needs to be a comprehensive approach in socializing this concept to the top management of the firms. And socializing this may need quite a time.

11.3.3 The Need for BCM in Contractors' Firms

All the respondents agreed that BCM implementation is highly needed for contractors. Their various reasons are as follows:

Respondent 1C Based on the nature of the business and the uncertainties in construction projects, I think this concept is needed. The knowledge from this concept is beneficial for the firms to become more prepared and resilient. As mentioned before, if BCM can help the firm to prepare for economic recession, then this will be very advantageous and highly needed.

Respondent 2C Yes, this concept is highly needed, particularly for the firms to coordinate their project sites when there are troubles/disasters occurring. I recommend that the contractors should have a specific BC plans for overcoming floods and earthquakes. Those events have occurred quite frequently these days.

Respondent 3C Considering that other firms from other countries have adopted this concept, I think we should start to adopt this also. And, if in Singapore (a neighbor of ours) and UK had shown that BCM has been regulated and considered as necessary for construction firms, then I think it is necessary for us too.

Respondent 4C I think so. A concept that assists the firms to be more prepared and aware of various threats should be considered. In the end, it is for the sake of the firm's continuity and sustainability. Implementing a concept in an integrated manner is beneficial for the firm to operate and coordinate effectively.

Respondent 5C I think BCM implementation is needed for contractors. By understanding and implementing it, they are able to plan and coordinate their resources for better resilience during crises.

11.3.4 BCM Certification

When asked about BCM certification, all respondents also agreed that the Indonesian contractors need to be BCM certified. Respondents 4C and 5C recommended the regulator to provide such requirements. Also, Respondents 2C and 3C mentioned the ISO as a benchmark for BCM certification. The interviewees provided their respective opinions below.

Respondent 1C Ideally, if there is an official BCM certification, then they should be BCM certified in order to regularly update and monitor the implementation.

Respondent 2C Yes, just like ISO, they should be BCM certified. Now that there is an ISO for BCM, this will make the certification process more approachable. A certification shows that the firm has implemented the concept the way it should be.

Respondent 3C If there is a BCM certification, then the firms should start to apply for it. I think the motivation for a BCM certification should be similar with those ISO certifications. Once your firm is certified, it shows a good credibility and image, thus provides higher opportunities to capture markets. If other global construction firms have been BCM certified, then the firms in Indonesia can compete globally by having BCM certification.

Respondent 4C If there is an official BCM certification provided in Indonesia, then I think this should be socialized (should be by the regulator, I think). They should have BCM certification if they are fully prepared to implement the concept thoroughly. What I mean by the term fully prepared is that they must be prepared financially, technically and culturally.

Respondent 5C I think the regulator must provide such requirements for the firms to follow. Therefore, the regulator should first understand the need for BCM implementation and certification. Then, once it is required, I think the firms will comply (like it or not).

11.3.5 Important Elements (BUs, CBFs, MBCOs, and Significant Crises with High Impacts) in Contracting Firms that Relates Mostly with BCM

In this section, the respondents were asked about the important elements in contracting firms that relate mostly with BCM, referring to the firm's BUs, CBFs, MBCOs, and the most significant crises which may highly impact the firms. Each respondent provided various answers based on their experiences and point of views. Regarding BUs in contracting firms which are essential, the project site unit, accounting, human resources department, contracts division were mentioned by Respondents 2C, 4C and 5C. Respondent 1C opined that this depends on the type of crises that occurred, and Respondent 3C viewed that all BUs are important due to their interdependencies.

In determining which CBFs are essential, functions for documentation, accounting, project assets, human resources management, and communication were all mentioned by the respondents. Moreover, the respondents viewed that securing all of the CBFs that were already mentioned should be the MBCOs of the firm. Also, Respondents 2C, 3C, 4C and 5C prioritized the safety of the firm's human resources as their top MBCO.

Lastly, regarding the most significant crises, economic/financial crisis and natural disasters were viewed to be mostly significant by the respondents. In addition, Respondent 4C added other crises such as riots and terrorism as most significant. Their responses are as follows:

Respondent 1C:

- *BU: I think it depends on the type of crises that occurred.*
- *CBF: The function is related to documentation, finance/accounting, project sites (project assets).*
- *MBCO: I think all of the CBFs should be secured, not lost/damaged, and communication between those divisions should be continued regularly.*
- *Economic recession is the most significant crisis. Impacts: downsizing and employee layoffs; bankruptcy.*

Respondent 2C:

- *BU: The projects site unit. They need to coordinate their resources, finances and operational activities.*
- *CBF: Project documentation (contract documents, reports, drawings); communication function (to coordinate between head office and project sites); accounting (payments and invoices).*
- *If any crisis occur, their MBCO should be the safety of their human resources (employees and laborers), the safety of their documentations and communications and their physical assets (heavy equipments, offices).*
- *Natural disasters: floods, earthquakes*

Respondent 3C:

- *BU: All BUs are important and related. So they are all essential in some ways.*

- *CBF: The firm's communication network and document storage should be secured effectively.*
- *MBCO: The human resources of the firm should be secured first, then followed by organized coordination (communication) and securing their materials and equipments in projects.*
- *Economic crisis. Therefore the firm needs business diversification as an alternative solution.*

Respondent 4C:

- *BU: Accounting department that is responsible for the firm's finance and also contracts division.*
- *CBF: The safety of the firm's data (reports, contracts, invoices/payment documents); the firm's communication network.*
- *MBCO: They should provide a safe condition for their employees/laborers, so they can survive and secure data and communication network, and they can continue to do minimum activities during the situation (depends on the crisis).*
- *A damaged head office (maybe due to disasters/riots/terrorism).*

Respondent 5C:

- *BU: Divisions that need a continuous flow of activities such as: accounting, HRD.*
- *CBF: HRD that manages their employees and laborers. The lives of the people are the main critical function for the firm (employee's emergency shelter), data storage should be functioning well, heavy equipment management (should have equipment/material warehouses to protect them during crises).*
- *MBCO: The employees/laborers should be safe (this is the minimum condition) and able to conduct their assigned tasks (or securing their tasks) in the minimum ways they can.*
- *Financial crisis (for example: no funding for the projects), force majeure.*

11.3.6 Additional Recommendations for BCM Implementation by Indonesian Contractors

This last section describes additional recommendations for BCM implementation by Indonesian contractor from the respondents:

Respondent 1C viewed that the full support from the firm's management and regulator are necessary for BCM implementation.

Respondent 1C In my view, BCM implementation should be fully supported by the regulator and the firm's top management. The regulator/ICA/NBCSD's roles are to: Raise the awareness of the importance of BCM to businesses; increase the supply of BCM specialists that can support contractors to become BCM ready; make BCM more accessible to the business community; and support contractor's efforts to become BCM certified through an incentive programme.

The firm's leader should also be able to identify and measure the impacts from certain threatening events toward their business, which can be used for business continuity planning. This plan should not only focus on the firm's location or physical assets, but also on the people and their daily operations. Throughout the planning process, communication and plan drilling/exercises should be conducted regularly by the employees for better awareness and preparedness toward unexpected events. This plan should also be regularly evaluated by the management.

Respondent 2C provided three factors that needed to be considered in implementing BCM, which are full support from the management, building awareness of BC plan to all employees, and maintaining the BC plan.

Respondent 2C In developing and applying the business continuity planning, other than providing steps to overcome disasters and testing the plans, factors that are also needed to be included are:

- *Full support and approval from the senior management.*

The senior management in the firm has the major role and responsibility for implementing BCM. When the unexpected events occur, they are responsible for the whole BC plan implementation and approvals.

- *Building the awareness towards the BC plan to all employees in the firm.*

Awareness towards the BC plan in the firm is necessary because the ability of the firm to overcome crises or recover/restore from certain unexpected situation depends on the employee's ability and preparedness. Specific training for BCM may be needed for some selected employees (who are championed to lead the BCM process in their department/division) in order to be able to lead and improve their commitment in business continuity planning.

- *Maintaining the plan, including updating it when necessary.*

The BC plan should always be evaluated, to see whether there are new updates from the BCM standards or BCM best practices worldwide.

Respondent 3C recommended strategies for BCM implementation that relate to the firm's human resources, the planning process, communication and attitude towards the implementation.

Respondent 3C Some strategic approaches that can be implemented by the firms in adopting BCM:

- *Human resources and responsibilities: Ensure that the BC team is led by business managers rather than technical/specialist managers; leadership rather than domination is necessary.*
- *BC planning and processes: Include strategic partnership beyond the organization's boundaries; Planning undertaken by functions/BUs with coordination supplied by the BC team to improve ownership of plans.*
- *Communications and structure: Use formal and informal communications management infrastructure to disseminate messages about BCM's importance; Tailor choice and use of media to improve the trajectory and understanding of*

BCM related communications; Appoint formal coordinators to underpin communications endeavours across the organizations.

- *Attitudes and ownership of the BC plan: Functions and departments must have (part) ownership of the planning process coupled with formal appraisals; Functions and departments must have an understanding of how crises and interruptions can threaten the organization's operations and advantages.*

Respondent 4C stressed that the firm's stakeholders should also be involved intensively throughout the implementation process. The stakeholders should be aware of this concept and able to implement it as well.

Respondent 4C I'm viewing that in implementing BCM, the firm should also assess its stakeholders, whether they are aware of BCM or not. The stakeholders should also be coordinated and communicated about the firm's BCM.

Some risks that may occur during crises that relate to the firm's stakeholders include the following:

- *Cascading failures from compounding disaster events (e.g. earthquake, tsunami);*
- *Failure of unexpectedly vulnerable support systems;*
- *Inability to get workers to their posts because of transportation infrastructure damage;*
- *Supply chain failure concentrated on suppliers and smaller firms without BC plan;*
- *Too-rigid BC plans left some firms unable to adapt to the shifting challenges produced by the crises; and*
- *Communications failures between public and private sectors and across national boundaries within large organizations.*

Some risk treatments that can help to overcome the risks above include the following:

- *Clearly identify all key support systems, work to reduce their vulnerability to likely hazards;*
- *Make provisions in BC plan for post-disaster transportation disruptions;*
- *Push BCP practices to suppliers;*
- *Ensure that the BC plan is flexible and is keyed to address common impacts and protect key processes; and*
- *Increase communications between private-sector entities across industries to improve collective action after a disaster.*

Last but not least, Respondent 5C opined that BCM implementation should be understood not only by the top management, but also by the firm's employees. This can be achieved by conducting relevant trainings, workshops or knowledge sharing about BCM.

Respondent 5C In adopting BCM as the firm's management system, it should require buy-in from senior management, and also require every employee to have an appropriate understanding of the policies and procedures relevant to them.

Relevant trainings, workshops or knowledge sharing about BCM should be conducted when BCM is initiated in the firm.

11.4 BCM Implementation Guidelines Development

After conducting the above case studies and interviews and analyzing the results, this study advanced to develop the BCM implementation guidelines (based on RO4). The guidelines were developed using a qualitative approach, where the data collected was used to generate ideas or framework and used for inductive reasoning. The sources for the guidelines include:

1. Results from the case studies and interviews—utilized for developing the Indonesian contractor's characteristic domain;
2. Literature on BCM level of preparedness (presented in Chaps. 3 and 8)—utilized for developing the BCM level of preparedness domain; and
3. Literature on BCM principles (presented in Chaps. 3 and 8)—utilized for developing the BCM principles domain.

11.4.1 BCM Implementation Guidelines Framework

The BCM implementation guidelines for Indonesian contractors were developed based on Fig. 8.4. The details of the BCM principle domains were based on Table 8.4, and the details of the BCM preparedness criteria domain (in general) were based on Table 8.5. For the Indonesian contractors' strategies and business value chains domain, the details were based on the results from the case studies and interviews (Sects. 11.2 and 11.3).

The guidelines consist of two aspects, which are:

1. The assessment process per BCM principle that provides results of the firm's BCM level of preparedness;
2. The recommended action plans for the firm's current BCM level of preparedness. These action plans consist of the technical aspects in implementing BCM (elaborated per BCM principle) and non-technical aspects that were related to Indonesian contractors' characteristics. The non-technical aspects were compiled from the OC, IF, case studies and interviews results.

11.4.2 Levels of Preparedness Development

In the proposed system, the level of BCM preparedness was grouped into four levels, which started from an undeveloped BCM, beginner level, moderate level and comprehensive level. These levels were adapted from various BCM level of preparedness studies (MOF-BC 2007; Lansley and McAtee 2009; Marsh Risk Consulting 2010; Smit 2005; ANAO 2009; SBF 2011). Currently, there is no official or standardized BCM level of preparedness yet. Therefore, this study summarized the existing levels from various institutions (which have similarities) and provided the four levels of preparedness. As described by Smit (2005), since a level of preparedness aims to provide a simplified and easily communicable reproduction of reality, these four levels of preparedness would assist the firm to understand its existing condition. The principle behind the different levels is that an organization develops new practices and processes, from which it learns, and from which it can subsequently optimize these practices and processes to move on to the next level.

Table 11.24 describes the four BCM levels of preparedness used for the study, which were based on the six BCM levels identified from various institutions. The detailed descriptions of each level were summarized from the references in the table. There are four main aspects that would describe each level, which are the deliverable (the outcome of BCM in the firm), management support, policies (policies about BCM), comprehension (the firms' understanding about BCM) and BCM principles (the practices that have been implemented by the firm).

11.4.3 Assessment Phase

In this phase, the guidelines provided descriptions (BCM practices) that needed to be assessed by the firm. The descriptions were compiled from the four main aspects that described the BCM level of preparedness. These descriptions would be rated and the inputs from the rating would provide the assessment result, which is the firm's current BCM level of preparedness. The assessment descriptions were grouped into six categories, based on the BCM principles, which are Risk analysis and review, Business impact analysis, Strategy analysis, BC plan development, Tests and exercises, and Programme management.

The rating for this assessment was based on the degree of implementation by the firm, which used a four-scale rating: 1 = not implemented; 2 = partially implemented 3 = largely implemented; and 4 = fully implemented (adopted from SPICE 2001; Guedria et al. 2009). Due to the difficulty faced by the assessor in making a fine judgment and assigning coherent numerical values for the rating process, using linguistic variables such as the proposed scale rating would be convenient for representing the assigned scores. In this study, the equation to find the BCM level of preparedness would be changed to a fuzzy version, which can be treated using fuzzy logic to obtain directly a linguistic qualification of the level of preparedness (Guedria et al. 2009; Ahmed and Capretz 2006). This is further described in Chap. 12. Table 11.25 shows the detailed descriptions (BCM practices) per BCM principle used for the assessment phase.

Table 11.24 BCM-KBDSS level of preparedness

Level number	BCM-KBDSS level of preparedness descriptions	MOF-BC (2007)	The Australian National Audit Office (2009)	Lansley and McAtee (2009)	Smit (2005)	Marsh Risk Consulting (2010)	Singapore Business Federation (2011)
1	*Undeveloped*						
Deliverable	There is no BCM in place		X	X	X	X	
Management support	BCM has not yet been recognized as strategically important by senior management	X		X		X	
Policies	There are no or minimum legal/regulatory requirements met providing protection for people and facilities	X		X		X	X
Comprehension	The firm has a minimal understanding of business continuity planning	X				X	X
BCM principle 1	Risk analysis and review have not been conducted			X		X	X
BCM principle 2	Business Impact Analysis (BIA) has not been conducted			X		X	X
BCM principle 3	Strategy development for business continuity has not been conducted			X		X	X
BCM principle 4	BC plan has not been developed			X		X	X
BCM principle 5	Tests and exercises for business continuity have not been conducted			X		X	X
BCM principle 6	Programme management for business continuity has not been conducted			X		X	X
2	*Beginner*						
Deliverable	The deliverable of the initiated stage is BCM as an initiative	X			X		
Policies	The firm exhibited a low level of awareness of policies and guidelines about BCM	X	X	X	X	X	

(continued)

Table 11.24 (continued)

Level number	BCM-KBDSS level of preparedness descriptions	MOF-BC (2007)	The Australian National Audit Office (2009)	Lansley and McAtee (2009)	Smit (2005)	Marsh Risk Consulting (2010)	Singapore Business Federation (2011)
Policies	The firm exhibited a low level of awareness of roles and responsibilities about BCM	X	X	X	X		
Management support	Participating business units (BUs) and departments have instituted a basic governance program, mandating at least limited compliance to the standardized BCM policy, practices and processes to which they have commonly agreed	X	X	X	X	X	X
Comprehension	The firm has a basic understanding of business continuity planning	X	X				X
BCM principle 1	Risk analysis and review have been documented	X	X		X	X	X
BCM principle 2	Business Impact Analysis (BIA) has been partially conducted	X	X		X	X	X
BCM principle 3	Strategy development for business continuity has been partially conducted	X	X		X	X	X
BCM principle 4	BC plan has not been developed	X	X		X	X	X
BCM principle 5	Tests and exercises for business continuity have not been conducted	X	X		X	X	X
BCM principle 6	Programme management for business continuity has not been conducted	X	X		X	X	X
3	*Moderate*						
Deliverable	A BCM framework is in place as a documented BC plan	X	X	X	X	X	X

Management support	Strong executive support for BCM in the firm	X		X	X		
Policies	The responsibility for BCM is covered at a sufficiently high level within the firm and an explicit BCM policy is in effect	X	X	X	X		
Comprehension	The firm has an advanced understanding of business continuity planning	X	X		X	X	X
BCM principle 1	Risk analysis and review have been documented	X	X		X	X	X
BCM principle 2	Business Impact Analysis (BIA) has been documented	X	X		X	X	X
BCM principle 3	Strategy development for business continuity has been documented	X	X		X	X	X
BCM principle 4	BC plan has been documented	X	X		X	X	X
BCM principle 5	Tests and exercises for business continuity have been partially conducted (not regularly)	X	X		X	X	X
BCM principle 6	Programme management for business continuity has been partially conducted	X	X		X	X	X
4	*Comprehensive*						
Comprehension	All BUs has a high degree of business continuity competency	X		X		X	X
Management support	Very strong executive support for BCM in the firm	X		X	X	X	
Policies	Staff responsible for BCM have a strong awareness of and compliance with core policy requirements, guidelines and procedures for business continuity planning	X	X	X	X	X	
Deliverable	If a firm has reached this level, it controls its existing BCM		X	X	X	X	

(continued)

Table 11.24 (continued)

Level number	BCM-KBDSS level of preparedness descriptions	MOF-BC (2007)	The Australian National Audit Office (2009)	Lansley and McAtee (2009)	Smit (2005)	Marsh Risk Consulting (2010)	Singapore Business Federation (2011)
BCM principle 1	Risk analysis and review have been documented	X	X		X	X	X
BCM principle 2	Business Impact Analysis (BIA) has been documented	X	X		X	X	X
BCM principle 3	Strategy development for business continuity has been documented	X	X		X	X	X
BCM principle 4	BC plan has been documented	X	X		X	X	X
BCM principle 5	Tests and exercises for business continuity have been conducted	X	X		X	X	X
BCM principle 6	Programme management for business continuity have been conducted and audited	X	X		X	X	X

Note: [X] means that the description is provided in the reference's BCM level of preparedness

Table 11.25 BCM practices for assessment (per BCM principle)

BCM practices for assessment
Risk analysis and review
1. The firm has documented its core business objectives and outputs
2. All areas of the firm's operational responsibility have been identified based on its functional organizational chart
3. Threats have been identified for each business unit (BU) of the firm
4. Detailed risk analyses have been conducted in each business unit (BU), complete with risk identification, risk category, and risk treatments
5. Possible disasters identification and key disasters scenarios development have been conducted in the firm
6. Reviewing each BU's critical risks and their impacts toward its operational processes and infrastructures has been implemented (reviews on which resources and physical assets—people, systems, equipment—are most critical for the organization)
7. Risk analyses and reviews process have been documented
Business impact analysis
1. Business functions and Minimum Business Continuity Objective (MBCO) for each business unit (BU) in the firm have been identified
2. Priorities for analyzing impact (from defined disasters) on each affected BU have been developed
3. Critical business functions (CBF) on each BU have been identified
4. Dependencies of each CBF have been identified
5. For each CBF the recovery time objective (RTO) and recovery point objective (RPO) have been established
6. Inventories have been conducted for each BU regarding its resource requirements and capabilities toward its CBFs
7. The Business Impact Analysis (BIA) process has been documented
Strategy analysis
1. The probable strategies for each affected Critical Business Function (CBF) have been determined
2. For each strategy, there are clearly defined roles and responsibilities
3. The defined strategy has also identified the internal and external linkages to the firm
4. The strategies for each affected CBF have been evaluated
5. The strategies for each affected CBF cover the emergency response and recovery response
6. Adequate strategies to maintain critical activities of the CBF in the event of a disruption have been selected and developed based on several considerations and arrangements
7. The recovery strategy selection has been consolidated and documented
BC plan development
1. The triggers and responses for each disaster have been identified
2. The command and control structure to respond to each disaster have been established
3. The BCM team and related suppliers have been coordinated and commitments have been finalized
4. A list of supporting documentation required to complete the Business Continuity (BC) processes have been prepared
5. The BC plan has been written based on selected disaster scenario, identified CBFs, and organization's recovery strategy
6. The BC plan also consists of emergency response plan (ERP) (to stabilize the situation following a disaster), setting Emergency operation center (EOC), and recovery and resumption of CBFs
7. The BC plan is confirmed and distributed within the firm

(continued)

Table 11.25 (continued)

BCM practices for assessment
Tests and exercises
1. The firm has established and conducted practices to operate BC plan in the form of tests and exercises to test and familiarize the staff about the continuity management arrangements in the event of a disruption
2. The firm has conducted assessments of the results from tests and exercises
3. The firm has prepared infrastructures to support tests and exercises of the BC plan
4. The firm has identified and implemented corrective actions from tests and exercises results
5. The firm has developed personnel assessment criteria for BC plan tests and exercises
6. The firm has identified the needs for further personnel development such as training and awareness programme
7. The firm has documented the whole BC plan tests and exercises processes
Programme management
1. The firm has the BCM framework fully developed
2. Staff responsible for BCM had a strong awareness of and compliance with core policy requirements, guidelines and procedures for BCM
3. The firm's management support BCM for implementation
4. The manager of each BU regularly manages and updates the BC plan
5. BCM in the firm is aligned with organization's operations
6. BCM review (BC plan reviews and its elements reviews) and audit are conducted regularly (minimum once a year)
7. The firm tracks the BCM trends and practices regularly

11.4.4 Action Plans per BCM Level of Preparedness

After conducting the assessment process and obtaining the current BCM level of preparedness, the guidelines provide the recommended action plans for the firm to consider. The action plans assist the firm to enhance its BCM level of preparedness. For example, if Firm A's assessment result was Level 1, then there would be recommended action plans that could improve the firm's BCM implementation into the next level (Level 2), and also into upper level/highest level (Level 1–Level 3; Level 1–Level 4). The action plans provide the technical and non-technical steps needed for each BCM principle.

The technical action plans consist of steps and practices that needed to be implemented, grouped into the six BCM principles. The knowledge base was acquired from the literature on BCM principle, supported by knowledge from the case studies and interview results. As for the non-technical action plans, the knowledge base were acquired from the study's fieldwork results, which were from the surveys, case studies and interview results. This second section consists of: recommended aspects to be considered for implementing the principle (based on case studies and interview results); Organizational culture (OC) attributes that can support the action plans (based on significant results of OC attributes from surveys); and several drivers to implement the BCM principle [based on significant Institutional Forces (IF) results from surveys]. The detailed recommended action plans per BCM level of preparedness can be seen in Appendix F—Analysis for Chap. 11.

11.5 Summary

In this chapter, the results of the case studies, interviews and BCM implementation guidelines development had been presented. The case studies were presented using various sources of evidence in exploring BCM implementation in the selected firms. The results from case studies and interviews were analyzed qualitatively to develop the BCM implementation guidelines. Elements of the guidelines are shown in Appendix F—Analysis for Chap. 11. Further discussions about the results from this chapter will be presented in Chap. 13.

Chapter 12
Data Analysis: BCM-KBDSS Development

12.1 Introduction

The final part of the study's data analysis is presented in this chapter. The first section elaborates the development of the BCM-KBDSS, where each element of the system will be described. Following that, the validation process of the BCM-KBDSS will be discussed and the results will be shown.

12.2 BCM-KBDSS Development

As described in Chap. 8, the BCM implementation guidelines for Indonesian contractors was developed into a KBDSS. This system provided the knowledge needed by the management team in developing BCP and the automation process in this system can be beneficial to facilitate a fast and effective decision making process. The BCM-KBDSS consisted of phases of development, started from knowledge base acquisition through the compiled data, rules development, frame and logical representation (inference engine), program algorithm used for the computing language, and finalized with validation (Mockler 1989; Turban 1995). Table 8.8 in Chap. 8 had described the variables needed for each of the phases.

12.2.1 Knowledge Base (KB) Development

The knowledge base (KB) for the BCM-KBDSS was grouped into two sections. The first section was for the assessment phase, where there were seven practices for each of the BCM principles provided. The second section was for the recommendations, in the form of action plans per BCM level of preparedness. Each level

L.S.R. Supriadi, L. Sui Pheng, *Business Continuity Management in Construction*, Management in the Built Environment, DOI 10.1007/978-981-10-5487-7_12

consisted of action plans for six BCM principles (technical and non-technical action plans). Section 11.4 had elaborated the details of the KB. All of the KB would be utilized for the BCM-KBDSS process that had been illustrated in Fig. 8.5.

12.2.2 Rules and Logics for BCM-KBDSS

Each of the process in the BCM-KBDSS contained rules, framework and logic representation. The KB was represented using these elements, where for each phase in the system: rules will be given for every decision situation; components such as inputs from users, processes after receiving inputs from users, knowledge to be used in the processes and the outputs were developed into a framework; and inference engine that controls the reasoning process was used for the framework. In this study, the reasoning process used was a forward chaining process, where it starts from utilizing the given knowledge (from the KB) and processed into deriving a solution or recommendation (Sudarto 2007; Mockler 1989).

12.2.2.1 The Use of Fuzzy Logic

The BCM-KBDSS used fuzzy logic for its inference engine, because this method had the ability to compute with words used for modeling qualitative human thought processes in the analysis of complex systems and decisions. The method also provided a simple way to extrapolate definite conclusions from vague and imprecise information (Liu and Lai 2009; Zadeh 1996; Dweiri and Kablan 2006). As discussed in Sect. 11.4.3 (Chap. 11), in finding the BCM level of preparedness based on the qualitative inputs from the assessment phase, fuzzy logic could obtain the linguistic qualification of the process to provide the results. The fuzzy set theory used in this method presented vague knowledge and allowed mathematical operators to be applied to its domain (Guedria et al. 2009; Ahmed and Capretz 2006; Xia et al. 2011).

The general process of fuzzy logic can be seen in Fig. 12.1. Firstly, a crisp set of input data is gathered and converted to a fuzzy set using fuzzy linguistic variables and membership functions. This step is known as fuzzification. Just like in human thinking, in fuzzy logic systems (FLS) linguistic variables are utilized to give a "value" to the element, some examples are *much, tall, cold,* etc. FLS require the linguistic variables to be related to their numeric values, their quantification and the connections between variables and the possible implications. In traditional sets, an element either belongs to the set or does not belong to the set {0,1}, while in fuzzy sets the degree to which the element belongs to the set is analyzed and it is called the membership degree, giving values in the range [0,1], where 1 indicates that the element belongs completely to the set. In FLS, the membership functions are utilized to find the degree of membership of the element in a given set (Mendel 1995; Ahmed and Capretz 2006; Ponce-Cruz and Ramirez-Figueroa 2010). The

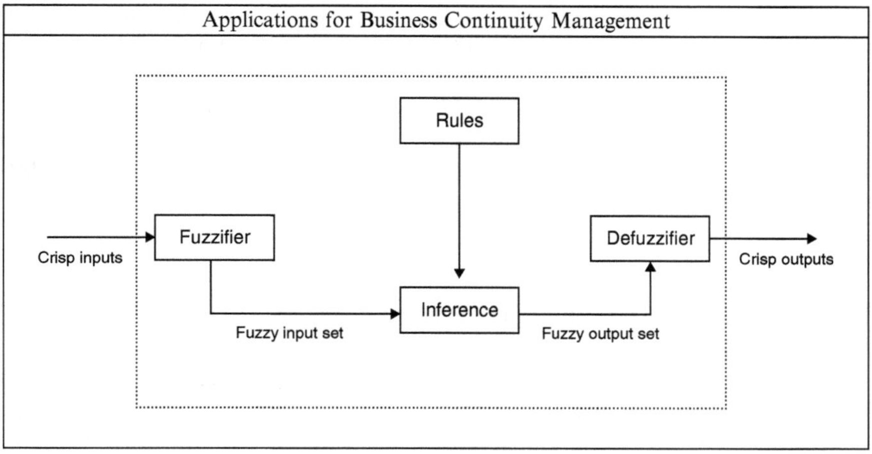

Fig. 12.1 A fuzzy logic system. Source: Adapted from Mendel (1995) and Ahmed and Capretz (2006)

next step after fuzzification is an inference which is made based on a set of rules. The rules used in the FLS can be in the form of the IF-THEN rule type or various fuzzy sets mathematical operations. Lastly, the resulting fuzzy output is mapped to a crisp output using the membership functions, in the defuzzification step (Mendel 1995; Ahmed and Capretz 2006; Dubios and Prade 1980; Cox 1998; Pedrycz and Gomide 2007).

In this study, triangular membership function was adopted for the fuzzification process. It has been widely known that this type of membership function is easy to use for information processing in a fuzzy environment (Pedrycz and Gomide 1998; Tah and Carr 2000; Xu et al. 2010). In this membership function, the triangular fuzzy set is utilized to quantify the qualitative information, using the equation below:

$$\mu_M(x) = \begin{cases} 0, x < a, \text{or } x > c \\ \dfrac{x-a}{b-a}, a \le x \le b \\ \dfrac{c-x}{c-b}, b < x \le c \end{cases} \tag{12.1}$$

The triangular fuzzy number $M = (a, b, c)$, where $a \le b \le c$; where $\mu_M(x)$ is the membership function of the imprecise numerical concepts, such as "close to b", "about b", or "approximately b" (Pedrycz and Gomide 1998; Lee 1990).

Furthermore, for the defuzzification process, this study adopted the centroid method. The method would find a point at which a vertical line would slice the aggregated fuzzy set into two equal masses, which represents the center of gravity of the fuzzy set (Negnetvitsky 2006; Lee 1990; Mendel 1995; Ahmed and Capretz 2006). The centroid method transformed the fuzzy numbers, into crisp numbers by assuming that fuzzy number, $D = (d_1, d_2, d_3)$, can be converted into the crisp number by using the equation below (Lee 1990; Mendel 1995):

$$x = (d_1 + d_2 + d_3)/3; \text{where x is the crisp number}\ldots \quad (12.2)$$

Due to its ease of calculation, the centroid method is widely used (Cox 1998; Chou and Chang 2008; Lam et al. 2010).

12.2.2.2 BCM-KBDSS Rules per Phase

The flowchart of the BCM-KBDSS's decision situation scenario is shown in Fig. 12.2. This was the first phase of developing the system, where every process and decision situation scenario that would be developed was structured. Each box described the process in the system, where it started from the user providing information about the firm that will be assessed. Following that, the system provided a brief description about BCM's definition and principles. The first phase, which is the assessment phase, would be the first decision scenario for the user to respond. There would be six BCM principles that needed to be rated based on the user's perspectives. After the inputs were acquired, the system provided the assessment results in the form of indexes and BCM level of preparedness. Moreover, the system showed the description about the user's BCM level of preparedness and provided the recommended action plans for each BCM principle that can be used by the user for further decision making. This process would be the BCM level of preparedness recommendation phase and the second decision situation scenario provided by the system. Both assessment and recommendation phases need to refer to rules using the fuzzy logic approach. The final process of the BCM-KBDSS was to provide the summary of the whole phases conducted by the user and the user could provide feedback regarding the results or the process of the BCM-KBDSS.

From the flowchart in Fig. 12.2, the system was developed into more details, particularly for the decision situation scenarios. In this phase, the concept and coding of the structured situation diagram were developed, as shown in Figs. 12.3 and 12.4 respectively. This diagram did not provide the rules yet, but showed the framework and coding that would be used for developing the dependency diagram (Mockler 1989; Turban 1995). As illustrated in these figures, descriptions of what would be conducted in the two decision situation scenarios (assessment and recommendation phases) were shown.

Furthermore, the coding for each description of the decision situation scenarios is illustrated in Fig. 12.4. The coding system would be used for developing rules and logic of the KBDSS.

After developing the structured situation diagram, the next phase was to develop the dependency diagram, as shown in Fig. 12.5. All of the rules needed for each structured situation would be described in this diagram.

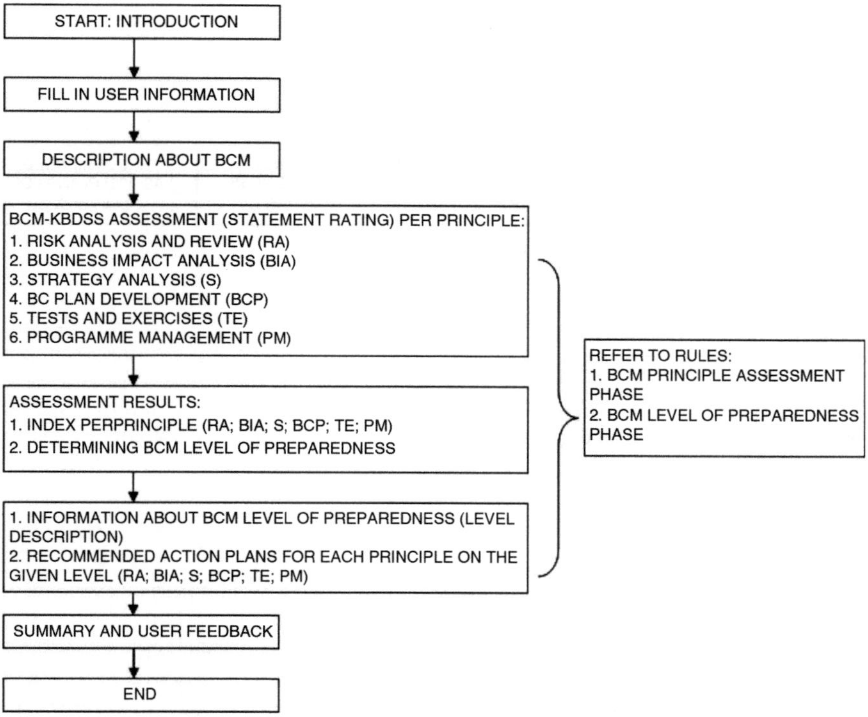

Fig. 12.2 BCM-KBDSS decision situation scenario flowchart

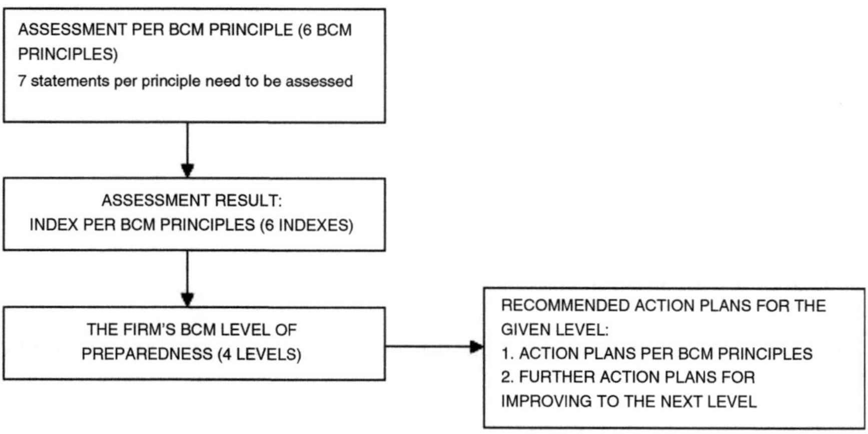

Fig. 12.3 Structured situation diagram (conceptual)

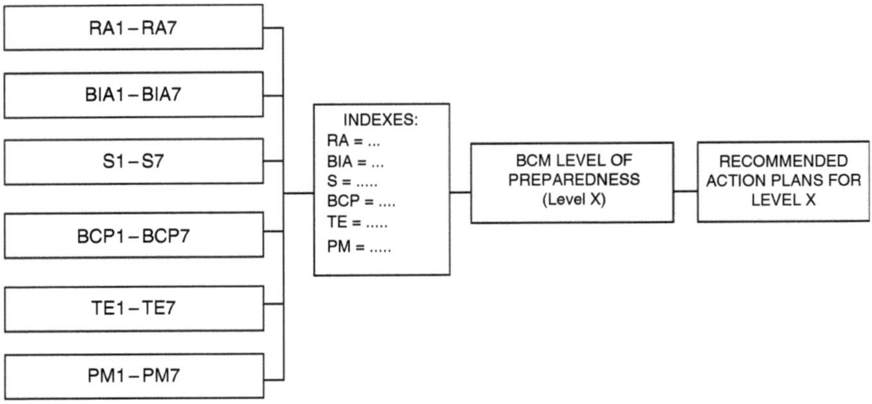

Fig. 12.4 Structured situation diagram (coding)

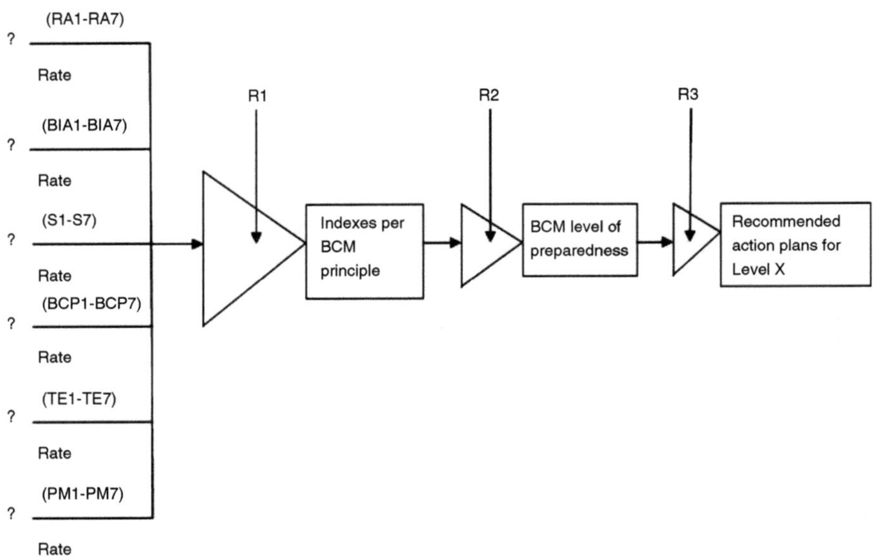

Fig. 12.5 BCM-KBDSS dependency diagram

Descriptions of symbols used in the dependency diagram are as follows (Mockler 1989):

- ? = Indicating the inputs needed from the user
- (...) = Subject of assessment
- Description below the line = Type of inputs from user to system (rating)
- R1, R2, R3 = Rules in the system process
- Triangle = Processes occurred in the system
- Box = Results from system (outputs)

- Description inside the box = Results description
- R1–R3 descriptions:
  - R1 =
    1. Determine the membership function per rating result
    2. Determine the total rating per BCM principle
    3. Determine the range number per BCM principle
    4. Determine the index per BCM principle
  - R2 = Determine the BCM level of preparedness
  - R3 = Provide the recommended action plans per BCM principle for the given level (Level X).

The details of each rule will be described in the next section.

Assessment Phase
The assessment phase of this system used fuzzy logic for conducting the rule of R1, which were (1) determining the membership function per rating result; (2) determining the total rating per BCM principle; (3) determining the range number per BCM principle; and (4) determining the index per BCM principle.

The fuzzification process was executed using linguistic variables as the rating descriptions with a four-scale rating: 1 = not implemented; 2 = partially implemented 3 = largely implemented; 4 = fully implemented (adopted from SPICE 2001; Guedria et al. 2009), as shown in Fig. 12.6. In defining the membership function (*a, b, c*) for each user responses (for each statement rating in each BCM principle), the system used the triangular membership function. Table 12.1 describes the membership function for each scale (1–4).

In determining the total rating for each BCM principle, the fuzzy rule in the form of mathematical operation, would be used as follows (Ahmed and Capretz 2006):

$$(a1 + \ldots + a7)/7, (b1 + \ldots + b7)/7, (c1 + \ldots + c7)/7 = A, B, C \ldots . \quad (12.3)$$

where:

a, b, c = membership function per statement responses (there are seven responses for each BCM principle)
A, B, C = total rating.

Following that, the defuzzification process would be started by determining the range number for each BCM principle. By adopting the centroid method, Eq. (12.4) below was used:

$$(A + B + C)/3 = Range\ number \ldots . \quad (12.4)$$

The index of each BCM principle was based on the range number result, by using IF-THEN rules as follows (Ahmed and Capretz 2006):

IF range number 0 to 0.165 THEN Index = 1

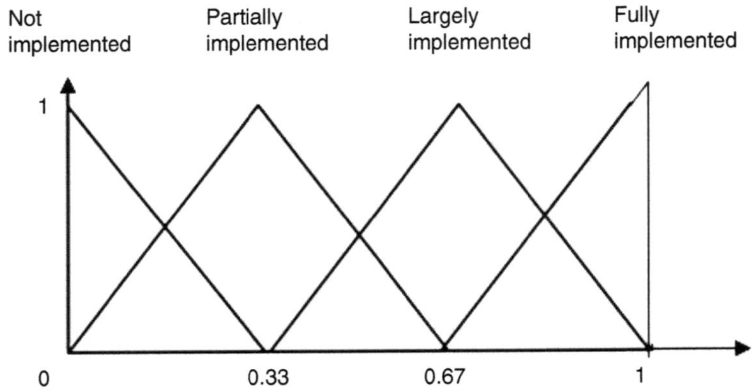

Fig. 12.6 Linguistic variables

Table 12.1 Triangular membership function

Scale	Membership function		
	a	b	c
1 = not implemented	0	0	0.33
2 = partially implemented	0	0.33	0.67
3 = largely implemented	0.33	0.67	1
4 = fully implemented	0.67	1	1

IF range number 0.166 to 0.495 THEN Index = *2*
IF range number 0.496 to 0.835 THEN Index = *3*
IF range number 0.836 to 1 THEN Index = *4*

This would be the first phase crisp output, which would be followed by determining the BCM level of preparedness (R2) as the final phase crisp output.

Recommendation Phase

The fuzzy logic approach used for the recommendation phase would continue to the defuzzification process to determine the BCM level of preparedness index (as the final crisp output). The fuzzy decision rule for this phase was achieved by using the two-input and one output rule base with the five-level hierarchical system (adopted from Zlateva et al. 2005, 2011; Ahmed and Capretz 2006). Every level provided one fuzzy logic subsystem (intermediate or final index) with two inputs, as shown in Fig. 12.7.

On the basis of the two inputs, the output mapping was defined in the truth value rules (IF-THEN rules) (Ahmed and Capretz 2006). There were 16 rules to be used to determine the output (as the intermediate and final index). The final index output would be the firm's BCM level of preparedness. Table 12.2 shows the truth value rules for providing the final crisp output of the system. The reference of the truth value rules were based on the four BCM levels of preparedness used for the study (see Table 11.24).

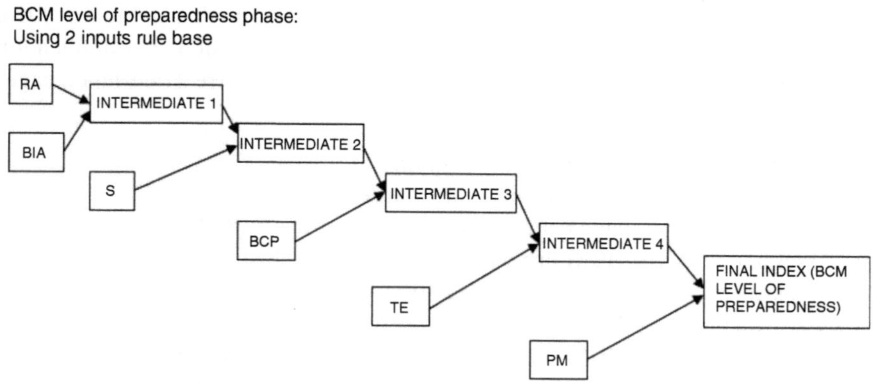

Fig. 12.7 Two-input and one output rule base

Table 12.2 Truth value rules

Truth value rules: IF input 1 = …. And input 2 = …. THEN output = ….			
No	Input 1	Input 2	Output (intermediate index and final index)
1	1	1	1
2	1	2	1
3	1	3	1
4	1	4	1
5	2	1	2
6	2	2	2
7	2	3	2
8	2	4	2
9	3	1	2
10	3	2	2
11	3	3	3
12	3	4	3
13	4	1	3
14	4	2	3
15	4	3	3
16	4	4	4

12.2.3 Design User Interface and BCM-KBDSS Compilation

In developing the BCM-KBDSS, the system was designed to have the capability to interface with the user. The user interface's objective was to provide a user friendly process in the system, starting from the introduction of the system, assessment phase, results and recommendation phase, summary and closing phase. The user interface should be easy for the user to understand and apply the BCM-KBDSS.

The design templates were developed using Microsoft Visual Studio 2010, where in this phase, considerations such as inserting descriptions per application pages, flows and command buttons were determined. There were 21 designed pages for the system as follows:

1. KBDSS introduction page ("Home" page)
2. User Information page (user fill in the data)
3. BCM introduction page (filled with descriptions about BCM)
4. Assessment page—BCM principle 1 (user rates each statement)
5. Assessment page—BCM principle 2 (user rates each statement)
6. Assessment page—BCM principle 3 (user rates each statement)
7. Assessment page—BCM principle 4 (user rates each statement)
8. Assessment page—BCM principle 5 (user rates each statement)
9. Assessment page—BCM principle 6 (user rates each statement)
10. Assessment result page (system provides the BCM level of preparedness, with index and level number)
11. Level description page
12. Recommended action plans page—BCM principle 1 (system provides the action plans)
13. Recommended action plans page—BCM principle 2 (system provides the action plans)
14. Recommended action plans page—BCM principle 3 (system provides the action plans)
15. Recommended action plans page—BCM principle 4 (system provides the action plans)
16. Recommended action plans page—BCM principle 5 (system provides the action plans)
17. Recommended action plans page—BCM principle 6 (system provides the action plans)
18. Additional recommendations page (applicable to each BCM principle from the Recommended action plans pages)
19. Summary page—result from clicking the "Summary" button
20. Feedback page (user input feedback for the system)
21. End page—result from clicking the "End" button.

The BCM-KBDSS consisted of groups of KB, fuzzy rules and guided user interface templates. All of these elements were incorporated in the system using MySQL and PHP scripting language softwares. The decision to proceed with the development of a web-based application was made to facilitate easy accessibility for all users (could be used via the firm's intranet), no software installation needed for users and quick process for updating the system (Wicht et al. 2011; Bisong et al. 2013).

12.2.4 BCM-KBDSS Prototype

The prototype of the BCM-KBDSS is illustrated from Figs. 12.8, 12.9, 12.10, 12.11, 12.12, 12.13, 12.14, 12.15, 12.16, 12.17, 12.18, 12.19, 12.20, 12.21, 12.22, 12.23, 12.24, 12.25, 12.26, 12.27 and 12.28. These figures show the application pages of each phase in the BCM-KBDSS. Figure 12.8 shows the introduction page, consisting of the system's general information and a link to start the assessment process. This page is the home screen page for the system.

Figure 12.9 shows the user information page, where the user needs to provide the firm's particulars.

Figure 12.10 shows the BCM introduction page, where BCM is briefly described based on its definition and main principles. The intention of this page is to provide general information to the user about BCM before starting the assessment.

Figures 12.11, 12.12, 12.13, 12.14, 12.15 and 12.16 show the assessment phase, where the user would rate all the statements for each BCM principle based on the existing conditions of the user's firm. There are six pages of assessments based on the six BCM principles.

After the user has given inputs on the assessment pages, the system would provide the assessment results, which consist of the indexes of the six BCM principles and the firm's BCM level of preparedness. The assessment result page would also provide links to the level description and the recommended action plans. This can be seen in Fig. 12.17.

Figure 12.18 shows the BCM level of preparedness description page. The description summarizes the resultant level based on the user's assessment process.

In Figs. 12.19, 12.20, 12.21, 12.22, 12.23 and 12.24, the recommended action plans based on the firm's current BCM level of preparedness are elaborated in details. These action plans are based on each of the BCM principles.

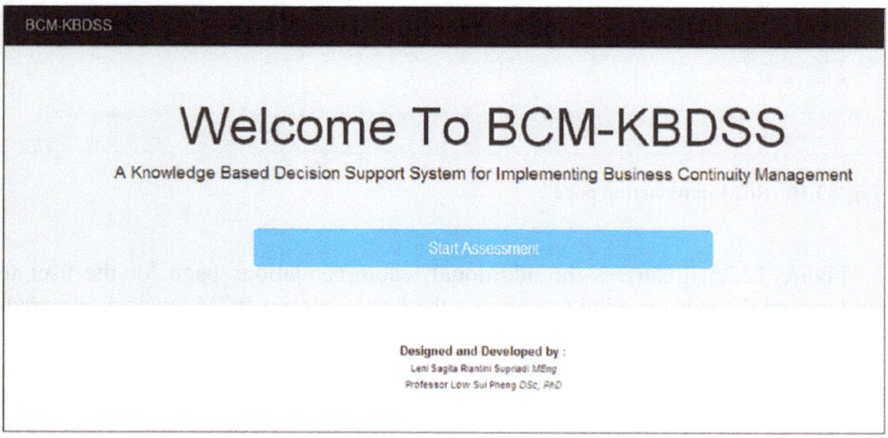

Fig. 12.8 BCM-KBDSS introduction page

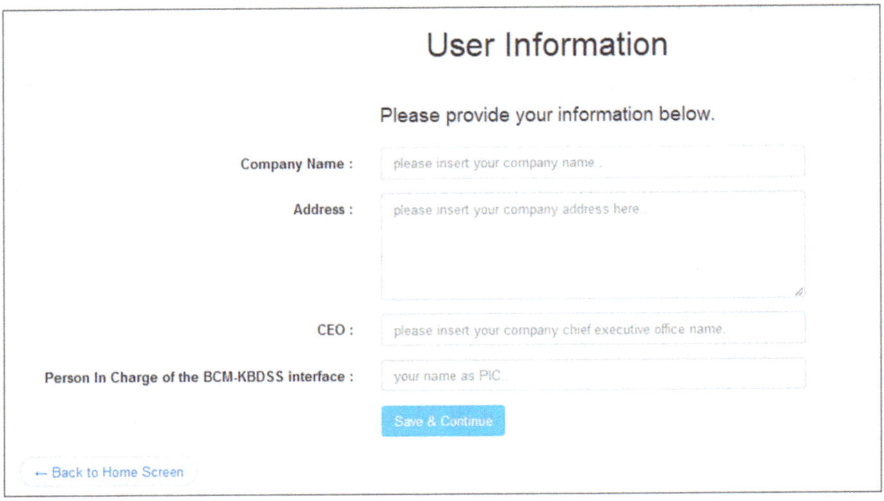

Fig. 12.9 User information page

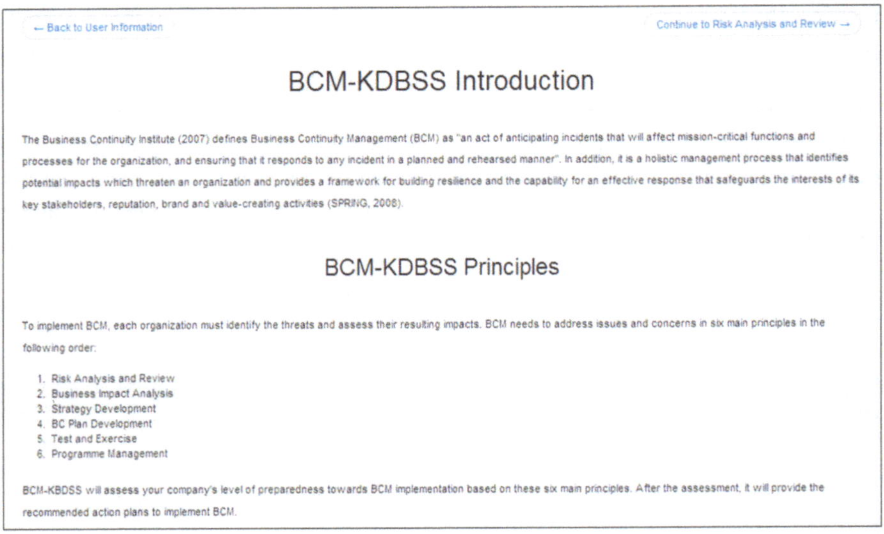

Fig. 12.10 BCM introduction page

Figure 12.25 illustrates the additional recommendations page for the user to understand the steps needed to improve the firm's current BCM level of preparedness to a higher level.

Moreover, the BCM-KBDSS is able to save all the assessments and results processes conducted by the user and these processes (inputs and outputs) can be viewed by the user through the summary page, as shown in Fig. 12.26. The summary page can also be viewed in a .pdf format.

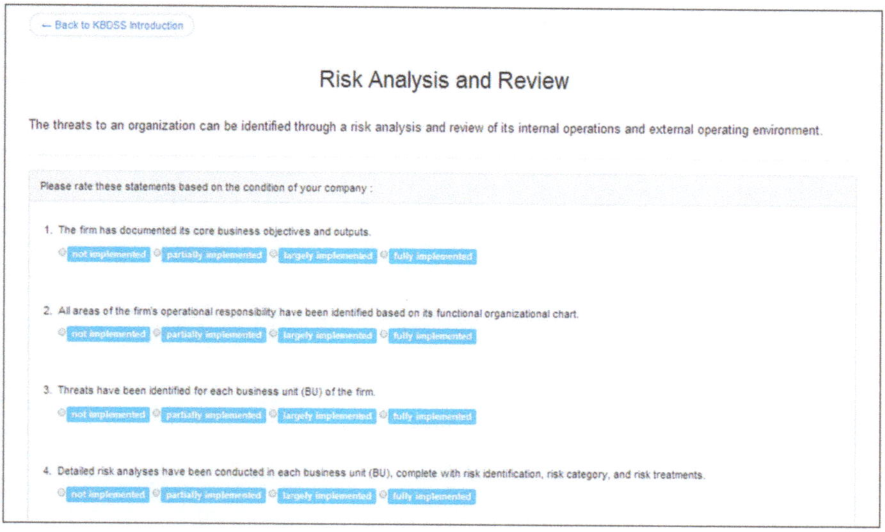

Fig. 12.11 Assessment page—BCM principle 1

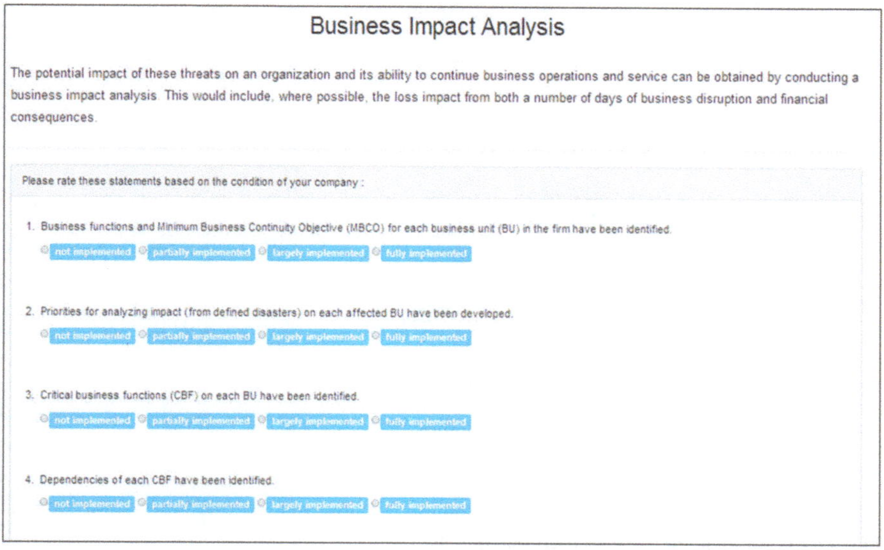

Fig. 12.12 Assessment page—BCM principle 2

The last two pages are the user's feedback page and the end page, which are shown in Figs. 12.27 and 12.28 respectively.

An example of the assessment and result processes in the BCM-KBDSS, which was summarized in .pdf form, can be seen in Appendix G—Analyses for Chap. 12. Furthermore, an example of the manual calculations, which used the fuzzy logic rules in Sect. 12.2.2.2 can also be seen in the same appendix section.

Fig. 12.13 Assessment page—BCM principle 3

Fig. 12.14 Assessment page—BCM principle 4

12.3 KBDSS Validation

The BCM-KBDSS prototype was validated in two stages, which were laboratory testing and field testing. Ten respondents were involved in this process where all of them were from different contractors in Indonesia. The validations were conducted within 3 months (September–November 2013).

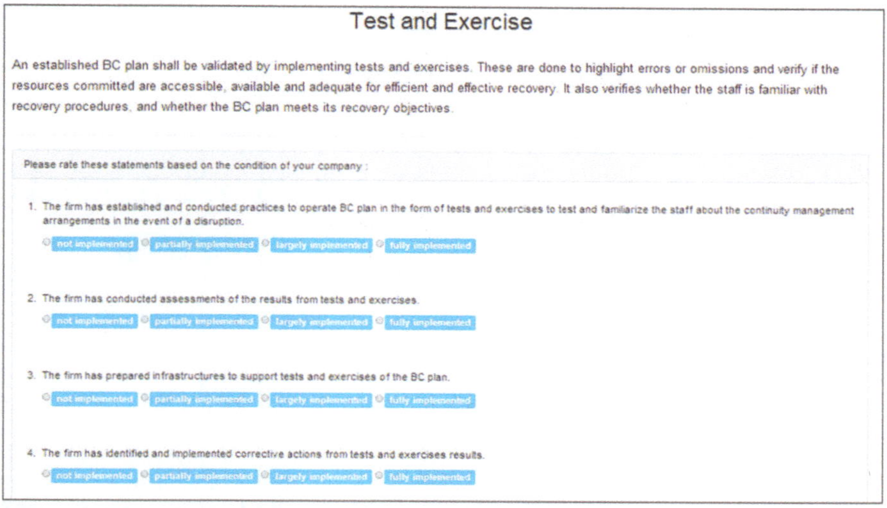

Test and Exercise

An established BC plan shall be validated by implementing tests and exercises. These are done to highlight errors or omissions and verify if the resources committed are accessible, available and adequate for efficient and effective recovery. It also verifies whether the staff is familiar with recovery procedures, and whether the BC plan meets its recovery objectives.

Please rate these statements based on the condition of your company :

1. The firm has established and conducted practices to operate BC plan in the form of tests and exercises to test and familiarize the staff about the continuity management arrangements in the event of a disruption.
   not implemented | partially implemented | largely implemented | fully implemented
2. The firm has conducted assessments of the results from tests and exercises.
   not implemented | partially implemented | largely implemented | fully implemented
3. The firm has prepared infrastructures to support tests and exercises of the BC plan.
   not implemented | partially implemented | largely implemented | fully implemented
4. The firm has identified and implemented corrective actions from tests and exercises results.
   not implemented | partially implemented | largely implemented | fully implemented

Fig. 12.15 Assessment page—BCM principle 5

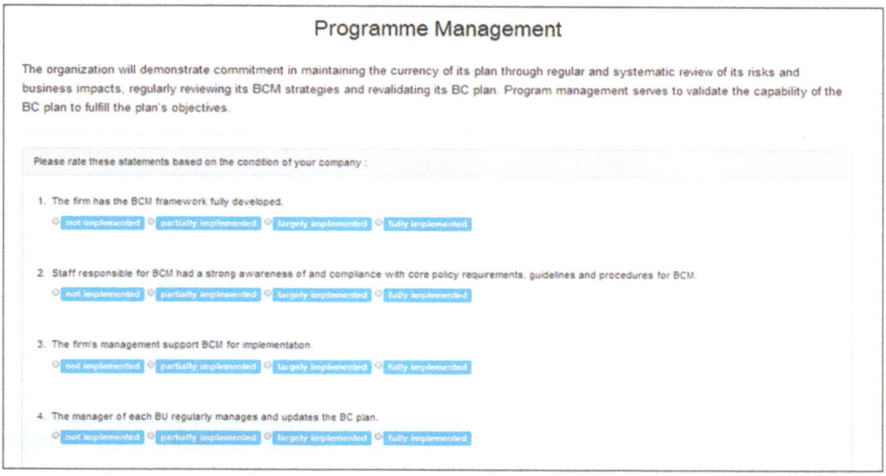

Programme Management

The organization will demonstrate commitment in maintaining the currency of its plan through regular and systematic review of its risks and business impacts, regularly reviewing its BCM strategies and revalidating its BC plan. Program management serves to validate the capability of the BC plan to fulfill the plan's objectives.

Please rate these statements based on the condition of your company :

1. The firm has the BCM framework fully developed.
   not implemented | partially implemented | largely implemented | fully implemented
2. Staff responsible for BCM had a strong awareness of and compliance with core policy requirements, guidelines and procedures for BCM.
   not implemented | partially implemented | largely implemented | fully implemented
3. The firm's management support BCM for implementation.
   not implemented | partially implemented | largely implemented | fully implemented
4. The manager of each BU regularly manages and updates the BC plan.
   not implemented | partially implemented | largely implemented | fully implemented

Fig. 12.16 Assessment page—BCM principle 6

For the laboratory testing, the BCM-KBDSS's logic flow and input-process-output were tested regularly. This was to see whether the KB-rules compilation was running as planned and whether there was any algorithm flaw. Once the system ran consistently, two respondents were invited to test the prototype for a predictive validation. This validation used test cases in which the results were known in order to obtain a statement of the applicability of the system by possible users (Imriyas 2006; O'Keefe et al. 1987). In the study, the two respondents were from the case studies firms (Firm A and Firm B), where the BCM preparedness descriptions of

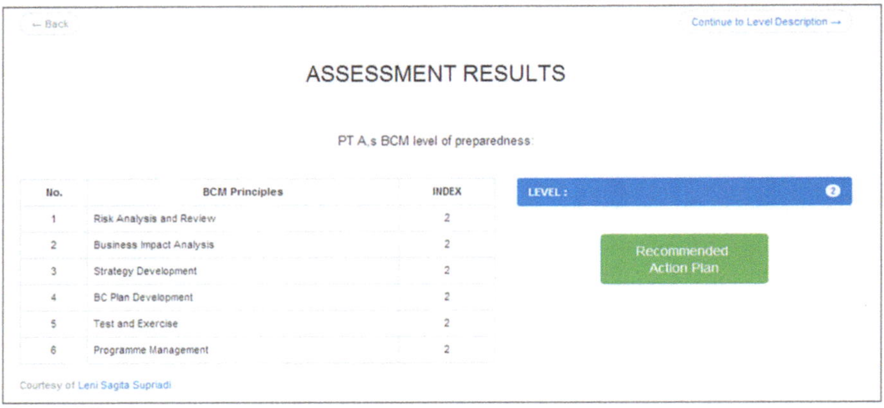

Fig. 12.17 Assessment result page

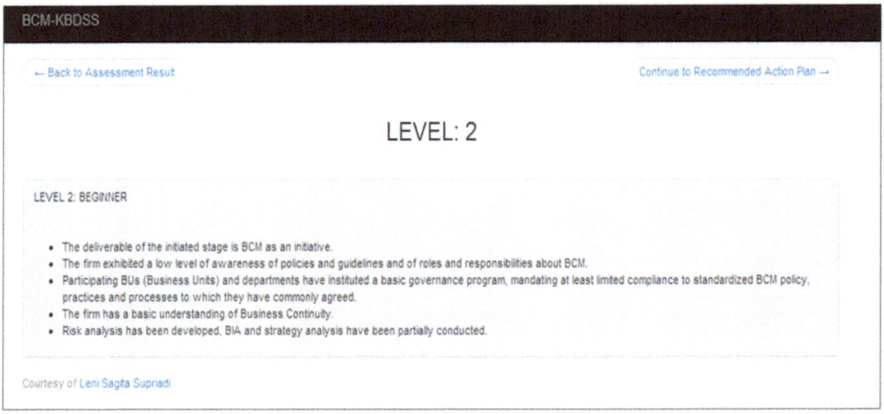

Fig. 12.18 Level description page

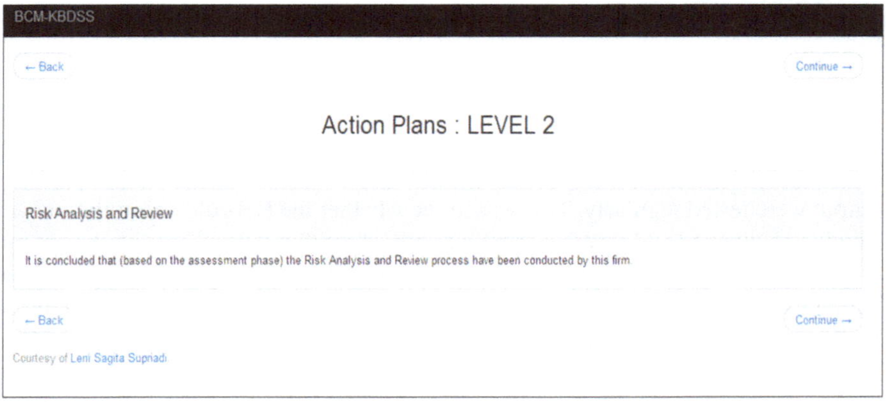

Fig. 12.19 Recommended action plans page—BCM principle 1

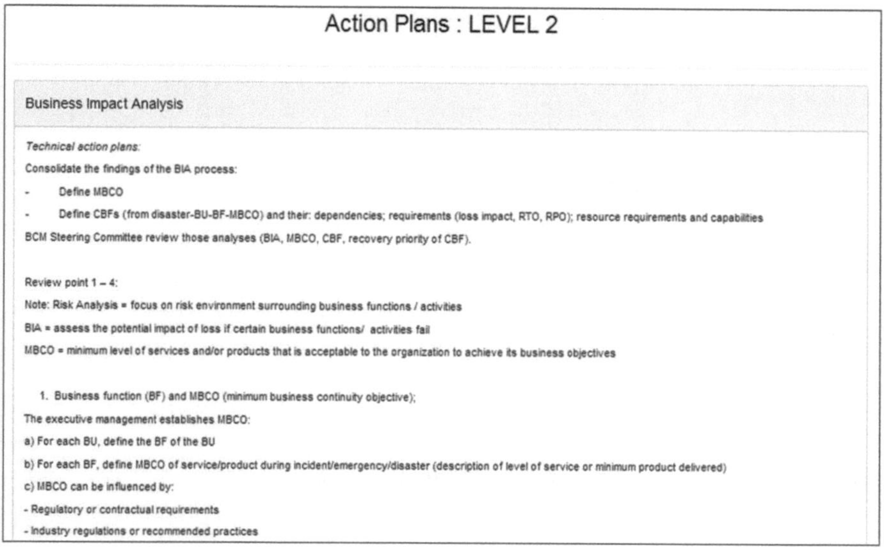

Fig. 12.20 Recommended action plans page—BCM principle 2

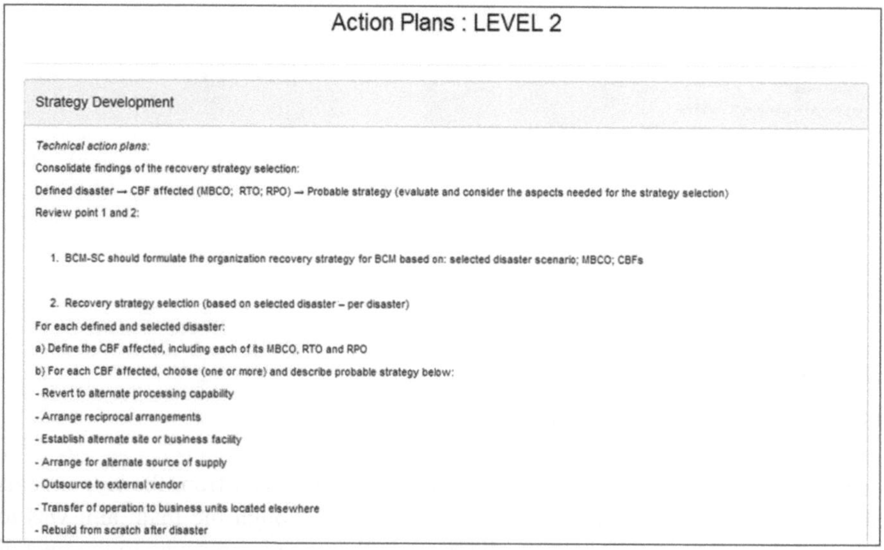

Fig. 12.21 Recommended action plans page—BCM principle 3

these firms were already known. This process used the qualitative approach (interviews) to derive the results.

After the laboratory testing, field testing was conducted to identify the BCM-KBDSS's general performance. Eight respondents participated in this phase, where they used the BCM-KBDSS prototype and provided feedbacks via a

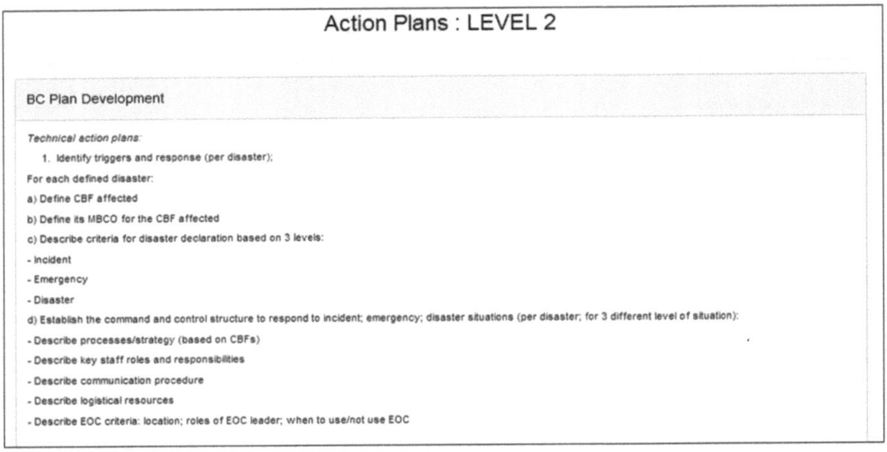

Fig. 12.22 Recommended action plans page—BCM principle 4

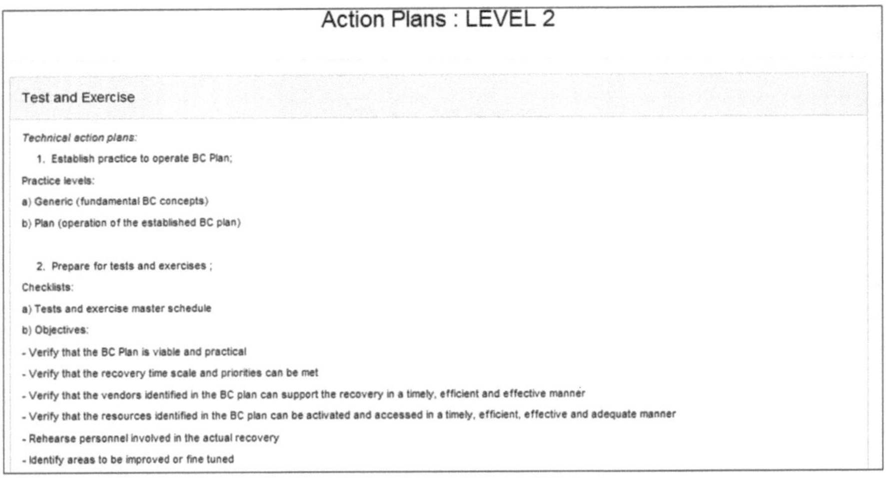

Fig. 12.23 Recommended action plans page—BCM principle 5

questionnaire. The questionnaire was designed to elicit data from the respondents about the system's strengths and weaknesses. Even though the main aim was to measure the utility of the system, other criteria that could contribute to the overall utility of the system were also identified. This process used the quantitative approach in the form of scoring or rating. The attributes needed to be assessed were based on these criteria (Bailey and Pearson 1983; Papamichail and French 2005):

1. Perceived utility—the user's judgment about the relevant balance between the cost and the considered usefulness of the system.

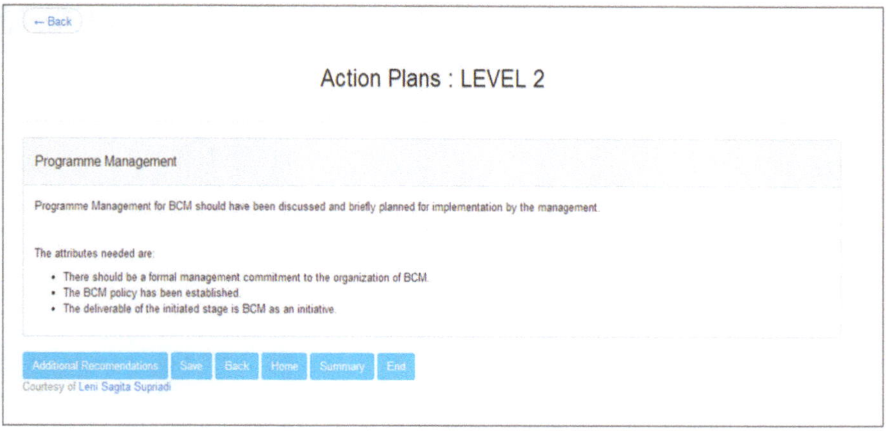

Fig. 12.24 Recommended action plans page—BCM principle 6

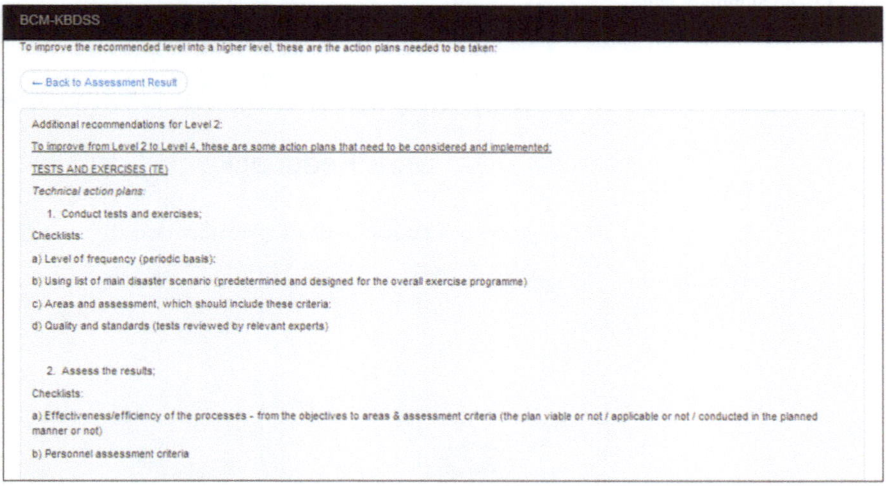

Fig. 12.25 Additional recommendations page

2. Relevance—the degree of congruence between what the user wants or requires and what the system provides.
3. Completeness—the comprehensiveness of the output information content.
4. Format of output—the layout design and display of the output contents.
5. Volume of output—the amount of information given to a user.
6. Ease of use—the amount of effort required by the user to take advantage of the tools provided by the system.
7. Ease of learning—the potential of a system to require minimal effort in learning how to use it.
8. Timeliness—the availability of the output information at a suitable time.

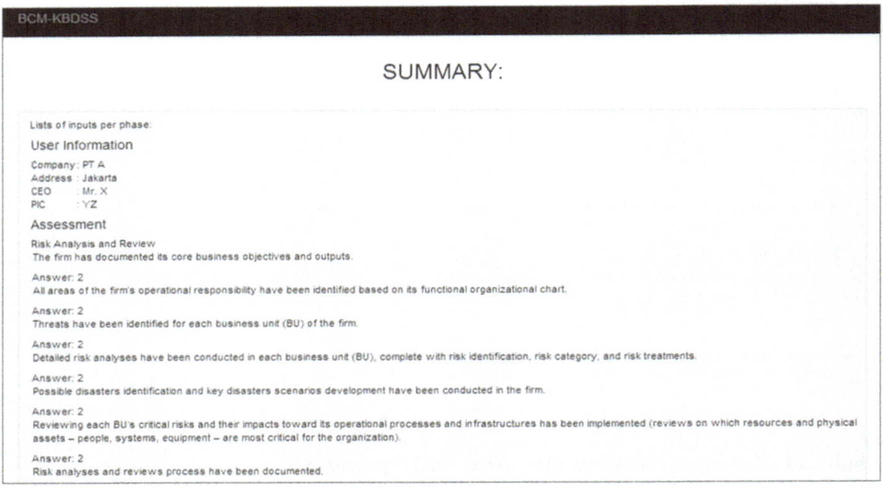

Fig. 12.26 Summary page

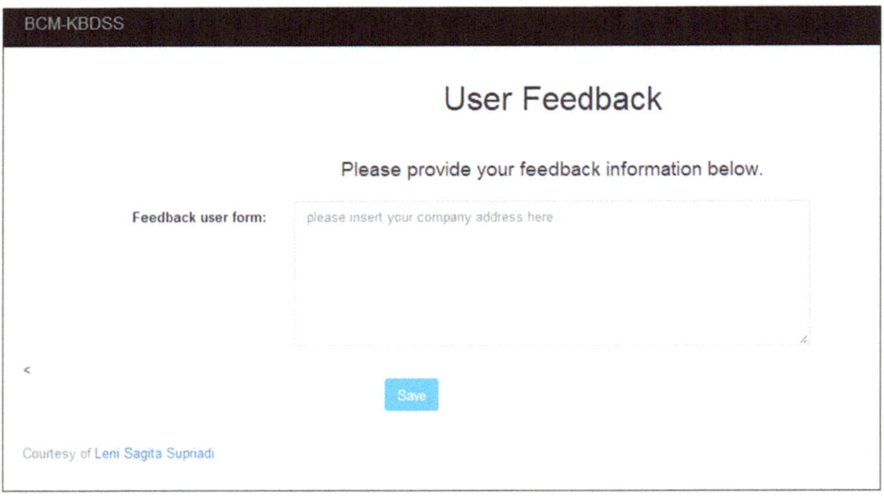

Fig. 12.27 Feedback page

9. Performance—the ability of the system to help a decision maker accomplish a task more effectively.
10. The benefits of the system.
11. The user's attitude towards the system.

Moreover, in order to measure the criteria, a 5-point Likert-type scale was used. The respondent had to state how much they agreed or disagreed with a statement on a scale from 1 (Strongly disagree); 2 (Disagree); 3 (Neutral); 4 (Agree); 5 (Strongly

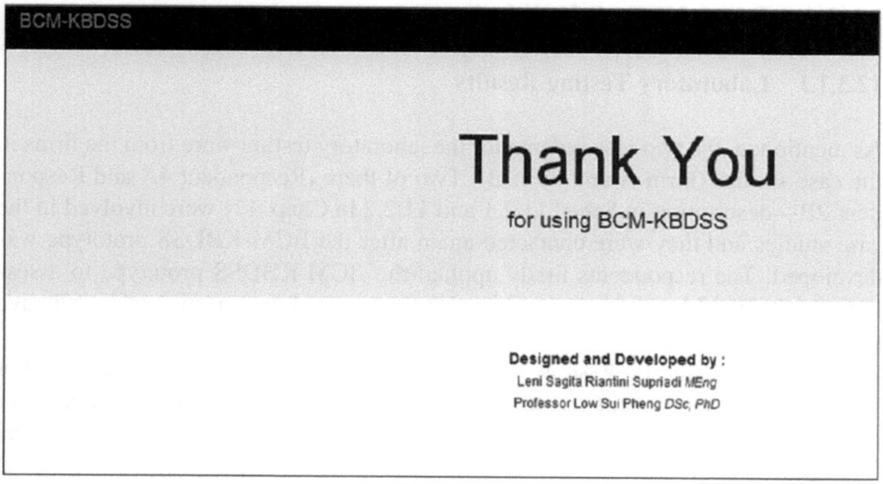

Fig. 12.28 End page

agree). The statements provided in the questionnaire were (Bailey and Pearson 1983; Papamichail and French 2005):

1. The layout of the information displayed in the BCM-KBDSS is straightforward and easy to understand (Format of output).
2. The BCM-KBDSS offers a structured and well-organized approach to assess the company's BCM level of preparedness (Perceived utility).
3. The BCM-KBDSS's ability to generate feasible recommendations is helpful (Performance).
4. The amount of information presented in the BCM-KBDSS is reasonable (Volume of output).
5. The BCM-KBDSS provides clear instructions and outputs (Completeness).
6. There were no technical problems when running the system (Ease of learning).
7. The processing speed of the system is fast enough (Timeliness).
8. I think that I would not need the support of a technical person to be able to use the system (Ease of use).
9. All the information (results and recommendations) provided by the BCM-KBDSS is useful and relevant (Relevance).
10. The BCM-KBDSS's action plans are beneficial (Benefit of the system).
11. My attitude towards the system is very positive (Attitude towards the system).

For both the testing processes, a face-to-face approach was used. Before the respondents used the BCM-KBDSS prototype, proper introduction about the system and the steps in using the system had been described by the research team. After completing the process in the BCM-KBDSS prototype, the respondents were given a questionnaire to assess the system's performance. The questionnaire form can be seen in Appendix G—Analysis for Chap. 12.

12.3.1 BCM-KBDSS Validation Results

12.3.1.1 Laboratory Testing Results

As mentioned, the two respondents for the laboratory testing were from the firms in the case studies (Firm A and Firm B). Two of them (Respondent 4A and Respondent 2B—description in Sects. 11.2.1 and 11.2.2 in Chap. 11) were involved in the case studies and they were contacted again after the BCM-KBDSS prototype was developed. The respondents firstly applied the BCM-KBDSS prototype to assess their firm's BCM level of preparedness, then answered the questionnaire and gave feedbacks in the form of interviews.

Figures 12.29 and 12.30 show the firms' BCM level of preparedness (screen shot pictures of the assessment results), based on their inputs using the BCM-KBDSS prototype. It turned out that the firms' BCM level of preparedness were Level 2 (Beginner) for Firm A and Level 3 (Moderate) for Firm B.

Furthermore, the results of the validation questionnaire (with 11 criteria to be rated) from both firms were described in Table 12.3. The 11 criteria were rated by the two respondents, followed by some comments during the interviews. The results showed that both respondents mostly agreed on the statements provided in the questionnaire, with 10 out of 11 statements in the scale range of "Neutral" to "Strongly agree". The only statement that they both disagreed on was regarding the system's ease of use ("I think that I would not need the support of a technical person to be able to use the system"). Both of them viewed that a technical person should be available to support the system in case technical issues or problems occurred.

The other result worth mentioning was the system's relevance with the current conditions of the firms. Both respondents strongly agreed that the results and

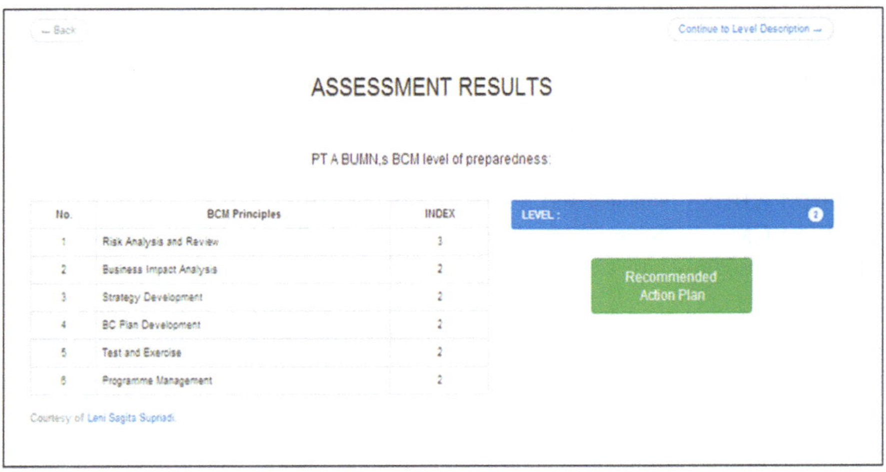

Fig. 12.29 BCM-KBDSS result of Firm A

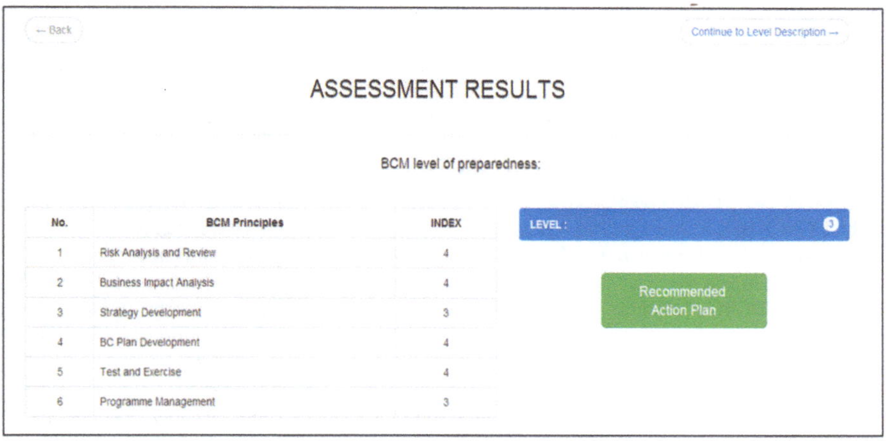

Fig. 12.30 BCM-KBDSS result of Firm B

recommendations provided by the system were relevant to their firms' current conditions. The action plans were viewed to be useful for their firms' further improvements on BCM level of preparedness.

Additional feedback was also obtained from the respondents. Table 12.4 elaborates the feedback on the printed results, the use of the BCM-KBDSS in their firms, and back-up for the BCM-KBDSS. From these results (questionnaire results and feedback), it appears that the laboratory testing for the BCM-KBDSS prototype was validated with positive feedback from the respondents. Some constructive feedback was also addressed for further enhancement of the system, such as more time needed to read and review the recommended action plans and the need for updates on the latest information about BCM best practices that could be included in the system.

12.3.1.2 Field Testing Results

Eight respondents were involved with the field testing process, where four persons worked in state-owned construction firms (Respondents 1, 3, 6 and 7) and the other four persons worked in private construction firms (Respondents 2, 4, 5 and 8). Their experiences of working in their firms ranged from 8 to 35 years, with various current job designations such as Project Managers, Senior Managers, Vice President and Directors.

During the validation process, the questionnaire was answered right after the respondent had finished using the BCM-KBDSS prototype. The results from the questionnaire were analyzed using descriptive statistics, in determining the mean scores of the responses of the 11 statements. Before analyzing the data collected, the reliability of the questionnaire was examined, and the Cronbach's Alpha

Table 12.3 Results from questionnaire's criteria (11 criteria)

Criteria to assess BCM-KBDSS	Respondent of Firm A	Respondent of Firm B
The layout of the information displayed in the BCM-KBDSS is straightforward and easy to understand	**Strongly agree** *The layout of the system is user friendly enough*	**Strongly agree** *The background template and layout are simple and understandable*
The BCM-KBDSS offers a structured and well-organized approach to assess the company's BCM level of preparedness	**Agree** *The system has a good structure of assessment and providing the results*	**Strongly agree** *The assessment process is well-structured, the result is given quickly, and the recommendations are provided in details*
The BCM-KBDSS's ability to generate feasible recommendations is helpful	**Agree** *The recommendations provide action plans that are reasonable. But the action plans should further be reviewed for better practical approach in the firm*	**Agree** *The recommended action plans help the user to decide and apply. I think one action plan can be generated into several more steps by the firm, depending on the approach they use. For example, a risk review in our firm may consist of reviewing the documentation back-ups and material-equipment warehouse in our main office and branch office. The steps may be a bit different due to different locations*
The amount of information presented in the BCM-KBDSS is reasonable	**Neutral** *The recommended action plans provide comprehensive steps. I think the user may need to spend more time focusing in reading and discussing it*	**Agree** *The summary page of the assessment and the results is very helpful to learn more about the information given by the tool*
The BCM-KBDSS provides clear instructions and outputs	**Agree** *I have no trouble in using the tool, because I only need to rate and follow the clear instructions. The rating may need a longer time because I need to review my firm's current condition regarding BCM preparedness*	**Agree** *I think the instructions given by the tool are clear enough and not ambiguous. The outputs are also clear*
There were no technical problems when running the system	**Neutral** *In my case, I have a bit of a glitch while running it. It turned out that my computer settings is an old format of Windows XP. I had to contact*	**Strongly agree** *I tried it on my PC and mobile device. It turned out good and I had no problem using it*

(continued)

Table 12.3 (continued)

Criteria to assess BCM-KBDSS	Respondent of Firm A	Respondent of Firm B
	the IT staff for re-setting it. In the future, I think before having this type of tool, our company should provide a more updated operating system (OS) for the company's PCs and laptops. Also, the tools should comply with the company's current OS	
The processing speed of the system is fast enough	**Neutral** *Due to the technical issue while using this tool, the speed of each page turning to the next page is quite slow*	**Agree** *The system provides a good processing speed*
I think that I would not need the support of a technical person to be able to use the system	**Disagree** *I think there should always be a technical team in charge of maintaining this tool. They should always stand by in case technical glitches occur*	**Disagree** *Even though the performance of the system is good and I had no problem in using it, a technical person who understands this tool should be provided. Therefore, whenever there are problems from this tool (the system crashed or pages did not respond), there is person in charge of repairing it*
All the information (results and recommendations) provided by the BCM-KBDSS is useful and relevant	**Strongly agree** *What the tool has given from the assessment result is relevant to my firm's current condition. We are in level 2 BCM preparedness. The action plans could assist us in implementing BCM principles (especially BIA and strategy development—as the early phase) more thoroughly*	**Strongly agree** *The result from the tool showed that my firm is in level 3 BCM preparedness. This is quite relevant because we have documented responses plan to overcome accidents, floods and earthquake. However, we do not have other responses plan for threats like terrorism or material loss. I think the action plans can be used for enhancing the current plans that we have, and for developing new responses plans for overcoming other threats*
The BCM-KBDSS's action plans are beneficial	**Agree** *The user can learn a lot from the recommendations and able to share with their colleagues*	**Agree** *Yes, I agree. The action plans are highly needed for the team who will develop BCM in my firm*

(continued)

Table 12.3 (continued)

Criteria to assess BCM-KBDSS	Respondent of Firm A	Respondent of Firm B
My attitude towards the system is very positive	**Agree** *I agree, the system provides a useful knowledge, particularly in learning more about BCM. BCM is still a new concept for my firm. Although we have implemented part of the principles (such as risk analysis and crises responses), but we have not adopted the whole concept*	**Agree** *I think this system is an effective approach to assess my firm's BCM preparedness, and also provides a detailed guideline for implementation*

Scale: Strongly disagree–Disagree–Neutral–Agree–Strongly agree; with comments (in *italics*)

coefficient for the questionnaire was 0.905. This showed that the internal consistency of the questionnaire was very good.

In answering the questionnaire, the user had to state how much they agreed or disagreed with a statement on a scale from 1 (Strongly disagree); 2 (Disagree); 3 (Neutral); 4 (Agree); and 5 (Strongly agree). Table 12.5 shows the questionnaire results from the respondents. The mean scores of the 11 statements ranged between 3.88 and 4.38. The mean scores of 10 (out of 11) criteria were >4.0, which were higher than the midpoint. This suggests that the system met the evaluation criteria stated in the questionnaire. The criterion which received the lowest mean score was "ease of use" (3.88). On the other hand, the respondents rated the system's "perceived utility", "performance", "completeness" and "ease of learning" with the highest mean scores. Moreover, all of the respondents agreed that the system provided benefits and their attitudes toward the system were positive, with mean scores of 4.25 and 4.12 respectively.

After responding to the questionnaire, the respondents had also given feedback regarding the system. Respondents 1 and 5 mentioned about commercializing the tool for firms to assess their BCM level of preparedness:

Respondent 1 I think this tool should be commercialized to assist companies to assess its BCM readiness.

Respondent 5 This tool should be used publicly (go public). Many firms would want to assess their BCM preparedness in this effective approach.

Respondents 3, 7 and 8 commented that the BCM-KBDSS prototype was useful for the continuity of their firm's business and could improve their firm's management performance:

Respondent 3 This is an excellent system and very useful for the continuity of our business and operation.

Table 12.4 Additional feedback (interviews)

Additional feedback for BCM-KBDSS	Respondent of Firm A	Respondent of Firm B
Printed results	*The additional printed format (in the form of pdf) for the summary of the assessment is highly needed for further discussion and review by the firm's management. Therefore, they can start to implement the recommended action plans as documented*	–
Using BCM-KBDSS in their firms	*We have our own knowledge management portal in our firm's website. It is used by the employees for enhancing their knowledge regarding their tasks. So far, many employees have used this tool and find it useful and informative for them. I think this type of tool (BCM-KBDSS) can be included in the portal as a section for learning and understanding about BCM. But of course, the top management should approve of this concept first*	*Up to now, we do not have this type of assessment tool applied in our firm. Most of the procedures or management tasks are all printed in big bundles of documents. I think developing this type of tool will be helpful and more effective than reading a thick bundle of papers. But we need to hire a permanent team for developing and maintaining this tool. This will need further consideration by the management (the cost of developing and maintaining such tool in the firm's online portal)*
Back-up for BCM-KBDSS	–	*I think, even if the firm has developed and maintained this type of tool, and the tool has successfully provided the knowledge needed to assess the firm's BCM preparedness and recommend the action steps, a hard copy of the detailed actions steps of BCM guidelines should always be prepared, as a back-up. But for applying the assessment process, a KBDSS should be more effective*

Respondent 7 This is a nice and useful application programme. Thank you for the opportunity to try it.

Respondent 8 This tool provides detail recommendations to improve the company's management.

Respondents 2 and 6 provided constructive feedback where they viewed the importance of understanding BCM before using the system:

Table 12.5 Questionnaire results from field testing

No	Criteria to assess	Mean	Std deviation
1	Format of output	4.12	0.991
2	Perceived utility	4.38	0.744
3	Performance	4.38	0.518
4	Volume of output	4.00	0.756
5	Completeness	4.38	1.061
6	Ease of learning the system	4.38	0.744
7	Timeliness	4.25	0.463
8	Ease of use	3.88	1.356
9	Relevance	4.00	0.756
10	Benefit of the system	4.25	0.886
11	Attitude towards the system	4.12	1.126

Respondent 2 When I utilize this tool, I still need some time to understand the terms. This concept is new for me. Therefore, I recommend that before using this tool, a practical overview about BCM should be conducted.

Respondent 6 I learn quite a lot about BCM from using this tool. But I'm not sure whether this will be useful in my firm, because we have not adopted BCM yet. Maybe this tool can be an introductory part for socializing BCM in our firm.

Last but not least, Respondent 4 suggested improvements on the system's interface:

Respondent 4 I think the system should provide information about the progress during the assessment phase (e.g. form/page 2 of 7). Overall, it is informative.

Based on the field testing results, the BCM-KBDSS prototype was viewed to be useful, with some strengths and weaknesses observed. Although the validation of the system was far from comprehensive, it yielded interesting results from several point of views. The feedback gained from the validation process could be used for further enhancement of the prototype.

12.4 Summary

This chapter provided descriptions about the development of the BCM-KBDSS, starting from the framework development, rules development, user interface designs and the system's finalization. As a prototype, the BCM-KBDSS was validated using two-phased tests (laboratory and field testing). Both the test results had shown that the system was viewed to be useful and provided important feedback for the system's further enhancement. A sample of a print-out (in .pdf format) for an Indonesian contracting firm from the validation exercise is shown in Appendix G. Findings and discussions related to this section will be further elaborated in Chap. 13.

Chapter 13
Findings and Discussion

13.1 Introduction

This chapter elaborates the findings from the data analyses results that were described in Chaps. 10, 11 and 12. Firstly, findings from surveys and case studies are explained and this is followed by discussion on BCM implementation guidelines development. These aspects are grouped into the findings and discussion section for RQ1. Secondly, findings from the BCM-KBDSS development and validation are addressed, where these aspects are part of the findings and discussion section for RQ2. The summary section diagrammatically illustrates all the findings that relate to the research questions (RQ1 and RQ2). This diagrammatic illustration updates the study's conceptual framework.

13.2 Findings and Discussion for RQ 1

13.2.1 Findings from Surveys

In order to achieve Research Objectives 1–3, questionnaires (please see Appendix B) were distributed to the sampling frame. The portions of the questionnaire that will be discussed here are Section B and C.

13.2.1.1 Questionnaire: Section B

This section describes the findings based on Table 8.1, where the constructs used for identifying the Indonesian contractor's knowledge about BCM were highlighted and discussed.

L.S.R. Supriadi, L. Sui Pheng, *Business Continuity Management in Construction*,
Management in the Built Environment, DOI 10.1007/978-981-10-5487-7_13

Types of Crises that Occurred in Contractors Within the Last 5 Years
Table 10.6 had earlier described the top seven crises (from 32 types of crises given in the questionnaire) that occurred in the firms within the last 5 years, which were: access/approval restriction or limitation; delays or uncertainty in resolving disputes; increase in prices of raw materials (unexpected price escalation); changes in regulations and statutory legislation; natural disasters; loss of management personnel or key staff; and lack of component workforce. These types of crises were similar with some actual crises that occurred and were documented in the past years in Indonesia, which had affected the Indonesian contractors. According to Suryadharma (2008), the 2008 financial crisis led to increase in the costs of raw materials such as steel and cement. This accounted for approximately 20% of an average project's cost structure. Moreover, natural disasters such as earthquakes, floods and tsunami that occurred had various impacts as well (MiyamotoINTL 2007; PTX 2008; Tambunan 2006; UNR/HC 2005; IFRC 2004; PTX 2008; Sutardi 2006).

Other listed crises were described by the respondents during the survey validation where they had experienced these crises in their firms. When they experienced the lack of component workforce, many project delays occurred due to that situation. An example of a crisis due to disputes was when the locals wanted to work in one of the respondent's projects as laborers and their demonstrations led to riots near the project site. The respondent's firm cannot hire the locals directly because they need to have the necessary skills. Consequently, the firm faced delays in the project, and they needed to resolve this with the locals and the community leaders.

Other respondent indicated that access restriction or limitation issue was caused mainly by the lack of good coordination between stakeholders. The example mentioned by the respondent relates to land acquisition for a road project. This process took a long duration and it was not resolved for months. The project was delayed until the land acquisition process was finished. The root of the problems was due to the land owners who disagreed with the compensation that they received from the project owner.

Furthermore, other respondent elaborated on how changes in regulations would have an impact towards project financing, which could lead to project delay or termination. Another respondent told his experience in overcoming haze and forest fires (as part of natural disasters), where the crises occurred annually and disrupted the firm's projects for 5–7 days. The impacts from these crises were high as these affected the cost of equipment leases and labor salaries. Lastly, a crisis such as the loss of management personnel was viewed to have a major impact on the firm because some periods of adjustment were needed in the firm. The new person who took up the management position needed to adjust to the work environment, job assignment, skills needed, coordination with other units, etc. Conflicts between personnel had occurred due to this adjustment period.

List of Impacts from Crises
Table 10.7 earlier showed the top and bottom three impacts of crises that were viewed as significant to the respondents. These are: delay in work/dissatisfied

customers; revenue impact; loss of productivity (the top three impacts); as well as huge data loss; building evacuation; and failure of few systems (the bottom three impacts).

An example of a crisis that led to the firm's revenue impact was the 2008 financial crisis. Due to the unfortunate financial condition for the contractors, they were more selective in managing their costs of contracts. Several infrastructure projects were also postponed due to the situation (Suryadharma 2008). Other impacts such as delays in work and loss of productivity could not be avoided when crises such as accidents and natural disasters occurred. Moreover, building evacuation and systems failure could have happened when floods and earthquakes occurred (Permana 2007; PTX 2008; MiyamotoINTL 2007; Tambunan 2006; UNR/HC 2005; IFRC 2004; Sutardi 2006).

One of the survey respondents suggested that the top three impacts have correlationships. If a crisis occurs, there would be disruptions that would lead to delays. Delays would lead to loss of productivity of the worker and consequently, more budget would be needed to finish the job, which had an impact on the firm's financial health, including its revenue. Another respondent opined that delays were unavoidable; therefore the important thing that a firm should consider is to minimize the period of delays for the project or activity to continue working. Minimizing the period of delays can be managed by adopting BCM in the firm (O'Hehir 1999; Health 1999).

Crisis SOP Elements in Firms

From the survey results, it was found that the respondents who had crisis SOP in their firms had various names for it, such as Crisis SOP, Crisis Management Plan, Business Continuity Plan, and Emergency Plan. This finding showed that the SOPs were named using different terms, and each SOP in each firm may not have the same contents. A Crisis Management Plan or Emergency Plan may not provide the same procedures as the Business Continuity Plan (BC Plan). As mentioned in Sect. 3.2.2, there are relationships between BCM and other concepts such as crisis management and emergency management. According to BCI (2010), BCM has strong links with crisis management through the incident management component. In the BCM context, incidents can come in different shapes and sizes and will typically invoke the BC plan. Crisis management is also seen as a response to non-physical as well as physical events. Emergency management also requires emergency planning that is usually included in incident management.

Even though BCM, crisis management and emergency management have strong relationships, several studies had made a distinction between them. BCM views that emergency planning is not only seen within the domain of services from the police, fire, ambulance and local authorities, but also for the organization in general. The company that adopts BCM would have a specific emergency response team that will coordinate with other external emergency response agencies. BCM focuses on the planning process of developing advance arrangements and procedures that enable an organization to respond to an event in such a manner that critical business functions can continue with planned levels of interruption or essential change.

Crisis management focuses on the immediate activities which need to be considered when the incident occurs. At most, the crisis management planning phase deals with the first couple of hours of the incident occurring, detailing who the key decision makers are, who will talk to the customers/clients/regulators and when this will be conducted (Smith 2003; Devlin 2007; Foster and Dye 2005).

The survey results also showed that only 21% of the respondents named their SOP as the BC Plan. This is probably because BCM is still a relatively newcomer to the business domain. However, aspects of BCM may have already been present in the firms, albeit under different names or terms (BCI 2010).

Furthermore, the survey validation interviews found that the respondents had a crisis SOP in place in their firms. However, they were not sure whether the crisis SOPs were developed holistically. This finding is in line with Supriadi and Low's (2012) findings where most of the contractors have provided relevant emergency responses for evacuating people during a crisis, but a detailed recovery procedure for their businesses to resume after the crisis and detailed responses for their business stakeholders had not been planned in advance. This suggests that there are still patches of responses that have not yet been integrated.

An example can be found in a case study in Australia, where similarly some of the large contractors had undertaken crisis management in an informal, fragmented fashion with few resources and little strategic guidance and support. This was seen as being insular and non-integrative, confined only to the senior management level and was limited in scope to issues such as safety, industrial relations and cost control (Loosemore and Teo 2006).

The survey results also confirmed that the clients and employees were the top priorities to communicate with during crises, followed by the firm's partners, regulator/government and vendor/supplier. As highlighted by Grunig (1992), when a firm is threatened by crises, the need for communication increases to some level where the use of communication management to assist in transformation and relationship with the environment is essential. Immediate and appropriate communication decision is crucial, particularly during the crisis response stage, which is characterized by short decision time, stress, complexity and uncertainty (Loosemore 1998a; More 1995). In this situation, the requirements of the various stakeholders in the firm should be considered, particularly for the protection of its employees and clients. Other parties such as investors and suppliers should also be in the communication loop due to their direct or indirect involvement with the firm (Elliott et al. 2002; Singapore Business Federation 2003).

Some studies showed that a crisis occurring in a construction firm would lead to various challenges for its management. It needs to reestablish an element of control and coordination in people's activities by attempting to communicate policies that might have been developed in isolation at a senior managerial level. Communication during this situation is not easy. Furthermore, a crisis at the project site level would be managed by focusing on the relationship between the site management and head office level, including the need to coordinate with subcontractors and suppliers (Mitroff and Pearson 1993; Loosemore and Teo 2006).

Regarding external coordination during crises, the survey results showed that the need for coordinating with the external agencies and government is essential. In addition, several respondents recommended the military and the Indonesian Contractors Association (ICA) to be the other external agencies to coordinate with during crises. These results were also supported by Mitroff and Pearson's (1993) study on crisis management, where parties who may affect or be affected by the crisis are one of the four main variables that should be considered in the crisis management plan. In this context, it means the inclusion of external coordination during crises.

BCM Knowledge

Section 10.3.2 described the survey results of the respondent's BCM general knowledge. 87.5% of the respondents did not know about BCM and only 12.5% of them knew about the concept. This was also confirmed by the respondents who validated the survey results. Most of the respondents who were interviewed also did not know about BCM. This result was in line with several studies which suggested that BCM is relatively new, particularly in the Asian region. In the construction sector, not many firms had implemented this concept. The main reason for this is a lack of awareness and lack of adequate personnel in their firms who can lead in BCM implementation. Similarly, a study in UK also found that less than 50% of the construction firms had a BC plan in place, and the drivers of implementing this concept were mostly due to regulations from the central government, insurers or auditors (Broughton 2005; Low et al. 2008). This result also confirmed that most of the respondents from the community of Indonesian contractors did not know about BCM. This study can therefore fill in the knowledge gaps about BCM awareness from the Indonesian contractors.

Table 10.9 earlier showed that from the respondents who knew about BCM, most of them had learned about it from their mentors or colleagues. This approach was also recommended by Power (1999), where the top management of the firm was viewed to be essential in initiating BCM. The role of the management board is to decide how broad or narrow the focus of the business continuity provisions is to be.

In addition, most of the respondents viewed that the main reason for their firms to implement BCM was to protect the firm and ensure its long-term survival. These results were also similar with the UK-based contractors that had adopted BCM. The firms viewed that the purposes for implementing BCM are to protect the firm by gaining resilience; to conduct part of risk management by a better understanding of the threats and activities at risk and the structured process for implementing measures to protect against these; and to enhance customer relations and improvements to supply chain (BSI Groups 2010). These purposes were similar with point (1), (2) and (5) in Table 10.10.

From the survey validation, most of the respondents viewed that the role of regulations in supporting BCM in Indonesia (which was ranked last as the reasons for implementing BCM) should be essential. They recommended that BCM should be regulated by the government in order to be widely implemented. Similarly,

Supriadi and Low (2012) had also suggested that full support should be obtained from the government for implementing BCM in the firm. Along with regulating its implementation, the government could develop good collaboration with the construction firms to foster the initiation process for BCP development. This is to leverage on the platform whereby the Indonesian government had developed various crisis responses for different regions of the country.

Effectiveness of BCM Implemented in Firms

Table 10.11 earlier showed that the respondents whose firms had implemented BCM had agreed that BCM provided effectiveness to their firms. They mostly agreed that BCM enables the organization to return to normal operations more quickly than otherwise would have been possible. This result is related to one of BCM's objectives, which is to develop a pragmatic, cost effective and operable recovery plan to enable the firm to achieve critical business processes during a major disruption to the firm's operations (O'Hehir 1999; Health 1999). The view that BCM can effectively reduce the impact of the disruption was ranked second for its effectiveness. Accordingly, this view is part of an effective BCM strategy, where it should be to minimize the impact of crises on the customers or clients (Smith 2003).

Moreover, the survey results also showed that most of the respondents whose firms had implemented BCM did not use any BCM standards. The results were also confirmed by the respondents who validated the survey results. These respondents recommended that the Indonesian contractors should develop their BCM based on BCM standards. They can undertake benchmarking with other countries that had implemented the concept. As mentioned earlier in Sect. 3.6.1, the standards from various countries have similar contents. The differences are on how the standards develop the detailed components in the BCM planning process. In general, all of these standards have common objectives, which are to guide the users to recover from any disasters that have occurred in their business environment and yet still continuously focus on the continuity of their business processes (Goh 2010; St-Germain et al. 2012).

Interests in BCM

Section 10.3.2 had earlier also found that there was a good level of interests from the survey respondents to learn more about BCM. This result was also similar with a majority of the respondents in China and Singapore, who had expressed interests in implementing BCM (Low et al. 2008). The respondents who were interviewed for validating the survey results also supported this result. They provided feedbacks on the parties that can provide the BCM workshop/training, such as ICA, NBCSD and the Public Works Ministry. In supporting this process, the firm's top management should be the main party to initiate and coordinate such an approach.

13.2.1.2 Questionnaire: Section C

This section describes the findings of the BCM principles for implementation in firms and earlier descriptions about Fig. 8.1 where the relationships between BCM-IF and BCM-OC were identified and discussed.

BCM Principles for Implementation in Firms

Table 10.12 earlier described the rankings of the BCM principles that had been implemented in the respondents' firms. The top five ranked BCM principles were from the Risk analysis and Strategy analysis principles. During the survey validation process, these results were also viewed to be significant by the respondents who confirmed that both analyses had been adopted in construction, especially for managing project risks. In addition, they viewed that both analyses are part of risk management, whom most of them have implemented in their firms.

In supporting this result, Sect. 3.2.2 earlier elaborated the relationships between BCM and risk management. BCM and risk management are related but also had distinctions. The risk analysis process in BCM focuses on addressing those risks which are deemed not acceptable to the firm. Subsequent BCM activities are aimed at establishing the appropriate measures to address these risks. It relegates BCM as part of risk treatment (Goh 2010). Moreover, comprehensive distinctions between BCM and risk management are shown earlier in Table 3.1.

This result also showed that the least implemented principles of BCM were Tests and exercises, Programme management, and BIA. These suggested that BCM had not been holistically implemented by the firms. Based on its definition, BCM should be developed and implemented in a holistic manner. BCM should be a system that develops a framework of protocols and sets of procedures and instructions which give structures, order and stability to the particular function being managed. It is a management system that also adopts the Plan-Do-Check-Act (PDCA) methodology for achieving continual improvement, where the BCM policy, objectives, processes and procedures are planned, implemented, assessed and reviewed regularly. PDCA is a key attribute within standards-based management systems that is widely used nowadays (SPRING 2008).

The results shown earlier in Table 10.13 suggested that there were correlations between the rankings of BCM implemented principles of private firms and the state-owned firms. The respondents who were interviewed during the validation stage agreed with these results where they viewed that both types of firms have similar traits. Both types of firms are operating in the same business environment, under the same ministry, construction board and association. Therefore, they appear to have similar sets of management and technical skills. Furthermore, both types of firms operate under the same legal framework with the same business functions to deliver the projects using the same types of contracting system (RI 1999, 2000; Raftery et al. 2004).

Institutional Forces (IF) that Support Implementation of BCM Principles
Various studies had found that a firm is supported and constrained by institutional forces. Liu et al. (2010) also observed that scholars have increasingly referred to the institutional theory as an important perspective for studies on organizational innovation or concept adoption. They argued that institutional pressures emanating from the environment and transmitted through operational channels can strongly affect firm's predisposition towards concept adoption. For this reason, this study investigated the motivation and current situations of implementing BCM from an institutional perspective. Different components of the institutional theory explain how these elements are created, diffused, adopted and adapted over space and time, and how they fall into decline and disuse. Collectively, this theory provides a framework to interpret the corresponding implementation issues.

The relationship between institutional forces (IF) and BCM principles is primarily about the contractor's perspective on BCM principles in the context of the institutional framework. It describes whether the contractor views the BCM principles as regulative, normative or cultural-cognitive forces towards implementation. The results from these analyses can derive the drivers and hindrances in implementing the BCM principles. These findings can be a good feedback for the Indonesian contractors in developing strategies for initiating BCM.

Section 10.3.4 earlier showed the meaningful factors from the institutional forces (IF) that support the implementation of BCM principles. Tables 10.14–10.18 earlier illustrated the most important IF factors for each BCM principle. These most important IF factors could further be categorized into variables and constructs of IF, based on Table 8.3 (see current Tables 13.1, 13.2, 13.3, 13.4, 13.5, 13.6, 13.7, 13.8, 13.9, 13.10 and 13.11). From here, the reasons and motivations for implementing each BCM principle could further be understood based on the IF variables and constructs.

Tables 13.1, 13.2 and 13.3 show the reasons and motivations for conducting risk analysis and review. It can be seen that risk analysis and review are mostly driven by normative forces (legitimacy and social influence), regulative forces (rules, laws and sanctions; gains, losses and consequences), and cultural-cognitive forces (shared understanding of compliance). Furthermore, the variables that are positioned as the highest loading in each factor of this principle are shared understanding of compliance of this principle and the legitimacy to implement this principle.

Table 13.1 The reasons and motivations for conducting risk analysis and cost benefit analysis

The reasons and motivations for conducting risk analysis and cost benefit analysis are mostly due to	Variables and constructs of Institutional Forces
• Awareness of potential risks • Part of the company culture	Shared understanding of compliance—cultural cognitive forces
• Fair procedures for better planning	Legitimacy—normative forces
• Management full support • Concern for reputation	Social influence—normative forces
• Cost impact	Rules, laws, sanctions—regulative forces

Table 13.2 The reasons and motivations for involving experts and BCM committee in risk review

The reasons and motivations for involving experts and BCM committee in risk review are mostly due to	Variables and constructs of Institutional Forces
• Appropriate and effective for better planning • A fair procedure for better planning in the organization	Legitimacy—normative forces
• Management full support • Concern for reputation	Social influence—normative forces
• Improve the company's procedure in preparing and handling crises	Gains, losses and consequences—regulative forces
• Non-compliance impact	Rules, laws, sanctions—regulative forces
• Awareness of potential risks • Comply with the regulation that is taken for granted • Part of the company culture	Shared understanding of compliance—cultural cognitive forces

Table 13.3 The reasons and motivations for conducting a detailed risk review

The reasons and motivations for conducting a detailed risk review are mostly due to	Variables and constructs of Institutional Forces
• Awareness of potential risks • Comply with the regulation that is taken for granted • Part of the company culture	Shared understanding of compliance—cultural cognitive forces
• Appropriate and effective for better planning	Legitimacy—normative forces
• Non-compliance impact	Rules, laws, sanctions—regulative forces
• Concern for reputation • Management full support	Social influence—normative forces

Table 13.4 The reasons and motivations for conducting BIA

The reasons and motivations for conducting BIA are mostly due to	Variables and constructs of Institutional Forces
• A fair procedure for better planning in the organization	Legitimacy—normative forces
• Management full support	Social influence—normative forces
• Improve the organization's procedures for facing crisis • It can be easily integrated with other management systems	Gains, losses and consequences—regulative forces
• Improve the employee's safety and welfare	Personal morality—normative forces
• Comply with the regulation that is taken for granted • Part of the company culture	Shared understanding of compliance—cultural cognitive forces
• Not implementing BIA can result in sanctions	Rules, laws, sanctions—regulative forces

Table 13.5 The reasons and motivations for involving experts, employees from related BUs and key staffs in BIA

The reasons and motivations for involving experts, employees from related BUs and key staffs in BIA are mostly due to	Variables and constructs of Institutional Forces
• Appropriate and effective for better planning • A fair procedure for better planning in the organization	Legitimacy—normative forces
• Awareness of potential risks • Comply with the regulation that is taken for granted • Part of the company culture	Shared understanding of compliance—cultural cognitive forces
• Concern for reputation • Management full support	Social influence—normative forces
• Improve the company's procedure in preparing and handling crises	Gains, losses and consequences—regulative forces
• Non-compliance impact	Rules, laws, sanctions—regulative forces

Table 13.6 The reasons and motivations for conducting strategy analysis for maintaining the operations of CBFs

The reasons and motivations for conducting strategy analysis for maintaining the operations of CBFs are mostly due to	Variables and constructs of Institutional Forces
• A fair procedure for better planning in the organization • Appropriate and effective for better planning in the organization	Legitimacy—normative forces
• Part of the awareness of potential impacts and losses • Part of the company culture	Shared understanding of compliance—cultural cognitive forces
• Management full support • Improve the company's reputation	Social influence—normative forces
• It can be easily integrated with other management systems	Gains, losses and consequences—regulative forces
• Non-compliance impact	Rules, laws, sanctions—regulative forces
• Improve the employee's health, safety and welfare	Personal morality—normative forces

Table 13.7 The reasons and motivations for determining staff to support the recovery strategy and providing training and awareness programme

The reasons and motivations for determining staff to support the recovery strategy and providing training and awareness programme are mostly due to	Variables and constructs of Institutional Forces
• A fair procedure for better planning in the organization	Legitimacy—normative forces
• Management full support • Improve the company's reputation	Social influence—normative forces
• Improve the organization's procedures for facing crisis • It can be easily integrated with other management systems	Gains, losses and consequences—regulative forces
• Part of the awareness of potential impacts and losses	Shared understanding of compliance—cultural cognitive forces

Table 13.8 The reasons and motivations for developing a detailed BC plan

The reasons and motivations for developing a detailed BC plan are mostly due to	Variables and constructs of Institutional Forces
• Appropriate and effective for better planning in the organization	Legitimacy—normative forces
• Improve the company's reputation	Social influence—normative forces
• Comply with the regulation that is taken for granted • Part of the awareness of potential impacts and losses • Part of the company culture	Shared understanding of compliance—cultural cognitive forces
• Non-compliance impact • Not implementing them can result in receiving sanctions	Rules, laws, sanctions—regulative forces

Table 13.9 The reasons and motivations for providing periodic tests and exercises to ensure that the BC plan is viable and workable

The reasons and motivations for providing periodic tests and exercises to ensure that the BC plan is viable and workable are mostly due to	Variables and constructs of Institutional Forces
• Appropriate and effective for better planning in the organization	Legitimacy—normative forces
• Improve the company's reputation • Management full support	Social influence—normative forces
• Improve the organization's procedures for facing crisis • Not providing periodic tests and exercises to ensure that the BC plan is viable and workable may lead to negative impact • It can be easily integrated with other management systems	Gains, losses and consequences—regulative forces
• Comply with the regulation that is taken for granted	Shared understanding of compliance—cultural cognitive forces

Table 13.10 The reasons and motivations for conducting BCM programme management

The reasons and motivations for conducting BCM programme management are mostly due to	Variables and constructs of Institutional Forces
• A fair procedure for better planning in the organization	Legitimacy—normative forces
• Management full support	Social influence—normative forces
• Part of the awareness of potential impacts and losses • Comply with the regulation that is taken for granted • Part of the company culture	Shared understanding of compliance—cultural cognitive forces
• Improve the organization's procedures for facing crisis • It can be easily integrated with other management systems	Gains, losses and consequences—regulative forces
• Improve the employee's health, safety and welfare	Personal morality—normative forces

Table 13.11 The reasons and motivations for conducting BCM training and awareness programmes for all staff and related external parties

The reasons and motivations for conducting BCM training and awareness programmes for all staff and related external parties are mostly due to	Variables and constructs of Institutional Forces
• A fair procedure for better planning in the organization • Appropriate and effective for better planning in the organization	Legitimacy—normative forces
• Management full support	Social influence—normative forces
• Comply with the regulation that is taken for granted • Part of the company culture	Shared understanding of compliance—cultural cognitive forces
• Not implementing them can result in receiving sanctions • Non-compliance impact	Rules, laws, sanctions—regulative forces
• Improve the employee's health, safety, and welfare	Personal morality—normative forces

Based on these findings, it can be seen that the firms would conduct risk analysis and risk review mostly because of their awareness of potential risks in their activities, and involving experts and BCM committee would be for the firm's legitimacy in implementing this principle. Other variables listed in the tables also correlate highly with these highest loading variables.

Tables 13.4 and 13.5 describe the reasons and motivations for implementing BIA. For this principle, the normative forces (legitimacy, social influence and personal morality) dominate the principle, followed by regulative forces (gains, losses and consequences; rules, laws and sanctions) and cultural cognitive forces (shared understanding of compliance). For conducting BIA, the variable that is positioned as the highest loading in each factor is legitimacy. The firms are observed to be implementing BIA and involving experts, employees from related BUs and key staffs in the process mostly for the legitimacy. BIA is opined to be a fair procedure, appropriate and effective for better planning in the firms.

Tables 13.6 and 13.7 explain the drivers for conducting strategy analysis. This principle is mostly driven by normative forces (legitimacy, social influence and personal morality), and also supported by regulative forces (gains, losses and consequences; rules, laws and sanctions) and cultural cognitive forces (shared understanding of compliance). Similar with BIA, in conducting strategy analysis, the variable that has the highest loading in each factor is legitimacy. This finding means that the firms would conduct strategy analysis for maintaining the operations of CBFs and that determining staff and providing training programme are mostly for legitimacy. Strategy analysis is viewed to be a fair procedure for better planning in the firms.

Table 13.8 shows the reasons and motivations for developing a detailed BC plan. This principle is also driven mostly by normative forces (legitimacy and social influence), followed by cultural cognitive forces (shared understanding of compliance) and regulative forces (rules, laws and sanctions). In implementing this principle, legitimacy is also the highest loading variable. This means that the firms would develop a detailed BC plan due to its appropriateness and effectiveness for better planning in the firms.

The drivers for conducting periodic tests and exercises to ensure that the BC plan is viable and workable are presented in Table 13.9. This principle is also driven mostly by the normative forces (legitimacy and social influence), followed by regulative forces (gains, losses and consequences) and cultural cognitive forces (shared understanding of compliance). The highest loading variable for this factor is also legitimacy. Based on this finding, it is viewed that the firms would provide periodic tests and exercises due to its legitimacy, where this principles is viewed as appropriate and effective for the firms.

Tables 13.10 and 13.11 present the reasons and motivations for conducting BCM programme management. Similar with the other BCM principles, this is mostly driven by normative forces (legitimacy, social influence and personal morality), followed by cultural cognitive forces (shared understanding of compliance) and regulative forces (gains, losses and consequences; rules, laws and sanctions). Legitimacy is also the highest loading variable for this principle. It appears that the firms would implement a BCM programme, complete with its training and awareness programme, mostly because they view that this principle is a fair procedure for better planning in their firms.

These findings showed that each BCM principle is motivated by a variety of IF combinations. A group of variables was analyzed and found to be highly correlated, where the most important and the least important factors were identified (Comrey and Lee 1992; Kline 1994). These findings were in line with what had been observed by Scott (1998, 2008), where in most institutional forms, not one single institutional force is at work, but varying combinations of IF. As an example in a stable social system, it was observed that practices that persist and are reinforced are due to them being taken for granted, normatively endorsed and backed by authorized powers. Furthermore, in some situations, one or another pillar would operate virtually alone and in other situations, a combination of pillars would assume dominance.

An interesting finding from these results is that all of the BCM principles are mostly driven by the normative forces, which is legitimacy. This variable has the highest loading in each factor for each principle. The views of implementing the BCM principles as appropriate, effective and fair procedures for better planning become the dominant motivation of the firms. They considered the principles as a value and expectations that could provide a certain legitimacy status by implementing it. The conception and controls from these normative forces focused more on a moral base which are likely to be internalized rather than through regulative controls (Scott 2008).

Only the risk analysis and review process is mostly driven by two variables, which are shared understanding of compliance of the principle and the legitimacy to conduct the principle. The risk analysis and review principle could be driven by cultural-cognitive forces because the firms might view this principle as a logical step granted that it is already part of the firm's culture. They had awareness of the potential risks if it is not implemented and they shared an understanding of the benefits of this principle for the firms (Low et al. 2010a; BSI Groups 2010; Scott 2008).

From the survey validation results, the respondents provided their views regarding the drivers related to the IF for implementing BCM. All of them supported the survey results, but they viewed that regulative forces would still dominate the way firms implement a system or a concept. Regulative forces such as regulation and sanctions from the top management of the firms would still be the reasons that motivate firms to implement a concept like BCM. These views are very much in line with the survey results, where regulative forces are still correlated with the highest loading variables. From the survey results, some BCM principles were also driven by regulative forces to meet the requirements from the firm's top level management which they considered as a rule that has legal sanctions, carried through in a coercive manner and to comply with expediency (BSI Groups 2010; Scott 2008).

Least Important IF Factors for Six BCM Principles

In Sect. 10.3.4, it had also been found earlier that there were two normative forces from the social influence variables that are viewed to be the least supporting factors for BCM implementation. From the analyses, the firm's competitiveness and stakeholders/clients' requirements were not viewed to be the main reasons for the firms to implement the whole suite of BCM principles.

These findings are different from what several studies had mentioned previously. BUCORIM (2008) for example, found that based on a survey of trends in business continuity, BCM has become mandatory due to the need to comply with the customers' mandate and regulatory requirements. The study also found that competitive edge drove the firms to develop a BC plan. Moreover, a report published by the Economist Intelligence Unit (EIU 2007) indicated that customers are the stakeholder viewed to be the most important factor in driving decisions about business continuity, with most citing them as a significant influence. The second most important influence over decisions about BCM is regulators (see Sect. 3.8).

Furthermore, the survey validation results also provided the respondents' views on the hindrance for implementing BCM within the context of institutional forces. Most of the respondents opined that cultural-cognitive forces may be the constructs that would least likely motivate firms to implement BCM. Their views were that the culture of discipline and awareness attitudes toward risk/crisis may not yet be dominant in the firms, particularly for low level employees in the firm. These views were not quite in line with what the survey results had shown, where shared understanding of compliance regarding implementing BCM principles was found to be one of the main motivations. Only one respondent supported the survey results, mentioning that the normative forces such as competitiveness and stakeholders' requirements may not be the main reason for firms to implement BCM. The respondent opined that the mindset of the firms would still be focusing on profitability, which relate to the mindset of gains and losses (regulative forces).

Based on these findings, it appears that the institutional framework used for explaining why construction firms do or do not wish to implement BCM principles is a useful platform in this study. This is similar to other studies that had utilized this framework for the same reason (Low et al. 2010b).

Organizational Culture Attributes (OC) Implemented for BCM Principles
The importance of organizational culture towards adopting a concept or system in the organization can be seen from studies which increasingly push organizational culture as the guide for organizational strategies. It is suggested that organizational culture can impact manager's ability to process information, rationalize and exercise discretion in their decision-making processes (Hofstede and Hofstede 2005). In a firm, identifying its organization culture provides practical utility. This process can identify whether the culture fits the strategies set out for the future. Cultural constraints determine which strategies are feasible for an organization and which are not (Hofstede and Hofstede 2005).

The relationship between organizational culture (OC) dimensions and BCM principles is generally to determine how the contractor views the BCM principles (in the context of organizational culture dimensions) regarding its level of influence (significant drivers) and level of hindrances for the firm. Furthermore, OC dimensions as the non-technical framework which affect any management system is an important element to be considered. Fulfilling the technical requirements of a management system is only one aspect, where the other aspect which focuses on non-technical attributes (such as OC dimensions) must not be overlooked. The non-technical attributes should be reviewed due to its role in helping to promote an integrative environment in the management system, which in this case is BCM (Low 1998; Kanter 1994).

As mentioned earlier in Sect. 8.3.2, several literatures had shown that OC dimensions are required in BCM. The OC attributes support the BCM implementation and could be applied to the six main BCM principles. Section 10.3.5 earlier explained the meaningful factors from the OC attributes that had been implemented by the responding Indonesian contractors that could support the BCM principles. Tables 10.19–10.28 earlier described the most implemented OC factors and the least implemented factors for each BCM principle. In these tables, the OC attributes were categorized based on its OC dimensions. Following these, the OC dimensions categorized in each BCM principle can be summarized in the form of a matrix, as shown in Table 13.12. From this table, the OC dimensions that were perceived as highly implemented in each BCM principle can be seen.

Based on Table 13.12, it appears that the Indonesian contractors had implemented the 14 OC dimensions in their organizations, and these dimensions would support the implementation of the respective BCM principles in the table. Knowing that they had these OC dimensions implemented in their organizations could drive BCM implementation to be executed more effectively. The non-technical attributes needed to support BCM implementation are already in place. However, the professional attribute seemed to be the least implemented attribute among Indonesian contractors, particularly in RA2, BIA2, S2 and PM2. Professionalism is essential for BCM implementation; but in this case, it is the least implemented attribute for these firms. This could be a hindrance for the firms to

Table 13.12 OC dimensions that have been implemented and support the BCM principles

Organizational culture dimensions	BCM principles										
	RA1	RA2	RA3	BIA1	BIA2	S1	S2	BCP1	TE1	PM1	PM2
Empowerment		X			X	X					
Team orientation		X			X		X		X		X
Developing employee's skills							X				X
A set of values										X	X
Coordination and integration	X		X			X		X	X	X	
Employee oriented	X										
Adaptability to change	X		X	X		X		X	X		
Setting standards and good performance								X		X	
Process oriented			X								
Customer orientation								X		X	
Reward orientation										X	
Power in organization										X	
Professional attribute		O			O		O				O
Open system		X									

X, opined to have been implemented and would mostly support the BCM principles; O, opined to have been least implemented for the BCM principles implement BCM effectively. Therefore, ways to improve the professionalism in the firms should be considered.

Detailed descriptions about each OC dimension based on Table 13.12 are as follows:

- Empowerment was opined to have been implemented and would mostly support: the involvement of experts and BCM committee in risk review (RA2); the involvement of experts, employees from related BUs and key staffs in BIA (BIA2); and strategy analysis for maintaining the operations of CBFs (S1). It had been identified that empowerment is a key part in BCP. This finding is in line with some studies which mentioned that BCP can be effective only if developed cooperatively, with the involvement of a wide range of individuals at various levels in the firms (Drennan and McConnell 2007; BCI 2010; McManus et al. 2008). By implementing empowerment in the aforementioned BCM principles, flexible and efficient decision making process, sharing authority while maintaining a strong chain of command, and designating responsibility for the BCP could run smoothly (Light 2008).
- Team orientation was viewed to have been implemented and would mostly support: the involvement of experts and BCM committee in risk review (RA2); the involvement of experts, employees from related BUs and key staffs in BIA (BIA2); the process determining staff to support the recovery strategy and providing training and awareness programme (S2); periodic tests and exercises to ensure that the BC plan is viable and workable (TE1); and BCM training and awareness programmes for all staff and related external parties (PM2). According to BCI (2010), cooperation among employees is essential and BCM principles should support team orientation in its process. This is in line with the survey results, where teamwork was viewed to be important in most of the BCM principles. The processes in RA2, BIA2, S2, TE1 and PM2 can be a good platform to encourage the employees and the related parties to get to know and understand each other's beliefs, cultures and practices (Drennan and McConnell 2007; Low et al. 2008a; Ministry of Manpower 2010; BCI 2010; Light 2008).
- The survey results showed that developing employee's skills was opined to have been implemented and would mostly support: the process determining staff to support the recovery strategy and providing training and awareness programme (S2); and BCM training and awareness programmes for all staff and related external parties (PM2). As described by several studies, the staff and the recovery team members should understand their roles and responsibilities when crises occurred. The need for training and awareness programme are necessary in developing the staff's skills and hence improving their performance (Ministry of Manpower 2010; BCMI 2011; BCI 2010; Light 2008). The survey results similarly showed that BCM training and awareness programme is the suitable platform to develop the employee's and other parties' skills regarding BCM.

- A set of values was viewed to have been implemented and would mostly support: BCM programme management (PM1); and BCM training and awareness programmes for all staff and related external parties (PM2). According to some studies, in conducting BCM programme management, there must be policies to guide BCM efforts, where this set of values should set out the organization's aims, principles and approach and would serve as the rationale and support for all BCM principles on an ongoing basis (Spring Singapore 2008; Denison 2000; Cheung et al. 2011). Therefore, having this OC dimension implemented in BCM programme management (PM1 and PM2) would be in line with what those studies had recommended.
- Coordination and integration were viewed to have been implemented and would mostly support: risk analysis and cost benefit analysis (RA1); a detailed risk review (RA3); strategy analysis for maintaining the operations of CBFs (S1); developing a detailed BC plan (BCP1); periodic tests and exercises to ensure that the BC plan is viable and workable (TE1); and BCM programme management (PM1). From this finding, it can be seen that in most of the BCM principles, coordination and integration are important attributes. In line with these results, Elliott et al. (2002) stated that coordination with the stakeholders is important during the planning and execution phase of BC. Moreover, an integrated approach within the business functions should be conducted in order to execute a coordinated response to a crisis situation. Developing and publicizing the BCM policy (by implementing RA1, RA3, S1, BCP1, TE1 and PM1) throughout the organization can be reinforced with suitable communications. Everyone involved in BCM should be informed and consulted.
- Employee oriented was found to have been implemented and would mostly support risk analysis and cost benefit analysis (RA1). This finding suggested the need to focus on the people element when conducting risk analysis and cost benefit analysis. As highlighted by BCI (2008), an effective BCM needs the people element to be involved actively. People are the principal asset of any business, because without them, the business will not function. BCP should consider that recovery planning is not just about a technical solution, but about people. Furthermore, understanding the people/staff's issues and needs during the relocation plan development is essential (McCrackan 2005; Low et al. 2008a).
- Adaptability to change was viewed to have been implemented and would mostly support: risk analysis and cost benefit analysis (RA1); a detailed risk review (RA3); BIA (BIA1); strategy analysis for maintaining the operations of CBFs (S1); developing a detailed BC plan (BCP1); as well as periodic tests and exercises to ensure that the BC plan is viable and workable (TE1). Similar to other OC dimensions such as coordination and integration, and team orientation, adaptability to change is an important dimension for most of the BCM principles. This finding is in line with what McManus et al. (2008) had stated, where the concept of adaptive capacity is at the core of the current organizational resilience methodology. Organizations that focus on their resilience in the face of disruption generally adopt adaptive qualities and proactive responses (Mallak

1998; Folke et al. 2002). Therefore, a firm that has the adaptability to change when implementing the BCM principles could become resilient more quickly.

- Setting standards and good performance was opined to have been implemented and would mostly support: developing a detailed BC plan (BCP1); and BCM programme management (PM1). BCM is a management process that must optimize cost efficiencies. Standards for BC procedures and processes should be set in order to develop a pragmatic, cost effective, and operable recovery plan; hence to enable the firm to run its critical business processes during disruptions. As a support, implementing and maintaining a robust exercise, rehearsal and testing programme could ensure that the business continuity capability is effective, up-to-date and fit-for-purpose. Moreover, during BC plan development, to be cost conscious is necessary (Smith 2003; Hiles 2007; Health 1999; O'Hehir 1999; Low et al. 2008a; Singapore Business Federation 2003; Pauchant et al. 1991). Based on this view, the finding for this dimension is in line with what the previous studies had recommended. Setting standards and good performance would be beneficial for the implementation of BCP1 and PM1 principles.
- Process oriented was viewed to have been implemented and would mostly support a detailed risk review (RA3). Conducting a detailed risk review means to examine and assess the availability of critical equipment, technology, and facilities for BU/CBF (including location of facilities, essential utilities and telecommunications, transportation to premises and physical security of premises). This process needs conformity to procedures and rules. Therefore, being process oriented is important while implementing this principle (Spring Singapore 2008; BCI 2010).
- The survey results described that customer orientation was viewed to have been implemented and would mostly support: developing a detailed BC plan (BCP1); and BCM programme management (PM1). Similar to the findings of another OC dimension, which is setting standards and good performance, customer orientation is essential for the implementation of BCP1 and PM1 principles. This means that the focus of continuity in customer service should be on defining what level of service must be maintained throughout a disaster, and what is required to achieve that level of service. In defining time lines for the resumption of support and services, and transparency of operations in a crisis, the impact on customers should also be considered (Singapore Business Federation 2003). Therefore, this dimension should be well considered for BC plan development and programme management. By demonstrating the importance of customers in BCM implementation, it can be used as a competitive advantage to gain new customers and to improve margins by using it as a "customer care" approach (BCI 2007a).
- Reward orientation was found to have been implemented and would mostly support BCM programme management (PM1). This finding is in line with the studies from BCI (2002), where it stated that performance-based rewards can ensure the active involvement of managers and staff at all levels of the organization, especially the operational middle management who has to implement and maintain BCM. Performance management and rewards are one of the

mechanisms that can exert influence upon what is seen as important and how it is to be done. In BCM programme management, the active involvement of managers and staff at various levels of the organization would be needed and necessary. Therefore, this dimension could definitely support the implementation of this principle.

- Power in organization was opined to have been implemented and would mostly support BCM programme management (PM1). This OC dimension is still related closely with the PM1 principle, because when implementing this, the role of the leader is very much needed. Other studies had stated that the development of BCM will be implemented successfully and continue to be successful as a programme when there is central control and coordination (McManus et al. 2008; BCI 2010; BCMI 2011). This means the quality of leadership and top management's commitment in the organization is essential.
- The survey results found that professional attribute/professionalism was viewed to have been the least implemented dimension for these BCM principles: the involvement of experts and BCM committee in risk review (RA2); the involvement of experts, employees from related BUs and key staffs in BIA (BIA2); determining staff to support the recovery strategy and providing training and awareness programme (S2); as well as BCM training and awareness programmes (PM2). This finding is interesting because from the survey analysis, this dimension was only found in the least implemented group of factors. From here, it appears that the firms had not emphasized fully on job competence (or not fully implemented the professional attribute) in its environment. The possible reason for this could be that the firms still adopt a parochial culture, where in hiring employees, the firms took the employee's social and family background into account as much as their job competence. Furthermore, the employees may feel that the organization's norms cover their behavior at home as well as on the job (Hofstede and Hofstede 2005). Therefore, based on this finding, when conducting RA2, BIA2, S2 and PM2, the firms might have implemented a parochial attribute than professional attribute. This finding is not quite in line with what several studies had recommended. The skill sets and competence of participants are essential to the success of BCM. Moreover, an independent expert or professional can be appointed as a supporting member in the BCM team, provided that the expert has complied with certain job competencies needed in developing BCM (Spring Singapore 2008; Low et al. 2008a).
- Open system was opined to have been implemented and would mostly support the involvement of experts and BCM committee in risk review (RA2). This finding is similar with what BCI (2008) suggested, where BCM should be using a people-focused approach. How the process of appointing staff with necessary skills for BCM development that involves the HR department, considering the impacts toward the staff during disruption or relocation process and managing how the people will fit into the BC process are needed in order to have an effective BCM.

In the survey validation results, when discussing the significant OC dimensions that may drive or hinder BCM implementation in contractors, the respondents supported the survey results. They viewed that communicating and coordinating, following procedures (process oriented), empowerment, adaptability to change, and power in organization have been implemented in the nature of their business. These are all drivers to support BCM implementation. However, they noticed that the culture of documentation (part of setting standard and good performance) was still a problem in their firms. One respondent explained that the skills for documenting reports by construction firm employees are still low. They are not used to write and document events or lessons learned from the project activities in a comprehensive manner. Another respondent stated that although the firm had provided template forms for progress reports or activity reports, the outcome depends on who would be disciplined and comprehensive enough to write the reports. This task seems to be easy, but not doing it properly may lead to problems. Based on this issue, the lesson learned is that when implementing BCM, the firms should select competent employees to conduct this documentation process.

From these findings, the respondents illustrated that the fourteen OC dimensions had been implemented by the Indonesian contractors, and these dimensions would support the implementation of the BCM principles, as summarized in Table 13.12. In other words, this study had identified the OC dimensions that could support BCM principles for implementation.

Each BCM principle is supported by a variety of OC attributes which had been implemented. These attributes are the common beliefs, pattern of behaviors and values which had been exercised among members of the organization (Kotler and Heskett 1992; Scholz 1987; Williams et al. 1993). The higher their correlations between the attributes, the most likely they would be grouped into one factor in the analysis (Comrey and Lee 1992; Kline 1994). From this study, the most implemented and the least implemented OC attributes/factors were identified.

13.2.2 Findings from Case Studies

Descriptive case studies were conducted based on two Indonesian contractors (state-owned and private-owned firms) to interpret the proposition based on RO4 and RQ2. The case studies explored the firms' characteristics and BCM implementation.

13.2.2.1 Organization Characteristics

Table 13.13 elaborates the findings from the case studies conducted in Firm A and Firm B. In the descriptions column, some literatures are cited to support the findings related to each of the organization characteristics.

Table 13.13 Organization characteristics of Firm A and Firm B

Organization characteristics	Firm A	Firm B	Descriptions
Organization objectives	Its mission is to make breakthroughs by evolving into an integrated infrastructure firm	The construction and development of premium, high-rise residential, commercial and office projects	Both firms' objectives supported the government in delivering more infrastructures in Indonesia. According to EIU (2010a), the Indonesian government had established a financing company to provide long-term financing and support for infrastructure schemes. Therefore, it is hoped that there would be an improvement in the quality of Indonesia's infrastructure
Organization products and service	Construction and energy	Building construction	Both firms' products are relevant to the Indonesian Government Regulation No. 28 Year 2000 regarding their business areas in construction services (RI 2000)
Organization geographic scale	• Cities in Indonesia and abroad • 45% of the projects were in west Indonesian islands (Sumatra, Kalimantan, Java and Bali) • 55% of the projects were in east Indonesian islands (Sulawesi, NTT, NTB, Papua)	• Operated only in Indonesia • 70% of the projects were delivered in west Indonesia (Sumatra, Kalimantan, Java and Bali) • 30% of the projects were in east Indonesia (Sulawesi, NTT, NTB and Papua)	• West Indonesia is still the main focus of development due to its more densely populated areas (EIU 2008) • Firm A focuses more on delivering projects in East Indonesia due to its demands for more infrastructure in that area (EIU 2010a)
Organization plans	• Revitalization and integration inside the firm • Short term: to focus on its internal growths and collaborate with other multi-national companies,	• Short term: establish a more synergistic approach to its repeat customers, and strengthening the company's internal foundation through the improvement of	• The plans are in line with NBCSD's (2004) recommendation to improve the firms' standards of competence that could help in delivering higher quality

(continued)

Table 13.13 (continued)

Organization characteristics	Firm A	Firm B	Descriptions
	respectively • Long term: to create new business expansion • Continue to implement a management system of Quality–Safety–Health–Environment (QSHE) as one of its differentiations	qualified human resources • Long term: focus on constructing buildings which have a relatively high investment values, and more emphasis on constructing green buildings in the future	construction products with better production processes • The firms' focus in internal growth is in line with the construction business value chain, including to create relationships with existing and prospective clients; maintaining effective human resource programs and policies; as well as maintaining its financial resources (Schaufelberger 2009)
Organization market and competitors	• It would increase its market in the private sector • It would continue to capture global market in the future • No single company has a dominant market share in Indonesia's construction industry	• It is focusing on domestic building projects • Its competitors were the local state-owned contractors, private firms and international firms working in Indonesia • The competition occurred in the tendering process (procurement phase)	A highly competitive industry is considered as a weakness for the contractors to expand their businesses. However, due to the current needs of infrastructures in many cities in Indonesia, there are still many opportunities for them to capture new markets (Suraji 2003; Raftery et al. 2004)
Organization clients	• It delivers projects to local (Indonesia) and international clients • The projects can be funded by the governments and private sector • It had arranged insurance between the clients and the firm for improved reliance on delivery. They have insurances for contractor all risk (CAR), erection all risk and machinery breakdown	• Its projects were all in Indonesia, but there were overseas companies from Singapore and Malaysia that became its customers • Its local clients were mostly government, state-owned firms and private firms (mostly banks and developers) • The insurance provided by the clients are contractor all risk (CAR) and equipment all risk (EAR)	• Both firms are still focusing on delivering projects in Indonesia. This is due to the fact that Indonesia's government provides its full effort to support infrastructure, residential and commercial development throughout the country by creating a more robust regulatory and legal framework (EIU 2010a; BMI 2009)

(continued)

Table 13.13 (continued)

Organization characteristics	Firm A	Firm B	Descriptions
			• Insurance policies by the clients were in line with the Construction Service Law No.18 Year 1999 (RI 1999)
MBCO for each CBF	The overview of the MBCO for the most influential CBFs includes • Managing and protecting all financial matters in all of the business value chains • Managing and protecting all of contracts and legal administration • Managing procurement for all projects (materials; men; machines) • Managing projects, documentation for completed projects and ongoing projects • Managing project sites (materials; equipment; facilities) and laborers • Maintaining the company's integrated management system; and managing and protecting the company's business development documentation • Managing the human resources in head office and projects (employee's wage and welfare)	The overview of the MBCO for the most influential CBFs includes • Managing legal and contractual matters • Managing relationships with the company's investors and clients • Managing the company's market forces and future business development • Managing the company's existing and current projects (head office—project site management) • Managing financial matters • Managing the company's human resources (employed and outsourced) • Managing equipment resources used by the company for all projects • Managing logistics for all projects • Managing safety for the human resources in the company • Managing the project during execution • Managing relationships with customers (before, during and after projects) • Managing all affairs related with the company (representing the	• Defining CBF and MBCO in the firm's business units are part of the BCM principles, particularly for Risk analysis (RA), BIA, Strategy Analysis (S), and BC Plan development • These elements would be applied in other BCM principles which are Tests and exercises (TE) and Programme management (PM) (see Tables 3.2 and 8.4)

(continued)

Table 13.13 (continued)

Organization characteristics	Firm A	Firm B	Descriptions
		company to external parties) • Managing the company's physical assets (head office, project office, resources)	
Internal and external coordination overview	• Coordination between head office and project-based office Regular reports on performance progress are conducted daily, weekly, monthly and annually Meetings; emails; intranet; video conference • Coordination between head office and governments; public; clients; suppliers (supply chain); sub-contractors Verbal: meetings Non-verbal: Reports, emails, letters • Coordination between project-base office and public; suppliers; sub-contractors Verbal: meetings Non-verbal: Reports, emails, letters	• Coordination between head office and project-based office Regular reports (daily, weekly, monthly) progress and performance reports were developed and distributed via regular meetings and emails Verbal: meetings Non-verbal: reports, notes, letters, emails, social media • Coordination between head office and governments; public; clients; suppliers (supply chain); sub-contractors Verbal: meetings; conference; workshops Non verbal: emails; letters Progress and performance reports to clients Other reports related to projects that need to be distributed/informed to government/public/supplier/sub-contractors Financial, technical and administration matters/coordination to suppliers and sub-contractors • Coordination between project-based	• These types of coordination are needed to be identified as part of BCM preparedness (BCIGPG 2002) • These types of coordination are part of the main principles of BCM, particularly on BC Plan development (Pitt and Goyal 2004; Elliott et al. 2002; BCI 2010)

(continued)

Table 13.13 (continued)

Organization characteristics	Firm A	Firm B	Descriptions
		office and public; suppliers; sub-contractors Verbal: meetings; conference; workshops Non verbal: emails; letters Progress and performance reports to suppliers and subcontractors Other reports related to projects that need to be distributed/informed to government/public/supplier/sub-contractors Financial, administration, and technical matters/coordination to suppliers and subcontractors	
List of important crises	• Changes in regulations and statutory legislation • Increase in price of raw materials (unexpected price escalation) • Shortage of key materials; the firm needs more time to coordinate for supplying the materials shortage • Subcontractor insolvency and conflicts at the project site • Weather issues/climate or natural disasters	• Action by environmentalist/pressure groups (protests) • Access/approval restriction or limitation • Delays or uncertainty in resolving disputes • Materials shortage • Increase in price of raw materials (unexpected price escalation) • Subcontractor insolvency • Serious accidents in a project • Natural disasters (earthquake, floods, tsunami, etc) • Loss of management personnel or key staff	The results from both firms are similar with the findings from the surveys (see Table 10.6)
List of crises impacts	• Loss of productivity • Wastage of materials • Delay in work/dissatisfied customers • Revenue impact	• Delays in work • Dissatisfied customers. Employees were not able to reach the office	The results from both firms are similar with the findings from the surveys (see Table 10.7)

(continued)

Table 13.13 (continued)

Organization characteristics	Firm A	Firm B	Descriptions
		• Revenue impact • Regarding the extent of resource loss, the firm viewed productivity losses, material wastes and equipment damages as the main losses	
Crises responses elements	• It has documented evacuation and communication procedures. However the further recovery and restoration procedure still need more coordination with the firm's head office • These plans are used for overcoming accidents, fires, earthquakes, floods and weather issues • Procedures for other types of crises have not been developed and documented • So far, the response plans for accidents in projects and floods have been implemented	• It has provided procedures for evacuation, communication during crisis, alternative facilities, recovery and restoration • These procedures are used for handling fires, accidents, earthquakes and floods • Procedures for other types of crises have not been developed and documented • Response plans for floods (for the head office) and accidents in projects have been implemented • However, it still needs to improve its communication procedures during floods, when some employees were still trapped in the floods and traffic when they do not receive the current information from the head office	• SOP for crisis management and crises responses had been developed in both firms. This is similar to the survey results (see Table 10.8) • The SOPs were not developed holistically and still need detailed responses that relate to various stakeholders (Supriadi and Low 2012)
Business environment and internal regulation that support crisis management in firms	• Its business regulation environment does not provide any support for overcoming crisis • Its internal regulation had provided SOPs for overcoming selected incidents and these SOPs provide coordination with external	• There is no business regulation that focuses mainly on overcoming crisis • It has its internal regulation and procedures (SOP) for overcoming selected incidents • These plans also need coordination	• The role of the management board is to decide how broad or constrained the focus of business continuity provision is to be. This would involve a consideration of the business processes that are to be covered by the

(continued)

Table 13.13 (continued)

Organization characteristics	Firm A	Firm B	Descriptions
	agencies • Every state-owned enterprise is supported financially by the government, where they would be supported to continue business as usual if any financial crisis occurred. This can be viewed as a support for business continuity • Within the reporting period, it has never received financial aid from the government or other facilities in terms of fiscal or financial matters	with external agencies such as local government, NBCSD, securities (police and army), hospitals, fire station and the local community • Regarding support for business continuity, there are regulations from the government that encourage the firm to comply with particular standards for safety, quality and environmental management. These standards may not focus specifically for business continuity • It needs to prepare internally by complying with a specific BCM international standards	continuity provision, and the extent to which external continuity services would be used. Without top down direction, support and ownership, success in both the BCM process and activating the BC Plan would be difficult, if not impossible (Power 1999) • Support and strong commitment from the senior management are needed and are part of the aspects for BCM preparedness (BCIGPG 2002) • Regulations to comply with safety, quality and environmental standards in construction have been established, but there is no regulation that is specifically established for implementing BCM (NBCSD 2004; Andi and Chandra 2007)
Lessons learned from crises	• Each business unit and function needs more coordination holistically. Defining the interdependencies of each BU is necessary for developing a holistic procedure to overcome crises (This is part of BCM principles: RA, BIA, S) • Communication procedures should be provided in details in order to reduce waiting time or delays (This is part of BCM	• Communication during crisis is essential. Firm B needs to review and improve its communication procedures. Regular communication drillings relating to certain crisis should be conducted (This is part of BCM principles: BIA, S, BCP) • Not all crises/threats have been analyzed by Firm B. It needs to have a more comprehensive plan for	All of the lessons learned from both firms would lead to the need for developing BCM, where they can be addressed by adopting the BCM principles (Supriadi and Low 2012)

(continued)

Table 13.13 (continued)

Organization characteristics	Firm A	Firm B	Descriptions
	principles: BIA, S, BCP)	overcoming crises (This is part of BCM principles: RA, BIA, S, BCP) • The role of top management is essential during crises, and the management should appoint a specific person for concentrate on developing the BC plans for various crises (This is part of BCM principles: RA, BIA, S, BCP, TE, PM; BCM initiation)	

13.2.2.2 BCM Preparedness of Firm A and Firm B

In Sect. 11.2.3, it was earlier seen that the case studies also provided descriptions about the firms' practices which relate to BCM. The tables in this section (namely Tables 11.10–11.23) illustrated some lists of activities that have/have not been implemented, based on each BCM principle. It had been found that the results of the case studies in this section are in line with the BCM-KBDSS results from Firm A and Firm B (see Sect. 12.3.1.1 and Figs. 12.29 and 12.30), where the BCM level of preparedness for Firm A and Firm B were level 2 (Beginner) and level 3 (Moderate), respectively.

To summarize the results, Firm A was in level 2, where:

- The deliverable of the initiated stage is BCM as an initiative.
- The firm exhibited a low level of awareness of policies and guidelines and of roles and responsibilities about BCM.
- Participating BUs and departments have instituted a basic governance program, mandating at least limited compliance to standardized BCM policy, practices and processes to which they have commonly agreed.
- The firm has a basic understanding of Business Continuity.
- Risk analysis has been developed. BIA and strategy analysis have been partially conducted.

Furthermore, Firm B was in level 3, where:

- A BCM framework is in place as a documented BC Plan.
- The responsibility for BCM is covered at a sufficiently high level within the organization and an explicit BCM policy is in effect.

- The firm has an advanced understanding of Business Continuity.
- BIA, Strategy analysis and BC plan have been documented, The tests and exercises have been planned and partially implemented.
- Top management strongly supports BCM.

A more comprehensive description about these BCM preparedness levels was presented earlier in Table 11.24.

13.2.2.3 Interviews with Experts for BCM Implementation

The second approach used to interpret the proposition based on RO4 and RQ2 was by interviewing five experts from the Indonesian construction industry. Their responses would be synthesized into the proposed BCM implementation guidelines for Indonesian contractors. The findings from the interviews are as follows:

The Benefits of Implementing BCM for Contractors
Section 11.3.1 has earlier described the benefits of implementing BCM for Indonesian contractors. Aspects such as the firm's resilience, competitive advantage, providing systematic data back-ups, economic recession preparedness, floods preparedness, securing the firm's assets, and better coordination with stakeholders were viewed as the benefits of implementing BCM. These aspects were also described as the benefits of BCM by other studies.

Tinston (2010) and BSI (2010) stated that BCM would help to improve resilience to disruptions. Regarding the gaining of competitive advantage by implementing BCM, BCI (2007a), Van Opstal (2007), and Council on Competitiveness (2006) supported this statement. Moreover, they viewed it as a contributor to profitability, shareholder value and competitiveness. Watkins (1997) had also opined that implementing BCM (particularly through BIA) could have positive implications especially in the event of a disaster, where the firm's assets can be secured. Raish, Statler and Burgi (2007) also found that protected revenue flows as a result of plans to protect key assets is one of the rewards of firm resiliency through BCM. Lastly, Hinton (2000) agreed that BCM could help to avoid losses of important data by providing systematic data back-ups.

The Drawbacks of Implementing BCM for Contractors
In Sect. 11.3.2, all respondents agreed that there would be drawbacks in implementing BCM in Indonesian contractors due to this not becoming the firm's main priority. Robinson (2009) also similarly opined that a challenge in implementing BCM could be due to the top management's view of BCM as a discretionary spending.

In addition, the respondents viewed this concept as a holistic and sophisticated concept to be implemented. Full support from the top management was their recommendation for implementing BCM. According to Continuity Central (2011), support from the top management is a big challenge, followed by the challenge of getting the wider organization to buy-in to business continuity.

The Need for BCM in Contractors

All of the respondents who were interviewed agreed that BCM implementation is needed for Indonesian contractors (see Sect. 11.3.3). This is similar to the findings from Supriadi and Low's (2012) study where this concept was needed for implementation by Indonesian contractors because there were many threats to a contractor by virtue of the nature of the business, particularly in Indonesia.

Other studies found the importance of implementing BCM, where it could be driven by the regulatory requirements (BUCORIM 2008; EIU 2007) and the customers' influence to implement BCM (EIU 2007). However, these types of requirements were not applied yet in Indonesia, particularly in the construction industry.

BCM Certification

Section 11.3.4 earlier described the views of the respondents regarding BCM certification. All of them agreed that the Indonesian contractors need to be BCM certified. Due to the support for BCM certification in this sector, the Indonesian contractors could refer to the existing BCM standards and certification from various countries, or using ISO's standards for BCM as the benchmark (Elliott et al. 2010; St-Germain et al. 2012; SPRING 2012).

The latest development of BCM standards is from ISO, where it has officially launched ISO 22301, "Societal security—Business continuity management systems—Requirements", the new international standard for Business Continuity Management System (BCMS). This standard is similar to the previous BCM standards, but it has some improvements for BCM implementation such as (St-Germain et al. 2012; SPRING 2012):

(a) Greater emphasis on setting the objectives, monitoring performance and metrics;
(b) Clearer expectations on management; and
(c) More careful planning for and preparing the resources needed for ensuring business continuity.

Furthermore, some respondents also recommended the regulator to provide the requirements for BCM certification. The Indonesian construction regulators could conduct benchmarking with other sectors or countries that have examples of regulators that provided BCM standards (Elliott et al. 2010).

Important Elements in Contractors that Relate Mostly with BCM

In Sect. 11.3.5, the respondents were earlier asked about the important elements in contractors that relate mostly with BCM, referring to the firm's BUs, CBFs, MBCO, and the most significant crises which might be highly impactful towards the firms. The interviews results were in line with the survey and case studies results (Tables 10.6, 11.2, 11.3, 11.6, 11.7). The findings particularly for Indonesian contractors (and for the contracting business in general) are:

(a) Regarding BUs in contractors, the project site unit, accounting, human resources department, and contracts division were viewed to be essential.

(b) In determining which CBFs are essential, functions for documentation, accounting, project assets, human resources management, and communication were all mentioned by the respondents.
(c) Lastly, regarding the most significant crises, economic/financial crisis and natural disasters were mostly viewed to be significant by the respondents.

Moreover, the essential BUs in contractors were all included in the primary and support activities of the construction business value chain (Schaufelberger 2009).

Additional Recommendations for BCM Implementation by Indonesian Contractors

Section 11.3.6 earlier explained the additional recommendations from the respondents for BCM implementation by Indonesian contractors. Table 13.14 describes the recommendations while relating them to some relevant aspects that are linked with BCM.

Table 13.14 Additional recommendations for Indonesian contractors to implement BCM

Additional recommendations for Indonesian contractors to implement BCM	Aspects related to BCM
Respondent 1C viewed that the full support from the firm's management and regulator are necessary for BCM implementation	BCM initiation phase; Management support and commitment to BCM
Respondent 2C provided three factors that needed to be considered in implementing BCM, which are full support from the management, building awareness of BC plan to all employees, and maintaining the BC plan	Management support and commitment to BCM; BCM awareness and training; BCM maintenance
Respondent 3C recommended strategies for BCM implementation that relate to the firm's human resources, the planning process, communication and attitude towards the implementation	BCM committee and roles; Communication and coordination in BCM; Exercises and tests for BCM; BC plan development
Respondent 4C opined that the firm's stakeholders should also be incorporated intensively throughout the implementation process. The stakeholders should be aware of this concept and able to implement it as well	BCM awareness and training; BCM maintenance; Communication for BCM during crises
Respondent 5C opined that BCM implementation should be understood by not only the top management, but also by the firm's employees. This can be achieved by conducting relevant trainings, workshops or knowledge sharing about BCM	Management support and commitment to BCM; BCM awareness and training

13.2.3 BCM Implementation Guidelines Development

Section 11.4 earlier described the BCM implementation guidelines development. These were developed in a qualitative approach where the data was collected to generate ideas or framework and used for inductive reasoning. Qualitative data emphasizes on determining the meaning of the data, where what is analyzed is not numbers, but texts. Therefore, these data must be handled systematically, starting from organizing and categorizing the data into concepts, establishing links between concepts, and refinement and development (Fellows and Liu 2003).

The development of the BCM implementation guidelines framework was conducted to fill in the knowledge gap of this study, where according to McKinsey (2013), before adopting and implementing a concept, an organization should also analyze its level of preparedness towards the concept. This type of assessment helps to identify strengths and areas for improvement and hence the organization can further invest its resources to implement the concept accordingly.

In addition, the framework was developed in the form of identifying the level of preparedness because this assessment had been proven to be an effective evaluation method. It can help the firm to verify what they had achieved related to the topic assessed and it can assist the firm in prioritizing the necessary improvement based on their assessment results (Peng et al. 2011; Stevanovic 2011). Therefore, the BCM implementation guidelines framework consists of the assessment of its level of preparedness and provides the guidelines or action plans based on its current level of preparedness.

13.3 Findings and Discussion for RQ 2

13.3.1 Findings from KBDSS Development

As explained earlier in Sect. 12.2, the BCM implementation guidelines for Indonesian contractors were developed into a KBDSS. This system is a supporting tool to assess the organization's level of preparedness and to provide the knowledge needed by the management team to develop BCM. The automation process in this system is beneficial for a fast and effective decision making process (Arain and Low 2006; Sudarto 2007). This section had comprehensively described the KBDSS development, starting from the knowledge acquisition from data analyses, rules and logic representation development, and decision framework development. All of these processes were structured, calculated and synthesized in the inference engine of the system. The tools used for developing the KBDSS were a DSS shell software and mathematical function software.

The system was named BCM-KBDSS, where it is the operational-prototype KBDSS. It had been successfully applied (in the validation process) and validated, but further improvements may still be considered based on future validation feedbacks. It has passed the developing KBDSS phase, partially operated (during the validation process), but not function as an operational KBDSS yet. Therefore, the BCM-KBDSS can still be grouped into the operational-prototype KBDSS category (Levitt 1987).

13.3.2 Findings from KBDSS Validation

Section 12.3 earlier illustrated the validation process of the BCM-KBDSS. The process consisted of two stages, which were laboratory testing and field testing (Borenstein 1998). The laboratory testing mainly tested the system's prototype development and utilized test cases for validating the system. The field testing was conducted afterwards to identify the system's general performance. This process utilized the rating approach to evaluate the system's performance attributes. Statistical software and DSS shell software were used for the validation process.

According to Borenstein (1998), validation is an important phase in developing a KBDSS. There is a need to validate complex model-based systems in order to ascertain what a system knows, knows incorrectly, or does not know. This process can be considered as a fundamental step for achieving a more scientific and effective computer based system.

From the validation results, it appeared that from both tests, the BCM-KBDSS was validated with positive feedback from the respondents. The system had also met the evaluation criteria stated in the questionnaire for field testing. The system's perceived utility, performance, completeness and ease of learning were viewed to be significant by the respondents. Furthermore, the respondents agreed that the system is beneficial and that they had positive attitude towards the system.

This validation is in line with McKinsey's (2013) view towards this type of supporting tool. It was viewed that a KBDSS provides an effective foundation for conversations and decision making process in an organization. It provides benefits such as: personal understanding of the topics in the system that relate to the organization and what improvements can be made; sharing opinions amongst the user; and access to the knowledge needed for the organization.

13.4 Summary

All the findings described in this chapter are summarized in Fig. 13.1. This diagram shows the conceptual framework of the study and the findings that are related to the study's research questions. All the findings answer the study's research questions (RQ1 and RQ2).

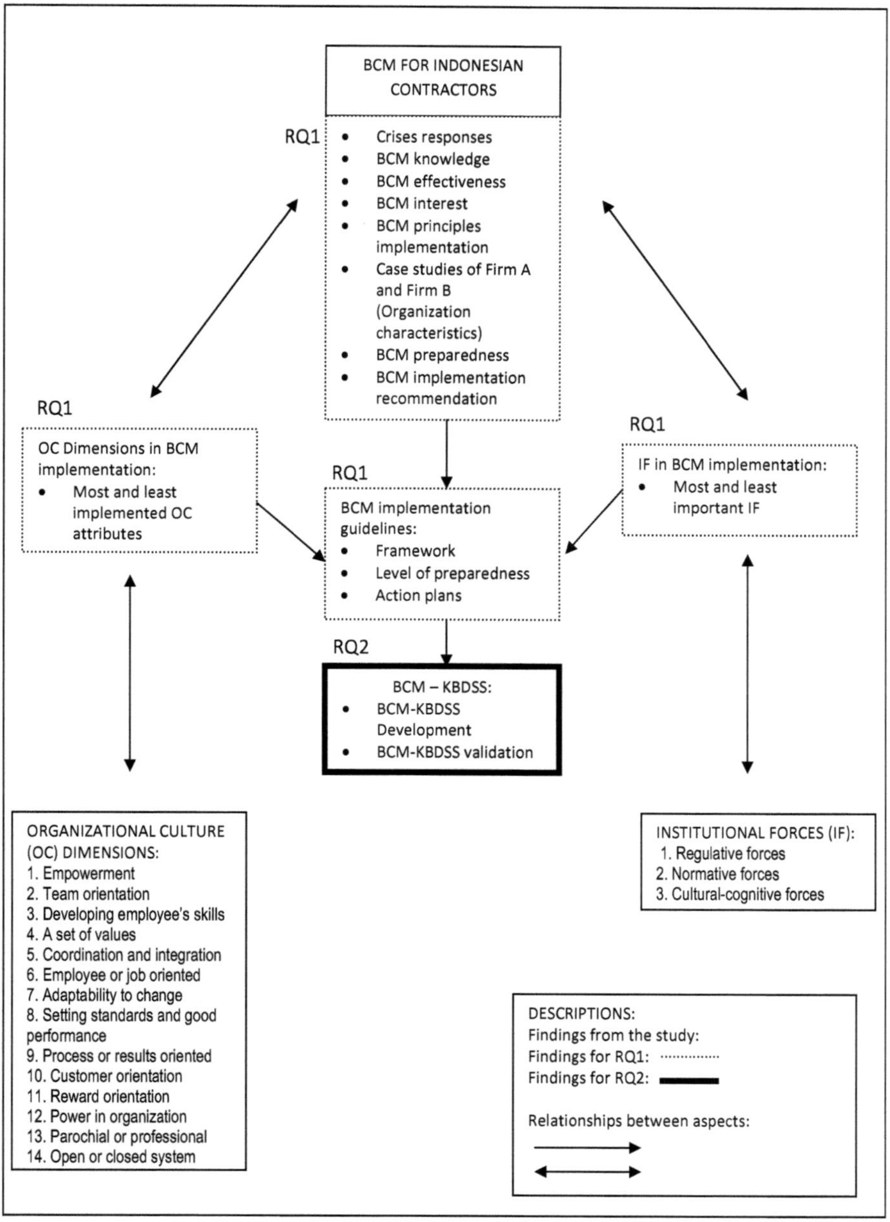

Fig. 13.1 Findings of the study

Chapter 14
Conclusions and Recommendations

14.1 Introduction

This final chapter describes the conclusions and recommendations of the study. Firstly, the summary of the study will be elaborated, followed by the conclusions of the study's research questions, research hypotheses and propositions. The following sections will explain the academic and practical contributions from the study. Afterwards, limitations of the research will be described with recommendations for future studies.

14.2 Summary

A crisis may give various consequences to an organization, whether financial, legal, or operational consequences. In order to overcome a crisis and to continue business as usual, an organization or firm must first have systematic ways and approaches in place. Although some organizations may survive such events due to perseverance, continuity of a business is primarily due to planning and preparation. One of the concepts that can be adopted to overcome crises is BCM, where it is not only focusing on overcoming any crises that occurred, but also considering thoroughly on how to sustain the business in order to obtain its goals and mission (Smith 2003). BCM provides a method for managing any disruption to ensure continuity of service when there is a disruption in the business.

In managing a crisis, decision making is considered as an important part of the process. Critical decisions such as task assignment, resource allocation, guideline to long-term decisions, training and the control capabilities of the organization are necessary for this situation (Yoon et al. 2008). As part of a decision making process in responding to crises, BCM can be designed into an effective model, which is by using computer applications to provide faster and reliable decisions. BCM can be

L.S.R. Supriadi, L. Sui Pheng, *Business Continuity Management in Construction*, Management in the Built Environment, DOI 10.1007/978-981-10-5487-7_14

automated by using a KBDSS. However, it should be pointed out that a KBDSS is not designed to make decisions for users, but rather it provides relevant information in an efficient and easy-to-access format that allows users to make more informed decisions (Arain and Low 2006).

One of the industries that should implement BCM is the construction industry, where it has an important role in a country's economic growth and development. Considering its characteristics and complexities, overcoming crises and threats in order to continue operations in this industry is necessary. The Indonesian construction industry is one of the important sectors in Indonesia. Its role can be seen from the major usage of domestic goods and services that contribute significantly to the country's GDP. This industry is also supported by and connected with a broad spectrum of the nation's legislation and agents. In delivering construction projects, a vast network of relationships between many parties is involved in the process, supported by regulations and management systems.

The country's regulations on construction services have provided for the types of services that can be carried out by the construction firms, types of firms, and the business areas of the services. Until now, the state-owned and large contractors are still dominating the construction market in Indonesia (Sutjipto 1991; NBCSD 2002; SCA 2012). Although there are still weaknesses and threats that are faced by Indonesian contractors, the strengths and opportunities of these firms are quite significant. Moreover, the growths of the country's economic performance and investment opportunities in the infrastructure sector have created a promising business environment for Indonesian contractors. BCM in the context of Indonesian contractors is therefore an important issue to address for the construction industry to continue to play an important role in the economic growth plans of Indonesia.

14.2.1 Research Problems

As firms located over a vast geographical area, which is known as the world's largest archipelago (Raftery et al. 2004), Indonesian contractors have also experienced various threats or crises that have significant impacts on their business activities. These crises had resulted in various levels of impacts, starting from disruptions to business activities, loss of potential markets, loss of productivity and profitability, to the extreme case such as bankruptcy of a firm.

Furthermore, considering the types of crises and the severe impacts that have occurred and have been experienced by Indonesian contractors, the existing responses made for these crises were not fully effective for safeguarding the business continuity of these firms. To become resilient and capable of providing an effective response to such threats, Indonesian contractors should start to adopt a systematic management concept in their organizations. BCM provides this framework, where based on its definition, it builds resilience and the capability for an effective response that safeguards the interests of its key stakeholders, reputation, brand and value creating activities (Elliott et al. 2010). There are several benefits that can be gained by Indonesian contractors in adopting the BCM concept, and developing BCM in an automated form using a KBDSS that can assist the firm in reaching an effective decision (Tinston 2010; BSI 2010; Singh et al. 2008).

14.2.2 Knowledge Gap

Although there are no studies yet on Indonesian contractors adopting BCM in their organizations, some existing studies on BCM implementation in other sectors and other construction firms in other countries, the importance of organizational culture and institutional perspectives in adopting a concept in an organization, and the benefits of a KBDSS may provide relevant findings for this study.

Adopting a new concept like BCM is not a straightforward process. There are issues to consider before implementing the concept into the firm. Previous studies had shown that the immediate motivation for a firm in adopting a concept or system comes from institutional forces (IF) and organizational culture (OC). Based on these considerations, before implementing BCM, the Indonesian contractor's OC and IF should be identified in order to determine whether these elements support or do not support BCM implementation.

Before adopting and implementing a concept, an organization should also analyze its level of preparedness towards the concept. This assessment helps to identify strengths and areas for improvement. Based on these analyses, the organization can further invest its resources to implement the concept accordingly. This process can be developed in the form of a KBDSS. The KBDSS can be used as a supporting tool to assess the organization's level of preparedness and to provide the knowledge needed by management to develop BCM.

14.2.3 Research Objectives and Research Designs Revisited

Based on the research problems and the knowledge gap, BCM is needed for implementation by Indonesian contractors in order to prepare them to overcome crises or threats. Therefore, the research questions for this study are:

1. What are the BCM implementation guidelines for different levels of preparedness for Indonesian contractors (RQ1)?
2. How can the BCM implementation guidelines for Indonesian contractors be automated as a KBDSS (RQ2)?

The research questions can be answered through the research objectives, which are:

1. To identify Indonesian contractors' knowledge about BCM (RO1).

2. To identify the significant drivers and hindrances from institutional forces for implementing BCM (RO2).
3. To identify the significant drivers and hindrances from organizational culture dimensions for implementing BCM (RO3).
4. To develop BCM implementation guidelines for different levels of preparedness for Indonesian contractors (RO4).
5. To automate BCM implementation guidelines for Indonesian contractors into a KBDSS (RO5).

In order to achieve the research objectives, extensive literature review on aspects related with the study were conducted. Chapters 2–7 reviewed the literature reviews on: the management of crisis—theories about management, organizational management and crisis management; BCM—its concept and implementation; OC and IF aspects in adopting a concept; mainstream theories implementation by contractors; overview of the Indonesian construction industry; and KBDSS—its formulation, development and validation. These literature reviews and their relationships formed the basis for developing the study's conceptual framework.

The detailed development of the conceptual framework was described in Chap. 8 and the framework was shown in Fig. 8.8. The framework consists mainly of two layers. The outer layer describes the constructs needed to understand the Indonesian contractor's knowledge of BCM (including BCM principles, BCM preparedness criteria, and the characteristics of Indonesian contractors) and the relationships between BCM, OC and IF. All of these aspects were used for developing the BCM implementation guidelines for Indonesian contractors. The second layer, which is the inner layer of the framework, describes the development of BCM implementation guidelines into a KBDSS. In this layer, the constructs needed for KBDSS development (knowledge base acquisition, rules and logic representation, system development, user interface, system prototype and validation) were provided.

Based on the conceptual framework and the research objectives, the research design of this study is summarized below:

- To achieve RO1–RO3, the survey method was used as the research design. In this case, it was used for understanding the Indonesian contractors' knowledge of BCM and gathering information on the significant drivers and hindrances in implementing BCM.
- To achieve RO4 and RO5, surveys and case studies were used as the research design. The surveys were utilized for obtaining knowledge regarding BCM implementation for Indonesian contractors from construction experts, and case studies were used to obtain more in-depth knowledge, particularly from the view of the management level in Indonesian construction firms.

14.3 Conclusions of the Research Questions

The research questions were answered based on the results and findings from the conceptual framework of the study.

14.3.1 Research Question 1

What are the BCM implementation guidelines for different level of preparedness for Indonesian contractors (RQ1)?

The BCM implementation guidelines for different level of preparedness for Indonesian contractors are guidelines developed for the Indonesian contractors to implement BCM in their organizations based on their respective BCM level of preparedness. The guidelines formed an assessment process that provide results of the firm's BCM level of preparedness and is followed by providing the recommended action plans. These guidelines are developed specifically for Indonesian contractors. Therefore, in developing these, four aspects were compiled in the guidelines, which are:

- Indonesian contractor's knowledge about BCM (RO1).
- The significant drivers and hindrances from institutional forces for implementing BCM (RO2).
- The significant drivers and hindrances from organizational culture dimensions for implementing BCM (RO3).
- The three domains of the BCM implementation guidelines framework, which are the BCM principles domain, BCM preparedness criteria domain, and the Indonesian contractors' strategies and business values chain domain (RO4).

The findings from these four aspects (or the four research objectives) are summarized as follows:

A. The identification of Indonesian contractors' knowledge about BCM (RO1)

In achieving this objective, Table 8.1 was utilized earlier where it provides three relevant constructs. The first construct viewed how the firms currently respond to crises based on their knowledge of the types of crises, standard operating procedures (SOP) on handling crises, and their coordination with the stakeholders during the crisis. The second construct identified the firms' knowledge about BCM and whether they have implemented it. The third construct identified the firms' knowledge about BCM principles and whether these principles have been implemented in their organizations. This last construct was viewed to be necessary because an organization may not know the formal concept of BCM, but may have implemented part of the BCM principles in its organization.

Using the proper research design and methodology, the results for this objective were addressed in Chaps. 10 and 13. The summary of these results are as follows:

1. Types of crises that occurred in contractors within the last 5 years

   Table 10.6 describes the top seven crises (from 32 types of crises given in the questionnaire) that occurred in the firms within the last 5 years, which were: access/approval restriction or limitation; delays or uncertainty in resolving disputes; increase in prices of raw materials (unexpected price escalation); changes in regulations and statutory legislation; natural disasters; loss of management personnel or key staff; and lack of component workforce. These types of crises were similar to some factual crises that have occurred and documented over the past years in Indonesia and were also validated by the respondents.
2. List of impacts from crises

   Table 10.7 provides the top three and the bottom three impacts of crises that were viewed as significant to the respondents. These are: delay in work/dissatisfied customers; revenue impact; loss of productivity (the top three impacts); and huge data loss; building evacuation; failure of few systems (the bottom three impacts). These results have also been validated.
3. Crisis SOP elements in firms

   The survey results found that the respondents who had crisis SOP in their firms had used various names for it, such as Crisis SOP, Crisis Management Plan, Business Continuity Plan, and Emergency Plan. This finding shows that the SOPs were named using different terms, and each SOP in each firm may not have the same contents. Furthermore, from the survey results, it was also shown that only 21% of the respondents named their SOP as a BC Plan. This is probably due to the fact that BCM is a relatively newcomer to the business discipline. However, aspects of BCM may have always been present in the firms, under different names or terms.
4. BCM knowledge

   Regarding the respondent's BCM general knowledge, the survey results showed that 87.5% of the respondents did not know about BCM and only 12.5% of them knew the concept. This result is in line with several studies that suggested that BCM is relatively new, particularly in the Asian region.
5. Effectiveness of BCM implemented in firms

   Based on Table 10.11, the results show that the respondents whose firms have implemented BCM had agreed that BCM provides effectiveness to their firms. They mostly agreed that BCM enables the organization to return to normal operations more quickly than otherwise would have been possible after a crisis. Moreover, the survey results also showed that most of the respondents whose firms have implemented BCM did not use any BCM standards. The results were also confirmed by the respondents who validated the survey results. These respondents recommended that the Indonesian contractors should develop their BCM based on BCM standards.
6. Interests in BCM

   It has been found that there is a good degree of interests from the respondents to learn more about BCM. The respondents who were interviewed for validating the survey results also supported this finding. They provided feedbacks on the parties that can provide the BCM workshop/training, which include the ICA,

NBCSD and the Public Works Ministry. In supporting this process, the firm's top management should be the main party to initiate and coordinate the approach.

7. BCM principles implementation in firms

   Table 10.12 describes the rankings of the BCM principles that have been implemented in the survey respondents' firms. The top five ranked BCM principles implemented are from the Risk analysis and Strategy analysis principles. During the survey validation process, these results were also viewed to be significant by the respondents. They confirmed that both analyses have been adopted in construction projects, especially for managing project risks. They also viewed that both analyses are part of risk management, which is mostly implemented in their firms. Moreover, this result also shows that the least implemented principles of BCM are Tests and exercises, Programme management, and BIA. These suggested that BCM have not been holistically implemented by the firms.

B. The identification of the significant drivers and hindrances from institutional forces for implementing BCM (RO2)

RO2 is obtained by utilizing the framework illustrated in Fig. 8.1, where the relationship between BCM and IF was identified. The relationship between institutional forces (IF) and BCM principles is primarily about the contractor's perspective on BCM principles in the context of the institutional framework. It describes whether the contractors view the BCM principles as regulative, normative or cultural-cognitive forces towards implementation. The results from these analyses can derive the drivers and hindrances in implementing the BCM principles. These findings provide good feedback for the Indonesian contractors in developing strategies for initiating BCM.

In determining the significant drivers and hindrances from IF for implementing BCM, Fig. 8.2 shows the variables that would need to be analyzed. The results for this objective are described in Sect. 10.3.4, which shows the meaningful factors from IF that support the implementation of BCM principles. The findings derived from these analyses were discussed in Chap. 13, where Tables 13.1–13.11 showed the most important IF variables and constructs that are the drivers for implementing the BCM principles. The findings are summarized as follows:

- The implementation of each BCM principle can be driven by a variety of IF combinations. This is in line with what Scott (1998, 2008) had observed, where in most institutional forms, not one single institutional force is at work, but varying combinations of IF.
- The implementation of all BCM principles is mostly driven by the normative forces, which is legitimacy. The views for implementing the BCM principles as being appropriate, effective and fair procedures for better planning become the dominant motivation for the firms.
- Risk analysis and review process is the only BCM principle that is mostly driven by two variables, which are shared understanding of compliance of the principle (cultural-cognitive forces) and the legitimacy to conduct the principle

(normative forces). This principle could be driven by cultural-cognitive forces because the firms might view it as a logical process and it is already part of the firms' culture.

- From the survey validation results, the respondents opined that although the regulative forces are not found to be dominant in the survey results, these forces would still drive the way firms implement a system. Regulative forces such as regulation and sanctions from the top management of the firms would still be the reasons for firms to implement a concept like BCM.
- There are two normative forces from the social influence variables which are found to be the least supporting factor for BCM implementation. The firms' competitiveness and stakeholders/clients' requirements are viewed not to be the main reasons for the firms to implement the entire BCM principles.
- The respondents from the survey validation results also opined that cultural-cognitive forces may be the constructs that would least likely motivate firms to implement BCM. Their views were that the culture of discipline and awareness attitudes toward risk/crisis may not yet be dominating in the firms, particularly for their low level employees. These views were not quite in line with what the survey results had shown, where shared understanding of compliance regarding the implementation of BCM principles was found to be one of the main motivations. However, this empirical finding can be a good feedback for the firms when initiating BCM.

C. The identification of the significant drivers and hindrances from organizational culture dimensions for implementing BCM (RO3)

Similar to RO2, RO3 is achieved by utilizing the framework in Fig. 8.1, where the relationship between BCM and OC was identified. The relationship between OC dimensions and BCM principles is generally to determine how the contractor views the BCM principles (in the context of OC dimensions) regarding its level of influence (significant drivers) and level of hindrances for the firm.

Several literatures have shown that OC attributes support BCM implementation and could be applied to the six BCM principles. In determining the significant drivers and hindrance from OC for implementing BCM, Fig. 8.3 presented the variables that would be analyzed. The results for this research objective were explained in Chap. 10, which shows the meaningful factors from OC attributes that have been implemented by the Indonesian contractors which could support the BCM principles. Furthermore, Chap. 13 discussed the findings from these analyses, where Table 13.12 illustrated the OC dimensions that have been implemented to support the BCM principles. From this table, the OC dimensions that were viewed as highly implemented in each BCM principle are presented. The findings from this section are as follows:

- It is suggested that Indonesian contractors have implemented the 14 OC dimensions in their organizations, and these dimensions could support the implementation of the relevant BCM principles.

- The OC dimension that is found to be the least implemented dimension in these firms is the professional attribute, particularly in supporting the BCM principles: the involvement of experts and BCM committee in risk review (RA2); the involvement of experts, employees from related BUs and key staffs in BIA (BIA2); determining staff to support the recovery strategy and providing training and awareness programme (S2); as well as BCM training and awareness programmes (PM2). This could be a hindrance for the firms to implement BCM effectively.
- The respondents from the survey validation results added another OC attribute that could become a hindrance, namely the culture of documentation (part of setting standards and good performance). They viewed that the construction firm employees have low and incomprehensive skills for documenting reports. Based on this issue, it is suggested that when implementing BCM, the firms should select competent employees to conduct this documentation process.

D. The development of BCM implementation guidelines for different levels of preparedness for Indonesian contractors (RO4)

In achieving RO4, Fig. 8.4 was utilized which shows the framework for developing the guidelines. The domains in this framework are the BCM principles domain (see Table 8.4), the BCM preparedness criteria domain (see Table 8.5), and the Indonesian contractors strategies and business value chain domain (based on surveys, case studies and interview results).

Furthermore, the two aspects that form the guidelines are explained in Chap. 11, starting from the assessment process per BCM principle that provided results of the firm's BCM level of preparedness to the recommended action plans for the respective BCM level of preparedness. Along with providing the technical aspects in implementing BCM (based on each BCM principle), the action plans also provide the non-technical aspects compiled from the results of the BCM-IF, BCM-OC, case studies and interview results.

There are several findings from this study area that are worth mentioning:

- The case studies have shown some similarities and differences between the two types of firms (namely Firm A which is a state-owned entity and Firm B which is a private entity), which provide an overview of the Indonesian contractors' characteristics. These studies also provide the following findings:
  1. Support and strong commitment from the firms' senior management are needed. A top-down direction is the preferable approach in initiating BCM.
  2. Regulations to comply with safety, quality and environmental standards in the construction sector have been established. However, there is no regulation that is specifically designed for implementing BCM in the construction sector.
  3. All the lessons learned about overcoming crises from both types of firms would lead to the need for adopting BCM, where they can be addressed by implementing the BCM principles.

- The results from interviewing the experts from the Indonesian construction industry are synthesized into the BCM implementation guidelines, which are summarized as follows:
    1. Aspects such as the firm's resilience, competitive advantage, providing systematic data back-ups, economic recession preparedness, floods preparedness, securing the firm's assets, and better coordination with stakeholders are viewed as the benefits of implementing BCM.
    2. BCM implementation is needed for Indonesian contractors, and the efforts to socialize and initiate it should be fostered by the top management and the regulators in the construction industry.
    3. BCM certification should be applied in this sector, and the firms could refer to the existing BCM standards from various countries or by using ISO's standards for BCM (ISO 22301).
    4. Relating to BUs in contractors, the project site unit, accounting, human resources departments, and contracts division are viewed to be essential.
    5. In determining which CBFs are essential, functions for documentation, accounting, project assets, human resources management, and communication were all mentioned by the respondents.
    6. The most significant crises for the construction business are economic/financial crisis and natural disasters.
- There are four BCM levels of preparedness used for the guidelines, which are Undeveloped, Beginner, Moderate and Comprehensive. Four main aspects that describe each level are: the deliverable (the outcome of BCM in the firm), management support, policies (policies about BCM), comprehension (the firms' understanding about BCM) and BCM principles (the practices that have been implemented by the firm) (see Table 11.24).
- The technical action plans consisted of steps and practices needed to be implemented, which were grouped into the six BCM principles. The knowledge base was acquired from the literature on BCM principles, and supported by knowledge from the case studies and interview results.
- As for the non-technical action plans, the knowledge base was acquired from the study's empirical findings, which were from the surveys, case studies and interview results. This second section consisted of: recommendations to be considered for implementing the principle (based on case studies and interview results); Organizational culture (OC) attributes that can support the action plans (based on significant OC attributes results from surveys); and several drivers to implement the BCM principle (based on significant Institutional Forces (IF) results from surveys.

14.3.2 Research Question 2

How can the BCM implementation guidelines for Indonesian contractors be automated as a KBDSS? (RQ2)

The BCM implementation guidelines for Indonesian contractors can be automated as a KBDSS by compiling the knowledge from the guidelines into a DSS shell, complete with rules, logic representation and user interface design. A summary of the processes for this development includes the following:

A. The automation of BCM implementation guidelines for Indonesian contractors using a KBDSS (RO5)

This final research objective is achieved as shown in Table 8.8 which highlights the variables needed for developing the BCM-KBDSS. The development phases started from the knowledge base (KB) acquisition through the compiled data, development of the rules and logical representation, user interface design and BCM-KBDSS compilation in a DSS shell, and finally validation. These results are summarized as follows:

- BCM-KBDSS used fuzzy logic for its inference engine because the assessment of the BCM level of preparedness is based on qualitative inputs using linguistic variables. Fuzzy logic ascertains the linguistic qualification of the process and provides the results. The fuzzy set theory used in this method presented vague knowledge and allowed mathematical operators to be applied to its domain (Guedria et al. 2009; Ahmed and Capretz 2006; Xia et al. 2011).
- The detailed process of developing the BCM-KBDSS started by developing the decision situation scenario, where every process and decision situation scenario in the system is structured. Thereafter, the system was developed into more details, where the concept and coding of the structured situation diagram were constructed. These diagrams were completed with rules in a detailed diagram named the dependency diagram. These diagrams formed the basis for incorporating all of the KB and relevant rules for the BCM-KBDSS in Chap. 12.
- In compiling the BCM-KBDSS in a DSS shell, the system is designed to achieve the capability to interface with the system's user in a user-friendly process. This is mainly to ensure the effectiveness of the system to be used by the user.
- The BCM-KBDSS is an operational-prototype KBDSS, which means that it has passed through the KBDSS development phase, partially operated (during the validation process), but not fully functioning as an operational KBDSS yet.
- The validation results showed that the BCM-KBDSS has met the evaluation criteria, where the system's perceived utility, performance, completeness and ease of learning are opined to be significant by the respondents. There were several constructive feedbacks that relate to the importance of understanding BCM before using the system and some improvements relating to the system's interface. Nonetheless, the respondents mostly agreed that the BCM-KBDSS is beneficial for guiding the Indonesian contractors to implement BCM.

14.4 Conclusions of the Research Hypotheses and Propositions

This section elaborates on the conclusions of the research hypotheses and propositions. Three hypotheses were tested and two propositions were provided.

14.4.1 Testing Hypotheses H1, H2 and H3

Hypothesis 1 (H1) *Indonesian contractors have not heard of BCM, but have emergency plans in place in their organizations.*
The survey results in Chap. 10 showed that a majority of the respondents have not heard of BCM (87.5%). Although most of them have not heard of BCM, Table 10.8 illustrates that the respondents have emergency plans in place in their firms, where a majority of them have emergency response, communication procedure, EOC, and restoration and recovery in place (with the percentages of 75%, 89.3%, 53.6% and 71.4% respectively). It appears that although the term BCM is new to the respondents, nevertheless some aspects of BCM (that relate to emergency plans) have been in place in their firms. These results are in line with what BCI (2010) suggested, where aspects of BCM may have always been present in the firms, under different names or terms. Based on these findings and consideration, this hypothesis (H1) is supported.

Hypothesis 2 (H2) *The institutional forces that consist of Regulative, Normative and Cultural-cognitive forces are the significant drivers for Indonesian contractors to implement BCM.*

- *Sub-Hypothesis 2.1 (H2.1): The regulative forces that consist of rules, laws, sanctions, gains, losses and consequences are significant drivers for Indonesian contractors to implement BCM.*
- *Sub-Hypothesis 2.2 (H2.2): The normative forces that consist of personal morality, social influence and legitimacy are significant drivers for Indonesian contractors to implement BCM.*
- *Sub-Hypothesis 2.3 (H2.3): The cultural-cognitive force that consists of shared understanding of compliance is a significant driver for Indonesian contractors to implement BCM.*

The study suggests that the meaningful factors from the institutional forces (IF) that support the BCM principles implementation consist of the three pillars, which are the regulative, normative and cultural-cognitive forces. These results in the study present a variety of IF combinations of the three pillars as the significant drivers to implement each BCM principle. These findings are in line with most institutional forms, where not one single institutional force is at work, but varying combinations of IF (Scott 1998, 2008). Based on these findings and consideration, Hypothesis 2 (H2) and its sub-hypotheses (H2.1, H2.2 and H2.3) are supported.

Hypothesis 3 (H3) *The organizational culture dimensions that consist of 14 attributes are the significant drivers for Indonesian contractors to implement BCM.*

The results in the study suggested that the Indonesian contractors have experienced the 14 OC dimensions in their firms, and these dimensions significantly drive the implementation of the relevant BCM principles. The descriptions of the OC dimensions that have been experienced and that support the relevant BCM principles are explained as follows:

- Empowerment was opined to have been implemented and would mostly support: the involvement of experts and BCM committee in risk review (RA2); the involvement of experts, employees from related BUs and key staffs in BIA (BIA2); and strategy analysis for maintaining the operations of CBFs (S1). (H3.1 is supported)
- Team orientation was viewed to have been implemented and would mostly support: the involvement of experts and BCM committee in risk review (RA2); the involvement of experts, employees from related BUs and key staffs in BIA (BIA2); the process determining staff to support the recovery strategy and providing training and awareness programme (S2); periodic tests and exercises to ensure that the BC plan is viable and workable (TE1); and BCM training and awareness programmes for all staff and related external parties (PM2). (H3.2 is supported)
- The survey results showed that developing employee's skills was opined to have been implemented and would mostly support: the process determining staff to support the recovery strategy and providing training and awareness programme (S2); and BCM training and awareness programmes for all staff and related external parties (PM2). (H3.3 is supported)
- A set of values was viewed to have been implemented and would mostly support: BCM programme management (PM1); and BCM training and awareness programmes for all staff and related external parties (PM2). (H3.4 is supported)
- Coordination and integration were viewed to have been implemented and would mostly support: risk analysis and cost benefit analysis (RA1); a detailed risk review (RA3); strategy analysis for maintaining the operations of CBFs (S1); developing a detailed BC plan (BCP1); periodic tests and exercises to ensure that the BC plan is viable and workable (TE1); and BCM programme management (PM1). (H3.5 is supported)
- Employee oriented was found to have been implemented and would mostly support risk analysis and cost benefit analysis (RA1). (H3.6 is supported)
- Adaptability to change was viewed to have been implemented and would mostly support: risk analysis and cost benefit analysis (RA1); a detailed risk review (RA3); BIA (BIA1); strategy analysis for maintaining the operations of CBFs (S1); developing a detailed BC plan (BCP1); and periodic tests and exercises to ensure that the BC plan is viable and workable (TE1). (H3.7 is supported)

- Setting standards and good performance was opined to have been implemented and would mostly support: developing a detailed BC plan (BCP1); and BCM programme management (PM1). (H3.8 is supported)
- Process oriented was viewed to have been implemented and would mostly support a detailed risk review (RA3). (H3.9 is supported)
- The survey results showed that customer orientation was viewed to have been implemented and would mostly support: developing a detailed BC plan (BCP1); and BCM programme management (PM1). (H3.10 is supported)
- Reward orientation was found to have been implemented and would mostly support BCM programme management (PM1). (H3.11 is supported)
- Power in organization was opined to have been implemented and would mostly support BCM programme management (PM1). (H3.12 is supported)
- Professional attribute was viewed to have been the least implemented dimension for these BCM principles: the involvement of experts and BCM committee in risk review (RA2); the involvement of experts, employees from related BUs and key staffs in BIA (BIA2); determining staff to support the recovery strategy and providing training and awareness programme (S2); and BCM training and awareness programmes (PM2). (H3.13 is not supported; the results of the study show that professional attribute is not a significant driver for Indonesian contractors to implement BCM.)
- Open system was opined to have been implemented and would mostly support the involvement of experts and BCM committee in risk review (RA2). (H3.14 is supported)

Based on these findings and considering that 13 out of the 14 sub-hypotheses are supported, it can be concluded that Hypothesis 3 (H3) appears to be primarily supported.

14.4.2 Propositions for RO4 and RQ2

To reiterate, Research Objective 4 states that: *A BCM implementation guidelines model for Indonesian contractors can be developed for different levels of preparedness based on the BCM principles and Indonesian contractor's characteristics.*

Based on the results in Chaps. 11 and 13, the BCM implementation guidelines model for Indonesian contractors had been developed for different levels of preparedness based on the BCM principles and Indonesian contractor's characteristics.

In response to Research Question 2, the study states that: *A BCM implementation guidelines model for Indonesian contractors can be automated into a KBDSS.*

Based on the results in Chaps. 12 and 13, a BCM implementation guidelines model for Indonesian contractors had been automated into a KBDSS, named the BCM-KBDSS. The BCM-KBDSS is an operational-prototype system and has been successfully validated.

14.5 Academic Contributions

This study provides several academic contributions based on its results and findings, which are:

- It has been found that there are no studies yet on Indonesian contractors adopting BCM in their organizations. Therefore, this study fills the knowledge gap. The findings from this study can be beneficial and insightful for the research and development of the Indonesian construction sector, particularly on knowledge about adopting and implementing management concepts in construction firms.
- The study provides new perspectives on how the institutional theory can be a useful platform for explaining why construction firms do or do not wish to implement the BCM principles. The relationship between BCM and IF has been established from this empirical research, where a variety of IF combinations of the three pillars are the significant drivers to implement each BCM principle. Different components of the institutional theory explain how these elements are created, diffused, adopted and adapted over time and how they fall into decline. Collectively, this theory provides a framework to interpret the corresponding implementation issues.
- Considering that currently the construction industry is lacking in organizational cultural studies (Coffey 2010), this study further contributes to this knowledge area. From the study, the most implemented and the least implemented OC attributes that relate to BCM implementation by Indonesian contractors were identified. This study surfaced the OC identifiers that could support BCM principles for implementation.
- The conceptual framework collates various schools of thoughts that relate to the management of crisis, BCM, construction management, institutional theory, organizational culture, and DSS development. It provides a framework for an organization to adopt a concept such as BCM. The various aspects that have been collated in this framework were found to be relevant for providing guidelines for implementing BCM. Moreover, even though construction firms have specific characteristics that differ from each other, it has been found that the management of crisis, construction management, institutional theory, organizational culture and DSS development are equally applicable to all of them, particularly in supporting BCM implementation.
- The development of the BCM-KBDSS in this study contributes to knowledge on automation in the construction industry. An automated system such as the BCM-KBDSS can be utilized for the management's decision making process when conducting business continuity planning. This computer-based system supports decision making by aiding knowledge storage, retrieval, transfer and application by facilitating knowledge access that benefits the management process. Moreover, it is always important to understand that this tool does not replace the BCM analysis phase, the BC planner, the selected methodology or the ongoing management commitment to the process. It is developed purely to assist with developing and maintaining business continuity programmes.

14.6 Practical Contributions

This study also provides practical contributions that are beneficial to the Indonesian contractors, which are:

- The study contributes a framework for understanding BCM for Indonesian contractors, through the cultural and institutional approach. The framework can be utilized as part of a concept adoption process for the organizations.
- The study also contributes knowledge on implementing BCM based on the technical and non-technical aspects. As mentioned by Low (1998), fulfilling technical requirements of a management system is one important aspect, but the other aspect which focuses on non-technical attributes (such as IF and OC dimensions) must not be overlooked. The non-technical attributes should also be reviewed due to its role in helping to promote an integrative environment in the management system, which in this case is BCM.
- The most practical contribution of this study is the BCM-KBDSS. The Indonesian contractors can gain more insights about BCM and able to promptly assess their level of BCM preparedness, followed by receiving recommendations regarding action plans or steps for developing BCM in their firms. In addition, the implementation guidelines provided by the BCM-KBDSS can assist the firms to develop better coordination with their stakeholders that lead to a competitive advantage and marketing value for the firms.

14.7 Limitations of the Research

The study is subject to a number of limitations relating to research methodology. The research team is fully aware of the limitations. Hence, every effort had been made to minimize the errors that may occur. The limitations of the study are as follows:

- The study has chosen surveys as the research design in order to understand the perceptions of the respondents. Their perceptions may be subjective views, which may or may not reflect reality. The personal bias of the respondents may reduce the objectivity of the responses, which is one of the weaknesses of using a survey. Nevertheless, in the absence of a better method, the survey has provided a better understanding of the relationships between BCM and aspects such as OC dimensions and institutional forces, and a framework that forms a basis for further studies.
- The total response rate from the surveys is <50%, which may not yield a very high degree of generalizability. However, despite its limitation, this study could still serve to provide meaningful evidence and insights into the research topic, which should be beneficial to future studies.

- This study has also conducted two case studies of Indonesian contractors as the research design. Ideally, an exploratory research should be conducted by collecting data in the form of observations, document analysis and interviews in the actual environment. For this study, interviews and document analysis had been conducted. However, due to legal concerns, time constraints and other practical limitations, this study only conducted observations to both firms for <3 weeks. Aspects or variables from the study that need longer observations to surface may not therefore be fully obtained. In overcoming this, an additional approach to gathering the data was conducted, which is by interviewing the relevant experts to verify and complement the results from the case studies.

14.8 Recommendations for Future Studies

The study presented in this book has developed BCM implementation guidelines in the form of a KBDSS which also identified the relationships between BCM implementation and the IF—OC attributes in Indonesian contractors. There are opportunities to extend this research further from various aspects. The recommendations for further studies are as follows:

- This study has provided the BCM implementation guidelines in an automated BCM-KBDSS. The future study can apply this tool to an Indonesian contractor and gain more detailed analyses by monitoring the BCM implementation in that specific organization over a period of time. Furthermore, this study can also measure whether the guidelines can improve the firm's BCM preparedness level. The research design that can be used for this study is a case study.
- In accordance with the scope of this study, the large contractors were chosen as the samples because these firms are mostly involved in major construction projects and are dominating the construction market in Indonesia. This study has not chosen SME contractors as the samples, and therefore the readiness of their IT infrastructure and human resources toward BCM-KBDSS have not been studied. Nonetheless, evaluating the Indonesian SME contractors' preparedness toward BCM and using BCM-KBDSS as the tool will be highly recommended for further studies. Surveys and case studies can similarly be used as the research design for future studies of SME contractors in Indonesia.
- Currently, this study has identified the significant OC attributes that support BCM implementation for Indonesian contractors. The next phase that can be conducted is to study whether the firms have actually implemented these OC attributes in their working environment. This future study should focus mainly on the organizational culture study in the construction firms. An ethnography study can be utilized for measuring this aspect, which is an ideal approach for in-depth cultural attributes studies.

Appendix A

Summary of OC dimensions from various studies

OC dimensions (keywords and indicators)	Hofstede OC dimension (2005)[a]	Denison (2000)	Cheung et al. (2011)[b]	Low (1998)	Osland et al. (2001), Luthans (2008)
1. Empowerment: Empowering individual to manage their own work; Decision making by individual; Value employee's ideas; Employee's input in major decisions; Employee participation in decision making (members' participation); Confidence in employee members	X Kotler and Heskett (1992), Schols (1987), Williams et al. (1993)	X	X Lynch (2006)	X	X
2. Team orientation: Cooperation among employees; Team contributions; Amicable opinions; Commitment to team	X Kotler and Heskett (1992), Schols (1987), Williams et al. (1993)	X	X Lynch (2006) Harrison (1972), Handy (1993)	X	
3. Developing employee's skills: Performance improvement; High expectations of performance		X	X	X	X

(continued)

L.S.R. Supriadi, L. Sui Pheng, *Business Continuity Management in Construction*, Management in the Built Environment, DOI 10.1007/978-981-10-5487-7

OC dimensions (keywords and indicators)	Hofstede OC dimension (2005)[a]	Denison (2000)	Cheung et al. (2011)[b]	Low (1998)	Osland et al. (2001), Luthans (2008)
4. A set of values: Clear goals-direction-approach-strategic intentions; Action = Goals; Shared visions		X	X Kotler and Heskett (1992), Schols (1987) and Williams et al. (1993) Hansen and Wernerfelt (1989)	X	X Lynch (2006)
5. Coordination and integration: Resolve problems effectively; Inter-department collaboration; Information sharing; Agreement on critical issues; Different functions work together; Decisions made by groups/individuals; Trust atmosphere; Managing conflict; Good communication	X	X	X Kotler and Heskett (1992), Schols (1987), Williams et al. (1993) Lynch (2006) Harrison (1972), Handy (1993)	X	
6. Employee/job oriented: Consider employee's welfare; Level of job pressures; Friendly workplace (warmth and support)	X Kotler and Heskett (1992), Schols (1987), Williams et al. (1993)			X	X
7. Adaptability to change: Creating change; Developing innovation-knowledge-capabilities; Innovation orientation; Employee's resistance to change		X	X Lynch (2006)	X	

(continued)

OC dimensions (keywords and indicators)	Hofstede OC dimension (2005)[a]	Denison (2000)	Cheung et al. (2011)[b]	Low (1998)	Osland et al. (2001), Luthans (2008)
8. Setting standards and good performance: Organizational structuring-meeting times-cost conscious (tight or loose control organization); Emphasize good performance; Set of performance standards	X		X Kotler and Heskett (1992), Schols (1987), Williams et al. (1993) Lynch (2006) Harrison (1972), Handy (1993)		X
9. Process/results oriented: Conformity to procedures and rules	X Kotler and Heskett (1992), Schols (1987), Williams et al. (1993)				X
10. Customer orientation: Focus on customer	X	X			
11. Reward orientation: Team and members accountability; Emphasize on reward; Performance-based rewards; Equitable reward			X Harrison (1972) and Handy (1993) Hansen and Wernerfelt (1989)		X
12. Power in organizations: Sources of power; expert (knowledge-based); leadership				X Kotler and Heskett (1992), Schols (1987) and Williams et al. (1993); Lynch (2006); Harrison (1972), Handy (1993)	X Lynch (2006)

(continued)

OC dimensions (keywords and indicators)	Hofstede OC dimension (2005)[a]	Denison (2000)	Cheung et al. (2011)[b]	Low (1998)	Osland et al. (2001), Luthans (2008)
13. Parochial/professional: Cover either social/family background plus job competence or job competence only	X				
14. Open/closed system: Organization open to newcomers or not; How people fit into the organization	X Kotler and Heskett (1992), Schols (1987) and Williams et al. (1993)				

Note: The references inside the matrix were related with the reference in its column header. Example: Lynch (2006)s OC is related with Osland et al. (2001) and Luthans (2008). It means that the value of Lynch's (2006) OC is similar with Osland et al. (2001) and Luthans' (2008) OC values
[a]Including studies from GLOBE's dimensions
[b]Including studies from: Bettinger (1989), Cameron and Quinn (1999)

Appendix B

Introduction

This questionnaire is part of a research study on Business Continuity Management (BCM) for Indonesian contractors. BCM provides a method for managing any disruption to ensure continuity of service when there is a disruption of business. Moreover, previous studies had shown that the immediate motivation for a firm in adopting a concept such as BCM comes from institutional forces and organizational culture.

The objectives of this questionnaire survey are (1) to collect information about BCM knowledge from Indonesian contractors, and (2) to collect information about factors that drives and hinders BCM implementation by Indonesian contractors. Based on these, the questionnaire comprises four sections. The results of this survey will be analyzed to achieve the objective of the study, which is to develop a BCM guideline model for Indonesian contractors.

L.S.R. Supriadi, L. Sui Pheng, *Business Continuity Management in Construction*, Management in the Built Environment, DOI 10.1007/978-981-10-5487-7

Section A: Firm Information

A1. Name: ____

A2. Contact number: ______________________

A3. Email address: ______________________

A4. Age: ____________

A5. Sex: Male / Female

A6. Education level:
☐ High School ☐ Diploma ☐ Bachelor degree ☐ Master degree

☐ Doctoral degree ☐ Others: ____________

A7. Company name: ____

A8. Job title: ____

A9. Department: ____

A10. Position Level:
☐ Board of Directors/CEO ☐ Senior Management/General Manager

☐ Manager/Assistant Manager ☐ Senior Executive/Sr. Staff ☐ Executive/Staff

☐ Junior Executive/Jr. Staff ☐ Others: __________________

A11. Professional membership: _____________________________________

A12. How long have you been employed in this firm:
☐ < 5 years ☐ 5 – 10 years ☐ 11 – 15 years ☐ 16 – 20 years ☐ > 20 years

A13. How long have you been in the construction industry:
☐ < 5 years ☐ 5 – 10 years ☐ 11 – 15 years ☐ 16 – 20 years ☐ > 20 years

A14. Some people spend most of the time in the office; some have to go to the project site frequently, while others have to be stationed at the project site. In one month, how do you spend your time?
_______ % in the office _______% at the project site _______% at _________
(please note that the percentages should add up to 100%)

A15. Firm type:
a) ☐ State-owned ☐ Private-owned
b) ☐Small ☐ Medium ☐ Large

Section B: Crises Response and General BCM Knowledge

This section asks for information about:

- The contractor's current response towards crises
- The respondent's BCM knowledge

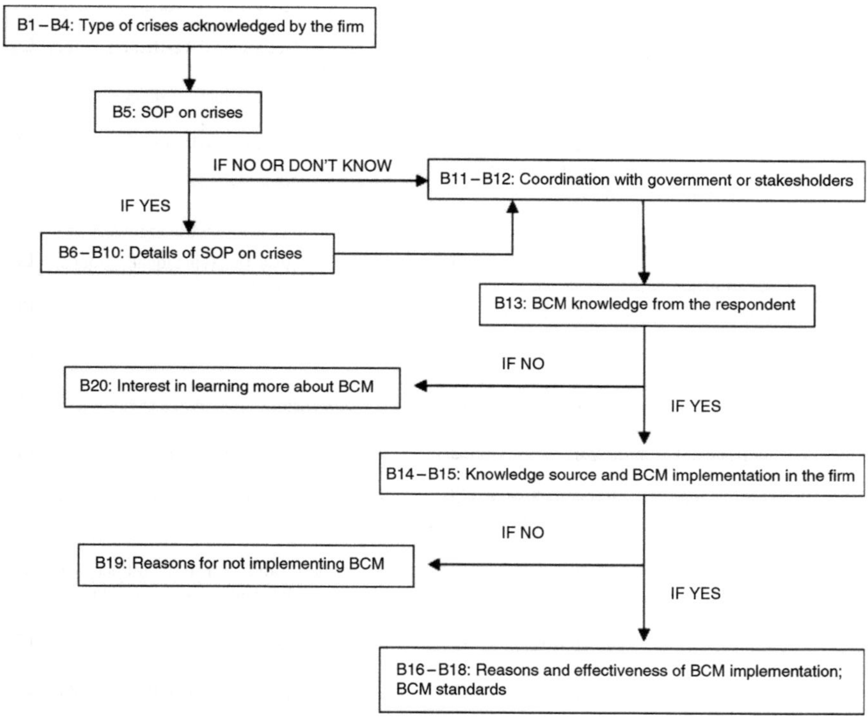

1. The contractor's current response towards crises

Based on the Oxford Dictionary (2006), the word crisis means a time of great danger, difficulty, or confusion when problems must be solved or important decisions must be made. A crisis can be a threat because it has the possibility of trouble, danger, or disaster. Moreover, Barton (1993) described crisis as an abnormal situation or perception which threatens or disrupts operations, staff, customers, or the reputation of the organization.

a. Type of crises acknowledged by the firm

B1. Number of significant business disruptions/crises in the last 5 years:

☐ No disruption ☐ Once ☐ 2 times ☐ 3 times

☐ > 3 times ☐ Don't know

B2. Causes of the significant business disruptions in the last 5 years (tick one or more):

➢ Loss of public goodwill, reputation, image due to malicious contamination	☐	➢ Theft	☐
➢ Action by environmentalist/pressure groups (protests)	☐	➢ Sabotage	☐
➢ Financial crisis	☐	➢ Access/approval restriction or limitation	☐
➢ Corruption scandal	☐	➢ Serious product defects or component failures	☐
➢ Political instability i.e. those leading to changes of project scope or cancellations, sanctions and embargoes, tighter exchange controls, repatriation of funds	☐	➢ Loss of confidential information	☐
➢ Changes in regulations and statutory legislation	☐	➢ Natural disasters (earthquake, floods, tsunami, etc)	☐
➢ Client insolvency leaves outstanding debts for work done	☐	➢ War	☐
➢ Delays or uncertainty in resolving disputes	☐	➢ Riot	☐
➢ Increase in price of raw materials (unexpected price escalation)	☐	➢ Terrorism	☐
➢ Shortage of key materials	☐	➢ Lack of component workforce	☐
➢ Material damages (during deliveries or faulty products from manufacturer)	☐	➢ Loss of management personnel or key staff	☐
➢ Subcontractor insolvency	☐	➢ Kidnap and ransom (effect on production and share price of loss of key personnel)	☐
➢ Breakdown of key construction plant	☐	➢ Strikes, labour disputes	☐
➢ Serious accidents in a project	☐	➢ Health issues (flu pandemic, SARS, etc)	☐
➢ Fire	☐	➢ No disruption occurred	☐
➢ Explosion	☐	➢ Others:____________________	☐

B3. Impacts from the significant business disruptions in the last 5 years (tick one or more):

- ➢ Loss of productivity ☐
- ➢ Delay in work/dissatisfied customers ☐
- ➢ Business closed temporarily/infrastructure down and unable to carry out business functions ☐
- ➢ Employees were not able to reach the office ☐
- ➢ Revenue impact ☐
- ➢ Huge data loss and client impact ☐
- ➢ Building had to be evacuated ☐
- ➢ Failure of few systems ☐
- ➢ Others: ____ ☐

B4. Effectiveness in dealing with the significant business disruptions in the last 5 years:
☐ Very effectively ☐ Somewhat effectively ☐ Not effectively ☐ Don't know

b. SOP on crises

B5. In responding to the crises/business disruptions, does the firm have any plans in place?
☐ Yes ☐ No ☐ Don't Know
If answered "Yes", please proceed to question B6 – B12.
If answered "No" or "Don't Know", please proceed to question B11 – B12.

B6. Does the plan have evacuation or emergency response procedure to stabilize the situation during crisis?
☐ Yes ☐ No ☐ Don't Know

B7. Does the plan have communication procedure during crisis/disruption?
☐ Yes ☐ No ☐ Don't Know

B8. Does the plan provide for procedures to move to alternative facilities or emergency operations center (EOC)?
☐ Yes ☐ No ☐ Don't Know

B9. Does the plan have recovery and restoration procedures to return of all business operations from temporary measures adopted during recovery to supporting normal business requirements after the disaster?
☐ Yes ☐ No ☐ Don't Know

B10. What are the names of the plan(s) to respond to crises? (please choose one or more):
☐ Crisis Standard Operation and Procedures (SOP)
☐ Crisis Management Plan
☐ Emergency Plan
☐ Business Continuity Plan
☐ Others: ____________________

c. Coordination with government or stakeholders

B11. In responding to crises, the firm establishes relationships with (please choose one or more):
☐ External agencies (police, fire department, hospitals, etc)
☐ Government
☐ Others: ____________________

B12. In any incident/disruption situations, the firm has communication procedures with (please choose one or more):
☐ Customers (clients)
☐ Partners
☐ Employees
☐ Regulatory authorities
☐ Suppliers / Vendors
☐ Others: ____________________

2. BCM knowledge

The Business Continuity Institute (Business Continuity Institute 2007b) defines Business Continuity Management (BCM) as an act of anticipating incidents that will affect mission-critical functions and processes for the organization, and ensuring that it responds to any incident in a planned and rehearsed manner.

Moreover, the Singapore Standard for BCM (Spring Singapore 2008) looked at this concept as a holistic management process that identifies potential impacts which threaten an organization and provides a framework for building resilience and the capability for an effective response that safeguards the interests of its key stakeholders, reputation, brand and value-creating activities.

The key objectives of an effective BCM strategy should be to ensure the safety of staff, maximize the defense of the organization's reputation and brand image, minimize the impact of business continuity events (including crises) on customers or clients, prevent impact beyond the organization, demonstrate effective and efficient governance to the media, markets and stakeholders, protect the organization's assets, and meet insurance, legal and regulatory requirements (Smith 2003).

The main process of a BCM is Business Continuity Planning (BCP). Business Continuity Planning (BCP) refers to the identification and protection of critical business processes and resources required to maintain an acceptable level of business, protection of such resources, and preparation of procedures to ensure the survival of the organization in times of business disruptions (Low et al. 2008a, b). This is regarded as the main process due to its vital output for the firm in handling disruptions and overcoming crises. This planning process will be followed by regular monitoring and updates.

B13. Do you know about the BCM concept before? (Yes/No)

☐ Yes ☐ No

If answered "Yes", please proceed to question B14 – B15.
If answered "No", please proceed to question B20.

B14. Where do you obtain the knowledge of this concept from? (please choose one or more):

☐ Information from the company
☐ Media (newspaper; magazine; television; internet)
☐ Academic reference (books; journals)
☐ Workshop/Seminar/Coursework
☐ Colleague/mentor
☐ Others: ____________________

B15. Is there any form of BCM implemented within your course of work/organization structure?

☐ Yes ☐ No

If answered "Yes", please proceed to question B16 – B18.
If answered "No" , please proceed to question B19.

B16. Why there is a need for BCM to be implemented into your course of work? (please choose one or more):

☐ Part of risk management
☐ Company requirements
☐ To protect revenues and minimize potential penalties
☐ To protect the firm and ensure long-term survival
☐ Response to past disruptions or crises
☐ Market practice within the industry
☐ Regulations within Indonesia
☐ Others: ____________________

B17. Please rate the following statements that relate to the effectiveness of BCM that are implemented in your firm:

	Disagree				Agree
1. It can effectively reduce the impact of the disruption	1	2	3	4	5
2. It enables continued delivery of key products and services without interruption to clients	1	2	3	4	5
3. It enables the organization to return to normal operations more quickly than otherwise would have been possible	1	2	3	4	5
4. It helps to cope with the immediate effects of an incident on employees	1	2	3	4	5
5. It supports employees after recovery	1	2	3	4	5

B18. Is the form of BCM implemented in your firm modeled after any relevant standard related to BCM?

☐ SS540: 2008 (Singapore Standards)	☐ NFPA1600: 2007 (US Standards)
☐ TR19: 2005 (Singapore Terms of Reference)	☐ ISO/IEC 27001: 2005
☐ BS 25999: 2006 (British Standards)	☐Others:________________________
☐ ANZ5050: 2009 (Australia/New Zealand Standards)	☐ None of the above

B19. What are the reasons for not implementing BCM in your firm? (please choose one or more):

☐ Waste of time/resources
☐ Not a priority
☐ Lack of awareness
☐ Deal with disruption as and when it happens
☐ Continuity issues are already covered in other plans (write the name of the plan: ________________________)
☐ Lack of support from management
☐ Not a statutory requirement
☐ Lack of expertise
☐ Too expensive and overly complex to develop
☐ Others: ____

B20. How interested would you be in obtaining more knowledge about BCM if the relevant assistance (financial budget; training sessions/workshops; others) were given?

Not interested				Very interested
1	2	3	4	5

CONTINUE TO SECTION C →

Section C: Significant Drivers and Hindrances for Implementing BCM for Indonesian Contractors

An organization/firm may not know or not have implemented the formal concept of BCM, but it may have implemented parts of BCM principles in its organization.

To implement BCM, each organization must identify the threats and assess their resulting impacts. BCM needs to address issues and concerns in six main principles in the following order: (1) Risk analysis and review; (2) Business Impact Analysis (BIA); (3) Strategy Analysis; (4) Development of BC plan; (5) Tests and exercises and (6) Programme management.

Please state your opinions on the questions related to these principles.

This section asks for the respondent's opinions about significant drivers and hindrances for implementing BCM for Indonesian contractors, based on BCM's main principles.

For each BCM principle, there are three (3) groups of questions, which are:

1. Questions on whether the principle is implemented or not.
2. Views on the principle (agree or disagree on the provided statements) (C(A)).
3. Factors that are viewed as drivers or hindrances in implementing the principle (C(B)).

The flow of the questions on Section C is as follows:

Risk Analysis and Review:
- C1
- C2A-C2B: *Conducting risk analysis and cost benefit analysis.*
- C3A-C3B: *The involvement of experts and BCM committee in risk review.*
- C4A-C4B: *Conducting a detailed risk review.*

↓

Business Impact Analysis:
- C5
- C6A-C6B: *Conducting business impact analysis.*
- C7A-C7B: *The involvement of experts, employees from related BUs and key staffs in BIA.*

↓

Strategy Analysis:
- C8
- C9A-C9B: *Conducting* strategy *analysis.*
- C10A-C10B: *Determining staff to support the recovery* strategy.

↓

Development of BC Plan:
- C11
- C12A-C12B: *Developing the detailed BC plan.*

↓

Tests and exercises
- C13
- C14A-C14B: *Providing periodic tests (to verify the capability of BC plan) and exercises (to train and condition BC team members*

↓

Programme Management:
- C15
- C16A-C16B: *Analyzing the ongoing efforts and activities to maintain the effectiveness of its BCM*
- C17A-C17B: *Conducting BCM training and awareness programmes.*

Risk Analysis and Review

The threats to an organization can be identified through a risk analysis and review of its internal operations and external operating environment.

C1. Is this principle being implemented in the firm?

☐ Yes ☐ No

C2. *Conducting risk analysis and cost benefit analysis.*

C2A. Please circle the number that indicates the extent to which you agree with each of the statement.

No.	Descriptions	1 (Strongly disagree); 2 (Disagree); 3 (Neutral); 4 (Agree); 5 (Strongly agree)				
1.	It can be easily integrated with other management systems in the organization that can provide positive gains for the organization	1	2	3	4	5
2.	It helps to improve the organization's procedures for facing crisis	1	2	3	4	5
3.	The cost for not conducting risk analysis and cost benefit analysis in the organization is not small	1	2	3	4	5
4.	There are strict regulations from the government or the organization's management about conducting risk analysis and cost benefit analysis on every planning process in the company. Not implementing them can result in receiving sanctions	1	2	3	4	5
5.	Analyzing internal and external risk may improve the employee's safety and welfare	1	2	3	4	5
6.	These analyses are insisted upon by the stakeholder and clients	1	2	3	4	5
7.	These analyses are conducted due to concern for reputation	1	2	3	4	5
8.	These analyses are conducted in order to increase the company's competitiveness	1	2	3	4	5
9.	Peer groups and the top management fully support the risk analysis and cost benefit analysis process	1	2	3	4	5
10.	These analyses are viewed as fair procedures for better planning in the organization	1	2	3	4	5
11.	These analyses are conducted in the organization (in the departments and business units) due to the awareness of potential risks that may occur during the operational process in the organization	1	2	3	4	5
12.	These analyses are already part of the company culture, where every plan has its risk and cost-benefit analysis in place	1	2	3	4	5

C2B. Please circle the number in both columns (Drivers and Hindrances) that indicates the extent to which you agree with each of the statement.

No.	Descriptions	1 (Strongly disagree); 2 (Disagree); 3 (Neutral); 4 (Agree); 5 (Strongly agree)									
		Drivers[a]					Hindrances[b]				
1.	The company has a high level of risk avoidance	1	2	3	4	5	1	2	3	4	5
2.	Unfamiliar situations should be managed and identified	1	2	3	4	5	1	2	3	4	5
3.	Following organizational procedures is essential	1	2	3	4	5	1	2	3	4	5
4.	The level of tolerance for ambiguity and chaos should be low	1	2	3	4	5	1	2	3	4	5
5.	During risk analysis and review, important decisions are made by groups or committee (as a consensus)	1	2	3	4	5	1	2	3	4	5

(continued)

No.	Descriptions	1 (Strongly disagree); 2 (Disagree); 3 (Neutral); 4 (Agree); 5 (Strongly agree)									
		Drivers[a]					Hindrances[b]				
6.	Good communication is essential and information sharing is encouraged	1	2	3	4	5	1	2	3	4	5
7.	The members of the committee or business units are able to reach agreement on critical issues	1	2	3	4	5	1	2	3	4	5
8.	The company should be open to alternative solutions	1	2	3	4	5	1	2	3	4	5
9.	The company should encourage creative and innovative ideas for determining the risk response and treatments	1	2	3	4	5	1	2	3	4	5
10.	The company should allocate sufficient resources for implementing innovative ideas	1	2	3	4	5	1	2	3	4	5
11.	The company should have a high level of cost-consciousness	1	2	3	4	5	1	2	3	4	5
12.	The company should provide a detailed set of performance standards that can support the analyses	1	2	3	4	5	1	2	3	4	5
13.	The company takes the responsibility for its employee's welfare (the risk treatments are considering this factor)	1	2	3	4	5	1	2	3	4	5
14.	The company has a main focus in meeting the client's needs and satisfaction. It views the analyses as part of the process to deliver products that meet the client's needs	1	2	3	4	5	1	2	3	4	5

[a]Drivers: In my opinion, this attribute is viewed as a supporting factor for conducting risk analysis and cost benefit analysis
[b]Hindrances: In my opinion, this attribute is not implemented in the company

C3. *The involvement of experts and BCM committee in risk review.*

C3A. Please circle the number that indicates the extent to which you agree with each of the statement.

No.	Descriptions	1 (Strongly disagree); 2 (Disagree); 3 (Neutral); 4 (Agree); 5 (Strongly agree)				
1.	It helps to improve the company's procedure in preparing and handling crises.	1	2	3	4	5
2.	Not involving the experts and BCM committee in risk review may lead to negative impact (not able to develop a rigorous risk analysis—more cost for risk treatment).	1	2	3	4	5
3.	It is insisted upon by the stakeholder and clients	1	2	3	4	5
4.	Involving the experts and BCM committee in risk review are for improving the company's reputation	1	2	3	4	5
5.	Involving the experts and BCM committee in risk review are for increasing the company's competitiveness in the industry	1	2	3	4	5

(continued)

No.	Descriptions	1 (Strongly disagree); 2 (Disagree); 3 (Neutral); 4 (Agree); 5 (Strongly agree)				
6.	Peer groups and the top management fully support the involvement of the experts and BCM committee in risk review process	1	2	3	4	5
7.	It is viewed as a fair procedure for better planning in the organization	1	2	3	4	5
8.	It is viewed as appropriate and effective for better planning (particularly for its risk management) in the organization	1	2	3	4	5
9.	It is done to comply with the regulation that is taken for granted	1	2	3	4	5
10.	It is part of the awareness of potential risks in the organization	1	2	3	4	5
11.	It is already part of the company culture	1	2	3	4	5

C3B. Please circle the number in both columns (Drivers and Hindrances) that indicates the extent to which you agree with each of the statement.

No.	Descriptions	1 (Strongly disagree); 2 (Disagree); 3 (Neutral); 4 (Agree); 5 (Strongly agree)									
		Drivers[a]					Hindrances[b]				
1.	The organization emphasizes on job competence only for each of the position in the company. Social and family backgrounds are not taken into account	1	2	3	4	5	1	2	3	4	5
2.	The company supports the member of the risk review team to participate in the decision-making process	1	2	3	4	5	1	2	3	4	5
3.	The member of the risk review team is enabled to decide or solve the problems within his/her sphere of responsibility or authority	1	2	3	4	5	1	2	3	4	5
4.	Amicable opinions and ideas exchange between members are facilitated by the company	1	2	3	4	5	1	2	3	4	5
5.	The experts and the BCM committee's commitment to the risk review team are highly valued by the company	1	2	3	4	5	1	2	3	4	5
6.	The company emphasizes on team contributions.	1	2	3	4	5	1	2	3	4	5
7.	The experts (from external party) and the BCM committee (from internal party) fit very well into the risk review team, and their opinions are appreciated	1	2	3	4	5	1	2	3	4	5

[a]Drivers: In my opinion, this attribute is viewed as a driver for supporting the experts and BCM committee to participate in risk review
[b]Hindrances: In my opinion, this attribute is not implemented in the company

C4. *Conducting a detailed risk review that examines and assesses the availability of critical equipment, technology, and facilities for BU/CBF (including location of facilities, essential utilities and telecommunications, transportation to premises and physical security of premises).*

C4A. Please circle the number that indicates the extent to which you agree with each of the statement.

No.	Descriptions	1 (Strongly disagree); 2 (Disagree); 3 (Neutral); 4 (Agree); 5 (Strongly agree)				
1.	It can be easily integrated with other management systems in the organization that can provide positive gains for the organization	1	2	3	4	5
2.	It helps to improve the organization's procedures for facing crisis	1	2	3	4	5
3.	Not conducting a detailed risk review may lead to negative impact (not able to develop a rigour risk analysis—more cost for risk treatment)	1	2	3	4	5
4.	It helps to improve the employees/worker's health, safety and welfare	1	2	3	4	5
5.	It is insisted upon by stakeholders, including clients	1	2	3	4	5
6.	It is conducted to sustain and improve the company's reputation	1	2	3	4	5
7.	It is conducted to increase the company's competitiveness	1	2	3	4	5
8.	Peer groups and the top management fully support the process	1	2	3	4	5
9.	It is viewed as fair procedures for better planning in the organization	1	2	3	4	5
10.	It is viewed as appropriate and effective for better planning (particularly for risk management) in the organization	1	2	3	4	5
11.	It is done to comply with the regulation that is taken for granted	1	2	3	4	5
12.	It is part of the awareness of potential risks in the organization	1	2	3	4	5
13.	It is already part of the company culture	1	2	3	4	5

C4B. Please circle the number in both columns (Drivers and Hindrances) that indicates the extent to which you agree with each of the statement.

No.	Descriptions	1 (Strongly disagree); 2 (Disagree); 3 (Neutral); 4 (Agree); 5 (Strongly agree)									
		Drivers[a]					Hindrances[b]				
1.	The company has a high level of risk avoidance	1	2	3	4	5	1	2	3	4	5
2.	Following organizational procedures is essential	1	2	3	4	5	1	2	3	4	5
3.	Unfamiliar situations are managed and identified	1	2	3	4	5	1	2	3	4	5
4.	Inter-departmental collaboration is encouraged for developing the detailed risk review	1	2	3	4	5	1	2	3	4	5
5.	Good communication is essential and information sharing is encouraged	1	2	3	4	5	1	2	3	4	5
6.	The company is able to create adaptive ways to meet changing needs, particularly to allocate resources/infrastructure when crises occur	1	2	3	4	5	1	2	3	4	5
7.	The company is able to allocate sufficient resources for implementing innovative ideas	1	2	3	4	5	1	2	3	4	5

(continued)

No.	Descriptions	1 (Strongly disagree); 2 (Disagree); 3 (Neutral); 4 (Agree); 5 (Strongly agree)									
		Drivers[a]					Hindrances[b]				
8.	The company has high level of cost-consciousness	1	2	3	4	5	1	2	3	4	5
9.	The company provides a detailed set of performance standards (regarding its equipments, technologies and facilities) that can support the risk review	1	2	3	4	5	1	2	3	4	5

[a]Drivers: In my opinion, this attribute is viewed as a supporting factor for conducting a detailed risk review that examines and assesses the availability of critical equipment, technology, and facilities for BU/CBF
[b]Hindrances: In my opinion, this attribute is not implemented in the company

Business Impact Analysis (BIA)

The potential impact of threats on an organization and its ability to continue business operations and service can be obtained by conducting a business impact analysis. This would include, where possible, the loss impact from both a number of days of business disruption and financial consequences.

C5. Is BIA being implemented in the firm?

☐ Yes ☐ No

C6. *Conducting business impact analysis (BIA) that focuses on assessing the impact of losses if the corresponding business operations and processes are disrupted.*

C6A. Please circle the number that indicates the extent to which you agree with each of the statement.

No.	Descriptions	1 (Strongly disagree); 2 (Disagree); 3 (Neutral); 4 (Agree); 5 (Strongly agree)				
1.	It can be easily integrated with other management systems in the organization that can provide positive gains for the organization	1	2	3	4	5
2.	It helps to improve the organization's procedures for facing crisis	1	2	3	4	5
3.	Not conducting BIA may lead to negative impact (not able to develop a rigorous analysis—more cost for recovery process)	1	2	3	4	5
4.	There are strict regulations from the government or the organization's management about conducting BIA in the company. Not implementing BIA can result in sanctions	1	2	3	4	5

(continued)

No.	Descriptions	1 (Strongly disagree); 2 (Disagree); 3 (Neutral); 4 (Agree); 5 (Strongly agree)				
5.	It may improve the employee's safety and welfare	1	2	3	4	5
6.	It is insisted upon by stakeholders and clients	1	2	3	4	5
7.	It is conducted to sustain and improve the company's reputation	1	2	3	4	5
8.	BIA is conducted to increase the company's competitiveness	1	2	3	4	5
9.	Peer groups and the top management fully support the BIA that focuses on assessing the impact of losses if the corresponding business operations and processes are disrupted	1	2	3	4	5
10.	BIA is viewed as fair procedures for better planning in the organization	1	2	3	4	5
11.	BIA is viewed as appropriate and effective for better planning in the organization	1	2	3	4	5
12.	It is done to comply with the regulation that is taken for granted	1	2	3	4	5
13.	It is part of the awareness of potential impacts and losses in the organization if any disruption occurs	1	2	3	4	5
14.	It is already part of the company culture	1	2	3	4	5

C6B. Please circle the number in both columns (Drivers and Hindrances) that indicates the extent to which you agree with each of the statement.

No.	Descriptions	1 (Strongly disagree); 2 (Disagree); 3 (Neutral); 4 (Agree); 5 (Strongly agree)									
		Drivers[a]					Hindrances[b]				
1.	The company has a high level of risk avoidance	1	2	3	4	5	1	2	3	4	5
2.	Unfamiliar situations are managed and identified	1	2	3	4	5	1	2	3	4	5
3.	Following organizational procedures is essential	1	2	3	4	5	1	2	3	4	5
4.	Important decisions are made by groups as a consensus	1	2	3	4	5	1	2	3	4	5
5.	Good communication is essential and information sharing is encouraged	1	2	3	4	5	1	2	3	4	5
6.	The members of the committee or business units are able to reach agreement on critical issues	1	2	3	4	5	1	2	3	4	5
7.	The company is open to alternative solutions	1	2	3	4	5	1	2	3	4	5
8.	The company encourages creative and innovative ideas for determining the responses due to business impacts	1	2	3	4	5	1	2	3	4	5
9.	The company is able to allocate sufficient resources for implementing innovative ideas	1	2	3	4	5	1	2	3	4	5
10.	The company has high level of cost-consciousness, particularly for analyzing the impacts due to disruptions	1	2	3	4	5	1	2	3	4	5
11.	The company provides a detailed set of performance standards (regarding its BUs) that could support the BIA	1	2	3	4	5	1	2	3	4	5

(continued)

No.	Descriptions	1 (Strongly disagree); 2 (Disagree); 3 (Neutral); 4 (Agree); 5 (Strongly agree)									
		Drivers[a]					Hindrances[b]				
12.	The company has many internal structuring (procedures) in the organization and considers that meeting times is essential	1	2	3	4	5	1	2	3	4	5
13.	The company takes the responsibility for its employee's welfare (the responses to losses from BIA are considering this factor)	1	2	3	4	5	1	2	3	4	5
14.	The company has a main focus in meeting the client's needs and satisfaction. It views the analysis as part of the process to deliver products that meet the client's needs	1	2	3	4	5	1	2	3	4	5

[a]Drivers: In my opinion, this attribute is viewed as a supporting factor for conducting business impact analysis (BIA) that focuses on assessing the impact of losses if the corresponding business operations and processes are disrupted
[b]Hindrances: In my opinion, this attribute is not implemented in the company

C7. *The involvement of experts, employees from related BUs and key staffs in BIA. BIA participants selected from related functional area/BU, including experts from the business, technology, financial, facility and legal domains. BIA will be conducted by BCM coordinator, committee, and experts.*

C7A. Please circle the number that indicates the extent to which you agree with each of the statement.

No.	Descriptions	1 (Strongly disagree); 2 (Disagree); 3 (Neutral); 4 (Agree); 5 (Strongly agree)				
1.	It helps to improve the company's procedure in preparing and handling crises	1	2	3	4	5
2.	Not involving experts, employees from related BUs and key staffs in BIA may lead to negative impact (not able to develop a rigorous BIA—more cost for impact recoveries)	1	2	3	4	5
3.	It is insisted upon by the stakeholder and clients	1	2	3	4	5
4.	It is for improving the company's reputation	1	2	3	4	5
5.	It is for increasing the company's competitiveness in the industry	1	2	3	4	5
6.	Peer groups and the top management fully support the involvement of experts, employees from related BUs and key staffs in BIA process	1	2	3	4	5
7.	It is viewed as a fair procedure for better planning in the organization	1	2	3	4	5
8.	It is viewed as appropriate and effective for better planning in the organization	1	2	3	4	5
9.	It is done to comply with the regulation that is taken for granted	1	2	3	4	5
10.	It is part of the awareness of potential risks in the organization	1	2	3	4	5
11.	It is already part of the company culture	1	2	3	4	5

C7B. Please circle the number in both columns (Drivers and Hindrances) that indicates the extent to which you agree with each of the statement.

No.	Descriptions	1 (Strongly disagree); 2 (Disagree); 3 (Neutral); 4 (Agree); 5 (Strongly agree) Drivers[a]					Hindrances[b]				
1.	The organization emphasize on job competence only for each of the position in the company. Social and family backgrounds are not taken into account	1	2	3	4	5	1	2	3	4	5
2.	The company supports the member of the BIA team to participate in the decision-making process	1	2	3	4	5	1	2	3	4	5
3.	The member of the BIA team is enabled to decide or solve the problems within his/her sphere of responsibility or authority	1	2	3	4	5	1	2	3	4	5
4.	Amicable opinions and ideas exchange between members are facilitated by the company	1	2	3	4	5	1	2	3	4	5
5.	The experts, employees from related BUs and key staff's commitment to the BIA team are highly valued by the company	1	2	3	4	5	1	2	3	4	5
6.	The company emphasize on team contributions.	1	2	3	4	5	1	2	3	4	5
7.	The experts (from external party) employees from related BUs and key staff (from internal party) fit very well into the BIA team, and their opinions are appreciated	1	2	3	4	5	1	2	3	4	5

[a]Drivers: In my opinion, this attribute is viewed as a driver for supporting the experts, employees from related BUs and key staffs to participate in BIA
[b]Hindrances: In my opinion, this attribute is not implemented in the company

Strategy Analysis

The organization determines the appropriate strategies to safeguard its interests. These strategies can be preventive or pre-emptive in nature.

C8. Is strategy analysis being implemented in the firm?

☐ Yes ☐ No

C9. *Conducting strategy analysis for maintaining the operations of CBFs that cover pre-incident preparedness, response and recovery.*

C9A. Please circle the number that indicates the extent to which you agree with each of the statement.

No.	Descriptions	1 (Strongly disagree); 2 (Disagree); 3 (Neutral); 4 (Agree); 5 (Strongly agree)				
1.	It can be easily integrated with other management systems in the organization that can provide positive gains for the organization	1	2	3	4	5
2.	It helps to improve the organization's procedures for facing crisis	1	2	3	4	5
3.	Not conducting strategy analysis may lead to negative impact (not able to develop a rigorous analysis—more cost for recovery process)	1	2	3	4	5
4.	There are strict regulations from the government or the organization's management about conducting strategy analysis in the company. Not implementing them can result in receiving sanctions	1	2	3	4	5
5.	It may improve the employee's health, safety and welfare	1	2	3	4	5
6.	It is insisted upon by stakeholders and clients	1	2	3	4	5
7.	It is conducted to sustain and improve the company's reputation	1	2	3	4	5
8.	It is conducted to increase the company's competitiveness	1	2	3	4	5
9.	Peer groups and the top management fully support the strategy analysis that covers pre-incident preparedness, response and recovery	1	2	3	4	5
10.	It is viewed as fair procedures for better planning in the organization	1	2	3	4	5
11.	It is viewed as appropriate and effective for better planning in the organization	1	2	3	4	5
12.	It is to comply with the regulation that is taken for granted	1	2	3	4	5
13.	It is part of the awareness of potential impacts and losses in the organization if any disruption occurs	1	2	3	4	5
14.	It is already part of the company culture	1	2	3	4	5

C9B. Please circle the number in both columns (Drivers and Hindrances) that indicates the extent to which you agree with each of the statement.

		1 (Strongly disagree); 2 (Disagree); 3 (Neutral); 4 (Agree); 5 (Strongly agree)									
No.	Descriptions	Drivers[a]					Hindrances[b]				
1.	The company is open to alternative solutions	1	2	3	4	5	1	2	3	4	5
2.	The company encourages creative and innovative ideas for determining the pre-incident preparedness, response and recovery	1	2	3	4	5	1	2	3	4	5
3.	The company is able to allocate sufficient resources for implementing innovative ideas	1	2	3	4	5	1	2	3	4	5
4.	Important decisions are made by groups as a consensus	1	2	3	4	5	1	2	3	4	5

(continued)

No.	Descriptions	1 (Strongly disagree); 2 (Disagree); 3 (Neutral); 4 (Agree); 5 (Strongly agree)									
		Drivers[a]					Hindrances[b]				
5.	Good communication is essential and information sharing is encouraged	1	2	3	4	5	1	2	3	4	5
6.	The members of the committee or business units are able to reach agreement on critical issues	1	2	3	4	5	1	2	3	4	5
7.	The company has high level of cost-consciousness, particularly for analyzing possible strategies for pre-incident preparedness, response and recovery	1	2	3	4	5	1	2	3	4	5
8.	The company provides a detailed set of performance standards (regarding its BUs) that can support the strategy analysis	1	2	3	4	5	1	2	3	4	5
9.	The company has many internal structuring (procedures) in the organization and considers that meeting times is essential (particularly on CBFs level of operation)	1	2	3	4	5	1	2	3	4	5
10.	The company values employee's ideas in analyzing possible strategies for pre-incident preparedness, response and recovery	1	2	3	4	5	1	2	3	4	5
11.	Employees' participation in decision-making process during strategy analysis is encouraged	1	2	3	4	5	1	2	3	4	5
12.	The company takes the responsibility for its employee's welfare (particularly for analyzing strategies for pre-incident preparedness, response and recovery)	1	2	3	4	5	1	2	3	4	5
13.	The company has a main focus in meeting the client's needs and satisfaction. It views the analysis as part of the process to deliver products that meet the client's needs	1	2	3	4	5	1	2	3	4	5
14.	The company has a high level of risk avoidance	1	2	3	4	5	1	2	3	4	5
15.	Unfamiliar situations are managed and identified	1	2	3	4	5	1	2	3	4	5
16.	Following organizational procedures is essential	1	2	3	4	5	1	2	3	4	5

[a]Drivers: In my opinion, this attribute is viewed as a supporting factor for conducting strategy analysis that covers pre-incident preparedness, response and recovery
[b]Hindrances: In my opinion, this attribute is not implemented in the company

C10. *Determining staff to support the recovery strategy and providing training and awareness programme.*

C10A. Please circle the number that indicates the extent to which you agree with each of the statement.

No.	Descriptions	1 (Strongly disagree); 2 (Disagree); 3 (Neutral); 4 (Agree); 5 (Strongly agree)				
1.	This process can be easily integrated with other management systems in the organization that can provide positive gains for the organization	1	2	3	4	5
2.	This process helps to improve the organization's procedures for facing crisis	1	2	3	4	5
3.	Not determining staff to support the recovery strategy and providing training and awareness programme may lead to negative impact (not able to develop a rigorous analysis—more cost for recovery process)	1	2	3	4	5
4.	It may improve the employee's welfare	1	2	3	4	5
5.	It is insisted upon by stakeholders and client.	1	2	3	4	5
6.	It is conducted to sustain and improve the company's reputation	1	2	3	4	5
7.	It is conducted to increase the company's competitiveness	1	2	3	4	5
8.	Peer groups and the top management fully support the process to determine staff to support the recovery strategy and provide training and awareness programme	1	2	3	4	5
9.	It is viewed as fair procedures for better planning in the organization	1	2	3	4	5
10.	It is viewed as appropriate and effective for better planning in the organization	1	2	3	4	5
11.	It is conducted to comply with the regulation that is taken for granted	1	2	3	4	5
12.	It is part of the awareness of potential impacts and losses in the organization if any disruption occurs	1	2	3	4	5
13.	It is already part of the company culture	1	2	3	4	5

C10B. Please circle the number in both columns (Drivers and Hindrances) that indicates the extent to which you agree with each of the statement.

No.	Descriptions	1 (Strongly disagree); 2 (Disagree); 3 (Neutral); 4 (Agree); 5 (Strongly agree)									
		Drivers[a]					Hindrances[b]				
1.	The organization emphasize on job competence only for each of the position in the company. Social and family backgrounds are not taken into account	1	2	3	4	5	1	2	3	4	5
2.	The company continually invests in the development of employee's skills in order to be competitive and able to meet on-going business needs	1	2	3	4	5	1	2	3	4	5
3.	Guidance for employee's performance improvement is provided in the company	1	2	3	4	5	1	2	3	4	5

(continued)

No.	Descriptions	1 (Strongly disagree); 2 (Disagree); 3 (Neutral); 4 (Agree); 5 (Strongly agree)									
		Drivers[a]					Hindrances[b]				
4.	A high degree of cooperation among employees (particularly among staffs for the recovery strategy) in working toward common goals	1	2	3	4	5	1	2	3	4	5
5.	Amicable opinions and ideas exchange between members are facilitated by the company	1	2	3	4	5	1	2	3	4	5
6.	The staffs to support the recovery strategy are highly valued by the company	1	2	3	4	5	1	2	3	4	5
7.	The company emphasize on team contributions	1	2	3	4	5	1	2	3	4	5
8.	The staffs to support the recovery strategy fit very well into the BCM team, and their opinions are appreciated	1	2	3	4	5	1	2	3	4	5

[a]Drivers: In my opinion, this attribute is viewed as a supporting factor for determining staff to support the recovery strategy and providing training and awareness programme
[b]Hindrances: In my opinion, this attribute is not implemented in the company

Development of BC Plan

A detailed business continuity plan should be formulated to indicate the resources and capabilities required of the organization to prepare, respond, and recover from potential threats.

C11. Is BC Plan being implemented in the firm?

☐ Yes ☐ No

C12. ***Developing the detailed BC plan (compiled from BCM principles 1,2,3), including its emergency response, EOC (Emergency Operations Center) plan, and identified CBFs (Critical Business Functions) with their RTOs (Recovery Time Objectives) and RPOs (Recovery Point Objectives). BC plan caters for 4 sets of activities (pre-incident preparation; response to incident/emergency/disaster; recovery and resumption of CBFs; restoration and return of all business operations from temporary measures adopted during recovery to supporting normal business requirements after disaster).***

C12A. Please circle the number that indicates the extent to which you agree with each of the statement.

No.	Descriptions	1 (Strongly disagree); 2 (Disagree); 3 (Neutral); 4 (Agree); 5 (Strongly agree)				
1.	This process can provide positive gains for the organization	1	2	3	4	5
2.	It helps to improve the organization's procedures for facing crisis	1	2	3	4	5
3.	Not examining the detailed BC plan may lead to negative impact (not able to develop a rigorous analysis—more cost for recovery process)	1	2	3	4	5
4.	There are strict regulations from the government or the organization's management about examining the detailed BC plan in the company. Not implementing them can result in receiving sanctions	1	2	3	4	5
5.	It may improve the employee's health, safety, and welfare	1	2	3	4	5
6.	It is insisted upon by stakeholders and clients	1	2	3	4	5
7.	It is conducted to sustain and improve the company's reputation	1	2	3	4	5
8.	It is conducted to increase the company's competitiveness	1	2	3	4	5
9.	Peer groups and the top management fully support the process to examine the detailed BC plan	1	2	3	4	5
10.	It is viewed as a fair procedure for better planning in the organization	1	2	3	4	5
11.	It is viewed as appropriate and effective for better planning in the organization	1	2	3	4	5
12.	It is conducted to comply with the regulation that is taken for granted	1	2	3	4	5
13.	It is part of the awareness of potential impacts and losses in the organization if any disruption occurs	1	2	3	4	5
14.	It is already part of the company culture	1	2	3	4	5

C12B. Please circle the number in both columns (Drivers and Hindrances) that indicates the extent to which you agree with each of the statement.

No.	Descriptions	1 (Strongly disagree); 2 (Disagree); 3 (Neutral); 4 (Agree); 5 (Strongly agree)									
		Drivers[a]					Hindrances[b]				
1.	The company has many internal procedures in the organization	1	2	3	4	5	1	2	3	4	5
2.	The company provides a detailed set of performance standards (regarding its BUs) that support the process of examining the detailed BC plan	1	2	3	4	5	1	2	3	4	5
3.	The company has high level of cost-consciousness, particularly for examining the detailed BC plan (compiled from BCM principle 1, 2, 3), including its emergency response, EOC plan, and identified CBFs with their RTOs and RPOs. Also, the company considers that meeting deadlines is essential (particularly on CBFs level of operation)	1	2	3	4	5	1	2	3	4	5
4.	Unfamiliar situations are managed and identified	1	2	3	4	5	1	2	3	4	5

(continued)

No.	Descriptions	1 (Strongly disagree); 2 (Disagree); 3 (Neutral); 4 (Agree); 5 (Strongly agree)									
		Drivers[a]					Hindrances[b]				
5.	Following organizational procedures is essential	1	2	3	4	5	1	2	3	4	5
6.	Important decisions are made by groups as a consensus	1	2	3	4	5	1	2	3	4	5
7.	Good communication is essential and information sharing is encouraged	1	2	3	4	5	1	2	3	4	5
8.	The members of the committee or business units are able to reach agreement on critical issues	1	2	3	4	5	1	2	3	4	5
9.	The company is able to create adaptive ways to meet changing needs	1	2	3	4	5	1	2	3	4	5
10.	The company is able to allocate sufficient resources for implementing innovative ideas	1	2	3	4	5	1	2	3	4	5
11.	The company takes the responsibility for its employee's welfare (particularly for examining the detailed BC plan)	1	2	3	4	5	1	2	3	4	5
12.	The manager in the company has a high concern for the procedures in the operation of the BUs	1	2	3	4	5	1	2	3	4	5
13.	The company has a main focus in meeting the client's needs and satisfaction. It views the BC plan review as part of the process to deliver products that meet the client's needs	1	2	3	4	5	1	2	3	4	5

[a]Drivers: In my opinion, this attribute is viewed as a supporting factor for developing the detailed BC plan (compiled from BCM principle 1, 2, 3), including its emergency response, EOC plan, and identified CBFs with their RTOs and RPOs
[b]Hindrances: In my opinion, this attribute is not implemented in the company

Tests and Exercises

An established BC plan shall be validated by implementing tests and exercises. These are done to highlight errors or omissions and verify if the resources committed are accessible, available and adequate for efficient and effective recovery. It also verifies whether the staff is familiar with recovery procedures, and whether the BC plan meets its recovery objectives.

C13. Are tests and exercises being implemented in the firm?

☐ Yes ☐ No

C14. ***Providing periodic tests (to verify the capability of BC plan) and exercises (to train and condition BC team members - to highlight any weaknesses in the operation and effectiveness of BC plan in following corrective actions) to ensure that the BC plan is viable and workable.***

C14A. Please circle the number that indicates the extent to which you agree with each of the statement.

No.	Descriptions	1 (Strongly disagree); 2 (Disagree); 3 (Neutral); 4 (Agree); 5 (Strongly agree)				
1.	These processes can be easily integrated with other management systems in the organization that can provide positive gains for the organization	1	2	3	4	5
2.	These processes help to improve the organization's procedures for facing crisis	1	2	3	4	5
3.	Not providing periodic tests and exercises to ensure that the BC plan is viable and workable may lead to negative impact (not able to develop a rigorous analysis—more cost for recovery process)	1	2	3	4	5
4.	There are strict regulations from the government or the organization's management about providing periodic tests and exercises. Not implementing them can result in receiving sanctions	1	2	3	4	5
5.	These processes may improve the employee's health, safety, and welfare	1	2	3	4	5
6.	These processes insisted upon by stakeholders and clients	1	2	3	4	5
7.	These processes are conducted to sustain and improve the company's reputation	1	2	3	4	5
8.	These processes are conducted to increase the company's competitiveness	1	2	3	4	5
9.	Peer groups and the top management fully support the periodic tests and exercises to ensure that the BC plan is viable and workable	1	2	3	4	5
10.	These processes are viewed as a fair procedure for better planning in the organization	1	2	3	4	5
11.	These processes are viewed as appropriate and effective for better planning in the organization	1	2	3	4	5
12.	These processes are conducted to comply with the regulation that is taken for granted	1	2	3	4	5
13.	These processes are part of the awareness of potential impacts and losses in the organization if any disruption occurs	1	2	3	4	5
14.	These processes are already part of the company culture	1	2	3	4	5

C14B. Please circle the number in both columns (Drivers and Hindrances) that indicates the extent to which you agree with each of the statement.

No.	Descriptions	1 (Strongly disagree); 2 (Disagree); 3 (Neutral); 4 (Agree); 5 (Strongly agree)									
		Drivers[a]					Hindrances[b]				
1.	The company emphasizes on good performance	1	2	3	4	5	1	2	3	4	5
2.	The company has high level of cost-consciousness, particularly for providing periodic tests and exercises to ensure that the BC plan is viable and workable. Also, the company considers that meeting times is essential (particularly on CBFs level of operation)	1	2	3	4	5	1	2	3	4	5
3.	Guidance for employee's performance improvement is provided in the company	1	2	3	4	5	1	2	3	4	5
4.	The company emphasizes on team accountability	1	2	3	4	5	1	2	3	4	5
5.	Different functions and units of the organization have a high capability to work together to achieve common goals	1	2	3	4	5	1	2	3	4	5
6.	The company can resolve internal problems effectively	1	2	3	4	5	1	2	3	4	5
7.	Inter-departmental collaboration in the company is highly encouraged	1	2	3	4	5	1	2	3	4	5
8.	Good communication is essential and information sharing is encouraged	1	2	3	4	5	1	2	3	4	5
9.	The members and BUs accept criticism and negative feedback	1	2	3	4	5	1	2	3	4	5
10.	The company is open to alternative solutions	1	2	3	4	5	1	2	3	4	5
11.	The company is able to allocate sufficient resources for implementing innovative ideas	1	2	3	4	5	1	2	3	4	5
12.	A high degree of cooperation among employees (particularly among staffs for the tests and exercises programmes) in working toward common goals	1	2	3	4	5	1	2	3	4	5
13.	The staffs commitment to be involved in the tests and exercises are highly valued by the company	1	2	3	4	5	1	2	3	4	5
14.	The company emphasizes on team contributions in the tests and exercises programmes	1	2	3	4	5	1	2	3	4	5
15.	Unfamiliar situations are managed and identified	1	2	3	4	5	1	2	3	4	5
16.	Following organizational procedures is essential	1	2	3	4	5	1	2	3	4	5

[a]Drivers: In my opinion, this attribute is viewed as a supporting factor for providing periodic tests and exercises to ensure that the BC plan is viable and workable
[b]Hindrances: In my opinion, this attribute is not implemented in the company

Programme Management

The organization will demonstrate commitment in maintaining the currency of its plan through regular and systematic review of its risks and business impacts, regularly reviewing its BCM strategies and revalidating its BC plan. Programme management serves to validate the capability of the BC plan to fulfill the plan's objectives. Validation aims to uncover flaws in the plan design, for example any inaccuracies and incompleteness of the design of the plan.

C15. Is programme management being implemented in the firm?

☐ Yes ☐ No

C16. ***Analyzing the ongoing efforts and activities to maintain the effectiveness of its BCM, including providing systematic training and awareness programmes to staffs.***

C16A. Please circle the number that indicates the extent to which you agree with each of the statement.

No.	Descriptions	1 (Strongly disagree); 2 (Disagree); 3 (Neutral); 4 (Agree); 5 (Strongly agree)				
1.	It can be easily integrated with other management systems in the organization that can provide positive gains for the organization	1	2	3	4	5
2.	It helps to improve the organization's procedures for facing crisis	1	2	3	4	5
3.	It may lead to negative impact (not able to develop a rigorous analysis—more cost for recovery process)	1	2	3	4	5
4.	There are strict regulations from the government or the organization's management the process. Not implementing them can result in receiving sanctions	1	2	3	4	5
5.	It may improve the employee's health, safety, and welfare	1	2	3	4	5
6.	It is insisted upon by stakeholders and clients	1	2	3	4	5
7.	It is conducted to sustain and improve the company's reputation	1	2	3	4	5
8.	It is conducted to increase the company's competitiveness	1	2	3	4	5
9.	Peer groups and the top management fully support the process	1	2	3	4	5
10.	It is viewed as a fair procedure for better planning in the organization	1	2	3	4	5
11.	It is viewed as appropriate and effective for better planning in the organization	1	2	3	4	5
12.	It is conducted to comply with the regulation that is taken for granted	1	2	3	4	5
13.	It is part of the awareness of potential impacts and losses in the organization if any disruption occurs	1	2	3	4	5
14.	It is already part of the company culture	1	2	3	4	5

C16B. Please circle the number in both columns (Drivers and Hindrances) that indicates the extent to which you agree with each of the statement.

No.	Descriptions	1 (Strongly disagree); 2 (Disagree); 3 (Neutral); 4 (Agree); 5 (Strongly agree)									
		Drivers[a]					Hindrances[b]				
1.	The leadership of the company's top management has a high influence on the company	1	2	3	4	5	1	2	3	4	5
2.	The company's operational programmes are highly supported by top management	1	2	3	4	5	1	2	3	4	5
3.	The organizational members are willing to accept the leadership and direction from the top management or qualified others	1	2	3	4	5	1	2	3	4	5
4.	The company emphasizes on good performance	1	2	3	4	5	1	2	3	4	5
5.	The company provides a detailed set of performance standards (regarding its BUs) that support the process of managing BCM programme	1	2	3	4	5	1	2	3	4	5
6.	The company has many internal structuring (procedures) in the organization	1	2	3	4	5	1	2	3	4	5
7.	The company has high level of cost-consciousness, particularly for managing BCM programme. Also, the company considers that meeting times is essential (particularly on CBFs level of operation)	1	2	3	4	5	1	2	3	4	5
8.	Important decisions are made by groups as a consensus	1	2	3	4	5	1	2	3	4	5
9.	Inter-departmental collaboration in the company is highly encouraged	1	2	3	4	5	1	2	3	4	5
10.	Good communication is essential and information sharing is encouraged	1	2	3	4	5	1	2	3	4	5
11.	Different functions and units of the organization have a high capability to work together to achieve common goals	1	2	3	4	5	1	2	3	4	5
12.	The company has a high degree of clear strategic intentions that are conveyed to make it clear how everyone can contribute	1	2	3	4	5	1	2	3	4	5
13.	The company provides a clear direction for employees in their work	1	2	3	4	5	1	2	3	4	5
14.	Actions are matched with the company's goals	1	2	3	4	5	1	2	3	4	5
15.	The company takes the responsibility for its employee's welfare (particularly for managing the BCM programme)	1	2	3	4	5	1	2	3	4	5
16.	The company has a main focus in meeting the client's needs and satisfaction. It views the BCM programme as part of the process to deliver products that meet the client's needs	1	2	3	4	5	1	2	3	4	5
17.	The company emphasizes on team accountability	1	2	3	4	5	1	2	3	4	5
18.	The company emphasizes on reward instead of punishment	1	2	3	4	5	1	2	3	4	5

(continued)

No.	Descriptions	1 (Strongly disagree); 2 (Disagree); 3 (Neutral); 4 (Agree); 5 (Strongly agree)									
		Drivers[a]					Hindrances[b]				
19.	The company recognize and reward members' performance	1	2	3	4	5	1	2	3	4	5
20.	Unfamiliar situations are managed and identified	1	2	3	4	5	1	2	3	4	5
21.	Following organizational procedures is essential	1	2	3	4	5	1	2	3	4	5

[a]Drivers: In my opinion, this attribute is viewed as a supporting factor for analyzing the ongoing efforts and activities to maintain the effectiveness of its BCM, including providing systematic training and awareness programmes to staffs
[b]Hindrances: In my opinion, this attribute is not implemented in the company

C17. ***Conducting BCM training and awareness programmes for all staff and related external parties.***

C17A. Please circle the number that indicates the extent to which you agree with each of the statement.

No.	Descriptions	1 (Strongly disagree); 2 (Disagree); 3 (Neutral); 4 (Agree); 5 (Strongly agree)				
1.	These programmes can be easily integrated with other management systems in the organization that can provide positive gains for the organization	1	2	3	4	5
2.	These programmes help to improve the organization's procedures for facing crisis	1	2	3	4	5
3.	Not conducting these programmes may lead to negative impact (not able to develop a rigorous analysis—more cost for recovery process)	1	2	3	4	5
4.	There are strict regulations from the government or the organization's management about conducting these programmes. Not implementing them can result in receiving sanctions	1	2	3	4	5
5.	These programmes may improve the employee's health, safety, and welfare	1	2	3	4	5
6.	These programmes are insisted upon by stakeholders and clients	1	2	3	4	5
7.	These programmes are conducted to sustain and improve the company's reputation	1	2	3	4	5
8.	These programmes are conducted to increase the company's competitiveness	1	2	3	4	5
9.	Peer groups and the top management fully support the programmes	1	2	3	4	5
10.	These programmes are viewed as a fair procedure for better planning in the organization	1	2	3	4	5

(continued)

No.	Descriptions	1 (Strongly disagree); 2 (Disagree); 3 (Neutral); 4 (Agree); 5 (Strongly agree)				
11.	These programmes are viewed as appropriate and effective for better planning in the organization	1	2	3	4	5
12.	These programmes are conducted to comply with the regulation that is taken for granted	1	2	3	4	5
13.	These programmes are part of the awareness of potential impacts and losses in the organization if any disruption occurs	1	2	3	4	5
14.	These programmes are already part of the company culture	1	2	3	4	5

C17B. Please circle the number in both columns (Drivers and Hindrances) that indicates the extent to which you agree with each of the statement.

		1 (Strongly disagree); 2 (Disagree); 3 (Neutral); 4 (Agree); 5 (Strongly agree)									
No.	Descriptions	Drivers[a]					Hindrances[b]				
1.	The company continually invests in the development of employee's skills in order to be competitive and be able to meet on-going business needs	1	2	3	4	5	1	2	3	4	5
2.	Guidance for employee's performance improvement is provided by the company	1	2	3	4	5	1	2	3	4	5
3.	The organization emphasizes on job competence only for each of the position in the company. Social and family backgrounds are not taken into account	1	2	3	4	5	1	2	3	4	5
4.	Amicable opinions and ideas exchange between members are facilitated by the company	1	2	3	4	5	1	2	3	4	5
5.	The staffs commitment to be involved BCM training and awareness programmes are highly valued by the company	1	2	3	4	5	1	2	3	4	5
6.	The company emphasizes on team contributions in the BCM training and awareness programmes	1	2	3	4	5	1	2	3	4	5
7.	The company provides a clear direction for employees in their work	1	2	3	4	5	1	2	3	4	5
8.	Actions are matched with the company's goals	1	2	3	4	5	1	2	3	4	5

[a]Drivers: In my opinion, this attribute is viewed as a supporting factor for conducting BCM training and awareness programmes for all staff and related external parties
[b]Hindrances: In my opinion, this attribute is not implemented in the company

Section D: Further Comments and Feedback Regarding the Survey

Do you have any additional comments regarding BCM implementation in your firm that were not covered in the above questions?

End of questionnaire
Thank you for your time and effort in responding to this questionnaire.
Your responses will be treated as confidential and will only be used for research purpose.

Appendix C

Questionnaire for RQ2: Researcher's Reference

Case Study

In-depth review of two (or more) Indonesian contractors (state and private-owned firms) regarding their BCM preparedness.

Company name: ______________________
☐ State-owned firm ☐ Private-owned firm

Data collected through checklists (documents review) and interviews.

Section A: Organization Characteristics

A1. What are the mission and vision of the firm?

A2. What are the firm's objectives?

A3. How does the firm achieve the objectives?

A4. What are the firm's products and services?

Products/ Services	Local/International scope	Types of projects	Product/service value (in USD or Rp)

L.S.R. Supriadi, L. Sui Pheng, *Business Continuity Management in Construction*, Management in the Built Environment, DOI 10.1007/978-981-10-5487-7

A5. What are the current direction and focus of the firm?

A6. What are the short and long term plans (in general) (any plans on these?; views on these matters)?

Descriptions	Short-term plans	Long-term plans
The firm's growth		
Downsizing		
Restructuring		
Acquisition		
Disposal		

A7. Are there any plans to develop a new product or service in the short term?

- What are the timescales for new products or services (in general)?

A8. What is the firm's operational geographic scale?

Location	Province—City	% of projects (or in numbers)
Indonesia		
Outside Indonesia		

A9. What is the geographic extent of a disruption (what types of disruptions that will affect the business (related to head office—project based office)?

A10. What is the extent of resource loss? (What types of resources that are considered to be a loss when disruptions occur?)

A11. What are the current market conditions of the firm's business?

- What are the expected market conditions of the firm's business (in 5 years/ 10 years)?

A12. Who are the firm's competitors? (local/international)

- What the competition looks like within the firm's sector?
- In which stage of the business process does the competition occur (with high impact/frequency)?

A13. What are the likely reactions of customers to the firm's operations being disrupted (describe any examples)?

A14. What are the competitor's reactions (in general/main reactions) when the firm is experiencing disruptions/crises:

- Taking advantage during difficulties (with example)
- Support one another (with example)
- Others

A15. Does the firm's business regulation environment provide any support for overcoming crisis? (in what form?)

A16. Does the firm's business regulation environment provide any support for business continuity? (in what form?)

A17. What are the types of suppliers involved in the firm's business?

Types of suppliers	Total suppliers

A18. What is the timescale for finding alternative suppliers?

- When does the firm decide to find alternative suppliers?

A19. What are the total customers/clients of the firm? (Local and International clients)

	Total customers/clients
Indonesia (local) clients	
International clients (insert name of country)	

A20. Do the customers/clients pay insurance for improved reliance on delivery?

- How are the insurance arrangements between the clients and the firm?

A21. How does the firm's organizational structure (for head office and project office) look like?

- Which BUs are involved in this business value chain; Which are the CBFs in this chain (related to what BUs)?

Business value chain	Business Unit(s) (BU)	Critical Business Functions (CBF)
Business development: • Creating relationships with existing and prospective customers • Obtaining the work • Market research		
Procurement: • Relationship with subcontractors • Relationship with suppliers • Efficiency and effectiveness of material purchasing procedures and management procedures		
Construction operations: • Cost and schedule estimation and control • Project management system • Quality management system • Safety management system		

(continued)

Business value chain	Business Unit(s) (BU)	Critical Business Functions (CBF)
Post-construction services: • Warranty management system • Customer relationship development program		
Firm infrastructure: • Adequacy and location of facilities and equipment • Efficiency and effectiveness of finance and accounting system • Information management system		
Human resources management: • Procedures for recruiting and developing employees • Working environment • Relationship with unions • Levels of employee motivation and job satisfaction		

A22. What is the MBCO for the defined CBFs when a crisis occurred? (Ex: CBF: …. – MBCO: …)

– Has this process of defining MBCO been conducted in the firm? (by the management?)

CBF	Minimum Business Continuity Objective (MBCO)

A23. How is the coordination between:

(a) Head office and project-base office, elements to review:

- How do they develop, communicate and distribute the reports on performance progress?
- What types of communication are there for this relationship? (verbal; non-verbal)

  – Who is in charge for the communication between them?

- How to allocate resources (from head office to project office; in the project office)?
- What types of infrastructure are there for coordination between head office and project-based office?

(b) Head office and governments; public; clients; suppliers (supply chain); sub-contractors

- What types of communication are there for these relationships? (verbal; non-verbal)

  – Who is in charge for the communication between them?

- What types of coordination/reports are there for these relationships?

  – How do they develop, communicate and distribute the reports?

- What are the resources needed for these relationships?
  – How to allocate resources between these relationships?
- What types of infrastructure are there for coordination between these relationships?

(c) Project-base office and public; suppliers; sub-contractors

- What types of communication are there for these relationships? (verbal; non-verbal)
  – Who is in charge for the communication between them?
- What types of coordination/reports are there for these relationships?
  – How do they develop, communicate and distribute the reports?
- What are the resources needed for these relationships?
  – How to allocate resources between these relationships?
- What types of infrastructure are there for coordination between these relationships?

Section B: BCM Preparedness Description

Questions for:
B1. BCM start-up phase or how to initiate BCM

Has the firm implemented this phase? (yes/no)

If yes:
What are the steps that are being implemented? (checklist)

Descriptions	Implemented or not
• The requirements for BC, taking into account on the firm's objectives; obligations—legal, regulatory, contractual	
• Identifying interests of key stakeholders: – What are their requirements and perceptions? – Who are they? (shareholders; employees; suppliers; regulators; financial investors; insurers; auditors; professional bodies; trade associations; government departments; competitors, the community; media and "vested interest" groups)	
• Scope of BC in terms of products and services	
• Accountability and responsibility for key areas were defined at the time that the framework was implemented	
• There are clearly defined and approved management processes to manage BC	
• The management and staff adopted an attitude to continuity management planning that ensures a positive control environment	
• The entity regularly communicates its vision, goals and objectives to staff	
• Others	

If No:
What are the reasons for not implementing this?
Do they have different methods for implementing this stage?
If the firm plans to implement this principle, what are the hindrances for implementing this phase (regarding its current condition)?: (check list)

Descriptions	A hindrance or not
– The entity has inadequate financial resources to implement the BCM	
– Designated user representatives are not promptly notified if a business disruption occurs	
– Others ...	

B2. Risk analysis and review

Has the firm implemented this principle? (yes/no)

If yes:
What are the steps that are being implemented? (checklist)

Descriptions	Implemented or not
• Threats identification in the firm	
• Management carefully analyse and assess risks and opportunities before authorizing new ventures or significant changes	
• Business units (BU) in organization (BU–BF–CBF)	
• Risk identification and treatment for each BU	
• Disaster identification (for key disaster scenario)—Disaster may be compiled from one or more identified risks	
• Risk review (per BU)	
• Others ...	

If No:
What are the reasons for not implementing this?
Do they have different methods for implementing this principle?
If the firm plans to implement this principle, what are the hindrances for implementing this phase (regarding its current condition)?: (check list)

Descriptions	A hindrance or not
– Contact details for interdependent entities are not regularly maintained	
– Lack of capacity to adequately implement a possible solution, such as remote access	
– No alternative contacts have been identified	
– No ongoing effort to minimize exposures to disasters and operations/ systems vulnerabilities	
– Others ...	

B3. Business Impact Analysis (BIA)

Has the firm implemented this principle? (yes/no)

If yes:
What are the steps that are being implemented? (checklist)

Descriptions	Implemented or not
• Defining business function	
• Defining MBCO (minimum business continuity objective)	
• Establish priority for analyzing impact (per disaster); the impact if key services and products are disrupted—for whatever reason	
• Establish CBF (for each BU)—what are the critical activities?	
• Defining dependencies of each CBF (each BF can span across one or more business operations)	
• Defining CBF requirements (this construct relates to previous CBF dependencies construct—above)	
• The BIA identifies the recovery timeframes of the critical business functions	
• Defining and reviewing resource requirements and capabilities (inventory for each BU)	
• Others ...	

If No:
What are the reasons for not implementing this?
Do they have different methods for implementing this principle?
If the firm plans to implement this principle, what are the hindrances for implementing this phase (regarding its current condition)?:

Descriptions	A hindrance or not
– Critical systems are not periodically evaluated and their minimum essential features cannot be provided for in a disaster	
– The entity experiences voluntary or involuntary separations of employment or relationships with any employees, suppliers, or other vendors between the occurrence of the disaster event and complete recovery	
– Daily transactions needed to reconstruct critical data are not rotated off-site with adequate frequency	
– Critical operations and systems documentation for each platform are not stored off-site	
– Replacement equipment is not readily available	
– Appropriately skilled IT personnel, or specialist equipment are not available	
– Ready access to public network following a disruption is not available	
– Lack of access to communications hardware (e.g. pager, fax, email)	
– Vital records are stored in a single location	
– Others ...	

B4. BC strategy

Has the firm implemented this principle? (yes/no)

If yes:
What are the steps that are being implemented? (checklist)

Descriptions	Implemented or not
• Recovery strategy selection (based on selected disaster—per disaster)	
• Recovery strategy must cover: people; premises; technology; information; supplies; stakeholders	
• In general you should consider four high level scenarios and what alternative working arrangements could be made if: – Cannot gain access to the building – A high percentage of the staff is unavailable – The ICT systems are unavailable – A key supplier/partner is disrupted	
• BCM Strategies must: – Recognise critical functions, dependencies and single points of failure – Enable organisation to perform critical activities – Allow decisions to be taken by responsible managers – Signed off by senior management	
• The continuity strategies that best meet the entity's needs have been implemented based on a cost-benefit analysis	
• Others . . .	

If No:
What are the reasons for not implementing this?
Do they have different methods for implementing this principle?
If the firm plans to implement this principle, what are the hindrances for implementing this phase (regarding its current condition)?: (check list)

Descriptions	A hindrance or not
– An insufficient number of qualified personnel are available to perform user tasks during the recovery phase	
– Personnel who play a role in recovery are unaware of their responsibilities and may not have been adequately trained to perform the recovery tasks	
– Staff support areas are not prepared to support the recovery operation	
– Lack of alternative processing facilities available as and when required	
– The organization lacks access to a fully configured second processing site sufficient in capacity to support data processing for critical business processes with critical application support needs	
– Critical users do not have the ability to reconstruct any lost work in-progress	
– Others . . .	

B5. BC plan development

Has the firm implemented this principle? (yes/no)

If yes:
What are the steps that are being implemented? (checklist)

Descriptions	Implemented or not
• Identify triggers and response (per disaster)	
• Establish the command and control structure to respond to incident; emergency; disaster situations (per disaster)	
• Prioritize activities; Time sequence of a BC plan for a selected disaster; Activities and tasks should be prioritized based on the time sequence	
• Coordinate and finalize commitment	
• Gather requirements (list of pre-incident measures)	
• Gather detailed requirements for each CBF	
• Checklists for writing the BC Plan (based from the tables and procedures/lists)	
• Confirm the BC plan	
• Distribute BC Plan; Not all BU require the entire BC Plan content; Based on need to know and need to hold basis	
• Incorporating communication management in the BC plan – Regularly update senior management – Keep the customers informed – Mechanisms to inform employees – Keep other stakeholders informed – Ensure media are briefed	
• Incorporating information management in the BC plan – Collate situation reports – Access to contact details – Access to staff records – Insurance policies, SLAs, contracts – Monitor the media – Maintain a log of decisions, activities and actions	
• The BC plan covers: critical products and services as specified in the scoping document; High level plans; Departmental plans; Unit plans	
• The BC plan is documented and endorsed	
• The BC plan is up-to-date	
• The BC plan is linked to the emergency management and incident management plans for the entity	
• The BC plan has been formally evaluated as part of the entity's overall corporate governance arrangements	
• Others ...	

If No:
What are the reasons for not implementing this?
Do they have different methods for implementing this principle?

If the firm plans to implement this principle, what are the hindrances for implementing this phase (regarding its current condition)?: (check list)

Descriptions	A hindrance or not
– The recovery plan will not cover any event which simultaneously renders both the primary and all alternative data/resource centre facilities inoperable	
– The recovery plan will not cover any event which simultaneously renders the data centre inoperable and the essential off-site storage inaccessible	
– Critical users do not have recovery plans developed to be able to proceed at the alternative processing facility	
– Others ...	

B6. Tests and exercises

Has the firm implemented this principle? (yes/no)

If yes:
What are the steps that are being implemented? (checklist)

Descriptions	Implemented or not
• Establish practice to operate BC Plan	
• Prepare for tests and exercises	
• Conduct test and exercises	
• Assess the results	
• Infrastructure to support tests and exercises	
• Identify and implement corrective actions	
• Considerations for implementing the test and exercise: – Risk, impacts and capabilities – Types of exercise to be used – Involvement of senior management – Process of delivering exercises – Relationship between exercising emergency plans and BCPs – Planning exercises which minimise the risk of disruption and the risk of an incident occurring as a direct result of the exercise is minimised	
• Others ...	

If No:
What are the reasons for not implementing this?
Do they have different methods for implementing this principle?
If the firm plans to implement this principle, what are the hindrances for implementing this phase (regarding its current condition)?: (check list)

Descriptions	A hindrance or not
– Testing and exercising does not involve interdependent entities	
– Periodic testing and exercising of the BC plan is not conducted	
– Others ...	

B7. BCM maintenance (programme management)

- Auditing
- Reviewing BC plan
- Has the firm implemented this principle? (yes/no)

If yes:
What are the steps that are being implemented? (checklist)

Descriptions	Implemented or not
• Align BCM with organization operations	
• Review key BCM elements by BCM Steering Committee	
• Review BC plan (minimum once a year by BCM SC)	
• Provide continuous training and awareness	
• Perform BCM audit; An internal audit or external review of the implemented framework has been undertaken	
• Track BCM trends and practices	
• The environment in which we operate is constantly changing so BCPs and BCM arrangements need reviewing. This involves the BCM team and author standing back and checking strategy on, say an annual basis, or after significant change using a formal process	
• Where changes are needed this will lead to re-writing, re-issuing and re-training and endorsement by management team	
• Others ...	

If No:
What are the reasons for not implementing this?
Do they have different methods for implementing this principle?
If the firm plans to implement this principle, what are the hindrances for implementing this phase (regarding its current condition)?: (check list)

Descriptions	A hindrance or not
– There is an insufficient number of personnel possessing the appropriate skills available to implement BC operations	
– There is inadequate maintenance of BC procedures	
– Others . . .	

Section C: The Firm's Practices (or Success Stories/Lessons Learned) Relating to Crises

C1. What are the types of crises that are acknowledged in the firm? (viewed as important and need to be prepared for this type of crisis) (tick from the list/can be more than one)

• Loss of public goodwill, reputation, image due to malicious contamination	□	• Theft	□
• Action by environmentalist/pressure groups (protests)	□	• Sabotage	□
• Financial crisis	□	• Access/approval restriction or limitation	□
• Corruption scandal	□	• Serious product defects or component failures	□
• Political instability i.e. those leading to changes of project scope or cancellations, sanctions and embargoes, tighter exchange controls, repatriation of funds	□	• Loss of confidential information	□
• Changes in regulations and statutory legislation	□	• Natural disasters (earthquake, floods, tsunami, etc.)	□
• Client insolvency leaves outstanding debts for work done	□	• War	□
• Delays or uncertainty in resolving disputes	□	• Riot	□
• Increase in price of raw materials (unexpected price escalation)	□	• Terrorism	□
• Shortage of key materials	□	• Lack of component workforce	□
• Material damages (during deliveries or faulty products from manufacturer)	□	• Loss of management personnel or key staff	□
• Subcontractor insolvency	□	• Kidnap and ransom (effect on production and share price of loss of key personnel)	□
• Breakdown of key construction plant	□	• Strikes, labour disputes	□
• Serious accidents in a project	□	• Health issues (flu pandemic, SARS, etc.)	□
• Fire	□	• No disruption occurred	□
• Explosion	□	• Others:_______________	□

C2. What type of crises that had occurred in the firm for the last: (can be based from the previous list)

- One year: ______________________________
- Three years: ______________________________
- 5–10 years: ______________________________

C3. What is/are the most significant crises that had occurred in the firm that resulted in highest impacts? (describe the case)

C4. What are the significant impacts from the crises that had occurred in the firm?

Can be chosen from the list:

- Loss of productivity
- Delay in work/dissatisfied customers
- Business closed temporarily/infrastructure down and unable to carry out business functions
- Employees were not able to reach the office
- Revenue impact
- Huge data loss and client impact
- Building had to be evacuated
- Failure of few systems

Crisis	Impacts

C5. Does the firm have any crises responses in place? (yes/no)

If yes:
What is the name of the plan?
What are the contents of the plan?
Does the plan have these procedures?: (yes/no/don't know + descriptions)

Description	Yes/No/don't know	Further comments (procedures; resources; infrastructures)
Evacuation procedure		
Communication procedure during crisis		
EOC/alternative facilities		
Recovery procedure		
Restoration procedure		

What type of crisis will the plan be used for?
Have the plans been implemented for a certain crisis? If yes, on which crisis? When? (general description about responding to the crisis)

If no:
What are the reasons that the firm does not have crisis response plans in place?
If crises occur, how will the firm respond and what will be the actions?

C6. What are the lessons-learned from the crises that occurred in the firm?

Survey:

Interviews with experts (academics; practitioners; governments in Indonesian construction industry) asking about BCM implementation for Indonesian contractors.

Interview with experts

(Confirming BCM implementation for Indonesian contractors—can be based on RQ1's results or part of the triangulation process for RQ1 and RQ2).

Questions about:

Section A: Knowledge of BCM in Construction

A1. Does the respondent know about BCM?
Do they think that the construction industry players (particularly the contractor firms) know about BCM?
Where do they know BCM from? (sources)

A2. What are the benefits of implementing BCM for contractors?

A3. What are the drawbacks of implementing BCM for contractors?

A4. Is BCM implementation in contractors highly needed or not? Why?

A5. Do the contractors need to be BCM certified? Why?

A6. In general, what are the drivers for implementing BCM in contractors? (general description)
Regulative forces (law/regulation/reward-sanction)
Normative forces (mimetic approach/association/peers/stakeholder/legitimacy)
Cultural-cognitive forces (habitual act/taken for granted)
A7. In general, what are the hindrances for implementing BCM in contractors? (general description)
Regulative forces (law/regulation/reward-sanction)
Normative forces (mimetic approach/association/peers/stakeholder/legitimacy)
Cultural-cognitive forces (habitual act/taken for granted)
A8. Important elements in contractors that relates mostly with BCM:
Which BUs in contractors (or their business value chain) are the essential elements for implementing BCM?
Which CBFs in contractors are the essential elements for implementing BCM (that need to be protected; that have high impacts to the business)?
What are the MBCOs for the CBFs mentioned above?
What are the most significant crises that may give the highest impact to the contractors?

A9. Organizational culture (OC) in contractors
In general, what are the significant organizational cultures (OC) that:

1. Drives the BCM implementation in contractors?
2. Hinders the BCM implementation in contractors? (can be described based on the list from each principle below)

- Risk analysis and review

  - Conducting risk analysis and cost benefit analysis

    Process or results oriented
    Coordination and integration
    Adaptability to change
    Setting standards
    Employee or job oriented
    Customer orientation

  - The involvement of experts and BCM committee in risk review.

    Parochial or professional
    Empowerment
    Team orientation
    Open or closed system

  - Conducting a detailed risk review that examines and assesses the availability of critical equipment, technology, and facilities for BU/CBF (including location of facilities, essential utilities and telecommunications, transportation to premises and physical security of premises).

    Process or results oriented
    Coordination and integration
    Adaptability to change
    Setting standards

- Business Impact Analysis (BIA)

  - Conducting business impact analysis (BIA) that focuses on assessing the impact of losses if the corresponding business operations and processes are disrupted.

    Process or results oriented
    Coordination and integration
    Adaptability to change
    Setting standards
    Employee or job oriented
    Customer orientation

  - The involvement of experts, employees from related BUs and key staffs in BIA. BIA participants selection from related functional area/BU, including experts from the business, technology, financial, facility and legal domains. BIA will be conducted by BCM coordinator, committee, and experts.

Parochial or professional
Empowerment
Team orientation
Open or closed system

- Strategy Analysis
  - Conducting strategy analysis for maintaining the operations of CBFs that covers pre-incident preparedness, response and recovery.

    Adaptability to change
    Coordination and integration
    Setting standards
    Empowerment
    Employee or job oriented
    Customer orientation
    Process or results oriented

  - Determining staff to support the recovery strategy and providing training and awareness programme.

    Parochial or professional
    Developing employee's skills
    Team orientation
    Open or closed system

- Development of BC plan
  - Formulating the detailed BC plan (compiled from BCM principles 1, 2, 3), including its emergency response, EOC (Emergency Operations Center) plan, and identified CBFs (Critical Business Functions) with their RTOs (Recovery Time Objectives) and RPOs (Recovery Point Objectives). BC plan caters for four sets of activities (pre-incident preparation; response to incident/emergency/disaster; recovery and resumption of CBFs; restoration and return of all business operations from temporary measures adopted during recovery to supporting normal business requirements after disaster).

    Setting standards
    Process or results oriented
    Coordination and integration
    Adaptability to change
    Employee or job oriented
    Customer orientation

- Tests and exercises
  - Providing periodic tests (to verify the capability of BC plan) and exercises (to train and condition BC team members—to highlight any weaknesses in the operation and effectiveness of BC plan with following corrective actions) to ensure that the BC plan is viable and workable.

    Setting standards
    Developing employee's skills
    Reward orientation
    Coordination and integration
    Adaptability to change
    Team orientation
    Process or results oriented

- Programme management
  - Analyzing the ongoing efforts and activities to maintain the effectiveness of its BCM, including providing systematic training and awareness programmes to staffs.

    Power in organization—leadership
    Setting standards
    Coordination and integration
    A set of values
    Employee or job oriented
    Customer orientation
    Reward orientation
    Process or results oriented

  - Conducting BCM training and awareness programmes for all staff and related external parties.

    Developing employee's skills
    Parochial or professional
    Team orientation
    A set of values

Section B: Recommendation for BCM Implementation (by Indonesian Contractors)

What are the recommendations for BCM implementation by Indonesian contractors, regarding these aspects:

B1. BCM initiation phase
B2. Management support and commitment to BCM
B3. Reward and recognition system

B4. Performance management system
B5. KPI for BCM
B6. Change management process
B7. Project management process
B8. MIS for BCM
B9. BCM awareness and training
B10. BCM committee and roles
B11. BCM budget
B12. BC plan development
B13. Legal issues in BCM
B14. Communication for BCM during crises
B15. Employing BCM professionals for developing BCM in the firm
B16. Exercises and tests for BCM
B17. BCM auditing
B18. BCM maintenance

Appendix D

Questions for Interviews

Case Study

In-depth review of two (or more) Indonesian contractors (state and private-owned firms) regarding their BCM preparedness.

Company name: ____________________

☐ State-owned firm ☐ Private-owned firm

Data collected through checklists (documents review) and interviews.

Section A: Organization Characteristics

A1. What are the mission and vision of the firm?
A2. What are the firm's objectives?
A3. How does the firm achieve the objectives?
A4. What are the firm's products and services?

Products/services	Local/international scope	Types of projects	Product/service value (in USD or Rp)

A5. What are the current direction and focus of the firm?
A6. What are the short and long term plans (in general) (any plans on these?; views on these matters)?

L.S.R. Supriadi, L. Sui Pheng, *Business Continuity Management in Construction*, Management in the Built Environment, DOI 10.1007/978-981-10-5487-7

Descriptions	Short-term plans	Long-term plans
The firm's growth		
Downsizing		
Restructuring		
Acquisition		
Disposal		

A7. Are there any plans to develop a new product or service in the short term?

- What are the timescales for new products or services (in general)?

A8. What is the firm's operational geographic scale?

Location	Province—city	% of projects (or in numbers)
Indonesia		
Outside Indonesia		

A9. What is the geographic extent of a disruption (what types of disruptions that will affect the business (related to head office—project based office)?

A10. What is the extent of resource loss? (What types of resources that are considered to be a loss when disruptions occur?)

A11. What are the current market conditions of the firm's business?

- What are the expected market conditions of the firm's business (in 5 years/10 years)?

A12. Who are the firm's competitors? (local/international)

- What the competition looks like within the firm's sector?
- In which stage of the business process does the competition occur (with high impact/frequency)?

A13. What are the likely reactions of customers to the firm's operations being disrupted (describe any examples)?

A14. What are the competitor's reactions (in general/main reactions) when the firm is experiencing disruptions/crises:

- Taking advantage during difficulties (with example)
- Support one another (with example)
- Others

A15. Does the firm's business regulation environment provide any support for overcoming crisis? (in what form?)

A16. Does the firm's business regulation environment provide any support for business continuity? (in what form?)

A17. What are the types of suppliers involved in the firm's business?

Types of suppliers	Total suppliers

A18. What is the timescale for finding alternative suppliers?

– When does the firm decide to find alternative suppliers?

A19. What are the total customers/clients of the firm? (Local and International clients)

	Total customers/clients
Indonesia (local) clients	
International clients (insert name of country)	

A20. Do the customers/clients pay insurance for improved reliance on delivery?

– How are the insurance arrangements between the clients and the firm?

A21. How does the firm's organizational structure (for head office and project office) look like?

– Which BUs are involved in this business value chain; Which are the CBFs in this chain (related to what BUs)?

Business value chain	Business Unit(s) (BU)	Critical Business Functions (CBF)
Business development: • Creating relationships with existing and prospective customers • Obtaining the work • Market research		
Procurement: • Relationship with subcontractors • Relationship with suppliers • Efficiency and effectiveness of material purchasing procedures and management procedures		
Construction operations: • Cost and schedule estimation and control • Project management system • Quality management system • Safety management system		
Post-construction services: • Warranty management system • Customer relationship development program		
Firm infrastructure: • Adequacy and location of facilities and equipment • Efficiency and effectiveness of finance and accounting system • Information management system		

(continued)

Business value chain	Business Unit(s) (BU)	Critical Business Functions (CBF)
Human resources management: • Procedures for recruiting and developing employees • Working environment • Relationship with unions • Levels of employee motivation and job satisfaction		

A22. What is the MBCO for the defined CBFs when a crisis occurred? (Ex: CBF: …. – MBCO: …)

– Has this process of defining MBCO been conducted in the firm? (by the management?)

CBF	Minimum Business Continuity Objective (MBCO)

A23. How is the coordination between:

(a) Head office and project-base office, elements to review:

- How do they develop, communicate and distribute the reports on performance progress?
- What types of communication are there for this relationship? (verbal; non-verbal)

  – Who is in charge for the communication between them?

- How to allocate resources (from head office to project office; in the project office)?
- What types of infrastructure are there for coordination between head office and project-based office?

(b) Head office and governments; public; clients; suppliers (supply chain); sub-contractors

- What types of communication are there for these relationships? (verbal; non-verbal)

  – Who is in charge for the communication between them?

- What types of coordination/reports are there for these relationships?

  – How do they develop, communicate and distribute the reports?

- What are the resources needed for these relationships?

  – How to allocate resources between these relationships?

- What types of infrastructure are there for coordination between these relationships?

(c) Project-base office and public; suppliers; sub-contractors

- What types of communication are there for these relationships? (verbal; non-verbal)
  - Who is in charge for the communication between them?
- What types of coordination/reports are there for these relationships?
  - How do they develop, communicate and distribute the reports?
- What are the resources needed for these relationships?
  - How to allocate resources between these relationships?
- What types of infrastructure are there for coordination between these relationships?

Section B: BCM Preparedness Description

Questions for:
B1. BCM start-up phase or how to initiate BCM
Has the firm implemented this phase? (yes/no)
If yes:
What are the steps that are being implemented?
If No:
What are the reasons for not implementing this?
Do they have different methods for implementing this stage?
If the firm plans to implement this principle, what are the hindrances for implementing this phase (regarding its current condition)?
B2. Risk analysis and review
Has the firm implemented this principle? (yes/no)
If yes:
What are the steps that are being implemented?
If No:
What are the reasons for not implementing this?
Do they have different methods for implementing this principle?
If the firm plans to implement this principle, what are the hindrances for implementing this phase (regarding its current condition)?
B3. Business Impact Analysis (BIA)
Has the firm implemented this principle? (yes/no)
If yes:
What are the steps that are being implemented?
If No:
What are the reasons for not implementing this?
Do they have different methods for implementing this principle?

If the firm plans to implement this principle, what are the hindrances for implementing this phase (regarding its current condition)?

B4. BC strategy

Has the firm implemented this principle? (yes/no)

If yes:

What are the steps that are being implemented?

If No:

What are the reasons for not implementing this?

Do they have different methods for implementing this principle?

If the firm plans to implement this principle, what are the hindrances for implementing this phase (regarding its current condition)?

B5. BC plan development

Has the firm implemented this principle? (yes/no)

If yes:

What are the steps that are being implemented?

If No:

What are the reasons for not implementing this?

Do they have different methods for implementing this principle?

If the firm plans to implement this principle, what are the hindrances for implementing this phase (regarding its current condition)?

B6. Tests and exercises

Has the firm implemented this principle? (yes/no)

If yes:

What are the steps that are being implemented?

If No:

What are the reasons for not implementing this?

Do they have different methods for implementing this principle?

If the firm plans to implement this principle, what are the hindrances for implementing this phase (regarding its current condition)?

B7. BCM maintenance (programme management)

- Auditing
- Reviewing BC plan

Has the firm implemented this principle? (yes/no)

If yes:

What are the steps that are being implemented?

If No:

What are the reasons for not implementing this?

Do they have different methods for implementing this principle?

If the firm plans to implement this principle, what are the hindrances for implementing this phase (regarding its current condition)?

Section C: The Firm's Practices (or Success Stories/Lessons Learned) Relating to Crises

C1. What are the types of crises that are acknowledged in the firm? (viewed as important and need to be prepared for this type of crisis)

C2. What type of crises that had occurred in the firm for the last:

- One year: ___
- Three years: ___
- 5 – 10 years: ___

C3. What is/are the most significant crises that had occurred in the firm that resulted in highest impacts? (describe the case)

C4. What are the significant impacts from the crises that had occurred in the firm?

Crisis	Impacts

C5. Does the firm have any crises responses in place? (yes/no)

If yes:

What is the name of the plan?

What are the contents of the plan?

Does the plan have these procedures?

Description	Yes/No/Don't know	Further comments (procedures; resources; infrastructures)
Evacuation procedure		
Communication procedure during crisis		
EOC/alternative facilities		
Recovery procedure		
Restoration procedure		

What type of crisis will the plan be used for?

Have the plans been implemented for a certain crisis? If yes, on which crisis? When? (general description about responding to the crisis)

If no:

What are the reasons that the firm does not have crisis response plans in place?

If crises occur, how will the firm respond and what will be the actions?

C6. What are the lessons-learned from the crises that occurred in the firm?

Survey:

Interviews with experts (academics; practitioners; governments in Indonesian construction industry) asking about BCM implementation for Indonesian contractors.

Interview with experts

(Confirming BCM implementation for Indonesian contractors—can be based on RQ1's results or part of the triangulation process for RQ1 and RQ2).

Questions about:

Section A: Knowledge of BCM in Construction

A1. Does the respondent know about BCM?

Do they think that the construction industry players (particularly the contractor firms) know about BCM?

Where do they know BCM from? (sources)

A2. What are the benefits of implementing BCM for contractors?

A3. What are the drawbacks of implementing BCM for contractors?

A4. Is BCM implementation in contractors highly needed or not? Why?

A5. Do the contractors need to be BCM certified? Why?

A6. In general, what are the drivers for implementing BCM in contractors? (general description)

- Regulative forces (law/regulation/reward-sanction)
- Normative forces (mimetic approach/association/peers/stakeholder/legitimacy)
- Cultural-cognitive forces (habitual act/taken for granted)

A7. In general, what are the hindrances for implementing BCM in contractors? (general description)

- Regulative forces (law/regulation/reward-sanction)
- Normative forces (mimetic approach/association/peers/stakeholder/legitimacy)
- Cultural-cognitive forces (habitual act/taken for granted)

A8. Important elements in contractors that relates mostly with BCM:

- Which BUs in contractors (or their business value chain) are the essential elements for implementing BCM?
- Which CBFs in contractors are the essential elements for implementing BCM (that need to be protected; that have high impacts to the business)?
- What are the MBCOs for the CBFs mentioned above?
- What are the most significant crises that may give the highest impact to the contractors?

A9. Organizational culture (OC) in contractors

In general, what are the significant organizational cultures (OC) that:

1. Drives the BCM implementation in contractors?

2. Hinders the BCM implementation in contractors? (can be described based on the six BCM principles)
   - Risk analysis and review
   - Business Impact Analysis (BIA)
   - Strategy Analysis
   - Development of BC plan
   - Tests and exercises
   - Programme management

Section B: Recommendation for BCM Implementation (by Indonesian contractors)

What are the recommendations for BCM implementation by Indonesian contractors, regarding these aspects:

B1. BCM initiation phase
B2. Management support and commitment to BCM
B3. Reward and recognition system
B4. Performance management system
B5. KPI for BCM
B6. Change management process
B7. Project management process
B8. MIS for BCM
B9. BCM awareness and training
B10. BCM committee and roles
B11. BCM budget
B12. BC plan development
B13. Legal issues in BCM
B14. Communication for BCM during crises
B15. Employing BCM professionals for developing BCM in the firm
B16. Exercises and tests for BCM
B17. BCM auditing
B18. BCM maintenance

Appendix E

Analysis for Chap. 10

EFA Results for IF-BCM Principles (Rotated Factor Matrix)

Risk Analysis

RA1: ***Conducting risk analysis and cost benefit analysis.***
KMO = 0.794
Bartlett's Test of Sphericity = 385.916; df = 66.000; sig .000

	Factor 1	Factor 2	Factor 3
Awareness of potential risks	0.873		
Fair procedures for better planning	0.858		
Cost impact	0.823		
Management full support	0.697		
Concern for reputation	0.658		
Part of the company culture	0.549		
Improve the employee's safety and welfare		0.867	
Not implementing them can result in receiving sanctions		0.809	
Improve the organization's procedures for facing crisis		0.651	
It can be easily integrated with other management systems		0.572	
Increase the company's competitiveness			0.887
Insisted upon by the stakeholder and clients			0.770

RA2: ***The involvement of experts and BCM committee in risk review.***
KMO = 0.766
Bartlett's Test of Sphericity = 311.029; 55.000; sig .000

L.S.R. Supriadi, L. Sui Pheng, *Business Continuity Management in Construction*, Management in the Built Environment, DOI 10.1007/978-981-10-5487-7

	Factor 1	Factor 2
Appropriate and effective for better planning	0.866	
Awareness of potential risks	0.828	
Management full support	0.806	
Concern for reputation	0.804	
Comply with the regulation that is taken for granted	0.709	
Improve the company's procedure in preparing and handling crises	0.701	
A fair procedure for better planning in the organization	0.639	
Non-compliance impact	0.623	
Part of the company culture	0.584	
It can be easily integrated with other management systems		0.834
Increase the company's competitiveness		0.828

RA3: ***Conducting a detailed risk review that examines and assesses the availability of critical equipment, technology, and facilities for BU/CBF (including location of facilities, essential utilities and telecommunications, transportation to premises and physical security of premises).***

KMO = 0.793

Bartlett's Test of Sphericity = 444.303; 78.000; sig .000

	Factor 1	Factor 2	Factor 3
Awareness of potential risks	0.869		
Appropriate and effective for better planning	0.818		
Comply with the regulation that is taken for granted	0.739		
Non-compliance impact	0.727		
Concern for reputation	0.677		
Management full support	0.592		
Part of the company culture	0.565		
Improve the employees/worker's health, safety and welfare		0.870	
A fair procedure for better planning in the organization		0.755	
Improve the organization's procedures for facing crisis		0.664	
It can be easily integrated with other management systems		0.559	
Increase the company's competitiveness			0.890
Insisted upon by the stakeholder and clients			0.818

Business Impact Analysis

BIA1: ***Conducting business impact analysis (BIA) that focuses on assessing the impact of losses if the corresponding business operations and processes are disrupted.***

KMO = 0.772

Bartlett's Test of Sphericity = 410.265; 91.000; sig .000

	Factor 1	Factor 2	Factor 3
A fair procedure for better planning in the organization	0.886		
Management full support	0.830		
Improve the organization's procedures for facing crisis	0.813		
Improve the employee's safety and welfare	0.763		
It can be easily integrated with other management systems	0.743		
Comply with the regulation that is taken for granted	0.716		
Not implementing BIA can result in sanctions	0.713		
Part of the company culture	0.679		
Appropriate and effective for better planning in the organization		0.917	
Part of the awareness of potential impacts and losses		0.897	
Non-compliance impact		0.798	
Improve the company's reputation		0.796	
Increase the company's competitiveness			0.792
Insisted upon by stakeholders and clients			0.733

BIA2: ***The involvement of experts, employees from related BUs and key staffs in BIA. BIA participants selected from related functional area/BU, including experts from the business, technology, financial, facility and legal domains. BIA will be conducted by BCM coordinator, committee, and experts.***

KMO = 0.647

Bartlett's Test of Sphericity = 160.987; 31.000; sig .000

	Factor 1	Factor 2
Appropriate and effective for better planning	0.913	
A fair procedure for better planning in the organization	0.913	
Awareness of potential risks	0.851	
Concern for reputation	0.808	
Management full support	0.798	
Improve the company's procedure in preparing and handling crises	0.789	
Non-compliance impact	0.751	
Comply with the regulation that is taken for granted	0.737	
Part of the company culture	0.658	
Increase the company's competitiveness		0.883
Insisted upon by the stakeholder and clients		0.842

Strategy Analysis

S1: ***Conducting strategy analysis for maintaining the operations of CBFs that cover pre-incident preparedness, response and recovery.***

KMO = 0.667

Bartlett's Test of Sphericity = 125.991; 21.000; sig .000

	Factor 1	Factor 2	Factor 3
A fair procedure for better planning in the organization	0.927		
Appropriate and effective for better planning in the organization	0.927		
Part of the awareness of potential impacts and losses	0.829		
Management full support	0.826		
Improve the company's reputation	0.784		
It can be easily integrated with other management systems	0.741		
Non-compliance impact	0.738		
Part of the company culture	0.652		
Improve the employee's health, safety and welfare	0.629		
Improve the organization's procedures for facing crisis		0.800	
Comply with the regulation that is taken for granted		0.756	
Not implementing them can result in receiving sanctions		0.746	
Increase the company's competitiveness			0.922
Insisted upon by stakeholders and clients			0.918

S2: ***Determining staff to support the recovery strategy and providing training and awareness programme.***

KMO = 0.697

Bartlett's Test of Sphericity = 424.482; 78.000; sig .000

	Factor 1	Factor 2	Factor 3
A fair procedure for better planning in the organization	0.876		
Management full support	0.865		
Improve the company's reputation	0.855		
Improve the organization's procedures for facing crisis	0.836		
Part of the awareness of potential impacts and losses	0.798		
It can be easily integrated with other management systems	0.775		
Appropriate and effective for better planning in the organization		0.934	
Improve the employee's welfare		0.809	
Comply with the regulation that is taken for granted		0.800	
Part of the company culture		0.764	
Not determining staff to support the recovery strategy and providing training and awareness programme may lead to negative impact		0.724	
Increase the company's competitiveness			0.899
Insisted upon by stakeholders and clients			0.830

BC Plan Development

BCP1: ***Developing the detailed BC plan (compiled from BCM principles 1,2,3), including its emergency response, EOC (Emergency Operations Center) plan,***

and identified CBFs (Critical Business Functions) with their RTOs (Recovery Time Objectives) and RPOs (Recovery Point Objectives). BC plan caters for 4 sets of activities (pre-incident preparation; response to incident/emergency/disaster; recovery and resumption of CBFs; restoration and return of all business operations from temporary measures adopted during recovery to supporting normal business requirements after disaster).

KMO = 0.720

Bartlett's Test of Sphericity = 443.693; 91.000; sig .000

	Factor 1	Factor 2	Factor 3
Appropriate and effective for better planning in the organization	0.877		
Part of the awareness of potential impacts and losses	0.844		
Improve the company's reputation	0.839		
Comply with the regulation that is taken for granted	0.816		
Part of the company culture	0.725		
Non-compliance impact	0.715		
Not implementing them can result in receiving sanctions	0.684		
A fair procedure for better planning in the organization		0.885	
Improve the employee's health, safety and welfare		0.848	
Improve the organization's procedures for facing crisis		0.812	
Provide positive gains for the organization		0.812	
Management full support		0.809	
Insisted upon by stakeholders and clients			0.857
Increase the company's competitiveness			0.854

Tests and Exercises

TE1: ***Providing periodic tests (to verify the capability of BC plan) and exercises (to train and condition BC team members—to highlight any weaknesses in the operation and effectiveness of BC plan in following corrective actions) to ensure that the BC plan is viable and workable.***

KMO = 0.644

Bartlett's Test of Sphericity = 483.243; 91.000; sig .000

	Factor 1	Factor 2	Factor 3
Appropriate and effective for better planning in the organization	0.885		
Improve the company's reputation	0.871		
Improve the organization's procedures for facing crisis	0.813		
Management full support	0.808		
Comply with the regulation that is taken for granted	0.790		

(continued)

	Factor 1	Factor 2	Factor 3
Not providing periodic tests and exercises to ensure that the BC plan is viable and workable may lead to negative impact	0.751		
It can be easily integrated with other management systems	0.597		
A fair procedure for better planning in the organization		0.925	
Part of the awareness of potential impacts and losses		0.819	
Improve the employee's health, safety, and welfare		0.810	
Part of the company culture		0.770	
Not implementing them can result in receiving sanctions		0.666	
Increase the company's competitiveness			0.859
Insisted upon by stakeholders and clients			0.812

Programme Management

PM1: ***Analyzing the ongoing efforts and activities to maintain the effectiveness of its BCM, including providing systematic training and awareness programmes to staffs.***

KMO = 0.740

Bartlett's Test of Sphericity = 398.949; 91.000; sig .000

	Factor 1	Factor 2	Factor 3
A fair procedure for better planning in the organization	0.889		
Management full support	0.823		
Part of the awareness of potential impacts and losses	0.807		
Improve the organization's procedures for facing crisis	0.805		
Comply with the regulation that is taken for granted	0.760		
It can be easily integrated with other management systems	0.750		
Improve the employee's health, safety, and welfare	0.697		
Part of the company culture	0.639		
Improve the company's reputation		0.874	
Appropriate and effective for better planning in the organization		0.868	
Non-compliance impact		0.789	
Not implementing them can result in receiving sanctions		0.700	
Insisted upon by stakeholders and clients			0.860
Increase the company's competitiveness			0.850

PM2: ***Conducting BCM training and awareness programmes for all staff and related external parties.***

KMO = 0.769

Bartlett's Test of Sphericity = 474.048; 91.000; sig .000

	Factor 1	Factor 2	Factor 3
A fair procedure for better planning in the organization	0.934		
Appropriate and effective for better planning in the organization	0.921		
Management full support	0.796		
Comply with the regulation that is taken for granted	0.771		
Not implementing them can result in receiving sanctions	0.710		
Part of the company culture	0.694		
Improve the employee's health, safety, and welfare	0.693		
Non-compliance impact	0.681		
Improve the organization's procedures for facing crisis		0.858	
Improve the company's reputation		0.820	
Part of the awareness of potential impacts and losses		0.810	
It can be easily integrated with other management systems		0.727	
Insisted upon by stakeholders and clients			0.825
Increase the company's competitiveness			0.741

EFA Results for OC—BCM Principles (Rotated Factor Matrix)

Risk Analysis and Review

RA1: ***Conducting risk analysis and cost benefit analysis.***

KMO = 0.809

Bartlett's Test of Sphericity = 425.433; 91.000; sig .000

	Factor 1	Factor 2	Factor 3
Open to alternative solutions	0.845		
Encourage creative and innovative ideas	0.838		
Allocate sufficient resources for implementing innovative ideas	0.742		
Good communication	0.719		
Able to reach agreement on critical issues	0.612		
The company takes the responsibility for its employee's welfare	0.588		
Unfamiliar situations should be managed and identified		0.772	
The level of tolerance for ambiguity and chaos should be low		0.734	
Following organizational procedures is essential		0.662	
Important decisions are made by groups or committee (as a consensus)		0.629	
A high level of cost-consciousness			0.865
Provide a detailed set of performance standards			0.687
Meeting the client's needs and satisfaction			0.649
High level of risk avoidance			0.495

RA2: ***The involvement of experts and BCM committee in risk review.***
KMO = 0.795
Bartlett's Test of Sphericity = 181.341; 21.000; sig .000

	Factor 1	Factor 2
The experts and the BCM committee's commitment to the risk review team are highly valued by the company	0.864	
The experts (from external party) and the BCM committee (from internal party) fit very well into the risk review team, and their opinions are appreciated	0.785	
The member of the risk review team is enabled to decide or solve the problems within his/her sphere of responsibility or authority	0.662	
Emphasizes on job competence		0.875
Emphasizes on team contributions		0.687
Amicable opinions and ideas exchange between members are facilitated		0.659
The company supports the member of the risk review team to participate in the decision-making process		0.625

RA3: ***Conducting a detailed risk review that examines and assesses the availability of critical equipment, technology, and facilities for BU/CBF (including location of facilities, essential utilities and telecommunications, transportation to premises and physical security of premises).***
KMO = 0.845
Bartlett's Test of Sphericity = 229.731; 36.000; sig .000

	Factor 1	Factor 2
Following organizational procedures is essential	0.816	
Good communication	0.781	
Unfamiliar situations are managed	0.771	
Allocate sufficient resources for implementing innovative ideas	0.588	
Inter-departmental collaboration is encouraged	0.514	
High level of cost-consciousness		0.883
Provides a detailed set of performance standards		0.757
Able to create adaptive ways to meet changing needs		0.688
High level of risk avoidance		0.644

Business Impact Analysis

BIA1: ***Conducting business impact analysis (BIA) that focuses on assessing the impact of losses if the corresponding business operations and processes are disrupted.***
KMO = 0.742
Bartlett's Test of Sphericity = 371.863; 91.000; sig .000

	Factor 1	Factor 2	Factor 3	Factor 4
Encourages creative and innovative ideas	0.883			
Open to alternative solutions	0.803			
Able to allocate sufficient resources for implementing innovative ideas	0.741			
Unfamiliar situations are managed		0.839		
Following organizational procedures is essential		0.763		
High level of risk avoidance		0.619		
The company takes the responsibility for its employee's welfare		0.602		
Important decisions are made by groups as a consensus		0.601		
The company provides a detailed set of performance standards			0.863	
Able to reach agreement on critical issues.			0.815	
Good communication			0.613	
The company has many internal structuring (procedures) in the organization and considers that meeting times is essential				0.730
The company has a main focus in meeting the client's needs and satisfaction				0.634
High level of cost-consciousness				0.588

BIA2: ***The involvement of experts, employees from related BUs and key staffs in BIA. BIA participants selected from related functional area/BU, including experts from the business, technology, financial, facility and legal domains. BIA will be conducted by BCM coordinator, committee, and experts.***

KMO = 0.736

Bartlett's Test of Sphericity = 123.195; 21.000; sig .000

	Factor 1	Factor 2
The company supports the member of the BIA team to participate in the decision-making process	0.884	
Emphasize on team contributions	0.881	
The experts, employees from related BUs and key staff's commitment to the BIA team are highly valued by the company	0.867	
The experts (from external party) employees from related BUs and key staff (from internal party) fit very well into the BIA team, and their opinions are appreciated		0.777
Amicable opinions and ideas exchange between members are facilitated		0.759
Emphasize on job competence		0.736
The member of the BIA team is enabled to decide or solve the problems within his/her sphere of responsibility or authority		0.678

Strategy Analysis

S1: ***Conducting strategy analysis for maintaining the operations of CBFs that cover pre-incident preparedness, response and recovery.***

KMO = 0.808

Bartlett's Test of Sphericity = 516.999; 120.000; sig .000

	Factor 1	Factor 2	Factor 3	Factor 4
Encourages creative and innovative ideas	0.839			
Open to alternative solutions	0.778			
Able to allocate sufficient resources for implementing innovative ideas	0.716			
Good communication	0.637			
The company values employee's ideas in analyzing possible strategies for pre-incident preparedness, response and recovery	0.601			
The members of the committee or business units are able to reach agreement on critical issues	0.590			
High level of cost-consciousness		0.831		
The company provides a detailed set of performance standards		0.720		
The company has many internal structuring (procedures) in the organization and considers that meeting times is essential (particularly on CBFs level of operation)		0.710		
The company has a main focus in meeting the client's needs and satisfaction		0.702		
High level of risk avoidance		0.604		
Unfamiliar situations are managed and identified			0.798	
Following organizational procedures is essential			0.753	
Important decisions are made by groups as a consensus			0.570	
The company takes the responsibility for its employee's welfare			0.557	
Employees' participation in decision-making process during strategy analysis is encouraged				0.920

S2: ***Determining staff to support the recovery strategy and providing training and awareness programme.***

KMO = 0.856

Bartlett's Test of Sphericity = 291.021; 28.000; sig .000

	Factor 1	Factor 2
A high degree of cooperation among employees	0.887	
The company continually invests in the development of employee's skills	0.798	
Guidance for employee's performance improvement is provided	0.774	
Amicable opinions and ideas exchange between members are facilitated	0.600	
The staffs to support the recovery strategy are highly valued		0.900

(continued)

	Factor 1	Factor 2
Emphasize on team contributions		0.715
Emphasize on job competence		0.621
The staffs to support the recovery strategy fit very well into the BCM team, and their opinions are appreciated		0.615

BC Plan Development

BCP1: ***Developing the detailed BC plan (compiled from BCM principles 1, 2, 3), including its emergency response, EOC (Emergency Operations Center) plan, and identified CBFs (Critical Business Functions) with their RTOs (Recovery Time Objectives) and RPOs (Recovery Point Objectives). BC plan caters for 4 sets of activities (pre-incident preparation; response to incident/emergency/disaster; recovery and resumption of CBFs; restoration and return of all business operations from temporary measures adopted during recovery to supporting normal business requirements after disaster).***

KMO = 0.776

Bartlett's Test of Sphericity = 371.318; 78.000; sig .000

	Factor 1	Factor 2	Factor 3
The company has a main focus in meeting the client's needs and satisfaction	0.820		
The company provides a detailed set of performance standards	0.762		
The company has many internal procedures in the organization	0.739		
Able to create adaptive ways to meet changing needs	0.707		
High level of cost-consciousness	0.665		
Able to allocate sufficient resources for implementing innovative ideas	0.590		
Able to reach agreement on critical issues	0.493		
Good communication		0.810	
Unfamiliar situations are managed and identified		0.780	
Following organizational procedures is essential		0.750	
The company takes the responsibility for its employee's welfare		0.709	
Important decisions are made by groups as a consensus		0.537	
The manager in the company has a high concern for the procedures in the operation of the BUs			0.871

Tests and Exercises

TE1: ***Providing periodic tests (to verify the capability of BC plan) and exercises (to train and condition BC team members—to highlight any weaknesses in the operation and effectiveness of BC plan in following corrective actions) to ensure that the BC plan is viable and workable.***

KMO = 0.842
Bartlett's Test of Sphericity = 568.317; 105.000; sig .000

	Factor 1	Factor 2	Factor 3	Factor 4
Different functions and units of the organization have a high capability to work together	0.870			
The members and BUs accept criticism and negative feedback	0.838			
High degree of cooperation among employees	0.711			
The company can resolve internal problems effectively	0.618			
Able to allocate sufficient resources for implementing innovative ideas	0.571			
Good communication		0.820		
Emphasizes on team contributions		0.760		
The staffs commitment to be involved in the tests and exercises are highly valued		0.691		
Open to alternative solutions		0.679		
High level of cost-consciousness			0.895	
Emphasizes on good performance			0.721	
Emphasizes on team accountability			0.561	
Guidance for employee's performance improvement is provided			0.455	
Unfamiliar situations are managed				0.911
Following organizational procedures				0.724

Programme Management

PM1: ***Analyzing the ongoing efforts and activities to maintain the effectiveness of its BCM, including providing systematic training and awareness programmes to staffs.***

KMO = 0.761
Bartlett's Test of Sphericity = 1135.873; 210.000; sig .000

	Factor 1	Factor 2	Factor 3	Factor 4	Factor 5
The company provides a detailed set of performance standards	0.770				
High level of cost-consciousness	0.728				
The company has many internal structuring (procedures)	0.718				
Main focus in meeting the client's needs and satisfaction	0.688				
High degree of clear strategic intentions	0.687				
Emphasizes on team accountability	0.677				

(continued)

	Factor 1	Factor 2	Factor 3	Factor 4	Factor 5
The leadership of the company's top management has a high influence on the company	0.592				
High capability to work together to achieve common goals	0.590				
Operational programmes are highly supported by top management		0.850			
Actions are matched with the company's goals		0.848			
Inter-departmental collaboration in the company is highly encouraged		0.729			
Members are willing to accept the leadership and direction from the top management or qualified others		0.704			
The company recognize and reward members' performance			0.838		
Emphasizes on good performance			0.824		
Emphasizes on reward instead of punishment			0.776		
Provides a clear direction for employees in their work			0.611		
Unfamiliar situations are managed and identified				0.828	
Following organizational procedures is essential				0.725	
Good communication				0.638	
The company takes the responsibility for its employee's welfare				0.576	
Important decisions are made by groups as a consensus					0.863

PM2: ***Conducting BCM training and awareness programmes for all staff and related external parties***

KMO = 0.853

Bartlett's Test of Sphericity = 254.053; 28.000; sig .000

	Factor 1	Factor 2
Continually invests in the development of employee's skills	0.866	
Guidance for employee's performance improvement is provided	0.835	
Amicable opinions and ideas exchange between members are facilitated	0.801	
Emphasizes on team contributions	0.721	
The staffs commitment to be involved BCM training and awareness programmes are highly valued by the company	0.672	
Actions are matched with the company's goals		0.857
Emphasizes on job competence		0.773
The company provides a clear direction for employees in their work		0.611

Means results per OC attributes—between perceived importance/ supporting factors versus perceived implementation:

- Coding based on the questions in the RQ1 questionnaire: CXBP = perceived importance/supporting factors; CXBI = perceived implementation.
- These figures illustrate the slightest difference between the perceived importance and the perceived implementation per OC attributes for each BCM principle.

RA1:

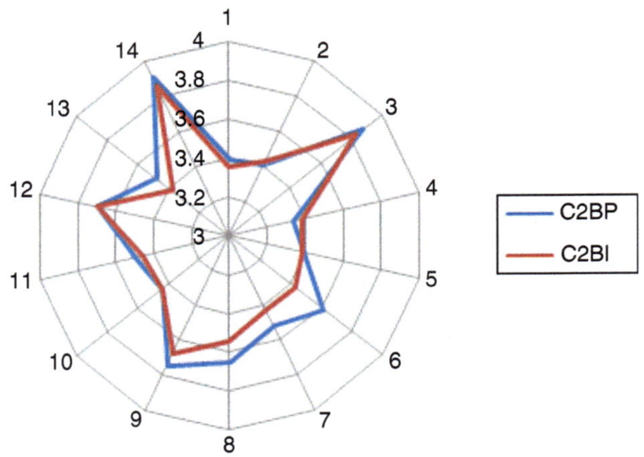

RA2:

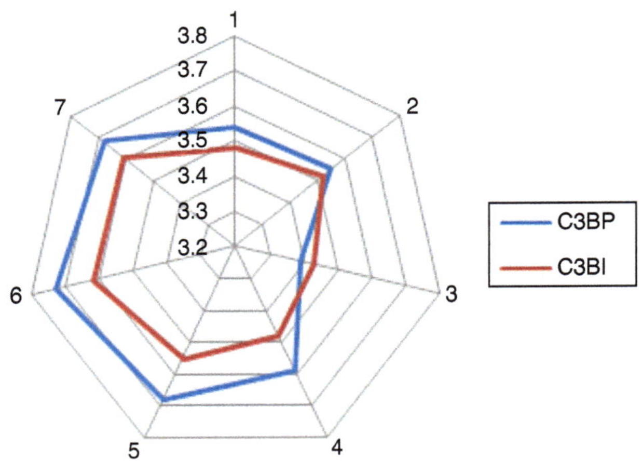

RA3:

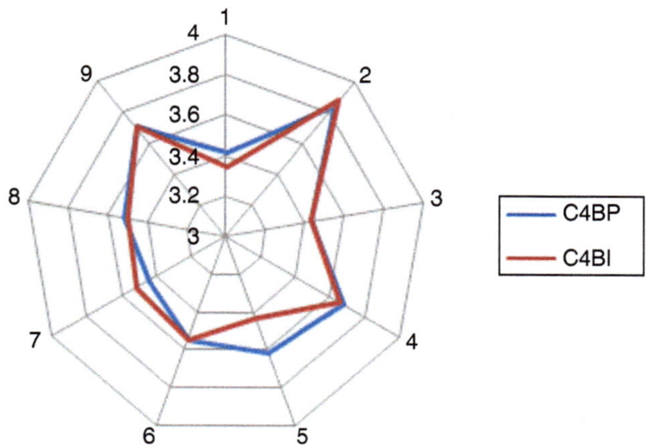
1
2
3
4
5
6
7
8
9
4
3.8
3.6
3.4
3.2
3
C4BP
C4BI

BIA1:

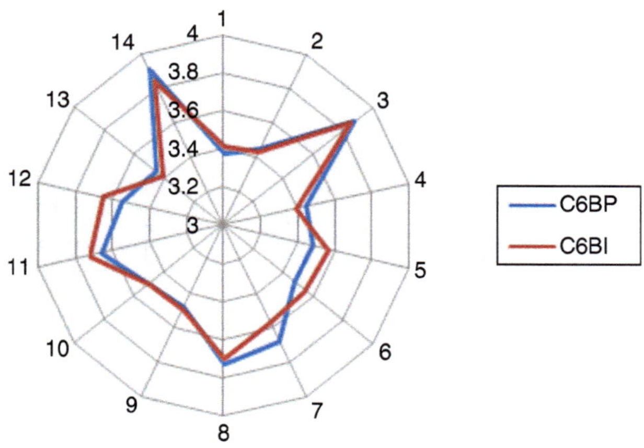
1
2
3
4
5
6
7
8
9
10
11
12
13
14
4
3.8
3.6
3.4
3.2
3
C6BP
C6BI

BIA2:

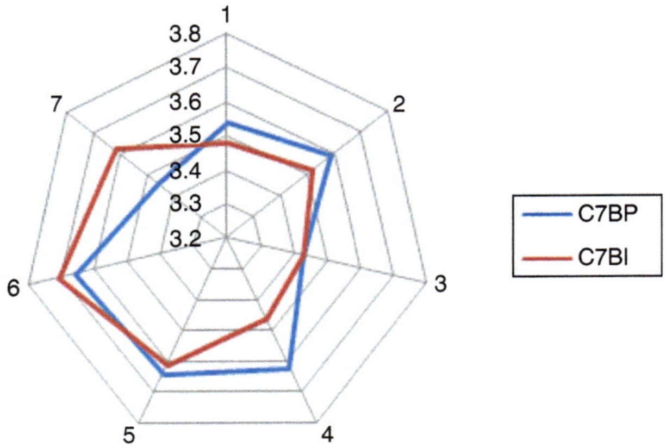
1
2
3
4
5
6
7
3.8
3.7
3.6
3.5
3.4
3.3
3.2
C7BP
C7BI

S1:

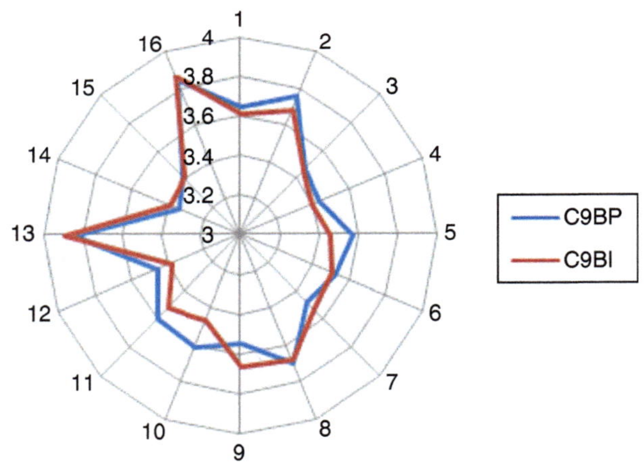
1
2
3
4
5
6
7
8
9
10
11
12
13
14
15
16
4
3.8
3.6
3.4
3.2
3
C9BP
C9BI

S2:

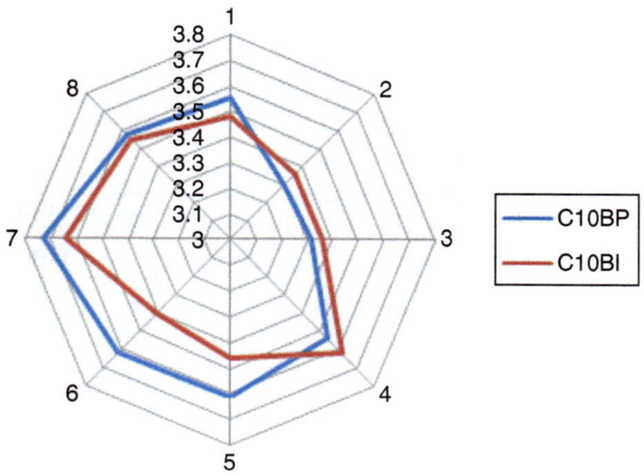
1
2
3
4
5
6
7
8
3.8
3.7
3.6
3.5
3.4
3.3
3.2
3.1
3
C10BP
C10BI

BCP1:

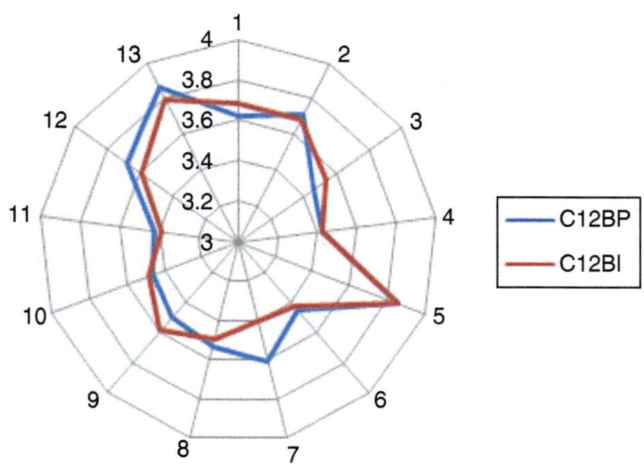
1
2
3
4
5
6
7
8
9
10
11
12
13
4
3.8
3.6
3.4
3.2
3
C12BP
C12BI

TE1:

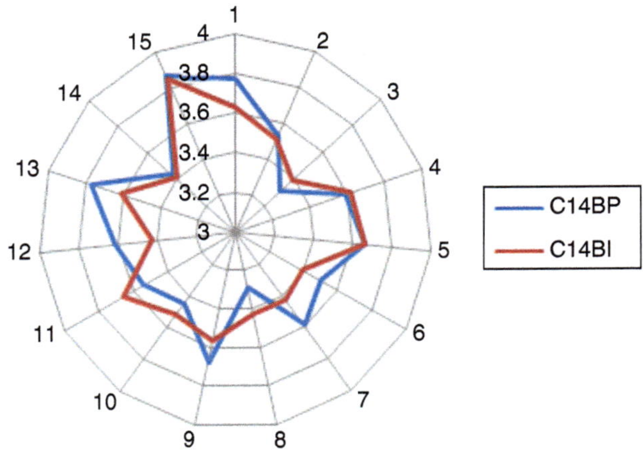
1
2
3
4
5
6
7
8
9
10
11
12
13
14
15
4
3.8
3.6
3.4
3.2
3
C14BP
C14BI

PM1:

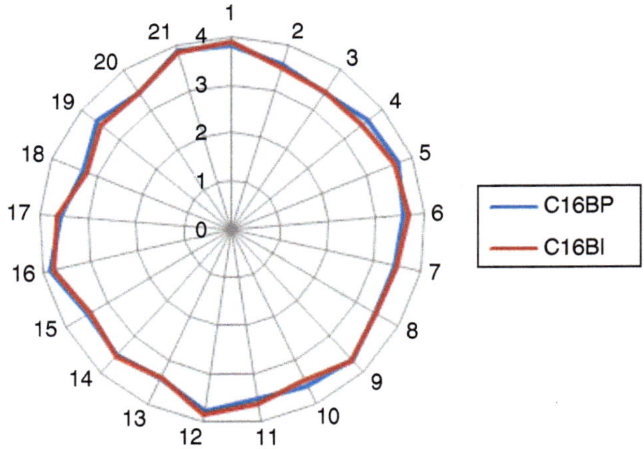
1
2
3
4
5
6
7
8
9
10
11
12
13
14
15
16
17
18
19
20
21
4
3
2
1
0
C16BP
C16BI

PM2:

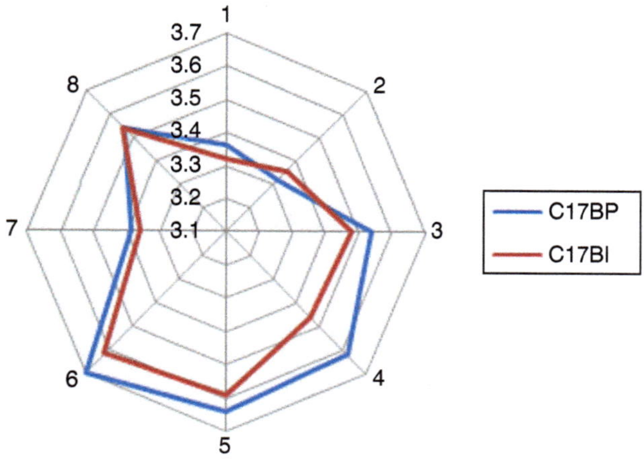

Appendix F

Analysis for Chap. 11

Action plans per BCM level of preparedness

Action Plans Level 1

Risk Analysis And Review (RA)

Technical action plans:

1. Threats identification in the firm;
   Identify threats/crisis related to the company, based on the categories (External environment; Public infrastructure; Internal environment OR Acts of nature; external, man-made event; internal unintentional event; internal intentional event)
2. Risk identification and treatment for each Business Unit (BU);
   Determine and rank risks and risk treatments for each BU.
   For each risk in each BU:

   (a) Define risk category, either beyond the organization's control or within the organization control
   (b) Define risk criteria (whether High-Medium-Low-none) for each of these six aspects: direct danger to health or life; inability of the organization to achieve its Minimum Business Continuity Objective (MBCO); non compliance of regulatory requirements; frequency of occurrence; financial lost; corporate image
   (c) Based on (a), (b) and (d), define and describe risk treatment for each of these four aspects: risk avoidance; risk reduction; risk transfer; risk acceptance
   (d) Define level of risk based on its frequency (High-Medium-Low (H-M-L)) and severity (H-M-L) (risk level analysis)—matrix form

L.S.R. Supriadi, L. Sui Pheng, *Business Continuity Management in Construction*, Management in the Built Environment, DOI 10.1007/978-981-10-5487-7

3. Disaster identification (for key disaster scenario)—Disaster may be compiled from one or more identified risks;

   Define a key disaster scenario. A disaster may be compiled from one or more identified risks (from point 2), not focusing on the BU yet, only the risks; For each disaster:

   (a) Define the level of severity in magnitude (can be in monetary losses; H-M-L) with description
   (b) Describe the results in severe to organization's ability to conduct business
   (c) Define the likelihood of occurrence (H-M-L) with description
   (d) Define whether the disaster can occur at worst possible time (yes or no and when)
   (e) Describe other risk (defined or not) involved due to the disaster

4. Risk review (per BU); For each BU:

   (a) Review its operational process, for the defined risk. Describe what will need to be controlled in that process regarding the defined risk:

       – Control to detect and inhibit risk occurrences (description)
       – Control to compensate for impact of risk occurrences (description)
       – Procedures for back-up and restoration of organization's vital records (description)

   (b) Review how each risk will have impacts toward the infrastructure of each BU. Describe these aspects:

       – Technology impacted
       – Technology needed
       – Equipment impacted
       – Facilities in BU impacted: Facilities; location of facilities; essential utilities and telecommunications; transportation to premises; physical security of premises.

5. Consolidate the findings of these risk analysis and review:

   – For each BU: defined risks, their treatments, their levels and review on the impacted operational processes and infrastructures.

   Defined key disaster scenario.

If the firm plans to implement this principle, consider these aspects for implementing this phase (regarding its current condition):

- Contact details for interdependent entities should regularly be maintained.
- A possible solution such as remote access should be considered.
- Alternative contacts should be identified.
- Effort in minimizing exposures to disasters and operations/systems vulnerabilities should be conducted.

Organizational culture attributes that can support this action plan:

1. In conducting risk analysis and cost-benefit analysis:
   - The firm should be open to alternative solutions.
   - The firm should encourage creative and innovative ideas.
   - The firm should be able to allocate sufficient resources for implementing innovative ideas.
   - Good communication is conducted.
   - The staffs and management are able to reach agreement on critical issues.
   - The firm has a high responsibility for its employee's welfare.
2. Involvement of experts and BCM committee:
   - The experts and BCM committee's commitment to this process are highly valued.
   - The experts and BCM committee should fit very well in the risk analysis team and their opinions are appreciated.
   - The member of the risk analysis team is enabled to solve the problems within their sphere of responsibility/authority.
3. In conducting a detailed risk review:
   - Organizational procedures are essential.
   - Unfamiliar situations should be managed and identified.
   - Good communication is conducted.
   - Inter-departmental collaboration is encouraged in this phase.
   - The firm should allocate sufficient resources for implementing innovative ideas.

   Several significant drivers to implement this principle:

- The firm's and its staff's awareness of their potential business risks.
- The awareness of this principle as a support for better organizational planning.
- The non-compliance impact of this principle is not small and may provide high costs.
- This principle provides effectiveness for better organizational planning.
- Top management fully supports this principle.
- This principle complies with the regulation.
- This principle is conducted to maintain/improve firm's reputation.
- This principle is already part of the company culture.
- This principle will complement/improve the firm's procedures.

Business Impact Analysis (BIA)

Note: Risk Analysis = focus on risk environment surrounding business functions/activities

BIA = assess the potential impact of loss if certain business functions/activities fail

MBCO = minimum level of services and/or products that is acceptable to the organization to achieve its business objectives

Technical action plans:

1. Business function (BF) and MBCO (minimum business continuity objective);
   The executive management establishes MBCO:

   (a) For each BU, define the BF of the BU
   (b) For each BF, define MBCO of service/product during incident/emergency/disaster (description of level of service or minimum product delivered)
   (c) MBCO can be influenced by:

   - Regulatory or contractual requirements
   - Industry regulations or recommended practices
   - Organizational vision, mission and policies
   - Highlights from audit reports
   - Case histories and past incidents
   - Risks due to existing environment and operations
   - Corporate governance and due diligence

2. Establish priority for analyzing impact (per disaster);
   For each disaster (from Risk Analysis and Review point 3): define what BUs are affected, where for each affected BU, define the BFs that will be affected. For each affected BF:

   (a) Choose and describe the impact on business:

   - Potential non compliance to external requirements and restrictions
   - Potential non compliance to organizational policies
   - Probable backlogs and resources needed to clear them
   - Impact due to concentration of critical staff, information or production assets at a particular location

   (b) Describe the operational constraints:

   - Minimum requirements to operate each business functions
   - Special requirements to operate each business functions
   - Maximum tolerable downtime within which systems, applications or functions must be recovered after a disaster has occurred

   (c) Rank the five categories of priority for analyzing the impact [rank each category 1 (Highest priority)–5 (Lowest priority)], which are:

   - Life safety and health
   - Compliance to regulatory requirements
   - Political and marketing standing
   - Financial position
   - Quality concerns

3. Establish CBF (for each BU);

   For each BU, define its BFs and its BFs' MBCO (point 1).

   For each BF, define disasters that may affect the BF. For each defined disaster:

   (a) Choose (one or more) and describe the significant impact:

   - Potential non compliance to external requirements and restrictions
   - Potential non compliance to organizational policies
   - Probable backlogs and resources needed to clear them
   - Impact due to concentration of critical staff, information or production assets at a particular location

   (b) Describe significant operational constraints:

   - Minimum requirements to operate each business functions
   - Special requirements to operate each business functions
   - Maximum tolerable downtime within which systems, applications or functions must be recovered after a disaster has occurred

   (c) Rank the five categories of priority for analyzing impact [rank each category 1 (Highest priority)–5 (Lowest priority)], which are:

   - Life safety and health
   - Compliance to regulatory requirements
   - Political and marketing standing
   - Financial position
   - Quality concerns

   (d) Based on (a) (b), (c): define whether the BF is critical or not (CBF: yes or no)

4. Dependencies of each CBF (each BF can span across one or more business operations); CBF requirements (this construct relates to previous CBF dependencies construct);

   a. Define and describe the CBF's dependencies (for each CBF):

   - In BU/Business process
   - Intra-department (to which department)
   - Inter-department (to which department)
   - On external parties (to which party)

   b. For each CBF, define and describe (based on 4a):

   - Potential loss impact if not recovered on time
   - Recovery time objective (RTO)
   - Recovery point objective (RPO)

   c. Resource requirements and capabilities (inventory for each BU).

      Defining inventories for each BU. For each CBF in the BU:

   - Describe the technology used
   - Describe the equipment used (computing/non-computing)

- Describe facilities used to support CBF
- Describe key staff to support CBF and its potential replacement (alternative)

BCM Steering Committee (BCM-SC) will review on: BIA, MBCO, CBF, Recovery priority of CBF.

Example of construction firm's CBFs and MBCOs:

CBF	Minimum Business Continuity Objective (MBCO)
Legal Unit (Contract Management)	Managing legal and contractual matters
System and Business Development Unit (Integrated management system and documentation)	Maintaining the company's integrated management system; Managing and protecting the company's business development documentation
Investors relations Unit (Relationship management)	Managing relationships with the company's investors and clients
Marketing Unit (Marketing Intelligence)	Managing the company's market forces and future business development
Project Development Unit (Project management)	Managing the company's existing and current projects (head office—project site management)
Accounting Unit (Financial Management)	Managing financial matters
Human Resources Unit (Human resources management)	Managing the company's human resources (employed and outsourced)
Equipment Unit (Equipment operations and maintenance)	Managing equipment resources used by the company for all projects
Logistic/Procurement Unit (Procurement management)	Managing logistics for all projects
Safety, Health, Environment Unit (Safety management and environmental management)	Managing safety for the human resources in the company
Project Control Unit (Project management)	Managing the project during execution; Managing project sites (materials, equipment, facilities) and labor
Customer Care Unit (Customer relationships)	Managing relationships with customers (before, during and after projects)
General Affairs Unit (Stakeholder management and customer relationships)	Managing all affairs related with the company (representing the company to external parties)
Property and Building Management Unit (Warranty, maintenance and internal asset management)	Managing the company's physical assets (head office, project office, resources)

***Additional recommendations regarding CBF and MBCO:*

1. These business functions below are viewed to be very important to construction business:

- Project documentation (contract documents, reports, drawings)
- Data storage should be functioning well
- Communication function (to coordinate between head office and project sites)

- Accounting (payments and invoices)
- HRD that manages their employees and labors. The lives of the people are the main critical function for the firm (employee's emergency shelter)
- Heavy equipment and materials management (should have equipment/material warehouses to protect them during crises)

2. Aspects to consider in determining the MBCOs:

- CBFs should be secured, not lost/damaged and communication between those divisions should be continued regularly
- The safety of human resources (employees and labors)
- The safety of documentations and communications
- The safety of physical assets (heavy equipments, offices)

Example of construction firm's internal (intra-department and inter-department) and external coordination:

Coordination	Description
Head office and project-base office	
How do they develop, communicate and distribute the reports on performance progress?	Regular reports (daily, weekly, monthly) progress and performance reports were developed and distributed via regular meetings and emails The reports are from project site office (by the project manager (PM)) to Head office (Project Development Unit)
What types of communication are there for this relationship? (verbal; non-verbal) Who is in charge for the communication between them?	Verbal: meetings Non-verbal: reports, notes, letters, emails, social media Persons in charge: PM and his/her administration team Project development and construction division
How to allocate resources (from head office to project office; in the project office)?	Permanent employees are assigned to projects Outsourced employees—based on the number needed, the employees will be outsourced from Head office or locals in the project
What types of infrastructure are there for coordination between head office and project-based office?	• Document storage • IT infrastructures (computers, telephone, facsimile, laptops, smartphones)
Head office and governments; public; clients; suppliers (supply chain); sub-contractors	
What types of communication are there for these relationships? (verbal; non-verbal) Who is in charge for the communication between them?	Verbal: meetings; conference; workshops Non verbal: emails, letters, reports Persons in charge: • External parties' representatives • Management representatives and Managers under Board of Directors
What types of coordination/reports are there for these relationships? How do they develop, communicate and distribute the reports?	• Progress and performance reports to clients • Other reports related to projects that need to be distributed/informed to government/public/supplier/sub-contractors • Financial, technical and administration

(continued)

Coordination	Description
	matters/coordination to suppliers and subcontractors These reports will be addressed by providing meetings or distributing them via mails/emails
What are the resources needed for these relationships? How to allocate resources between these relationships?	Human resources and documentation are needed Scheduled coordination between parties should be developed
What types of infrastructure are there for coordination between these relationships?	• Document storage • IT infrastructures (computers, telephone, facsimile, laptops)
Project-base office and public; suppliers; sub-contractors	
What types of communication are there for these relationships? (verbal; non-verbal) Who is in charge for the communication between them?	Verbal: meetings; conference; workshops Non verbal: emails, letters, reports Persons in charge: External parties representatives PM for the project and the resident engineer and administration teams
What types of coordination/reports are there for these relationships? How do they develop, communicate and distribute the reports?	• Progress and performance reports to suppliers and subcontractors • Other reports related to projects that need to be distributed/informed to government/public/supplier/sub-contractors • Financial, administration, and technical matters/coordination to suppliers and subcontractors These reports will be addressed by providing meetings or distributing them via mails/emails
What are the resources needed for these relationships? How to allocate resources between these relationships?	Human resources and documentation are needed Scheduled coordination between parties should be developed
What types of infrastructure are there for coordination between these relationships?	• Document storage • IT infrastructures (computers, telephone, facsimile, laptops)

If the firm plans to implement this principle, consider these aspects for implementing this phase (regarding its current condition):

- Critical systems are periodically evaluated and their minimum essential features should be provided for in a disaster.
- The firms should have good coordination between employees, suppliers/vendors to overcome the occurrence of a disaster and during recovery.
- Daily transactions that consist of critical data should be rotated off-site with adequate frequency.
- Critical operations and systems documentation should also be stored off-site (as back-ups).
- Appropriately skilled IT personnel or equipment specialists should be available.

- Ready access to public network following a disruption should be available.
- Access to communications hardware (e.g. fax, email, pager, etc.) should be prepared.
- Vital records should be stored in multiple locations.
- Each business unit and function needs more coordination holistically. Defining the interdependencies of each BU is necessary for developing a holistic procedure to overcome crises.
- Communication procedures should be provided in details in order to reduce waiting time or delays.

Organizational culture attributes that can support this action plan:

1. In conducting Business Impact Analysis (BIA):
   - The firm should encourage creative and innovative ideas.
   - The firm should be open to alternative solutions.
   - The firm should be able to allocate sufficient resources for implementing innovative ideas.
2. Involvement of BIA experts and employees:
   - The firm should support the member of the BIA team to participate in the decision-making process.
   - The firm should emphasize on team contributions.
   - The BIA team is highly valued by the firm.

Several significant drivers to implement this principle:

- The awareness of this principle as a support for better organizational planning.
- This principle provides effectiveness for better organizational planning.
- Top management fully supports this principle.
- This principle will complement/improve the firm's procedures.
- The firm's and its staff's awareness of their potential business risks.
- This principle can improve the employee's safety and welfare.
- This principle is conducted to maintain/improve firm's reputation.
- This principle can be easily integrated with the firm's other management systems.
- This principle complies with the regulation.
- Not implementing this principle may result in receiving regulative sanctions.
- The non-compliance impact of this principle is not small and may provide high costs.
- This principle is already part of the company culture.

Strategy Analysis(S)

Technical action plans:

1. BCM-SC should formulate the organization recovery strategy for BCM based on: selected disaster scenario; MBCO; CBFs
2. Recovery strategy selection (based on selected disaster—per disaster)

   For each defined and selected disaster:

(a) Define the CBF affected, including each of its MBCO, RTO and RPO
(b) For each CBF affected, choose (one or more) and describe probable strategy below:

   – Revert to alternate processing capability
   – Arrange reciprocal arrangements
   – Establish alternate site or business facility
   – Arrange for alternate source of supply
   – Outsource to external vendor
   – Transfer of operation to business units located elsewhere
   – Rebuild from scratch after disaster
   – Do Nothing
   – Other strategy

(c) For each chosen strategy, describe strategy evaluation below:

   – Costs to implement the strategy
   – Availability of alternate strategies
   – Costs of alternate strategies
   – Ease of implementing strategies
   – Comparison of the time needed to re-establish the CBF by the strategy and alternate strategies
   – Potential security breaches or control lapses due to the atypical measures associated with the strategy
   – Long term costs to maintain the strategy

(d) For each chosen strategy, describe strategy selection consideration below:

   – Skills set required by supporting staff
   – Technology and equipment
   – Facilities
   – Offsite storage and alternate site(s)
   – Alternate processing capabilities
   – Alternate processing facilities (either gained through acquisitions, mutual agreement, outsourcing to external vendors or manual workarounds)
   – Criteria for outsourcing to external vendors for alternate facilities (review of existing capacity, costs, location, technology and capacity of vendor's facilities)
   – Non-technology continuity issues

If the firm plans to implement this principle, consider these aspects for implementing this phase (regarding its current condition):

- A sufficient number of qualified personnel should be available to perform user tasks during the recovery phase.
- Personnel who play a role in recovery should be aware of their responsibilities and should have been adequately trained to perform the recovery tasks.
- Staff support areas should be prepared to support the recovery operation.
- Alternative processing facilities should be available as and when required.
- The organization should have access to a fully configured second processing site sufficient in capacity to support data processing for critical business processes with critical application support needs.
- Critical users (from the CBFs) should have the ability to reconstruct any lost work in-progress.

Organizational culture attributes that can support this action plan:

1. In conducting strategy analysis:
   - The firm should encourage creative and innovative ideas.
   - The firm should be open to alternative solutions.
   - The firm should be able to allocate sufficient resources for implementing innovative ideas.
   - Good communication is conducted.
   - The firm should value employee's ideas.
   - The staffs and management are able to reach agreement on critical issues.
2. In determining staff to support recovery:
   - A high degree of cooperation among employees should be conducted.
   - The firm should invest in the development of employee's skills.
   - The firm should provide guidance for employee's performance improvement.
   - Amicable opinions and ideas exchange between employees are facilitated by the firm.

Several significant drivers to implement this principle:

- The awareness of this principle as a support for better organizational planning.
- This principle provides effectiveness for better organizational planning.
- Top management fully supports this principle.
- The firm's and its staff's awareness of their potential business impacts.
- This principle is conducted to maintain/improve firm's reputation.
- This principle will complement/improve the firm's procedures for facing crises.
- This principle can be easily integrated with the firm's other management systems.
- The non-compliance impact of this principle is not small and may provide high costs.
- This principle is already part of the company culture.
- This principle can improve the employee's safety and welfare.

BC Plan Development (BCP)
Not Applicable
Tests and Exercises (TE)
Not applicable
Programme Management (PM)
Not applicable

Action Plans Level 2

Risk Analysis and Review (RA)

It is concluded that (based on the assessment phase) the Risk Analysis and Review process have been conducted by this firm.

Business Impact Analysis (BIA)

Technical action plans:

Consolidate the findings of the BIA process:

- Define MBCO
- Define CBFs (from disaster-BU-BF-MBCO) and their: dependencies; requirements (loss impact, RTO, RPO); resource requirements and capabilities

BCM Steering Committee reviews those analyses (BIA, MBCO, CBF, recovery priority of CBF).

Review point 1–4:

Note: Risk Analysis = focus on risk environment surrounding business functions/activities

BIA = assess the potential impact of loss if certain business functions/activities fail

MBCO = minimum level of services and/or products that is acceptable to the organization to achieve its business objectives

1. Business function (BF) and MBCO (minimum business continuity objective);
   The executive management establishes MBCO:
   (a) For each BU, define the BF of the BU
   (b) For each BF, define MBCO of service/product during incident/emergency/disaster (description of level of service or minimum product delivered)
   (c) MBCO can be influenced by:
      - Regulatory or contractual requirements
      - Industry regulations or recommended practices
      - Organizational vision, mission and policies
      - Highlights from audit reports
      - Case histories and past incidents
      - Risks due to existing environment and operations
      - Corporate governance and due diligence

2. Establish priority for analyzing impact (per disaster);

   For each disaster (from Risk Analysis and Review point 3): define what BUs are affected, where for each affected BU, define the BFs that will be affected. For each affected BF:

   (a) Choose and describe the impact on business:

      - Potential non compliance to external requirements and restrictions
      - Potential non compliance to organizational policies
      - Probable backlogs and resources needed to clear them
      - Impact due to concentration of critical staff, information or production assets at a particular location

   (b) Describe the operational constraints:

      - Minimum requirements to operate each business functions
      - Special requirements to operate each business functions
      - Maximum tolerable downtime within which systems, applications or functions must be recovered after a disaster has occurred

   (c) Rank the five categories of priority for analyzing the impact [rank each category 1 (Highest priority)–5 (Lowest priority)], which are:

      - Life safety and health
      - Compliance to regulatory requirements
      - Political and marketing standing
      - Financial position
      - Quality concerns

3. Establish CBF (for each BU);

   For each BU, define its BFs and its BFs' MBCO (point 1).

   For each BF, define disasters that may affect the BF. For each defined disaster:

   (a) Choose (one or more) and describe the significant impact:

      - Potential non compliance to external requirements and restrictions
      - Potential non compliance to organizational policies
      - Probable backlogs and resources needed to clear them
      - Impact due to concentration of critical staff, information or production assets at a particular location

   (b) Describe significant operational constraints:

      - Minimum requirements to operate each business functions
      - Special requirements to operate each business functions
      - Maximum tolerable downtime within which systems, applications or functions must be recovered after a disaster has occurred

(c) Rank the five categories of priority for analyzing impact [rank each category 1 (Highest priority)–5 (Lowest priority)], which are:

- Life safety and health
- Compliance to regulatory requirements
- Political and marketing standing
- Financial position
- Quality concerns

(d) Based on (a), (b), (c): define whether the BF is critical or not (CBF: yes or no)

4. Dependencies of each CBF (each BF can span across one or more business operations); CBF requirements (this construct relates to previous CBF dependencies construct);

(a) Define and describe the CBF's dependencies (for each CBF):

- In BU/Business process
- Intra-department (to which department)
- Inter-department (to which department)
- On external parties (to which party)

(b) For each CBF, define and describe (based on 4a):

- Potential loss impact if not recovered on time
- Recovery time objective (RTO)
- Recovery point objective (RPO)

(c) Resource requirements and capabilities (inventory for each BU).
Defining inventories for each BU. For each CBF in the BU:

- Describe the technology used
- Describe the equipment used (computing/non-computing)
- Describe facilities used to support CBF
- Describe key staff to support CBF and its potential replacement (alternative)

BCM Steering Committee (BCM-SC) will review on: BIA, MBCO, CBF, Recovery priority of CBF.

Example of construction firm's CBFs and MBCOs:

CBF	Minimum Business Continuity Objective (MBCO)
Legal Unit (Contract Management)	Managing legal and contractual matters
System and Business Development Unit (Integrated management system and documentation)	Maintaining the company's integrated management system; Managing and protecting the company's business development documentation

(continued)

CBF	Minimum Business Continuity Objective (MBCO)
Investors relations Unit (Relationship management)	Managing relationships with the company's investors and clients
Marketing Unit (Marketing Intelligence)	Managing the company's market forces and future business development
Project Development Unit (Project management)	Managing the company's existing and current projects (head office—project site management)
Accounting Unit (Financial Management)	Managing financial matters
Human Resources Unit (Human resources management)	Managing the company's human resources (employed and outsourced)
Equipment Unit (Equipment operations and maintenance)	Managing equipment resources used by the company for all projects
Logistic/Procurement Unit (Procurement management)	Managing logistics for all projects
Safety, Health, Environment Unit (Safety management and environmental management)	Managing safety for the human resources in the company
Project Control Unit (Project management)	Managing the project during execution; Managing project sites (materials, equipment, facilities) and labor
Customer Care Unit (Customer relationships)	Managing relationships with customers (before, during and after projects)
General Affairs Unit (Stakeholder management and customer relationships)	Managing all affairs related with the company (representing the company to external parties)
Property and Building Management Unit (Warranty, maintenance and internal asset management)	Managing the company's physical assets (head office, project office, resources)

***Additional recommendations regarding CBF and MBCO*:

1. These business functions below are viewed to be very important to construction business:
   - Project documentation (contract documents, reports, drawings)
   - Data storage should be functioning well
   - Communication function (to coordinate between head office and project sites)
   - Accounting (payments and invoices)
   - HRD that manages their employees and labors. The lives of the people are the main critical function for the firm (employee's emergency shelter)
   - Heavy equipment and materials management (should have equipment/material warehouses to protect them during crises)
2. Aspects to consider in determining the MBCOs:
   - CBFs should be secured, not lost/damaged and communication between those divisions should be continued regularly
   - The safety of human resources (employees and labors)

- The safety of documentations and communications
- The safety of physical assets (heavy equipments, offices)

Example of construction firm's internal (intra-department and inter-department) and external coordination:

Coordination	Description
Head office and project-base office	
How do they develop, communicate and distribute the reports on performance progress?	Regular reports (daily, weekly, monthly) progress and performance reports were developed and distributed via regular meetings and emails. The reports are from project site office (by the project manager (PM)) to Head office (Project Development Unit).
What types of communication are there for this relationship? (verbal; non-verbal) Who is in charge for the communication between them?	Verbal: meetings Non-verbal: reports, notes, letters, emails, social media Persons in charge: PM and his/her administration team; Project development and construction division
How to allocate resources (from head office to project office; in the project office)?	Permanent employees are assigned to projects; Outsourced employees—based on the number needed, the employees will be outsourced from Head office or locals in the project.
What types of infrastructure are there for coordination between head office and project-based office?	• Document storage • IT infrastructures (computers, telephone, facsimile, laptops, smartphones)
Head office and governments; public; clients; suppliers (supply chain); sub-contractors	
What types of communication are there for these relationships? (verbal; non-verbal) Who is in charge for the communication between them?	Verbal: meetings; conference; workshops Non verbal: emails, letters, reports Persons in charge: • External parties' representatives • Management representatives and Managers under Board of Directors
What types of coordination/reports are there for these relationships? How do they develop, communicate and distribute the reports?	• Progress and performance reports to clients; • Other reports related to projects that need to be distributed/informed to government/public/supplier/sub-contractors; • Financial, technical and administration matters/coordination to suppliers and subcontractors. These reports will be addressed by providing meetings or distributing them via mails/emails.
What are the resources needed for these relationships? How to allocate resources between these relationships?	Human resources and documentation are needed. Scheduled coordination between parties should be developed.
What types of infrastructure are there for coordination between these relationships?	• Document storage • IT infrastructures (computers, telephone, facsimile, laptops)

(continued)

Coordination	Description
Project-base office and public; suppliers; sub-contractors	
What types of communication are there for these relationships? (verbal; non-verbal) Who is in charge for the communication between them?	Verbal: meetings; conference; workshops Non verbal: emails, letters, reports Persons in charge: External parties representatives; PM for the project and the resident engineer and administration teams.
What types of coordination/reports are there for these relationships? How do they develop, communicate and distribute the reports?	• Progress and performance reports to suppliers and subcontractors; • Other reports related to projects that need to be distributed/informed to government/public/supplier/sub-contractors; • Financial, administration, and technical matters/coordination to suppliers and subcontractors. These reports will be addressed by providing meetings or distributing them via mails/emails.
What are the resources needed for these relationships? How to allocate resources between these relationships?	Human resources and documentation are needed. Scheduled coordination between parties should be developed.
What types of infrastructure are there for coordination between these relationships?	• Document storage • IT infrastructures (computers, telephone, facsimile, laptops)

If the firm plans to implement this principle, consider these aspects for implementing this phase (regarding its current condition):

- Critical systems are periodically evaluated and their minimum essential features should be provided for in a disaster.
- The firms should have good coordination between employees, suppliers/vendors to overcome the occurrence of a disaster and during recovery.
- Daily transactions that consist of critical data should be rotated off-site with adequate frequency.
- Critical operations and systems documentation should also be stored off-site (as back-ups).
- Appropriately skilled IT personnel or equipment specialists should be available.
- Ready access to public network following a disruption should be available.
- Access to communications hardware (e.g. fax, email, pager, etc.) should be prepared.
- Vital records should be stored in multiple locations.
- Each business unit and function needs more coordination holistically. Defining the interdependencies of each BU is necessary for developing a holistic procedure to overcome crises.
- Communication procedures should be provided in details in order to reduce waiting time or delays.

Organizational culture attributes that can support this action plan:

1. In conducting Business Impact Analysis (BIA):
   - The firm should encourage creative and innovative ideas.
   - The firm should be open to alternative solutions.
   - The firm should be able to allocate sufficient resources for implementing innovative ideas.
2. Involvement of BIA experts and employees:
   - The firm should support the member of the BIA team to participate in the decision-making process.
   - The firm should emphasize on team contributions.
   - The BIA team is highly valued by the firm.

   Several significant drivers to implement this principle:

- The awareness of this principle as a support for better organizational planning.
- This principle provides effectiveness for better organizational planning.
- Top management fully supports this principle.
- This principle will complement/improve the firm's procedures.
- The firm's and its staff's awareness of their potential business risks.
- This principle can improve the employee's safety and welfare.
- This principle is conducted to maintain/improve firm's reputation.
- This principle can be easily integrated with the firm's other management systems.
- This principle complies with the regulation.
- Not implementing this principle may result in receiving regulative sanctions.
- The non-compliance impact of this principle is not small and may provide high costs.
- This principle is already part of the company culture.

Strategy Analysis(S)

Technical action plans:

Consolidate findings of the recovery strategy selection:

Defined disaster → CBF affected (MBCO; RTO; RPO) → Probable strategy (evaluate and consider the aspects needed for the strategy selection)

Review point 1 and 2:

1. BCM-SC should formulate the organization recovery strategy for BCM based on: selected disaster scenario; MBCO; CBFs
2. Recovery strategy selection (based on selected disaster—per disaster)

   For each defined and selected disaster:

   (a) Define the CBF affected, including each of its MBCO, RTO and RPO
   (b) For each CBF affected, choose (one or more) and describe probable strategy below:

- Revert to alternate processing capability
- Arrange reciprocal arrangements
- Establish alternate site or business facility
- Arrange for alternate source of supply
- Outsource to external vendor
- Transfer of operation to business units located elsewhere
- Rebuild from scratch after disaster
- Do Nothing
- Other strategy

(c) For each chosen strategy, describe strategy evaluation below:

- Costs to implement the strategy
- Availability of alternate strategies
- Costs of alternate strategies
- Ease of implementing strategies
- Comparison of the time needed to re-establish the CBF by the strategy and alternate strategies
- Potential security breaches or control lapses due to the atypical measures associated with the strategy
- Long term costs to maintain the strategy

(d) For each chosen strategy, describe strategy selection consideration below:

- Skills set required by supporting staff
- Technology and equipment
- Facilities
- Offsite storage and alternate site(s)
- Alternate processing capabilities
- Alternate processing facilities (either gained through acquisitions, mutual agreement, outsourcing to external vendors or manual workarounds)
- Criteria for outsourcing to external vendors for alternate facilities (review of existing capacity, costs, location, technology and capacity of vendor's facilities)
- Non-technology continuity issues

If the firm plans to implement this principle, consider these aspects for implementing this phase (regarding its current condition):

- A sufficient number of qualified personnel should be available to perform user tasks during the recovery phase.
- Personnel who play a role in recovery should be aware of their responsibilities and should have been adequately trained to perform the recovery tasks.
- Staff support areas should be prepared to support the recovery operation.
- Alternative processing facilities should be available as and when required.
- The organization should have access to a fully configured second processing site sufficient in capacity to support data processing for critical business processes with critical application support needs.

- Critical users (from the CBFs) should have the ability to reconstruct any lost work in-progress.

  Organizational culture attributes that can support this action plan:

1. In conducting strategy analysis:
   - The firm should encourage creative and innovative ideas.
   - The firm should be open to alternative solutions.
   - The firm should be able to allocate sufficient resources for implementing innovative ideas.
   - Good communication is conducted.
   - The firm should value employee's ideas.
   - The staffs and management are able to reach agreement on critical issues.
2. In determining staff to support recovery:
   - A high degree of cooperation among employees should be conducted.
   - The firm should invest in the development of employee's skills.
   - The firm should provide guidance for employee's performance improvement.
   - Amicable opinions and ideas exchange between employees are facilitated by the firm.

   Several significant drivers to implement this principle:

- The awareness of this principle as a support for better organizational planning.
- This principle provides effectiveness for better organizational planning.
- Top management fully supports this principle.
- The firm's and its staff's awareness of their potential business impacts.
- This principle is conducted to maintain/improve firm's reputation.
- This principle will complement/improve the firm's procedures for facing crises.
- This principle can be easily integrated with the firm's other management systems.
- The non-compliance impact of this principle is not small and may provide high costs.
- This principle is already part of the company culture.
- This principle can improve the employee's safety and welfare.

BC Plan Development (BCP)

Technical action plans:

1. Identify triggers and response (per disaster);
   For each defined disaster:
   (a) Define CBF affected
   (b) Define its MBCO for the CBF affected
   (c) Describe criteria for disaster declaration based on three levels:

- Incident
- Emergency
- Disaster

(d) Establish the command and control structure to respond to incident; emergency; disaster situations (per disaster; for three different level of situation):

- Describe processes/strategy (based on CBFs)
- Describe key staff roles and responsibilities
- Describe communication procedure
- Describe logistical resources
- Describe EOC criteria: location; roles of EOC leader; when to use/not use EOC

2. Prioritize activities;
   Time sequence of a BC plan for a selected disaster:

   (a) Immediate response to an incident; emergency; or disaster
   (b) Subsequent recovery and resumption of CBF (selected strategies)
   (c) Restoration from temporary measures to return to business operations
   Activities and tasks should be prioritized based on the time sequence above.

3. Coordinate and finalize commitment;
   Each BU should have BCM team/personnel in their CBFs.
   Each BU should be tasked in the following:

   (a) Appoint an appropriate staff to participate in the organization BCM efforts
   (b) Establish prior arrangement with its suppliers on supply of raw materials, equipment and spares to support CBFs
   (c) Contribute to the content and approved portions of the BC plan pertaining to their business operations

4. Gather detailed requirements for each CBF;

   (a) List the pre-incident measures based on each risk:

      1. Define the CBFs affected
      2. Describe the chosen risk treatment (risk avoidance; risk reduction; risk transference; risk acceptance)

   (b) For each CBF from each defined disaster, describe recovery strategy (Recovery Strategy Development)
   (c) For each CBF from each defined disaster, describe IT & IS requirements:

      - IT systems and application programmes
      - Vital records
      - Electronic backup and storage
      - Manual procedures

(d) For each CBF from each defined disaster, describe communication requirements:

- Procedures
- Tools/media
- Stakeholders to communicate with
- Facilities for communications

(e) For each CBF from each defined disaster, describe security and controls needed during recovery:

- Information processing and information security (information processing, documentation, storage and retrieval, and access control)
- Physical access security and control
- Access and control

(f) For each CBF from each defined disaster, describe inventory lists:

- List of computer items (computer software requirements; communication requirements; computer hardware; computer data files formats)
- List of non computer items (hardcopy forms and documents; office equipment; file storage locations; transportation and logistics; Funding and source of funding)
- List of vital records (operations records; management records; accounting and financial records; intellectual property records)

(g) For each CBF from each defined disaster, describe critical item lists:

- Grab list (small items to grab during disaster evacuation)
- Priority Salvage list (hand-carried items)

(h) For each CBF from each defined disaster, describe list of names from related CBFs for BC operations:

- Disaster declaration officer
- Plan execution teams (Damage Assessment Team (DAT); Emergency Response and Management Team; Recovery Team; Salvage Team; Restoration Team)
- Damage Assessment Team (Produce: Report that identifies the disrupted operations, estimates the downtime, and recommends the next course of action) - Staff with skills in testing CBFs, safety regulations and procedures, and technical specialists)

(i) For each CBF from each defined disaster, describe assembly area and personnel safety

(j) For each CBF from each defined disaster, describe hazardous material handling:

- Purchase or hire of appropriate equipment and facilities to handle the hazardous materials

- Training of personnel to operate and maintain these equipment and facilities
- Procedures to operate and maintain these equipment and facilities

(k) For each CBF from each defined disaster, describe the medical attention needed:

- Immediate first aid and medical treatment
- Isolation and decontamination
- Seeking of professional advice and assistance

(l) For each CBF from each defined disaster, list and describe related contact list needed:

- Personnel contact list (telephone call and notification tree)
- Vendors contact list
- Stakeholders contact list (informed at the earliest possible moment) (Ex: regulatory agencies, media, key customers, board of Directors, shareholders)

(m) For each CBF from each defined disaster, describe infrastructure requirements:

1. General equipment requirements:
   - Replacement of equipment (costs, availability, delivery time)
   - Supplies to operate equipment
   - Spares to operate equipment
   - Customization of equipment and their configurations
   - Protection of equipment
   - Documentation to set up, operate and maintain equipment
2. Additional equipment requirements (for supporting recovery operation)
3. Critical equipment and supplies lead times
4. Computing equipment requirements (such as PC, video, voice and data communications equipment, network connectivity)
5. Communication requirements:
   - Voice communications (telephone lines; voice mail; automated voice response; call center)
   - Data communications (IS mainframe database; LAN; WAN; Email etc.)

(n) For each CBF from each defined disaster, describe facilities needed for:

1. EOC, with characteristic descriptions:
   - Activation time
   - Distance
   - Means of communications
   - Space

- Availability of extended time
- Accessibility to basic amenities (ex: toilets, cafeteria)
- Security
- Alternate Site requirements (work area requirements) (office space, furniture, equipment and supplies, power supply, AC, Telecommunications, Transportation and storage logistics, food and lodging)
- Facility protection (for on-site facilities)
- Off-site storage (for back up of vital records, etc.)

2. Restore and return after disaster (determine when and how on the descriptions below):

   - Restoration of non critical or business functions that became critical because of the recovery time has exceeded its tolerable downtime
   - Salvage operations
   - Renovation, retrofitting or rebuilding of facilities
   - Preparation of facilities
   - Transportation
   - Audit and corporate governance requirements needed during restoration and return to normal

5. Checklists for writing the BC Plan (based on the tables and procedures/lists);

   (a) Policy (by the Executive Management) on:

   1. Priorities of actions:

      - Saving and preserving human lives should overrule all other considerations
      - Reducing the impact of the risk occurrence from escalating to a higher level
      - Salvaging and restarting affected business operations and functions

   2. Emergency response:

      - Stabilizing the situation immediately following an emergency declaration
      - Deciding on the disaster declaration after the initial damage assessment
      - Communicating with internal and external parties
      - Interacting with external agencies and regulatory authorities
      - Ensuring security of personnel, information and physical premise

   (b) BC Plan structure:

   1. ER Plan activation:

      - Detection and determination
      - Emergency Response
      - Disaster notification and declaration

- Controls and security
- Activation of EOC
- Activation of alternate site
- Including Disaster Recovery Plan (DRP) for IT systems

2. Recovery and resumption operations:

- Recovery of time sensitive CBFs
- Resumption of CBFs at alternate sites
- Crisis communications

(c) Procedures: (Policy–Process–Procedure)
Main set of procedures in a BC Plan:

a. Initial Damage Assessment (guiding the assessment of the extent of the damage when an incident occurs)
b. Damage Assessment
c. Emergency Response Procedures, which should:

1. Respond to the list of potential emergencies drawn up
2. Escalate response to the appropriate level if an incident situations worsens
3. Stabilize the situation immediately following a disaster declaration, which caters:

- Summon resources to minimize loss of human lives
- Provide an action plan for site safety, security of personnel, information and assets
- Assemble the relevant resources at the affected site
- Minimize or contain the disaster
- Perform initial and full damage assessment
- Coordinate salvage operations
- Set up and operate the EOC

d. Crisis Communications (parties involved: Executive Management; Employees and their immediate family members; Customers; Shareholders; Vendors; Media; Authorities)
e. Coordination with External Agencies

6. Confirm the BC plan;
Confirm this to: Management; Personnel involved with the plan
By:

- Briefings
- Verification (by the BU and its recovery teams)
- Final management endorsement

7. Distribute BC Plan; Not all BU require the entire BC Plan content; Based on need to know and need to hold basis

If the firm plans to implement this principle, consider these aspects for implementing this phase (regarding its current condition):

- The recovery plan should cover any event that simultaneously renders both the primary and all alternative data/resource center facilities inoperable.
- The recovery plan should cover any event which simultaneously renders the data centre inoperable and the essential off-site storage inaccessible.
- Critical users (from CBFs) should have recovery plans developed to be able to proceed at the alternative processing facility.

Some strategic approaches that can be implemented by the firms in developing BC Plan:

- Human resources and responsibilities: Leadership rather than domination is necessary.
- BC planning and processes: Include strategic partnership beyond the organization's boundaries; Planning undertaken by functions/BUs with coordination supplied by the BC team to improve ownership of plans.
- Communications and structure: Use formal and informal communications management infrastructure to disseminate messages about BCM's importance; Tailor choice and use of media to improve the trajectory and understanding of BCM related communications; Appoint formal coordinators to underpin communications endeavours across the organizations.
- Attitudes and ownership of the BC plan: Functions and departments must have (part) ownership of the planning process coupled with formal appraisals; Functions and departments must have an understanding of how crises and interruptions can threaten the organization's operations and advantages.
- The firm should also assess its stakeholders, whether they are aware of BCM or not. The stakeholders should also be coordinated and communicated about the firm's BCM.

Some risks that may occur during crises that relate to the firm's stakeholders:

- Cascading failures from compound disaster events (e.g. earthquake, tsunami);
- Failure of unexpectedly vulnerable support systems;
- Inability to get workers to their posts because of transportation infrastructure damage;
- Supply chain failure concentrated on suppliers and smaller firms without BC plan;
- Too-rigid BC plans left some firms unable to adapt to the shifting challenges produced by the crises;
- Communications failures between public and private sectors and across national boundaries within large organizations.

Some risk treatments that can overcome the risks above:

- Clearly identify all key support systems, work to reduce their vulnerability to likely hazards;

- Make provisions in BC plan for post-disaster transportation disruptions;
- Push BCP practices to suppliers;
- Ensure that the BC plan is flexible and is keyed to address common impacts and protect key processes;
- Increase communications between private-sector entities across industries to improve collective action after a disaster.

Organizational culture attributes that can support this action plan:
In developing the detailed BC plan:

- The firm should have a main focus in meeting the client's needs and satisfaction.
- The firm should have a detailed set of related performance standards.
- The firm should have relevant internal procedures, particularly for documentation process.
- The firm should be able to create adaptive ways to meet changing needs.
- The firm should have a high-level of cost consciousness.
- The firm should be able to allocate sufficient resources for implementing innovative ideas.
- The staffs and management are able to reach agreement on critical issues.

Several significant drivers to implement this principle:

- This principle provides effectiveness for better organizational planning.
- The firm's and its staff's awareness of their potential business impacts.
- This principle is conducted to maintain/improve firm's reputation.
- This principle complies with the regulation.
- This principle is already part of the company culture.
- The non-compliance impact of this principle is not small and may provide high costs.
- Not implementing this principle may result in receiving regulative sanctions.

Tests and Exercises (TE)

Technical action plans:

1. Establish practice to operate BC Plan;
   Practice levels:
   (a) Generic (fundamental BC concepts)
   (b) Plan (operation of the established BC plan)
2. Prepare for tests and exercises;
   Checklists:
   (a) Tests and exercise master schedule
   (b) Objectives:
      - Verify that the BC Plan is viable and practical

- Verify that the recovery time scale and priorities can be met
- Verify that the vendors identified in the BC plan can support the recovery in a timely, efficient and effective manner
- Verify that the resources identified in the BC plan can be activated and accessed in a timely, efficient, effective and adequate manner
- Rehearse personnel involved in the actual recovery
- Identify areas to be improved or fine tuned

(c) Methods used:

- Unannounced or surprised
- Announced or scheduled
- Walkthrough
- Functional or specific business units
- Simulations

(d) Key resources for test and exercises:

- Availability of personnel
- Equipment
- Facilities

(e) Budget for tests and exercises

If the firm plans to implement this principle, consider these aspects for implementing this phase (regarding its current condition):

- Testing and exercising should involve interdependent entities.
- Periodic testing and exercising of the BC plan should be conducted.

Organizational culture attributes that can support this action plan:
In providing periodic tests and exercises on BC Plan:

- The firm should have its Business Units with a high capability to work together.
- The employees (and management) should be able to accept criticism and negative feedback.
- A high degree of cooperation among employees should be implemented.
- The firm should be able to resolve internal problems effectively.
- The firm should be able to allocate sufficient resources for implementing innovative ideas.

Several significant drivers to implement this principle:

- This principle provides effectiveness for better organizational planning.
- This principle is conducted to maintain/improve firm's reputation.
- This principle will complement/improve the firm's procedures.
- Top management fully supports this principle.
- This principle complies with the regulation.
- The non-compliance impact of this principle is not small and may provide high costs.

- This principle can be easily integrated with the firm's other management systems.

Programme Management (PM)

Programme Management for BCM should have been discussed and briefly planned for implementation by the management.

The attributes needed are:

- There should be a formal management commitment to the organization of BCM.
- The BCM policy has been established.
- The deliverable of the initiated stage is BCM as an initiative.

***Additional recommendations to implement Programme Management*:

The regulator/ICA/NBCSD's roles are to:

- Raise the awareness of the importance of BCM to businesses;
- Increase the supply of BCM specialists that can support contractors to become BCM ready; make BCM more accessible to the business community; and
- Support contractor's efforts to become BCM certified through an incentive programme.

The firm's leader should also be able to identify and measure the impacts from certain threatening events toward their business, which can be used for business continuity planning. This plan should not only focus on the firm's location or physical assets, but also on the people and their daily operations. Throughout the planning process, communication and plan drilling/exercises should regularly be conducted by the employees for better awareness and preparedness toward unexpected events. This plan should also be regularly evaluated by the management.

In developing and applying the business continuity planning, other than providing steps to overcome disasters and testing the plans, factors that are also needed to be included are:

- Full support and approval from the senior management.

  The senior management in the firm has the major role and responsibility for implementing BCM. During the unexpected events occur, they are responsible for the whole BC plan implementation and approvals.
- Building the awareness towards the BC plan to all employees in the firm.

  Awareness towards BC plan in the firm is necessary because the ability of the firm to overcome crises or recover/restore from certain unexpected situation depends on the employee's ability and preparedness. A specific training regarding BCM may be needed for some selected employees (who are championed to lead the BCM process in their department/division) in order to be able to lead and improve their commitment in business continuity planning.
- Maintaining the plan, including updating it when necessary.

BC plan should always be evaluated, to see whether there are new updates from the BCM standards or BCM best practices worldwide.

Action Plans Level 3

Risk Analysis and Review (RA)

It is concluded that (based on the assessment phase) the Risk Analysis and Review process have been conducted by this firm.

Business Impact Analysis (BIA)

It is concluded that (based on the assessment phase) the Business Impact Analysis process have been conducted by this firm.

Strategy Analysis (S)

It is concluded that (based on the assessment phase) the Strategy Development process has been conducted by this firm.

BC Plan Development (BCP)

It is concluded that (based on the assessment phase) the BC Plan Development process has been conducted by this firm.

Some strategic approaches that can be implemented by the firms in developing BC Plan:

- Human resources and responsibilities: Leadership rather than domination is necessary.
- BC planning and processes: Include strategic partnership beyond the organization's boundaries; Planning undertaken by functions/BUs with coordination supplied by the BC team to improve ownership of plans.
- Communications and structure: Use formal and informal communications management infrastructure to disseminate messages about BCM's importance; Tailor choice and use of media to improve the trajectory and understanding of BCM related communications; Appoint formal coordinators to underpin communications endeavours across the organizations.
- Attitudes and ownership of the BC plan: Functions and departments must have (part) ownership of the planning process coupled with formal appraisals; Functions and departments must have an understanding of how crises and interruptions can threaten the organization's operations and advantages.
- The firm should also assess its stakeholders, whether they are aware of BCM or not. The stakeholders should also be coordinated and communicated about the firm's BCM.

Some risks that may occur during crises that relate to the firm's stakeholders:

- Cascading failures from compound disaster events (e.g. earthquake, tsunami);
- Failure of unexpectedly vulnerable support systems;
- Inability to get workers to their posts because of transportation infrastructure damage;

- Supply chain failure concentrated on suppliers and smaller firms without BC plan;
- Too-rigid BC plans left some firms unable to adapt to the shifting challenges produced by the crises;
- Communications failures between public and private sectors and across national boundaries within large organizations.

Some risk treatments that can overcome the risks above:

- Clearly identify all key support systems, work to reduce their vulnerability to likely hazards;
- Make provisions in BC plan for post-disaster transportation disruptions;
- Push BCP practices to suppliers;
- Ensure that the BC plan is flexible and is keyed to address common impacts and protect key processes;
- Increase communications between private-sector entities across industries to improve collective action after a disaster.

<u>Tests and Exercises (TE)</u>
Technical action plans:

1. Conduct tests and exercises;
   Checklists:

   (a) Level of frequency (periodic basis):

      – Discrete/individual level (CBF tested individually)
      – Integrated level (all CBFs tested together)

   (b) Using list of main disaster scenario (predetermined and designed for the overall exercise programme)
   (c) Areas and assessment, which should include these criteria:

      – Operation of the CBFs during tests and exercises conditions
      – Backup, offsite storage and retrieval of information supporting the CBFs
      – Critical processes supporting the CBFs
      – Staff and recovery teams supporting the CBFs
      – Alternate site(s), if they form part of the recovery strategy
      – Internal and external interfaces

   (d) Quality and standards (tests reviewed by relevant experts)

2. Assess the results;
   Checklists:

   (a) Effectiveness/efficiency of the processes—from the objectives to areas and assessment criteria (the plan viable or not/applicable or not/conducted in the planned manner or not)
   (b) Personnel assessment criteria:

- Attendance and preparedness (at their designated recovery locations and respective tasks)
- Roles, responsibilities and tasks execution
- Leadership, coordination and control (in accomplishing their team tasks within tight time constraints)
- Feedback from Participants (including from external parties)

3. Infrastructure to support tests and exercises;
   Checklists:

   (a) Technology and equipment (verification)

   - Timing for setup and operation of equipment
   - Vital records and resources (availability)

   (b) Verifying facilities:

   - Timing for setup and operation of facilities
   - Alternate site (at least tested once)

   (c) Criteria facilitating for alternate site: (whether complied/accomplished or not)

   - Logistics and transportation
   - Utilities
   - Telecommunications
   - Operation

4. Identify and implement corrective actions;
   Checklists:

   (a) From tests → identify corrective actions (per areas of assessment)
   (b) Implement corrective actions:

   - Post test and exercise (scheduled post test) for reviewing the recommendations
   - Waivers from unrectified recommendation and corrective action
   - Documentation for the whole process

If the firm plans to implement this principle, consider these aspects for implementing this phase (regarding its current condition):

- Testing and exercising should involve interdependent entities.
- Periodic testing and exercising of the BC plan should be conducted.

Organizational culture attributes that can support this action plan:
In providing periodic tests and exercises on BC Plan:

- The firm should have its Business Units with a high capability to work together.
- The employees (and management) should be able to accept criticism and negative feedback.

- A high degree of cooperation among employees should be implemented.
- The firm should be able to resolve internal problems effectively.
- The firm should be able to allocate sufficient resources for implementing innovative ideas.

Several significant drivers to implement this principle:

- This principle provides effectiveness for better organizational planning.
- This principle is conducted to maintain/improve firm's reputation.
- This principle will complement/improve the firm's procedures.
- Top management fully supports this principle.
- This principle complies with the regulation.
- The non-compliance impact of this principle is not small and may provide high costs.
- This principle can be easily integrated with the firm's other management systems.

Programme Management (PM)
Technical action plans:

1. Align BCM with organization operations; BCM can be aligned with the organization's operations (checklists);
   BCM can be aligned with the organization's operations via (checklists):
   (a) Policy
   (b) Structure and organization
   (c) Maintaining BCM planning methodology (ensuring a standardized process to update processes and to upkeep the program)
   (d) BC Plan maintenance → should be reviewed and revised to maintain its relevancy. Ongoing activities for BC plan review include:
      – Changes to the threat and risk profile
      – Upgrading and downgrading of CBFs
      – Strategy realignment of organization
      – Gaps in recovery procedures discovered during Tests and Exercises
   (e) Tests and exercises (From discrete to integrated tests)
   (f) BC Plan validation
   (g) Exploration of alternatives
   (h) Other things to consider:
      1. A permanent BCM Steering Committee (BCM SC), ensuring:
         - Risk analysis and review
         - BIA
         - Strategy
         - BC Plan
         - Tests and exercises

2. Organization BCM Coordinator → keep in tune with the latest BCM trends and development. Relevant activities:

   – Maintain a constant lookout for changes to the business operations and environment.
   – Initiate and propose BCM projects to the BCM SC when new risks arise
   – Ensure that test and exercises of the BC Plan are carried out on periodic and systematic basis and changes required are updated in the BC plan
   – Ensure that all staff in the organization are appropriately trained and informed of the latest BCM practices.

3. Review and updates on (for each CBFs):

   – Infrastructure
   – Technology and equipment
   – Facilities

4. BCM manual is created, with guidelines and instructions governing BCM activities. Two main topics:

   – The scope and conduct of Risk Analysis and Review, BIA, Strategy selection and maintenance of the BC plan
   – The policies, processes and procedures governing BCM activities in the organization

2. Review key BCM elements by BCM-SC (checklists);

   (a) Risks, Business Impacts and Recovery Strategies. Reviews on:

   – Environmental and operational risks
   – The list of risk acceptance
   – Risks and their impact on CBFs
   – Recovery strategies

   (b) MBCO review, to reflect:

   – Current regulatory requirements
   – Business operations
   – Environmental conditions
   – When significant changes necessitate a review of this objective

   (c) Reviewing roles and responsibilities of the following:

   – BCM-SC
   – Organization BCM coordinator
   – BU BCM coordinator
   – Disaster declaration offices
   – Teams involved in executing the BC plan

3. Review BC plan (minimum once a year by BCM-SC) (Key aspects of review—Checklists);

   (a) Content:

      – Contain a set of well defined and sequenced procedures to guide various recovery activities such that the number of decision to be made is minimized.
      – Specify the resources needed to carry out the activities specified by each procedure
      – Specify the tasks to be accomplished by each designated team
      – Specify the critical data to be recovered at each phase of the recovery

   (b) Currency of plan: (maintained with independent review)

      – BC Plan should be related to the organization's structure (support the BCM needs of different BU)
      – BC Plan should be tested and maintained on a predefined basis

   (c) Plan documentation maintenance:

      – BC Plan documentation maintained using automated means. Benefits: helps to reduce errors and ensure consistency and quality of the BC Plan
      – Using BC Plan software which supports inputs from various BCM teams → consolidate into the final BC document

   (d) Review vendors contracts: (should include BCM requirements clauses therein)

      – Materials
      – Outsourcing services
      – Telecommunications
      – Power supply
      – Utilities

   (e) Review reports. All reviews and findings should be all documented and reported to BCM SC.

4. Provide continuous training and awareness
   These are based on (checklists);

   (a) Levels

      – Basic → all staff
      – Management → Management staff
      – Specialized → Specific staff (IT support, recovery operations staffs, staff handling hazardous materials)

   (b) Assessments (at 2 level):

      – Programme level (the programme drawn up should meet the needs of BU supporting CBFs)
      – People level (assessing participants whether meeting the objectives set forth)

(c) Coverage:

1. Appointment holders (staff holding key BCM responsibilities); can be accomplished by:

   - Attending BCM planning conferences and seminars
   - Participating in BCM user groups and associations
   - Perusing BCM publications
   - Enrolling in formal BCM courses that are internationally recognized and accredited

2. All staff, trained by stipulated activities:

   - Evacuation and assembly
   - Activation of alarm
   - Emergency response
   - Reporting to the appropriate authority to handle the emergency situation

3. Specific staff (directly involved in the recovery operations; including external parties + staffs handling hazardous materials)

5. Perform BCM audit;

(a) Annual internal BCM audit
(b) External parties supporting the CBFs (review their capability to support the organization's CBFs during recovery from disaster)

6. Track BCM trends and practices
This can be done by (checklists):

(a) BCM meeting, to discuss:

1. Proposal for new BC initiatives
2. Status of BC initiatives being implemented
3. Changes and corresponding implications for BC planning and the BC plan:

   - Business and business operations
   - Industry practices
   - Laws or regulations

4. Schedule of tests and exercises and their coordination
5. Review of tests and exercises results and corrective actions
6. Review and revision:

   - Recovery time requirements of each CBF (RTO and RPO)
   - Strategy
   - BC Plan

(b) Participation in industry BCM activities, such as:

- Membership in industry BCM interest group
- Attendance at international or regional BCM conferences
- Organizing committee of conferences and seminars
- Presentation of paper at conferences and seminars

(c) Tracking industry development (related to BCM - Industry)
(d) Internal update and awareness via:

- Publishing periodic newsletter
- Organizing internal seminars or conferences
- Providing BCM update to the Executive Management

If the firm plans to implement this principle, consider these aspects for implementing this phase (regarding its current condition):

- There should be a sufficient number of personnel possessing the appropriate skills available to implement BC operations.
- There should be adequate maintenance of BC procedures.

***Additional recommendations to implement Programme Management:*
The regulator/ICA/NBCSD's roles are to:

- Raise the awareness of the importance of BCM to businesses;
- Increase the supply of BCM specialists that can support contractors to become BCM ready; make BCM more accessible to the business community; and
- Support contractor's efforts to become BCM certified through an incentive programme.

The firm's leader should also be able to identify and measure the impacts from certain threatening events toward their business, which can be used for business continuity planning. This plan should not only focus on the firm's location or physical assets, but also on the people and their daily operations. Throughout the planning process, communication and plan drilling/exercises should regularly be conducted by the employees for better awareness and preparedness toward unexpected events. This plan should also be regularly evaluated by the management.

In developing and applying the business continuity planning, other than providing steps to overcome disasters and testing the plans, factors that are also needed to be included are:

- Full support and approval from the senior management.
  The senior management in the firm has the major role and responsibility for implementing BCM. During the unexpected events occur, they are responsible for the whole BC plan implementation and approvals.
- Building the awareness towards the BC plan to all employees in the firm.
  Awareness towards BC plan in the firm is necessary because the ability of the firm to overcome crises or recover/restore from certain unexpected situation depends on the employee's ability and preparedness. A specific training regarding BCM may be needed for some selected employees (who are championed to

lead the BCM process in their department/division) in order to be able to lead and improve their commitment in business continuity planning.

- Maintaining the plan, including updating it when necessary.

  BC plan should always be evaluated, to see whether there are new updates from the BCM standards or BCM best practices worldwide.

Organizational culture attributes that can support this action plan:

1. In analyzing ongoing process of BCM:
   - The firm should have a detailed set of related performance standards.
   - The firm should have a high-level of cost consciousness.
   - The firm should have relevant internal procedures, particularly for documentation process.
   - The firm should have a main focus in meeting the client's needs and satisfaction.
   - The firm should have a high degree of clear strategic intentions.
   - The firm should emphasize on team accountability.
   - The leadership of the firm's top management has a high influence.
   - The firm should have its Business Units with a high capability to work together.
2. In conducting BCM training and awareness programme:
   - The firm should continually invest in the development of its employee's skills.
   - The firm should provide guidance for employee's performance improvement.
   - Amicable opinions and ideas exchange between employees are facilitated by the firm.
   - The firm emphasizes on team contributions.
   - The employee's commitment for this programme is highly valued by the firm.

Several significant drivers to implement this principle:

- The awareness of this principle as a support for better organizational planning.
- Top management fully supports this principle.
- This principle provides effectiveness for better organizational planning.
- The firm's and its staff's awareness of their potential business impacts.
- This principle will complement/improve the firm's procedures for facing crises.
- This principle complies with the regulation.
- Not implementing this principle may result in receiving regulative sanctions.
- This principle can be easily integrated with the firm's other management systems.
- This principle is already part of the company culture.
- This principle can improve the employee's safety and welfare.
- The non-compliance impact of this principle is not small and may provide high costs.

Action Plans Level 4

Risk Analysis and Review (RA)

It is concluded that (based on the assessment phase) the Risk Analysis and Review process have been conducted by this firm.

Business Impact Analysis (BIA)

It is concluded that (based on the assessment phase) the Business Impact Analysis process have been conducted by this firm.

Strategy Analysis (S)

It is concluded that (based on the assessment phase) the Strategy Development process have been conducted by this firm.

BC Plan Development (BCP)

It is concluded that (based on the assessment phase) the BC Plan Development process have been conducted by this firm.

Tests and Exercises (TE)

It is concluded that (based on the assessment phase) the Tests and Exercises process have been conducted by this firm.

Programme Management (PM)

Need to continue to implement the continuous improvement process for BCM of this firm.

Full support from the Management for:

- Periodic BCM review
- Periodic BCM audit
- Regularly conducting training and awareness programme on BCM
- Regularly tracking and updating BCM news and development

Appendix G

Analysis for Chap. 12

Example of manual calculations to determine the BCM level of preparedness of a firm using the fuzzy logic rules

		Membership function: (Triangular form)		
RatingScale:		a	b	c
1	not implemented	0	0	0.33
2	partially implemented	0	0.33	0.67
3	largely implemented	0.33	0.67	1
4	fully implemented	0.67	1	1

IF range number 0 to 0.165 THEN Index = 1
IF range number 0.166 to 0.495 THEN Index = 2
IF range number 0.496 to 0.835 THEN Index = 3
IF range number 0.836 to 1 THEN Index = 4
(INDEX FOR RA, BIA, S, BCP, TE AND PM)

L.S.R. Supriadi, L. Sui Pheng, *Business Continuity Management in Construction*, Management in the Built Environment, DOI 10.1007/978-981-10-5487-7

Calculation example:					
RISK ANALYSIS (RA) ASSESSMENT					
Statement rating:	RATING (scale 1 - 4)		Membership function		
			a	b	c
RA1	2		0	0.33	0.67
RA2	2		0	0.33	0.67
RA3	3		0.33	0.67	1
RA4	2		0	0.33	0.67
RA5	1		0	0	0.33
RA6	3		0.33	0.67	1
RA7	4		0.67	1	1
Total rating			0.19	0.475714	0.762857
Range number			0.47619		
RA index			2		

BUSINESS IMPACT ANALYSIS (BIA) ASSESSMENT					
Statement rating:	RATING (scale 1 - 4)		Membership function		
			a	b	c
BIA1	1		0	0	0.33
BIA2	1		0	0	0.33
BIA3	2		0	0.33	0.67
BIA4	2		0	0.33	0.67
BIA5	3		0.33	0.67	1
BIA6	3		0.33	0.67	1
BIA7	1		0	0	0.33
Total rating			0.094286	0.285714	0.618571
Range number			0.332857		
BIA Index			2		

STRATEGY DEVELOPMENT (S) ASSESSMENT					
Statement rating:	RATING (scale 1 - 4)		Membership function		
			a	b	c
S1	2		0	0.33	0.67
S2	2		0	0.33	0.67
S3	2		0	0.33	0.67
S4	3		0.33	0.67	1
S5	3		0.33	0.67	1
S6	2		0	0.33	0.67
S7	1		0	0	0.33
Total rating			0.094286	0.38	0.715714
Range number			0.396667		
S Index			2		

BC PLAN (BCP) DEVELOPMENT ASSESSMENT					
Statement rating:	RATING (scale 1 - 4)		Membership function		
			a	b	c
BCP1	1		0	0	0.33
BCP2	1		0	0	0.33
BCP3	2		0	0.33	0.67
BCP4	1		0	0	0.33
BCP5	3		0.33	0.67	1
BCP6	3		0.33	0.67	1
BCP7	1		0	0	0.33
Total rating			0.094286	0.238571	0.57
Range number			0.300952		
BCP Index			2		

TESTS AND EXERCISE (TE) ASSESSMENT				
Statement rating:	RATING (scale 1 - 4)	Membership function		
		a	b	c
TE1	2	0	0.33	0.67
TE2	2	0	0.33	0.67
TE3	2	0	0.33	0.67
TE4	2	0	0.33	0.67
TE5	1	0	0	0.33
TE6	2	0	0.33	0.67
TE7	3	0.33	0.67	1
Total rating		0.047143	0.331429	0.668571
Range number		0.349048		
TE index		2		

PROGRAMME MANAGEMENT (PM) ASSESSMENT				
Statement rating:	RATING (scale 1 - 4)	Membership function		
		a	b	c
PM1	2	0	0.33	0.67
PM2	2	0	0.33	0.67
PM3	1	0	0	0.33
PM4	1	0	0	0.33
PM5	2	0	0.33	0.67
PM6	2	0	0.33	0.67
PM7	3	0.33	0.67	1
Total rating		0.047143	0.284286	0.62
Range number		0.317143		
PM index		2		

RA INDEX	2
BIA INDEX	2
S INDEX	2
BCP INDEX	2
TE INDEX	2
PM INDEX	2
BCM LEVEL OF PREPAREDNESS:	2

Truth value rule: IF input 1 .... And input 2 ...... THEN OUTPUT

No.	Input 1	Input 2	Output (Intermediate index and BCM level of preparedness)
1	1	1	1
2	1	2	1
3	1	3	1
4	1	4	1
5	2	1	2
6	2	2	2
7	2	3	2
8	2	4	2
9	3	1	2
10	3	2	2
11	3	3	3
12	3	4	3
13	4	1	3
14	4	2	3
15	4	3	3
16	4	4	4

12.3 KBDSS Validation—Questionnaire Format

This survey is conducted to evaluate the BCM-KBDSS that you have used.

Please kindly rate the following statements below based on your perspectives.

	Disagree				Agree
The layout of the information displayed in the BCM-KBDSS is straightforward and easy to understand	1	2	3	4	5
The BCM-KBDSS offers a structured and well-organized approach to assess the company's BCM level of preparedness	1	2	3	4	5
The BCM-KBDSS's ability to generate feasible recommendations is helpful	1	2	3	4	5
The amount of information presented in the BCM-KBDSS is reasonable	1	2	3	4	5
The BCM-KBDSS provides clear instructions and outputs	1	2	3	4	5
There were no technical problems when running the system	1	2	3	4	5
The processing speed of the system is fast enough	1	2	3	4	5
I think that I would not need the support of a technical person to be able to use the system	1	2	3	4	5
All the information (results and recommendations) provided by the BCM-KBDSS is useful and relevant	1	2	3	4	5
The BCM-KBDSS's action plans are beneficial	1	2	3	4	5
My attitude towards the system is very positive	1	2	3	4	5

Additional feedback:

____

____

____

____

____

References

Abdul-Aziz, A. R., Jaafar, M., Nuruddin, A. R., & Lai, S. W. (2010). Using institutional theory and resource-based perspective to aid transformation of housing-related public enterprises in Malaysia. *Habitat International, 34*(2), 196–203.

Abeysekera, V. (2003). *Understanding culture in an international construction context*. Rotterdam: International Council for Research and Innovation in Building and Construction.

Agustinus, P., & Luhur, F. (2008). *Indonesia's Ponzi economy: Does financial crisis give a lesson*. Retrieved from http://mpra.ub.uni-muenchen.de/6776/

Ahmed, F., & Capretz, L. (2006). A decision support tool for assessing the maturity of the software product line process. *International Journal of Computing and Information Sciences, 4*(3), 97–113.

Aldrich, H. E., & Fiol, C. M. (1994). Fools rush in? The institutional context of industry creation. *Academy of Management Review, 19*(4), 645–670.

Alkass, S., & Aronian, A. (1993). Computer-aided equipment selection for transporting and placing concrete. *Journal of Construction Engineering and Management, 119*(3), 445–465.

Alpaslan, C. M., Green, S. E., & Mitroff, I. I. (2009). Corporate governance in the context of crises: Towards a stakeholder theory of crisis management. *Journal of Contingencies and Crisis Management, 17*(1), 38–49.

Alvesson, M., & Thompson, P. (2006). Post-bureaucracy. In S. Ackroyd, R. Batt, P. Thompson, & P. S. Tolbert (Eds.), *The Oxford handbook of work and organization*. Oxford: Oxford University Press.

Amirkhanian, S. N., & Baker, N. J. (1992). Expert system for equipment selection for earth-moving operations. *Journal of Construction Engineering and Management, 118*(2), 318–331.

ANAO. (2000). *Business continuity management keeping the wheels in motion*. Canberra: Australian National Audit Office.

ANAO. (2009). *Business continuity management; Building resilience in public sector entities – Better practice guide*. Barton: Australian National Audit Office (ANAO).

Anderson, P., Meyer, A., Eisenhardt, K., Carley, K., & Pettigrew, A. (1999). Applications of complexity theory to organizational science. *Organization Science, 10*(3), 216–376.

Andi, & Chandra, H.P. (2007). Budaya Kualitas pada Proyek Konstruksi. In *Konstruksi: Industri, Pengelolaan dan Rekayasa* (pp. 85–96). Bandung: Institut Teknologi Bandung.

Andrews, A. Y., & Hirsch, M. (1983). Ambushes, shootouts, and knights of the roundtable: The language of corporate takeovers. In R. Rondy, F. G. Morgan, & T. Dandridge (Eds.), *Organizational symbolism*. Greenwich, CT: JAI Press.

Angehrn, A. A. (1991). Modeling by example: A link between users, models and methods in DSS. *European Journal of Operational Research, 55*(3), 296–308.

L.S.R. Supriadi, L. Sui Pheng, *Business Continuity Management in Construction*, Management in the Built Environment, DOI 10.1007/978-981-10-5487-7

Angehrn, A. A., & Jelassi, T. (1994). DSS research and practice in perspective. *Decision Support Systems, 12*(4–5), 267–275.

ANZS. (2009). *Business continuity – Managing disruption related risk*. Barton: New Zealand Standard (ANZS).

Applewhite, P. B. (1965). *Organizational behaviour*. Englewood Cliffs, NJ: Prentice Hall.

APRA. (2005). *Prudential standard APS 222 business continuity management*. The Australian Prudential Regulation Authority (APRA): Sydney.

Arain, F. M., & Low, S. P. (2006). Knowledge-based decision support system for management of variation orders for institutional building projects. *Automation in Construction, 15*(3), 272–291.

Arinze, B., Igbaria, M., & Young, L. F. (1992). A knowledge based decision support system for computer performance management. *Decision Support Systems, 8*(6), 501–515.

Arnott, D., & Pervan, G. (2008). Eight key issues for the decision support systems discipline. *Decision Support Systems, 44*(3), 657–672.

AS/NZS. (2004). *Business continuity management HB 221:2004*. Sydney: Standards Australia International Ltd.

Asemi, Safari, & Zavareh. (2011). The role of management information system (MIS) and decision support system (DSS) for manager's decision making process. *International Journal of Business and Management, 6*(7), 164–173.

Aziz. (1994). *Pemrograman sistem pakar*. Jakarta: PT. Elex Media Computindo.

Bailey, J. E., & Pearson, S. W. (1983). Development of a tool for measuring and analyzing computer user satisfaction. *Management Science, 29*, 530–545.

Bajcsy, P., Kooper, R., & Lee, S.-C. (2010). Understanding documentation and reconstruction requirements for computer-assisted decision processes. *Decision Support Systems, 50*, 316–324.

Balci, O., & Sargent, R. G. (1981). A methodology for cost-risk analysis in the statistical validation of simulation models. *Communication ACM, 24*(11), 190–197.

Barthforth, S., Duncan, R., & Miller, C. (1999). A literature review on studies in culture: A pluralistic concept. In S. Ogunlana (Ed.), *Profitable partnering in construction procurement* (pp. 533–542). London: E and FM Spon.

Barton, L. (1993). *Crisis in organizations: Managing and communicating in the heat of chaos*. Cincinnati, OH: South-Western Publishing Co.

Bashiri, M., & Badri, H. (2011). A group decision making procedure for fuzzy interactive linear assignment programming. *Expert Systems with Applications, 38*(5), 5561–5568.

Baum, J. A., & Oliver, C. (1991). Institutional linkages and organizational mortality. *Administrative Science Quarterly, 36*, 187–218.

BCI. (2002). *Development tools for best practices in BCM: BS25999 and other standards*. London: Business Continuity Institute (BCI).

BCI. (2006). *Indonesian construction project data*. Building and Construction Interchange Asia (BCI): Indonesia.

BCI. (2007a). *Business continuity institute good practice guidelines 2008*. London: Business Continuity Institute (BCI).

BCI. (2007b). *Corporate excellence in business continuity management: Current thinking*. Retrieved from www.thebci.org

BCI. (2009). *BCI and BSI join forces on BCM in the USA*. Retrieved from www.businessstandards.com/Articles/08101010_NEWS_BCIBSI

BCI. (2010). *The business continuity institute good practice guidelines 2010 global edition; A management guide to implementing global good practice in business continuity management*. London: Business Continuity Institute (BCI).

BCMI. (2011). *Business as usual*. Retrieved from http://www.bcmpedia.org/wiki/'Business_as_usual%3F

Bertalanffy, L. (1968). *General system theory: Foundations, development, applications*. New York: Braziller.

Bettinger, C. (1989). Use corporate culture to trigger high performance. *Journal of Business*, 38–42.

Bisong, A. Y., Ononse, E., Oluwabanke, F., Enifome, O., Jimi, A., & Olusola, M. (2013). Framework model for a soil suitability decision support system for crop production in Nigeria. *American Journal of Engineering Research, 2*(6), 9–13.

Blau, P., & Scott, W. (1966). *Formal organizations*. London: Routledge.

Bloomberg. (2012). *Foreign exchange rate*. Retrieved from http://www.bloomberg.com/quote/USDIDR:IND

BMI. (2009). *Indonesia business forecast report*. London: Business Monitor International.

Bofa, M. L. (2012). *Indonesia's widening current account and fiscal deficits are concerning*. Retrieved from http://brazilianbubble.com/indonesia-widening-current-account-and-fiscal-deficits/

Boje, M., Fedor, B., & Rowland, M. (1982). Myth making: A qualitative step in OD interventions. *Journal of Applied Behavioural Science, 18*, 17–28.

Borenstein, D. (1998). Towards a practical method to validate decision support systems. *Decision Support Systems, 23*(3), 227–239.

Bossuyt, D. L. V. (2008). *Mechanical engineering design across cultures: A method of designing for cultures*. Oregon: Oregon State University.

Brake, T., Medina-Walker, D., & Walker, A. (2002). *Doing business internationally: The guide to cross-cultural success* (2nd ed.). New York: McGraw-Hill.

Bresnen, M., & Marshall, N. (2000). Partnering in construction: A critical review of issues, problems and dilemmas. *Construction Management and Economics, 18*(2), 229–237.

Brikci, N., & Green, J. (2007). *A guide to using qualitative research methodology*. London: Medecins Sans Frontieres.

Broadbent, D. (1979). *Contingency planning*. Manchester: National Computing Centre.

Brochner, J., Josephson, P. E., & Kadefors, S. (2002). Swedish construction culture, quality management and collaborative practice. *Building Research and Information, 30*(6), 392–400.

Brodbeck, P. W. (2002). Complexity theory and organization procedure design. *Business Process Management Journal, 8*(4), 377–402.

Brooks, N. (2010). *Putting the business into continuity management: Using BCM software to manage risk and resilience*. UK: Sungard Availability Services Ltd.

Broughton, T. (2005). *Industry ill-prepared for disasters*. Retrieved from http://www.lexisnexis.com

BSI. (2006). *BS25999-1 code of practice for business continuity management and BS25999-2 specification for business continuity management*. London: British Standards Institute (BSI).

BSI. (2010a). *Construction company uses BS25999 certification to demonstrate best practice*. Retrieved from http://www.talkingbusinesscontinuity.com

BSI. (2010b). *How a leading construction company uses certification to BC 25999 as an additional insurance policy*. Retrieved from http://www.talkingbusinesscontinuity.com/bs25999-implementation-case-studies/construction-company-uses-bs-25999-certification-to-demonstrate-best-practice.aspx

BUCORIM. (2008). *Risk management, business continuity, and disaster recovery – Industry overview*. Waverley: BUCORIM Inc.

Busemeyer, J. R., & Pleskac, T. J. (2009). Theoretical tools for understanding and aiding dynamic decision making. *Journal of Mathematical Psychology, 53*(3), 126–138.

Buyukozkan, & Feyzioglu. (2004). A fuzzy-logic-based decision-making approach for new product development. *International Journal of Production Economics, 90*, 27–45.

Cameron, K. S., & Quinn, R. E. (1999). *Diagnosing and changing organizational culture*. Englewood Cliffs, NJ: Prentice Hall.

Carlisle, H. M. (1976). *Management concepts and situations*. Chicago, IL: Science Research Associates Inc.

CCDG. (2005). *Response to exposure draft of proposed amendments to IFRS 3 business combinations*. Singapore: CCDG.

Certo, S. C. (1997). *Modern management, diversity, quality ethic and the global environment*. Englewood Cliffs, NJ: Prentice Hall.
Chan, F. T. S., Chan, H. K., Chan, M. H., & Humphreys, P. K. (2006). An integrated fuzzy approach for the selection of manufacturing technologies. *International Journal of Advanced Manufacturing Technology, 27*, 747–758.
Chang, Y. H., & Yeh, C. H. (2001). Evaluating airline competitiveness using multiattribute decision making. *Omega, 29*, 405–415.
Checkland, P. (1981). *Systems thinking, systems practice*. New York: Wiley.
Chen, T. C. (2000). Extensions of the TOPSIS for group decision-making under fuzzy environment. *Fuzzy Sets and Systems, 114*, 1–9.
Chen, S. J., & Hwang, C. I. (1992). *Fuzzy multiple attribute decision making: Methods and applications*. New York: Springer-Verlag.
Chen, P., & Partington, D. (2006). Three conceptual levels of construction project management work. *International Journal of Project Management, 24*(5), 412–421.
Chen, M. F., Tzeng, G. H., & Ding, C. G. (2008). Combining fuzzy AHP with MDS in identifying the preference similarity of alternatives. *Applied Soft Computing, 8*, 110–117.
Cheung, S. O., Wong, P. S. P., & Wu, A. W. Y. (2011). Towards an organizational culture framework in construction. *International Journal of Project Management, 29*(1), 33–44.
Chisholm, B. (2010). *Motivating knowledge workers inside small french start-up organizations*. Paris: Grenoble Ecole de Management.
Chou, S. Y., & Chang, Y. H. (2008). A decision support system for supplier selection based on a strategy-aligned fuzzy SMART approach. *Expert Systems with Applications, 34*(4), 2241–2253.
Cillier, P. (1998). *Complexity and postmodernism*. London: Routledge.
CIRC. (2001). *Construct for excellence*. Construction Industry Review Committee (CIRC): Hong Kong.
Clarkson, M. B. E. (1995). A stakeholder framework for analyzing and evaluating corporate social performance. *Academy of Management Review, 20*, 92–117.
Cleland, D., & King, W. (1983). *Systems analysis and project management* (3rd ed.). New York: McGraw-Hill.
Clemens, R. T. (1996). *Making hard decisiosn: An introduction to decision analysis*. Belmont: Wadsworth.
CMfEA. (2009). *Trade and investment news*. Jakarta: Ministry of Finance.
Coffey, W. V. (2005). *The organizational culture and effectiveness of companies involved in public sector housing construction in Hong Kong*. Perth: Curtin University.
Coffey, V. (2010). *Understanding organisational culture in the construction industry*. Abingdon, UK: Spon Press.
Cole, R. E. (1989). *Strategies for learning: Small-group activities in American, Japanese, and Swedish Industry*. Berkeley: University of California Press.
Cole, R. E. (1999). *Managing quality fads: How American business learned to play the quality game*. New York: Oxford University Press.
Coleman, J., & Henry, J. (1999). What enables self organizing behavior in business. *Emergence, 1*(1), 33–49.
Collier, P. M. (2009). *Fundamentals of risk management for accountants and managers. London*: Butterworth-Heinemann.
Competitiveness-Council. (2006). *The value of resilience* (Council on competitiveness report). UK: Council on Competitiveness.
Comrey, A. L., & Lee, H. B. (1992). *A first course in factor analysis* (2nd ed.). Hillsdale, NJ: Lawrence Erlbaum Associates.
Continuity-Central. (2011a). *Stephenson resilience launches new version of business resilience intelligence tool*. Retrieved from http://www.continuitycentral.com/news05762.html
Continuity-Central. (2011b). *Top BCM challenges*. Retrieved from http://www.continuitycentral.com/feature0864.html

Continuity-Insights. (2009). *Special report on software tools*. Retrieved from http://www.continuityinsights.com/articles/special-report-on-software-tools
Coombs, W. T. (2007). *Ongoing crisis communication, planning, managing and responding*. Newbury Park, CA: Sage Publications.
COOP-Systems. (2010). *Two new challenges for BCM software: Delivering BCMS management process and management system auditing*. Retrieved from http://www.coop-systems.com/
Cox, E. (1998). *The fuzzy systems handbook*. Boston: AP Professional.
Crabb, I. (2011). *Why business continuity software sceptics are right, and why they are wrong*. Retrieved from http://www.continuitycentral.com/feature0850.html
Creswell, J. W. (2009). *Research design: Qualitative, quantitative, and mixed methods approaches* (3rd ed.). Thousand Oaks, CA: Sage Publications.
CSS. (2007). *Business continuity management*. Retrieved from http://www.ukresilience.info/upload/assets/www.ukresilience.info/ep_chap_06.pdf
Cunningham, W. G., & Gresso, D. W. (1994). *Cultural leadership: The culture of excellence in education*. Boston: Allyn and Bacon.
Daft, R. L. (1998). *Organization theory and design* (6th ed.). Cincinnati, OH: Southwestern College Publishing.
Dahl, S.-Å., & Nesheim, T. (1998). Downsizing strategies and institutional environments. *Scandinavian Journal of Management, 14*(3), 239–257.
Davis, G. B., & Olson, M. H. (1985). *Management information systems, conceptual foundation, structure and development*. New York: McGraw Hill.
de Winter, J. C. F. (2013). Using the students t-test with extremely small sample sizes. *Practical Assessment, Research & Evaluation, 18*(10), 1–12.
Deal, T. E., & Kennedy, A. A. (1982). *Corporate cultures the rites and rituals of corporate life*. Reading, MA: Addison-Wesley.
DeGroot, M. H. (1970). *Optimal statistical decisions*. New York: McGraw-Hill.
Denison, D. R. (2000). Organizational culture: Can it be a key lever for driving organizational change? In S. Cartwright & C. I. Cooper (Eds.), *The handbook of organizational culture*. London: Wiley.
Devlin, E. S. (2007). *Crisis management planning and execution*. New York: Auerbach Publications.
DiMaggio, P. (1988). Interest and agency in institutional theory. In L. G. Zucker (Ed.), *Institutional patterns and organizations: Culture and environment*. Cambridge, MA: Ballinger.
DiMaggio, P. J., & Powell, W. W. (1983). The iron cage revisited: Institutional isomorphism and collective rationality in organizational fields. *American Sociological Review, 48*, 147–160.
Directors and Boards. (2006). Boardroom briefing business continuity and disaster recovery. *Directors and Boards Magazine*.
Doll, W. E. (1987). *Foundations for a post-modern curriculum*. Paper presented at the Annual Conference of the American Educational Research Association.
Drennan, L. T., & McConnell, A. (2007). *Risk and crisis management in the public sector*. Basingstoke, UK: Routledge.
Dubois, D., & Prade, H. (1980). *Fuzzy sets and systems: Theory and applications*. New York: Academic Press.
Dunteman, G. H. (1989). *Principal components analysis*. Newbury Park, CA: Sage Publication.
Dweiri, F. T., & Kablan, M. M. (2006). Using fuzzy decision making for the evaluation of the project management internal efficiency. *Decision Support Systems, 42*, 712–726.
EBCC. (2007). *Business continuity checklist*. Retrieved from http://www.eternity.ca/articles/Eternity%20Business%20Continuity%20Checklist.pdf
Edwards, L. (1995). *Practical risk management in the construction industry*. London: Thomas Telford.
Egan, J. (1998). *Rethinking construction*. London: HMSO.
EIU. (2007). *Business resilience; Ensuring continuity in a volatile environment*. London: The Economist Intelligence Unit (EIU).

EIU. (2008). *Country profile 2008 – Indonesia*. London: The Economist Intelligence Unit.
EIU. (2010a). *Indonesia business environment – Infrastructure*. Retrieved from http://country.eiu.com
EIU. (2010b). *Indonesia business environment: The labour market*. Retrieved from http://country.eiu.com
EIU. (2010c). *Indonesia: Macroeconomic environment*. Retrieved from http://country.eiu.com
EIU. (2011). *Indonesia: Business environment at a glance*. Retrieved from http://country.eiu.com
EIU. (2012). *Country report: Indonesia*. London: The Economist Intelligence Unit.
Elliott, D., Swartz, E., & Herbane, B. (1999). Reducing the impact of the unexpected. *The Antidote, 21*.
Elliott, D., Swartz, E., & Herbane, B. (2002). *Business continuity management: A crisis management approach*. London: Routledge.
Elliott, D., Swartz, E., & Herbane, B. (2010). *Business continuity management; A crisis management approach* (2nd ed.). London: Routledge.
Elzarka, H. M., & Bell, L. C. (1995). Object-oriented methodology for materials-management systems. *Journal of Construction Engineering and Management, 121*, 438–445.
Eom, S., & Min, H. (1999). The contributions of multi-criteria decision making to the development of decision support system subspecialties: An empirical investigation. *Journal of Multi-Criteria Decision Analysis, 8*, 239–255.
Fazlollahi, B., Parikh, M. A., & Verma, S. (1997). Adaptive decision support systems. *Decision Support Systems, 20*(4), 297–315.
Fellows, R., & Liu, A. (2003). *Research methods for construction*. London: Blackwell Publishing.
Fellows, R. F., & Seymour, D. E. (2002). *Preface to perspectives on culture in construction*. Rotterdam: International Council for Research and Innovation in Building and Construction.
Ferre, R. (2000). *Building confidence in business continuity*. London: Institute of Directors Guidebook.
Fink, S. (1986). *Crisis management: Planning for the inevitable*. New York: American Management Association.
Finlay, P. (1989). *Introducing decision support systems*. Oxford: NCC Blackwell.
Firdausy, C. M. (2002). *The impact of the regional economic crisis on employment and an evaluation of public work programmes in Indonesia*. Indonesia: EADN.
Firm A. (2012). *Annual performance report*. Jakarta: Firm A.
Firm B. (2013). *Portfolio report: Year 2013*. Jakarta: Firm B.
Firm Y. (2011). *Firm A website*. Retrieved from http://www.wika.co.id/ina/company/?act=organitation
Firm Z. (2011). *Firm B website*. Retrieved from http://www.totalbp.com/project_list.htm
Flynn, C., & Heitzmann, D. (2008). Tragedy at Virginia Tech: Trauma and its aftermath. *The Counseling Psychologist, 36*, 479–489.
FMG. (2003). *Improving risk quality to drive value*. Oxford Metrica. Retrieved from www.oxfordmetrica.com/pdf/OMRiskQualityReport.pdf
Folke, C. (2002). *Resilience and sustainable development: Building adaptive capacity in a world of transformations*. Stockholm: Environmental Advisory Council to the Swedish Government.
Foster, S. P., & Dye, K. (2005). Building continuity into strategy. *Journal of Corporate Real Estate, 7*(2), 105–119.
French, J., & Raven, B. H. (1959). *Studies of social power*. Ann Arbor, MI: Institute for Social Research.
French, S., & Turoff, M. (2007, March). Decision support system. *Communications of the ACM, 50*.
Galbraith, J. R. (1973). *Designing complex organisations*. New York: Addison-Wesley.
Gale, A. W. (1992). The construction industry's male culture must feminize if conflict is to be reduced: The role of education as a gatekeeper to a male construction industry. In P. Fenn & R. Gameson (Eds.), *Construction conflict management and resolution*. London: E and FM Spon.

Gass, S. I. (1983). Decision-aiding models: Validation, assessment and related issues for policy analysis. *Operations Research, 31*(4), 603–631.

Geldermann, J., Bertsch, V., Treitz, M., French, S., Papamichail, K. N., & Hamalainen, R. (2009). Multi-criteria decision support and evaluation of strategies for nuclear remediation management. *Omega, 37*, 238–251.

Geraldi, J. G., Lee-Kelley, L., & Kutsch, E. (2009). The Titanic sunk, so what? Project manager response to unexpected events. *International Journal of Project Management, 28*, 547–558.

Ghosh, A., & Ramakrishnan, U. (2012). *Current account deficits: Is there a problem?* (International Monetary Fund (IMF) report). Retrieved from http://www.imf.org/external/pubs/ft/fandd/basics/current.htm

Gigliotti, R., & Ronald, J. (1991). *Emergency planning for maximum protection*. New York: Butterworth-Heinemann.

Goerner, S. (1994). The physics of evolution: From chaos to evolution and deep ecology. *World Futures, 42*, 193–214.

Goh, M. H. (2010). *A manager's guide to Singapore standard SS540 for business continuity management*. Singapore: Business Continuity Management Institute.

Goldberg, E. M. (2008). Sustainable utility business continuity planning: A primer, an overview and a proven culture-based approach. *The Electricity Journal, 21*(10), 67–74.

Goldberg, J., & Markoczy, L. (2000). Complex rhetoric and simple games. *Emergence, 2*(1), 72–100.

Gordon, J. R. (1993). *A diagnostic approach to organizational behavior*. Upper Saddle River, NJ: Prentice Hall.

Gorry, G. A., & Morton, M. (1971). A framework for management information system. *Sloan Management Review, 13*, 55–70.

Griffin, G. S. (1996). *Complexity theory in construction management*. Unpublished Dissertation, Colorado State University, Fort Collins, CO.

Griffith, A. (1999). Developing an integrated quality, safety and environmental management system. *Construction Information Quarterly, Construction Papers, 108*, 6–18.

Griffith, A. (2011). *Integrated management systems for construction; Quality, environment and safety*. Essex: Pearson Education Limited.

Grunig, J. E. (1992). *Excellence in public relations and communication management*. Mahwah, NJ: Lawrence Erlbaum Associates.

Guedria, W., Chen, D., & Naudet, Y. (2009). A maturity model for enterprise interoperability. In R. Meersman, P. Herreo, & T. Dillon (Eds.), *OTM 2009 workshops, LNCS 5872* (pp. 216–225). Berlin: Springer.

Hair, J. F., Anderson, R. E., Tatham, R. L., & Black, W. C. (1995). *Multivariate data analysis with readings*. Upper Saddle River, NJ: Prentice Hall.

Hair, J. F., Celsi, M. W., Money, A. H., Samouel, P., & Page, M. J. (2011). *Essentials of business research methods* (2nd ed.). New York: M.E. Sharpe.

Hall, M. A., & Jaggar, D. M. (1997). *Accommodating cultural differences in international construction project procurement arrangements*. Paper presented at the CIB W92 Conference.

Hallgren, M., Wilson, L., & T. (2008). The nature and management of crises in construction projects: Projects-as-practice observations. *International Journal of Project Management, 26*, 830–838.

Hamilton, A. (1997). *Management by projects: Achieving success in a changing world*. London: Thomas Telford.

Han, S. H., Kim, D. Y., Kim, H., & Jang, W.-S. (2008). A web-based integrated system for international project risk management. *Automation in Construction, 17*(3), 342–356.

Handy, C. (1993). *Understanding organizations*. Harmondsworth: Penguin.

Hansen, G. S., & Wernerfelt, B. (1989). Determinants of firm performance: The relative importance of economic and organizational factors. *Strategic Management Journal, 10*, 399–411.

Harald, S. H., Heaton, D. P., & Alexander, C. N. (1999). Evolution of organizations in the new millennium. *Leadership and Organizational Development Journal, 20*(4), 198–207.

Harris, M. (2004). *Culture, people, nature: An introduction to anthropology*. Boston: Allyn and Bacon.

Harrison, R. (1972, May–June). How to describe your organization. *Harvard Business Review*, 110–128.

Haykin, S. (1999). *Neural networks*. Upper Saddle River, NJ: Prentice Hall.

Health, R. (1999). A crisis management perspective of business continuity. In A. Hiles & P. Barnes (Eds.), *The definitive handbook of business continuity management* (pp. 43–54). Chichester: Wiley.

Hendrickson, C. (1989). *Project management for construction. Fundamental concepts for owners, engineer, architects, and builders*. Singapore: Prentice Hall.

Herbane, B., Elliott, D., & Swartz, E. (1997). Contingency and continua: Achieving excellence through business continuity planning. *Business Horizons, 40*(6), 19–25.

Herbane, B., Elliott, D., & Swartz, E. (2004). Business continuity management: Time for a strategic role? *Long Range Planning, 37*, 435–457.

Herlijanto, J. (2004). *The May 1998 riots and the emergence of Chinese Indonesian: Social movements in the post-Soeharto era*. Paper presented at the 18th Conference of International Association of Historians of Asia (IAHA), Taipei, Taiwan.

Hermann, C. F. (1963). Some consequences of crisis which limit the viability of organizations. *Administrative Science Quarterly, 8*(25), 61–82.

Hiles, A. (Ed.). (2007). *The definitive handbook of business continuity management (BCM)*. London: Wiley.

Hill, L. A. (1992). *Becoming a manager: Mastery of a new identity*. Boston: Harvard Business School Press.

Hillebrandt, P. M. (1984). *Analysis of the British construction industry*. London: MacMillan Publishers.

Hillebrandt, P. M. (1988). *Analysis of the British construction industry*. London: MacMillan Press Ltd.

Hillebrandt, P., & Cannon, J. (1990). *The modern construction firm*. London: The Macmillan Press Ltd.

Hinton, W. (2000). *Data protection and disaster recovery*. Retrieved from www.msiservice.com/uploads/1084916515.pdf

Hodge, B. J., Anthony, W. P., & Gales, L. M. (2003). *Organization theory: A strategic approach* (6th ed.). Upper Saddle River, NJ: Prentice-Hall.

Hofsede, G., Hofstede, G. J., & Minkov, M. (2010). *Cultures and organizations; Software of the mind*. New York: McGraw Hill.

Hofstede, G., & Hofstede, G. J. (2005). *Cultures and organizations; Software of the mind*. New York: McGraw-Hill.

Holland, T. P. (Ed.). (1995). *Encyclopedia of social work* (Vol. 2). Washington, DC: NASW Press.

Holland, J. H., & Miller, J. H. (1991). Artificial adaptive agents in economic theory. *American Economic Review, 81*(2), 365–370.

Holland, T. P., & Petchers, M. K. (Eds.). (1987). *Encyclopedia of social work* (Vol. 2). Silver Spring, MD: NASW.

House, R. J., Hanges, P. J., Javidan, M., Dorfman, P., & Gupta, V. (Eds.). (2003). *GLOBE, cultures, leadership, and organizations: GLOBE study of 62 societies*. Newbury Park, CA: Sage.

House, R. J., Hanges, P. J., Javidan, M., Dorfman, P. W., & Gupta, V. (2004). *Culture, leadership and organizations – The globe study of 62 societies*. Thousand Oaks, CA: Sage Publications.

House, R. J., Hanges, P. J., Ruiz-Quintanilla, S. A., Dorfman, P. W., Javidan, M., & Dickson, M. (2010). *Cultual influences on leadership and organizations: Project Globe*. Retrieved from http://www.thunderbird.edu/wwwfiles/sites/globe/pdf/process.pdf

Hughes, M. A., Price, R. L., & Marrs, D. W. (1986). Linking theory construction and theory testing: Models with multiple indicators of latent variables. *Academy of Management Review, 11*(1), 128–144.

Hwang, C. I., & Yoon, K. (1981). *Multiple attribute decision making: Method and applications, a state-of-the-art survey*. New York: Springer.

IBI. (2000). *Indonesia*. Essex: Walden Publishing Ltd.

ICA. (2012). *Indonesian contractors association (ICA) membership report*. Jakarta: ICA.

ICG. (2002). *Indonesia: Resources and conflict in Papua*. Jakarta: International Crisis Group (ICG).

IFRC. (2004). *Indonesia: Floods emergency response appeal*. Jakarta: International Federation of Red Cross and Red Crescent Societies (IFRC).

Imriyas, K. (2006). *A fuzzy knowledge-based system for premium-rating of workers' compensation insurance for construction projects*. Unpublished PhD Dissertation, National University of Singapore, Singapore.

INTERCEP. (2007). *The business case for preparedness*. New York: New York University (NYU).

Israel, G. D. (2013). *Determining sample size*. Gainsville, FL: Agriculture Education and Communication Department, University of Florida.

Jackson, S., & Browne, J. (1992). AI-based decision support tool for strategic decision making in the factory of the future. *Computer Integrated Manufacturing Systems, 5*(2), 83–90.

Janakiraman, V. S., & Sarukeshi, K. (Eds.). (2001). *Decision support systems*. New Delhi: Prentice-Hall of India Private Limited.

Javidan, M., House, R. J., Dorfman, P. W., Hanges, P. J., & De Luque, M. S. (2006). Conceptualizing and measuring culture and their consequences: A comparative review of Globe's and Hofstede's approaches. *Journal of International Business Studies, 37*, 897–914.

Kanter, R. M. (1983). *The change masters*. New York: Simon and Schuster.

Kanter, R. M. (1994). *The change masters: Corporate entrepreneurs at work*. New York: Routledge.

Karakasidis, R. (1997). A project planning process for business continuity, information management, and computer security. In D. Elliott, E. Swartz, & B. Herbane (Eds.), *Business continuity management: A crisis management approach* (pp. 78–109). London: Routledge.

Kartasasmita, G. (2000). *Globalization and the economic crisis: The Indonesian story*. Paper presented at the FASID/GSAPS/WIAPS Joint ADMP Workshop, Japan.

Kaya, İ., & Kahraman, C. (2010). Development of fuzzy process accuracy index for decision making problems. *Information Sciences, 180*(6), 861–872.

Keeney, R. L., & Raiffa, H. (1993). *Decisions with multiple objectives: Preferences and value tradeoffs*. New York: Cambridge University Press.

Kelly, S. (1999). *The complexity advantage: How the science of complexity can help your business achieve peak performance*. New York: Business Week Books.

Kerr, D. (2004). Factors influencing the development and adoption of knowledge based decision support systems for small, owner-operated rural business. *Artificial Intelligence Review, 22*, 127–147.

Kirst-Ashman, K. (2000). *Human behavior, communities, organizations, and groups in the macro social environment: An empowerment approach*. Belmont, CA: Wadsworth.

Kline, P. (1994). *An easy guide to factor analysis*. London: Routledge.

Knight, R., & Pretty, D. (1996). *The impact of catastrophe on shareholders value*. Oxford: Templeton College of Oxford University.

Knight, R. F., & Pretty, D. J. (2005). *Protecting value in the face of mass fatality event*. Oxford Metrica. Retrieved from www.oxfordmetrica.com/pdf/OMMassFatalitiesBriefing.pdf

Koontz, H. (1980). The management theory jungle revisited. *The Academy of Management Review, 5*(2), 175–187.

Koontz, H., & Weihrich, H. (1990). *Essentials of management* (5th ed.). New York: McGraw-Hill.

Kotler, J., & Heskett, J. (1992). *Corporate culture and performance*. London: Macmillan.

Kuong, J., & Isaacson, J. (1986). *How to prepare an EDP plan for business contingency*. Wellesley Hills: Management Advisory Publications.

Lam, K. C., Tao, R., & Lam, M. C. K. (2010). A material supplier selection model for property developers using fuzzy principal component analysis. *Automation in Construction, 19*, 608–618.

Lansley, & McAtee. (2009). *What is best practice business continuity management?* Retrieved from www.bcminabox.com

Larasati, D., & Tsunemi, W. (2009). *Evaluation study on existing condition of indonesian construction industry: How to improve performance and the competitiveness.* Paper presented at the International Seminar of Society for Social Management System, Kochi, Japan.

Latham, M. (1994). *Constructing the team.* London: HMSO.

Lavender, S. (1996). *Management for the construction industry.* Essex: Addison Wesley Longman Limited.

Lawrence, T. B. (1999). Institutional strategy. *Journal of Management, 25*(2), 161–187.

Lawrence, P. R., & Lorsch, J. W. (1967). *Organization and environment.* Cambridge, MA: Harvard University Press.

Lawrence, R., & Lorsch, J. W. (1986). *Organization and environment; Managing differentiation and integration.* Boston: Harvard Business School Press.

Lawrence, P. R., Barnes, L. B., & Lorsch, J. W. (Eds.). (1976). *Organizational behavior and administration; Cases and readings* (3rd ed.). Chicago, IL: Richard D. Irwin, Inc.

Leavitt, H. J. (1964). *New perspectives in organization research.* New York: Wiley.

Lee, C. (1990). Fuzzy logic in control systems: Fuzzy logic parts I, II. *IEEE Transaction on Systems, Man, and Cybernetics, 20*, 404.

Lee, S. H. (2009, May 22). Distressed in a foreign land? You've got a friend. *The Straits Times,* A24.

Lee, J.-W., & McKibbin, W. J. (2007). *Domestic investment and external imbalances in East Asia.* Organisation for Economic Cooperation and Development (OECD) Report. Retrieved from http://www.oecd.org/home/0,2987,en_2649_201185_1_1_1_1_1,00.html

Leidner, D. E., & Kayworth, T. (2006). Review: A review of culture in information systems research: Toward a theory of information technology culture conflict. *MIS Quarterly, 30*, 357–399.

Levitt, R. E., & Maher, M. L. (1987). Expert system in construction. In M. L. Maher (Ed.), *Expert system for civil engineers* (pp. 85–112). New York: ASCE.

Lewin, A. Y., Long, C. P., & Carroll, T. N. (1999). The co-evolution of new organizational forms. *Organization Science, 10*(5), 535–550.

Light, P. C. (2008). *Predicting organizational crisis readiness: Perspectives and practices toward a pathway to preparedness.* New York: New York University (NYU).

Lingard, H., & Rowlinson, S. M. (1998). Behaviour-based safety management in Hong Kong's construction industry: The results of a field study. *Construction Management and Economics, 16*(4), 481–488.

Lissack, M. R. (1997). Of chaos and complexity: Managerial insights from a new science. *Management Decision, 35*, 205–219.

Littrell, R. F. (2008). *Managing culturally diverse stakeholders in international projects: The problem of planning*. Retrieved from www.rij.eng.uerj.br/scientific/2008/sp083-02.pdf

Liu, A. M. M. (1999). Culture in the Hong Kong real-estate profession: A trait approach. *Habitat International, 23*(3), 413–425.

Liu, A. M. M. (2002). *Keys to harmony and harmonic keys.* Rotterdam: International Council for Research and Innovation in Building and Construction.

Liu, A. M. M., & Fellows, R. F. (1999a). Cultural issues. In S. M. Rowlinson & P. McDermott (Eds.), *Procurement systems: A guide to best practice in construction* (pp. 141–162). London: E and FM Spon.

Liu, A. M. M., & Fellows, R. F. (1999b). The impact of culture on project goals. In S. Ogunlana (Ed.), *Profitable partnering in construction procurement* (pp. 523–532). London: E and FM Spon.

Liu, K., & Lai, J. H. (2009). Decision-support for environmental impact assessment: A hybrid approach using fuzzy logic and fuzzy analytic network process. *Expert Systems with Applications, 36*, 5119–5135.

Liu, H., Ke, W., Wei, K. K., Gu, J., & Chen, H. (2010). The role of institutional pressures and organizational culture in the firm's intention to adopt internet-enabled supply chain management systems. *Journal of Operations Management, 28*(5), 372–384.

Lock, D. (1992). *Project management*. London: Gower.

Longman (Ed.). (1978). *Longman dictionary of contemporary english*. London: Longman Group Limited.

Loosemore, M. (1998a). Organisational behaviour during a construction crisis. *International Journal of Project Management, 16*(2), 115–121.

Loosemore, M. (1998b). Reactive crisis management in construction projects – Patterns of communication and behavior. *Journal of Contingencies and Crisis Management, 6*(1), 23–34.

Loosemore, M. (1998c). The three ironies of crisis management in construction projects. *International Journal of Project Management, 16*(3), 139–144.

Loosemore, M., & Teo, M. M. M. (2006). Crisis preparedness of construction companies. *Journal of Management in Engineering, 16*(5), 60–65.

Louis, M. R. (1980). Surprise and sense making: What newcomers experience in entering organizational settings. *Administrative Science Quarterly, 25*, 226–251.

Low, S. P. (1998). Managing total service quality: A systemic view. *Managing Service Quality, 8* (1), 34–45.

Low, S. P., & Shi, Y. (2002). An exploratory study of Hofstede's cross-cultural dimensions in construction projects. *Management Decision, 40*(1), 7–16.

Low, S. P., Ho, D. K. H., & Yeap, S. A. (1999). Crisis management: A survey of property development firms. *Property Management, 17*(3), 231–251.

Low, S. P., Liu, J., & Sio, S. (2008a). Attitude towards business continuity management. *The Singapore Engineer*, 36–44.

Low, S. P., Liu, J. Y., & Kumaraswamy, M. (2008b). Business continuity management in construction: A regional perspective. *The Singapore Engineer*, 35–40.

Low, S. P., Liu, J. Y., & Sio, S. (2010a). Business continuity management in large construction companies in Singapore. *Disaster Prevention and Management, 19*(2), 219–232.

Low, S. P., Liu, J. Y., & Kumaraswamy, M. (2010b). Institutional compliance framework and business continuity management in Mainland China, Hong Kong SAR and Singapore. *Disaster Prevention and Management, 19*(5), 596–614.

Luce, R. D. (2000). *Utility of gains and losses*. Mahwah, NJ: Lawrence Erlbaum Associates.

Luthans, F. (1992). *Organizational behavior*. New York: McGraw Hill.

Luthans, F. (2008). *Organizational behavior*. New York: McGraw Hill.

Lynch, R. (2006). *Corporate strategy*. Harlow, Essex: Prentice Hall Education.

Magnusson, P., Wilson, R. T., Srdan, Z., Zhou, J. X., & Westjohn, S. A. (2008). Breaking through the cultural clutter: A comparative assessment of multiple cultural and institutional frameworks. *International Marketing Review, 25*, 183–200.

Maleki, A. (2010). *The effect of culture and political structure on participatory policy analysis*. Delft University of Technology: Netherlands.

Mallak, L. (1998). Putting organizational resilience to work. *Industrial Management, 40*(6), 8–13.

Maloney, W. F., & Federle, M. O. (1991). Organizational culture and management. *Journal of Management and Engineering, 7*(1), 1–25.

Mamadani, E. H. (1974). Applications of fuzzy algorithm for simple dynamic plant. *IEE Proceedings, 121*, 1585–1588.

Mamadani, E. H., & Assilian, S. (1975). An experiment in linguistic synthesis with a fuzzy logic controller. *International Journal of Man-Machine Studies, 7*, 1–13.

Marsh. (2007). *Risk management and insurance training essential: Business continuity management training*. Melbourne: VMIA.

Marsh. (2010). *Business continuity preparedness review and maturity assessment*. Canada: Marsh Risk Consulting.
MAS. (2003). *Business continuity management guidelines*. Monetary Authority of Singapore (MAS): Singapore.
Mason, M. (2008). Complexity theory and the philosophy of education. *Educational Philosophy and Theory, 40*(1), 4–18.
McCabe, S. (2010). *Corporate strategy in construction; Understanding today's theory and practice*. London: Wiley-Blackwell.
McCrackan, A. (2005). *Practical guide to business continuity assurance*. London: Artech House Inc.
McKinsey. (2013). *The organizational capacity assessment tool (OCAT): 2.0 – Social sector practice*. New York: McKinsey & Company.
McManus, S., Seville, E., Vargo, J., & Brunsdon, D. (2008). Facilitated process for improving organizational resilience. *Natural Hazards Review, 9*, 81–90.
McMillan, E. (2006). *Complexity, organizations and change an essential introduction*. London: Routledge.
Mead, R. (2000). *Cases and projects in international management: Cross-cultural dimensions*. Oxford: Blackwell.
Mendel, J. (1995). Fuzzy logic systems for engineering: A tutorial. *Proceedings of the IEEE, 83* (3), 345–377.
Mercer, D. (1999). Organizational futures: Unprepared for the surprises to come. *Management Decision, 37*(5), 411–416.
Meyer, J. W., & Scott, R. W. (1983). *Organizational environments: Ritual and rationality*. Beverly Hills, CA: Sage.
Miles, M. B., & Huberman, A. M. (1994). *Qualitative data analysis: An expanded sourcebook*. Thousand Oaks, CA: Sage Publications.
Millar, D. P., & Health, R. L. (2003). *Responding to crisis, a rhetorical approach to crisis communication*. Mahwah, NJ: Lawrence Erlbaum Associates.
Mintzberg, H. (1987). Crafting strategy. *Harvard Business Review, 65*, 66–75.
Mintzberg, H. (2009). *Managing*. San Francisco, CA: Berrett-Koehler Publishers.
Miser, J. H., & Quade, E. S. (1988). Validation. In H. J. Miser (Ed.), *Handbook of system analysis – Craft issues and procedural choices*. Chichester: Wiley.
Mitleton-Kelly, E. (2004). Ten principles of complexity and enabling infrastructures. In E. - Mitleton-Kelly (Ed.), *Complex systems and evolutionary perspectives on organizations: The application of complexity theory to organizations* (pp. 23–50). London: Pergamon.
Mitroff, I., & Kilmann, H. (1976). On organizational stories: An approach to the design and of organizations through myths and stories. In H. Kilmann, R. Pondy, & P. Slevin (Eds.), *The management of organization design* (pp. 189–207). New York: Elsevier-North Holland.
Mitroff, I., & Pearson, C. (1993). *Crisis management: A diagnostic guide for improving your organization's crisis preparedness*. San Francisco: Jossey-Bass Publishers.
MiyamotoINTL. (2007). *2007 West Sumatra Earthquake: Reconnaissance report*. Sacramento, CA: Miyamoto International.
Mockler, J. R. (1989). *Knowledge based system for management decision*. Singapore: Prentice Hall.
MOF-BC. (2007). *Report on the cross government review of business continuity management*. British Columbia: Ministry of Finance – British Columbia (MOF-BC).
Mohan, S. (1990). Expert system application in construction management and engineering. *Journal of Engineering and Management, 116*(1), 87–99.
Molinier, L. (2009, September/October). BCM Bureau. *Continuity Magazine*. BCI.
MOM. (2010). Safeguard your business through fostering a harmonious, cohesive and resilient workplace. *Entrepreneurs Digest*.
More, E. (1995). Crisis management and communication in Australian organizations. *Australian Journal of Communication, 22*, 31–47.

Morrison, K. (2002). *School leadership and complexity theory*. London/New York: Routledge/Falmer.

Myerson, R. B. (1991). *Game theory: Analysis of conflict*. Cambridge, MA: Harvard University Press.

Naoum, S. (2001). *People and organizational management in construction*. London: Thomas Telford.

NBCSD. (2002a). *Certification and Registration Procedure for National Construction Service Firms*. Statute SK 75/KPTS/LPJK/D/X/2002. Indonesia: NBCSD.

NBCSD. (2002b). *Certification and registration procedure for national construction service firms*. Jakarta: NBCSD.

NBCSD. (2004). *Country report Indonesia*. Paper presented at the 10th Asia Construct, Bali.

Negnevitsky, M. (2006). *Artificial intelligence: A guide to intelligent systems*. New York: Pearson Education.

Newcombe, R., Langford, D., & Fellows, R. (1990). *Construction management*. London: Mitchell.

Newell, A. (1990). *Unified theories of cognition*. Cambridge, MA: Harvard University Press.

NFPA. (2007). Standard on disaster/emergency management and business continuity programs – NFPA 1600. Quincy, MA: National Fire Protection Association.

Noyes, C. M. (1992). *Creativity, change and culture: An investigation into the relationship between organizational culture and innovation*. Ann Arbor: UMI Research Press.

NYU. (2006). *New York's small and medium sized business are not prepared for emergencies. Why?* New York: New York University (NYU).

O'Hehir, M. (1999). What is a business continuity planning (BCP) strategy. In A. Hiles & P. Barnes (Eds.), *The definitive handbook of business continuity management* (pp. 25–54). Chichester: John Wiley and Sons Ltd.

O'Keefe, R. M., Balci, O., & Smith, E. P. (1987). Validating expert system performance. *IEEE Expert, 2*(4), 81–89.

O'Leary, T. J., Goul, M., Moffitt, K. E., & Radwan, A. E. (1990). Validating expert systems. *IEEE Expert, 5*(3), 51–58.

Ofori, G. (1990). *The construction industry: Aspects of its economics and management*. Singapore: Singapore University Press.

Ogbonna, E., & Harris, L. C. (2000). Leadership style, organizational culture and performance: Empirical evidence from UK companies. *International Journal of Human Resource Management, 11*(4), 766–788.

Oliver, C. (1991). Strategic responses to institutional process. *Academy of Management Review, 16*(1), 145–179.

Olum, Y. (2004). *Modern management theories and practices*. Paper presented at the 15th East African Central Banking Seminar. Kenya School of Monetary Studies, Kenya.

Opstal, D. V. (2007). *Transform. The resilient economy: Integrating competitiveness and security*. Council on Competitiveness, UK.

Osland, J. S., Kolb, D. A., & Rubin, I. M. (2001). *Organizational behavior: An experiential approach*. Englewood Cliffs, NJ: Prentice Hall.

Owens, H. D., & Philippakis, A. S. (1995). Inductive consistency in knowledge-based decision support systems. *Decision Support Systems, 13*(2), 167–181.

Oxford University Press. (2006). *Oxford Dictionary*. Cambridge: OUP.

Pallant, J. (2001). *SPSS survival manual: A step by step guide to data analysis using SPSS*. Buckingham: Open University Press.

Papamichail, K. N., & French, S. (2005). Design and evaluation of an intelligent decision support system for nuclear emergencies. *Decision Support Systems, 41*, 84–111.

Paraskevas, A. (2006). Crisis management or crisis response system? A complexity science approach to organizational crises. *Management Decision, 44*(7), 892–907.

Parsons, T. (1960). *Structure and process in modern societies*. Glencoe, IL: Free Press.

Parthasarathi, P. (2005). *Operational risk management – Business continuity planning DBS.CO.IS Audit. No. 19/31.02.03/2004-05*. Mumbai: Reserve Bank of India.

Pascale, R. T., Millemann, M., & Gioja, L. (2000). *Surfing the edge of chaos: The laws of nature and the new laws of business*. New York: Crown Business.

Patton, M. Q. (2002). *Qualitative research and evaluation methods*. Thousand Oaks, CA: Sage.

Pauchant, T., & Mitroff, I. (1992). *Transforming the crisis prone organization*. San Francisco: Jossey-Bass Inc..

Pauchant, T. C., Mitroff, I. I., & Lagadec, P. (1991). Toward a systemic crisis management strategy: Learning from the best examples in the US, Canada and France. *Industrial Crisis Quarterly, 5*, 209–232.

Pearl, J. (1988). *Probabilistic reasoning in intelligent systems: Networks of plausible inference*. San Francisco: Morgan Kauffman.

Pedrycz, W., & Gomide, F. (1998). *An introduction to fuzzy sets: Analysis and design*. Cambridge, MA: MIT Press.

Pedrycz, W., & Gomide, F. (2007). *Fuzzy systems engineering toward human-centric computing*. Hoboken, NJ: John Wiley and Sons, Inc.

Peel, M. (1993). *Introduction to management: A guide for better business performance*. London: Pitman Publishing.

Peng, J., Zhang, G., Chen, R., & Tan, Y. (2011). Impacts of essential elements of management on IT application maturity – A perspective from firms in China. *Decision Supports Systems, 51*(1), 88–98.

Permana, I. E. (2007). *Construction safety practices in Batam, Indonesia*. Unpublished Master Degree Thesis. Universiti Teknologi Malaysia, Malaysia.

Pfeffer, J. (1981). Management as symbolic action: The creation and maintenance of organizational paradigms. In L. L. Cummings & B. M. Staw (Eds.), *Research in organizational behavior*. Greenwich, CT: JAI Press.

Phua, F. T. T. (2007). Does senior executives' perception of environmental uncertainty affect the strategic functions of construction firms? *International Journal of Project Management, 25*(8), 753–761.

Pitt, M., & Goyal, S. (2004). Business continuity planning as a facilities management tool. *Facilities, 22*, 87–99.

Ponce-Cruz, P., & Ramirez-Figueroa, F. D. (2010). *Intelligent control systems with LabVIEW*. London: Springer-Verlag.

Porter, M. E. (1991). Towards a dynamic theory of strategy. *Strategic Management Journal, 12* (Winter), 95–117.

Porter, M. E., & Millar, V. E. (1985). How information gives you competitive advantage. *Harvard Business Review, 63*(4), 149–160.

Power, P. (1999). Business continuity management – Preventing chaos in a crisis, management action notes. In D. Elliott, E. Swartz, & B. Herbane (Eds.), *Business continuity management: A crisis management approach* (pp. 54–57). Abingdon: Routledge.

Preece, A. D. (1990). Towards a methodology for evaluating expert systems. *Expert Systems, 7*(4), 215–223.

Pritchard, J. A. T. (1976). *Contingency planning*. Manchester: NCC.

PTX. (2008). *PT X emergency procedural report*. Jakarta: PTX.

Puterman, M. I. (1994). *Markov decision processes: Discrete stochastic dynamic programming*. New York: John Wiley and Sons Inc.

PWC. (2009). Embedding BCM in the organization's culture. *Business Continuity Journal, 3*(3).

Quarantelli, E. I. (1988). Disaster crisis management: A summary of research findings. *Journal of Management Studies, 25*(4), 373–386.

Raftery, J., Chiang, Y. H., & Anson, M. (2004). *The construction sector in Asian economies*. London: Spon Press.

Raisch, W., & Statler, M. (2006). *Insurance incentives for corporate preparedness*. New York: New York University (NYU).

Raisch, W., Statler, M., & Burgi, P. (2007). *Mobilizing corporate resources to disasters: Toward a program for action*. New York: New York University (NYU).

Raman, K. S., & Phoon, C. K. (1990). Decision support systems in Singapore: Issues in their management and their beneficial contributions. *Information and Management, 18*(3), 153–165.
Raykov, T., & Marcoulides, G. A. (2006). *A first course in structural equation modeling*. Mahwah, NJ: Lawrence Erlbaum Associates.
Raymond, M. (1990). *Information systems*. New York: Macmillan Publishing Company.
Republic of Indonesia Government Regulation No. 28 Year 2000 on Construction Service Business and Roles (2000).
Republic of Indonesia Regulation No. 18 Year 1999 on Construction Services (1999).
Robinson, J. (2009, September/October). BCM Bureau. *Continuity Magazine*. BCI.
Rogers, E. M. (1995). *Diffusion of innovations*. New York: The Free Press.
Root, F. R. (1994). *Entry strategies for international markets*. New York: Lexington Books.
Root, D. S. (2001). *The influence of professional and occupational cultures on project relationships mediated through standard forms and conditions of contract*. Unpublished Thesis. University of Bath, Bath, UK.
Rowlinson, S. M. (2001). Matrix organizational structure, culture and commitment: A Hong Kong public sector case study of change. *Construction Management and Economics, 19*, 669–673.
Rowlinson, S. M., & Root, D. S. (1996). *The impact of culture on project management*. Hong Kong: British Council.
Saaty, T. I. (1988). *Multicriteria decision making: The analytic hierarchy process*. Pittsburgh: RWS Publication.
Santos, M. (1998). Simple, yet complex. *CIO Enterprise Magazine*.
Saparini, H. (2009). *Policy response to overcome crisis: A lesson from indonesian case*. Paper presented at the International Conference on Re-regulating Global Finance in the Light of the Global Crisis, Beijing.
SBF. (2003). *Business continuity management*. Retrieved from http://www.hbiz.com.sg/html/BizContinuityManagement_200304.htm
SBF. (2010). *BCM diagnostic tool*. Retrieved from http://www.bcm.org.sg/Links/BCMDiagnosticTool.aspx
SCA. (2007). *Statistical Yearbook of Indonesia*. Jakarta: Statistical Central Agency (SCA) (Badan Pusat Statistik).
SCA. (2009). *Statistical Yearbook of Indonesia*. Jakarta: Statistic Central Agency (*Badan Pusat Statistik*).
SCA. (2012). *Statistical Yearbook of Indonesia*. Jakarta: Statistic Central Agency (*Badan Pusat Statistik*).
Schaufelberger, J. (2009). *Construction business management*. Upper Saddle River, NJ: Pearson.
Schein, E. (1985). *Organizational culture and leadership: A dynamic view*. San Francisco: Jossey Bass.
Schermerhorn, J. R. (1996). *Management*. New York: Wiley.
Schmucker, K. J. (1984). *Fuzzy sets, natural language computations and risks analysis*. Rockville, MD: Computer Science Press.
Schols, C. (1987). Corporate culture and strategy – The problem of strategic fit. *Long Range Planning, 20*(4), 78–87.
Schuwer, R. V., & Kusters, R. J. (1993). Application areas and added value of knowledge base systems. *Information and Management, 24*(2), 83–92.
Schwartz, S. H. (1994). A theory of cultural values and some implications for work. *Applied Psychology International Review, 48*(1), 23–47.
Scott, W. R. (1987). The adolescence of institutional theory. *Administrative Science Quarterly, 32*, 493–511.
Scott, W. R. (1992). *Organizations: Rational, natural, and open systems*. Upper Saddle River, NJ: Prentice-Hall.
Scott, W. R. (1998). *Organizations; Rational, natural, and open systems* (4th ed.). Upper Saddle River, NJ: Prentice Hall.
Scott, W. R. (2001). *Institutions and organizations*. Thousand Oaks, CA: Sage.

Scott, W. R. (2004). Institutional theory. In G. Ritzer (Ed.), *Encyclopedia of social theory*. Thousand Oaks, CA: Sage.

Scott, J. E. (2007). Mobility, business process management, software sourcing, and maturity model trends: Propositions for the IS organization of the future. *Information Systems Management, 24*(2), 139–145.

Scott, W. R. (2008). *Institutions and organizations; Ideas and interests* (3rd ed.). Los Angeles: Sage Publications.

Sen, A., & Biswas, G. (1985). Decision support systems: An expert systems approach. *Decision Support Systems, 1*(3), 197–204.

Seymour, D., & Rooke, J. (1995). The culture of the industry and the culture of research. *Construction Management and Economics, 13*(6), 511–523.

Sheffi, Y. (2006, February 19). Waiting for the next 'big one'. *Boston Globe*, 11.

Sherman, H., & Schultz, R. (1998). *Open boundaries*. Cambridge, MA: Perseus Books.

Shrivastava, P., & Mitroff, I. (1988). Strategic management of corporate crises. In D. Elliott, E. Swartz, & B. Herbane (Eds.), *Business continuity management: A crisis management approach* (pp. 74–79). London: Routledge.

Shrivastava, P., & Siomkos, G. (1989). Disaster containment strategies. *The Journal of Business Strategy, 10*(5), 26–30.

Siehl, C., & Martin, J. (1981). *Learning organizational culture*. Stanford, CA: Stanford University.

Simon, H. (1997). *Administrative behavior: A study of decision-making processes in administrative organizations* (4th ed.). New York: The Free Press.

Singapore Business Federation. (2006). *A flu pandemic business continuity guide for SMEs*. Singapore: Spring Singapore.

Singh, M., Singh, P., & Singh, S. B. (2008). Decision support system for farm management. In *Proceeding of World Academy of Science, Engineering and Technology* (Vol. 29).

Smircich, L. (1983). Concepts of culture and organizational analysis. *Administrative Science Quarterly, 28*, 339–358.

Smit, N. (2005). *Business continuity management – A maturity model*. Unpublished Thesis. University Rotterdam, Rotterdam.

Smith, D. (2003). Business continuity and crisis management. *Management Quarterly Magazine*, 27–33.

SPICE. (2001). *ISO/IEC 15504 Software process improvement and capability determination model*. Limerick: International Organization for Standardization and International Electrotechnical Commission.

SPRING. (2008). *Singapore standard: Business continuity management (BCM) SS540:2008*. Singapore: Spring Singapore Standard.

SPRING. (2009). *Flu pandemic business continuity guide*. Singapore: SPRING Singapore.

SPRING. (2012). *SS ISO 22301:2012; Societal security – Business Continuity Management System –Requirement*. Singapore: SPRING Singapore.

Srinivas, V., & Shekar, B. (1997). Strategic decision-making processes: Network-based representation and stochastic simulation. *Decision Support Systems, 21*(2), 99–110.

Sriraj, P. S., & Khisty, C. J. (1999). Crisis management and planning using systems methodologies. *Journal of Urban Planning and Development, 125*(3), 121–133.

Stacey, R. (1996). *Complexity and creativity in organizations*. San Francisco: Berrett-Koehler.

Stacey, R. D. (2000). *Strategic management and organizational dynamics*. Englewood Cliffs, NJ: Prentice Hall.

Stair, R. M. (1992). *Principles of information systesm: A managerial approach*. Boston: Boyd and Fraser Publishing Company.

Starr, R., Newfrock, J., & Delurey, M. (2004). Enterprise resilience: Managing risk in the networked economy. *Strategy+Business*. Retrieved from http://www.bah.com

Stengel, R. F. (1986). *Stochastic optimal control: Theory and application*. New York: John Wiley and Sons Inc.

Stevanovic, B. (2011). Maturity models in information security. *International Journal of Information and Communication Technology Research, 1*(2), 44–47.

St-Germain, R., Aliu, F., Lachapele, E., & Decoteam, P. D. (2012). *Societal security: Business continuity management systems*. Montreal: Professional Evaluation and Certification Board.

Strang, D., & Meyer, J. W. (1993). Institutional conditions for diffusion. *Theory and Society, 22* (487–511).

Suchman, M. C. (1995). Managing legitimacy: Strategic and institutional approaches. *Academy of Management Review, 20*, 571–610.

Sudarto. (2007). *Penggunaan Knowledge Based Management System untuk Meningkatkan Kinerja Perusahaan Jasa Konstruksi di Indonesia.* Unpublished PhD Dissertation, University of Indonesia, Jakarta.

Supriadi, L., & Low, S. P. (2012a). *Analysis of business continuity management for Indonesian contractors using cultural and institutional approach.* Paper presented at International Construction Business and Management Symposium, Kuala Lumpur, Malaysia, 11 pages (in - CD-Rom proceedings).

Supriadi, L., & Low, S. P. (2012b). Business continuity management for Indonesian contractors. *Management, Procurement and Law, 165*, MPO (in press).

Suraji, A. (2003). *Peta Kesiapan Industri Jasa Konstruksi Menuju Liberalisasi Perdagangan Jasa Konstruksi*. Paper presented at the *National Seminar Peran Jasa Industry Era Otonomi Daerah dan AFTA/AFAS*, Jakarta.

Suraji, A., Krisnandar, D., & Mulyo, S. S. (2006). *The Indonesian construction: Now and future*. National Board of Construction Services Development Indonesia (NBCSD): Jakarta.

Suryadharma, A. (2008, August 13). Indonesia's construction sector facing tough challenges. *The Jakarta Post*. Retrieved from www.thejakartapost.com

Sutardi. (2006). *Action report towards flood disasters reduction: Indonesian case*. Tokyo: International Flood Network Action.

Sutjipto. (1991). *Strategi Industri Jasa Konstruksi Nasional dalam Era Globalisasi*. Jakarta: Jakarta Press.

Sutton, R. S., & Barto, A. G. (1998). *Reinforcement learning*. Cambridge, MA: MIT Press.

Swartz, E., Elliott, D., & Herbane, B. (1995). Out of sight, out of mind: The limitations of traditional information systems planning. *Facilities, 13*, 15–21.

Tadie, J. (2008). *Flammable cities: Fire, urban environment, and culture in history*. Paper presented at the Conference at the German Historical Institute (GHI), Washington, DC.

Tah, J. H. M., & Carr, V. (2000). A proposal for construction project risk assessment using fuzzy logic. *Construction Management Economics, 18*(4), 491–500.

Tambunan, T. (2006). *The likely economic impact from the Yogyakarta and central Java earthquake*. Jakarta: Kadin Indonesia-Jetro.

Tan, W. (2008). *Practical research methods* (3rd ed.). Singapore: Prentice Hall.

Tengblad, S. (2012). *The work of managers: Towards a practice theory of management.* New York: Oxford University Press.

Teo, P. (2011). *Business continuity management for quantity surveyors in Singapore*. Unpublished B.Sc. (Project and Facilities Management) Dissertation, National University of Singapore, Singapore.

Thayer, W. (1998). An inside look at the scary world of food safey scares. *Frozen Food Age, 46*(1), 12–38.

Thompson, J. D. (1967). *Organizations in action*. New York: McGraw Hill.

Thompson, B. (2004). *Exploratory and confirmatory factor analysis; Understanding concepts and applications*. Washington, DC: American Psychological Association.

Tinston, C. (2010). *Construction firm cements one of the industry's most effective business continuity management plans*. Retrieved from http://www.24dash.com/news/housing

Tirtosudarmo, R. (2005, June 29–30). *Indonesia: Decentralization and horizontal inequality.* Paper presented at the Center for Research on Inequality, Human Security, and Ethnicity (CRISE) Workshop, Oxford, UK.

Tjandra, I. K. (2004). *Organisational learning in construction: The case of construction firms in Jakarta*. Unpublished PhD Dissertation, National University of Singapore, Singapore.

Toda, M., Hiraishi, K., Shintani, T., & Katayama, Y. (1991). Information structuring and its implementations on a research decision support system. *Decision Support Systems, 7*(2), 169–184.

Todini, E. (1999). An operational decision support system for flood risk mapping, forecasting and management. *Urban Water, 1*, 131–143.

Tridata. (2009). *Mass shootings at Virginia Tech; Addendum to the report of the review panel*. Virginia, US: Tridata Corporation.

Trompenaars, F. (1993). *Riding the waves of culture: Understanding cultural diversity in business*. London: The Economist Books.

Turban, E. (1995). *Decision support system and expert system*. Singapore: Prentice Hall.

Ulrich, W. (1983). *Critical heuristics of social planning: A new approach to practical philosophy*. Berne: Paul Hapt.

UNR/HC. (2005). *Situation report – Indonesia*. United Nations Resident/Humanitarian Coordination (UNR/HC): Indonesia. in Indonesia.

Vaidya, O. S., & Kumar, S. (2006). Analytic hierarchy process: An overview of applications. *European Journal of Operational Research, 169*, 1–29.

Vancoppenolle, G. (1999). What are we planning for? In A. Hiles & P. Barnes (Eds.), *The definitive handbook of business continuity management* (pp. 3–23). Chichester: John Wiley and Sons.

VDNA. (2010). *Provisions and guidelines for business continuity management*. Bank Van De Nederlandse Antillen (Central Bank): Willemstad.

VMIA. (2007). *Business continuity management toolkit*. Melbourne: Victorian Managed Insurance Authority (VMIA).

VT. (2010). *Emergency response plan*. Blacksburg, Virginia: Virginia Polytechnic Institute and State University.

Wang, S. Q. (2001). ESSCAD: Expert system integrating construction scheduling with CAD drawing. *Construction Informatics Digital Library, 28*, 1–28.

Watkins, J. (1997). Justifying the contingency Plan. *Disaster Recovery Journal*. Retrieved from www.drj.com/new2dr/w2_011.htm

Watson, T. J. (2001). *In search of management: Culture, chaos and control in managerial work*. London: Thomson Learning.

Wertheim, E. G. (2001). *Schools of historical thought and their components by decade*. Retrieved from http://www.scribd.com/doc/6926402/Historical-Background-of-Organizational-Behavior

Whitten, P., & Hunter, D. E. (1987). *Anthropology: Contemporary perspective*. Boston: Little Brown.

Wicht, A., Meixner, G., & Klein, U. (2011). Design and prototypical development of a web based decision support system for early detection of sepsis in hematology. *EICS4Med Proceedings*, 69–74.

Wilkins, A., & Martin, J. (1980). *Organizational legends*. Stanford, CA: Stanford University.

Williams, A., Dobson, P., & Walters, M. (1993). *Changing culture – New organizational approaches*. London: London Institute of Personnel Management.

World Bank. (2012). *World development indicators and global development finance*. Retrieved from http://databank.worldbank.org/ddp/home.do?Step=12&id=4&CNO=2

Xia, B., Chan, A. P. C., & Yeung, J. F. Y. (2011). Developing a fuzzy multi-criteria decision-making model for selecting design-build operational variations. *Journal of Construction Engineering and Management, 137*(12), 1176–1184.

Xu, Y., Chan, A. P. C., & Yeung, J. F. Y. (2010). Developing a fuzzy risk allocation model for PPP projects in China. *Journal of Construction Engineering and Management, 136*(8), 894–903.

Yen, J., & Langari, R. (1999). *Fuzzy logic intelligence, control and information*. New York: Prentice Hall.

Yin, R. K. (2003). *Applications of case study research* (2nd ed.). Los Angeles: Sage Publications.
Yoon, S. W., Velasquez, J. D., Partridge, B. K., & Nof, S. Y. (2008). Transportation security decision support system for emergency response: A training prototype. *Decision Support Systems, 46*, 139–148.
Zadeh, I. A. (1965). Fuzzy sets. *Information and Control, 8*, 338–353.
Zadeh, I. A. (1996). Fuzzy Logic; computing with words. *IEEE Transactions on Fuzzy Systems, 4* (2), 103–111.
Zhong, Y., & Low, S. P. (2009). Managing crisis response communication in construction projects – From a complexity perspective. *Disaster Prevention and Management, 18*(3), 270–282.
Zlateva, P., Kirov, G., & Stoyanov, K. (2005). Fuzzy logic application for eco-tourism potential assessment of villages. *Automatics and Informatics, 4*, 20–23.
Zlateva, P., Pashova, L., Stoyanov, K., & Velev, D. (2011). *Fuzzy logic model for natural risk assessment in SW Bulgaria*. Paper presented at 2nd International Conference on Education and Management Technology; IPEDR Vol. 13, IACSIT Press, Singapore, pp. 109–118.

MIX
Papier aus verantwortungsvollen Quellen
Paper from responsible sources
FSC® C105338

If you have any concerns about our products,
you can contact us on
ProductSafety@springernature.com

In case Publisher is established outside the EU,
the EU authorized representative is:
Springer Nature Customer Service Center GmbH
Europaplatz 3, 69115 Heidelberg, Germany

Printed by Libri Plureos GmbH
in Hamburg, Germany